Texts in Statistical Science

Applied
Nonparametric
Statistical Methods

Fourth Edition

Peter Sprent and Nigel C. Smeeton

Chapman & Hall/CRC
Taylor & Francis Group
Boca Raton   London   New York

Chapman & Hall/CRC is an imprint of the
Taylor & Francis Group, an informa business

CHAPMAN & HALL/CRC
Texts in Statistical Science Series

Series Editors

Bradley P. Carlin, *University of Minnesota, USA*
Julian J. Faraway, *University of Bath, UK*
Martin Tanner, *Northwestern University, USA*
Jim Zidek, *University of British Columbia, Canada*

Chapman & Hall/CRC
Taylor & Francis Group
6000 Broken Sound Parkway NW, Suite 300
Boca Raton, FL 33487-2742

© 2007 by Taylor & Francis Group, LLC
Chapman & Hall/CRC is an imprint of Taylor & Francis Group, an Informa business

Library of Congress Cataloging-in-Publication Data

Sprent, Peter.
  Applied nonparametric statistical methods. -- 4th ed. / Peter Sprent and Nigel C. Smeeton.
      p. cm. -- (Chapman & Hall/CRC texts in statistical science series)
    Includes bibliographical references and index.
    ISBN-13: 978-1-58488-701-0 (alk. paper)
    ISBN-10: 1-58488-701-X (alk. paper)
    1. Nonparametric statistics. I. Smeeton, N. C. II. Title. III. Series.

QA278.8.S74 2006
519.5'4--dc22   2006037003

Visit the Taylor & Francis Web site at
http://www.taylorandfrancis.com

and the CRC Press Web site at
http://www.crcpress.com

Contents

Preface

Applied Nonparametric Statistical Methods first appeared in 1989. Major developments in computing, especially for exact permutation tests, inspired a revised second edition in 1993. The third edition in 2001 reflected not only further advances in computing, but a widening of the scope of nonparametric or distribution-free methods and a tendency for these to merge with, or to be used in conjunction with, techniques such as exploratory data analysis, robust estimation and semiparametric methods. This trend has continued, being evident especially in computer intensive methods to deal with both intractable analytic problems and in processing large data sets.

This new edition reflects these developments while retaining features that have met a continuing positive response from readers and reviewers.

Nonparametric methods are basically tools for statistical analyses, but data collection and the interpretation of analyses are interrelated parts of the statistician's role. As in the third edition we comment, where appropriate, on all these aspects, some of which do not always receive the attention they deserve in undergraduate mainstream or in service courses in statistics.

Our approach is midway between a bare description of techniques and a detailed exposition of the theory, often illustrating key practical points by examples. We keep mathematics to the minimum needed for a clear understanding of scope and limitions.

We have two aims. One is to provide a textbook for those making first contact with nonparametric methods at the undergraduate level. This may be in mainstream statistics courses, or in service courses for students majoring in other disciplines. The second is to make the basic methods readily available to specialist workers, managers, research and development staff, consultants and others working in various fields. Many of them may have an understanding of basic statistics but only a limited acquaintance with nonparametric methods, yet feel these may prove useful in their work. The format we have adopted makes the book suitable not only as a class text, but also for self-study or use as a reference manual.

To meet our aims the treatment is broad rather than deep. We believe this to be a fruitful approach at the introductory stage. Once one has a broad overview of nonparametrics a more advanced study of topics of special interest becomes appropriate. Because fields of interest will vary from person to person, this second phase is best tackled by attending courses on, or referring to the literature that deals in depth with, aspects of particular interest. We give references to books and papers where more advanced treatments of many topics can be found.

Popular features of earlier editions are retained, including a formal structure for most examples, lists of potential fields of application and a selection of exercises at the end of each chapter.

There has been a substantial reordering of topics and new material has been added. The former Chapter 1 has been split into two, with consequent renumbering of later chapters. Chapter 1 now gives a brief summary of some relevant general statistical concepts, while Chapter 2 introduces ideas basic to nonparametric or distribution-free methods. The new Chapters 3 to 7 broadly cover the content of the former Chapters 2 to 6, but with many changes in emphasis and removal of some material on designed experiments to a new Chapter 8, and some on analysis of survival data to an extended treatment in Chapter 9. Designed experiments, particularly those with a factorial treatment structure, are handled in an up-to-date way in Chapter 8. Chapters 10 to 14 are revisions of the former Chapters 7 to 11. Chapter 15 is new and introduces a few of the many important modern developments, most of the applications being computer intensive.

As in earlier editions we have not included tables of quantiles, critical values, etc. relevant to basic nonparametric procedures. Modern software has made many of these somewhat redundant, but those who need such tables will find them in many standard collections of statistical tables. We give references as needed to relevant specialized tables. Solutions to selected exercises are given in an appendix.

We are grateful to many readers and reviewers of the earlier editions who have made constructive comments about content and treatment of particular topics. Their input triggered several major changes in the present edition. We thank Edgar Brunner and Thomas P. Hettmansperger for drawing our attention to a number of papers dealing with interactions in a nonparametric context, and Joseph Gastwirth for alerting us to many recent practical developments and Nick Cox for providing a Stata program for simulating runs distributions. We renew thanks to those whose assistance was acknowledged in earlier editions — to Jim McGarrick for useful discussions on physiological measurements — to Professor Richard Hughes for advice on the Guillain–Barré syndrome — to Timothy P. Davis and Chris Theobald who supplied data sets for examples. We are grateful to Cyrus Mehta and Cytel Software for providing us with complementary software and manuals for StatXact 7 and to Shashi Kumar for help with technical problems with embedding fonts in diagrams.

P. Sprent
N.C. Smeeton

CHAPTER 1

SOME BASIC CONCEPTS

1.1 Basic Statistics

We assume most readers are familiar with the basic statistical notions met in introductory or service courses in statistics of some 20 hours duration. Nevertheless, those with no formal statistical training should be able to use this book in parallel with an introductory statistical text. Rees (2000) adopts a straightforward approach. Some may prefer a more advanced treatment, or an introduction that emphasizes applications in a discipline in which they are working.

Readers trained in general statistics, but who are new to nonparametric methods will be familiar with some of the background material in this chapter. However, we urge them at least to skim through it to see where we depart from conventional treatments, and to learn how nonparametric procedures relate to other approaches. We explain the difference between parametric and nonparametric methods and survey some general statistical notions that are relevant to nonparametric methods. We also comment on good practice in applied statistics.

In Chapter 2 we use simple examples to illustrate some basic nonparametric ideas and introduce some statistical notions and tools that are widely used in this field. Their application to a range of problems is covered in the remaining chapters.

1.1.1 Parametric and Nonparametric Methods

The word *statistics* has several meanings. It is used to describe a collection of data, and also to designate operations that may be performed with that primary data. The simplest of these is to form *descriptive* statistics. These include the mean, range, or other quantities to summarize primary data, as well as preparing tables or pictorial representations (e.g., graphs) to exhibit specific facets of the data. The scientific discipline called *statistics*, or *statistical inference*, uses observed data — in this context called a *sample* — to make inferences about a larger potentially observable collection of data called a *population*. We explain the terms *sample* and *population* more fully in Section 1.2

We associate *distributions* with populations. Early in their careers statistics students meet families of distributions such as the *normal* and *binomial* where individual members of the family are distinguished by assigning specific values to entities called *parameters*.

The notation $N(\mu, \sigma^2)$ denotes a member of the normal, or Gaussian, family with mean μ and variance σ^2. Here μ and σ are parameters.

The *binomial* family depends on two parameters, n and p, where n is the total number of observations and p is the probability associated with one of two possible outcomes at any observation. Subject to certain conditions, the number of occurrences, r, where $0 \leq r \leq n$, of that outcome among n observations, has a binomial distribution with parameters n and p. We call this a $B(n, p)$ distribution.

Given a set of independent observations, called a random sample, from some population with a distribution that is a member of a family such as the normal or binomial, *parametric statistical inference* is often concerned with testing hypotheses about, or estimation of, unknown parameters.

For a sample from a normal distribution the sample mean is a point (i.e., a single value) estimate of the parameter μ. Here the well-known t-test provides a measure of the strength of the evidence provided by a sample in support of an *a priori* hypothesized value μ_0 for the distribution, or population, mean. We may also obtain a *confidence interval*, a term we explain in Section 1.4.1, for the "true" population mean.

When we have a sample of n observations from a $B(n, p)$ distribution with p unknown, if the event with probability p is observed r times an appropriate estimate of p is $\hat{p} = r/n$. We may want to assess how strongly sample evidence supports an *a priori* hypothesized value p_0, say, for p, or to obtain a confidence interval for the population parameter p.

Other well-known families of distributions include the uniform (or rectangular), multinomial, Poisson, exponential, gamma, beta, Cauchy and Weibull distributions. This list is not exhaustive and you may not be, and need not be, familiar with all of them.

It may be reasonable on theoretical grounds, or on the basis of past experience, to assume that observations come from a particular family of distributions. Also experience, backed by theory, suggests that for many measurements inferences based on the assumption that observations form a random sample from some normal distribution may not be misleading, even if the normality assumption is incorrect. A theorem called the *central limit theorem* justifies such a use of the normal distribution especially in what are called *asymptotic approximations*. We often refer to these in this book.

Parametric inference may be inappropriate or even impossible. For example, records of examination results may only give the numbers of candidates in banded and ordered grades designated Grade A, Grade B, Grade C, etc. Given these numbers for pupils from two different schools, we may want to know if they indicate a difference in performance between those schools that might be attributed to different methods of teaching, or to the ability of one school to attract more able pupils. There is no obvious family of distributions that provides our data, and there are no clearly defined *parameters* about which we can make inferences. Two terms are in common use for the type of inferences we may then make. They are either described as *nonparametric* or as *distribution-free*. There is sometimes an incorrect belief among nonstatisticians that these terms refer to the data.

There is not unanimity among statisticians in their use of the terms *nonparametric* and *distribution-free*. This is of no great consequence in practice, and to some extent simply reflects historical developments.

There is not even universal agreement about what constitutes a parameter. Quantities such as μ, σ^2 appearing in the density functions for the normal family are unquestionably parameters. The term is often used more widely to describe any population characteristic within a family such as a mean, median, moment, quantile, or range. In Chapter 8 and elsewhere we meet situations where we have observations composed of a deterministic and random element where we want to estimate constants occuring in the deterministic element. Such constants are also sometimes called parameters.

In nonparametric, or distribution-free, methods we often make inferences about parameters in this wider sense. The important point is that we do not assume our samples are associated with any prespecified *family* of distributions. In this situation the name *distribution-free* is more appropriate if we are still interested in parameters in the broader senses mentioned above.

Some procedures are both distribution-free and nonparametric in that they do not involve parameters even in the broader use of that term. The above example involving examination grades falls into this category.

Historically, the term *nonparametric* was in use before *distribution-free* became popular. There are procedures for which one name is more appropriate than the other, but as in many areas of statistics, terminology does not always fit procedures into watertight compartments. A consequence is the spawning of hybrid descriptions such as *asymptotically distribution-free* and *semiparametric* methods. There is even some overlap between descriptive statistics and inferential statistics, evident in a practice described as *exploratory data analysis*, often abbreviated to EDA. We shall see in Section 2.4, and elsewhere, that sensible use of EDA may prove invaluable in selecting an appropriate technique, parametric or nonparametric, for making statistical inferences.

Designating procedures as *distribution-free* or *nonparametric* does not mean they are assumption free. In practice we nearly always make some assumptions about the underlying population distribution. For example, we may assume that it is continuous and symmetric. These assumptions do not restrict us to a particular family such as the normal, but they exclude both discrete and asymmetric distributions. Given some data, EDA will often indicate whether or not an assumption such as one of symmetry is justified.

Many *nonparametric* or *distribution-free* procedures involve, through the test statistic, distributions and parameters (often the normal or binomial distributions). This is because the terms refer *not* to the test statistic, but to the fact that the methods can be applied to samples from populations having distributions only specified in broad terms, e.g., as being continuous, symmetric,

identical, differing only in medians, means, etc. The distribution of the appropriate test statistic is the same no matter what the population distribution
may be, providing only that it satisfies the broad-term specification. There is
a grey area between what is clearly distribution-free and what is parametric
inference. Some of the association tests described in Chapters 12 and 13 fall
in this area.

1.1.2 The Use of Nonparametric Methods

Some parametric tests do not depend critically on the correctness of an assumption that samples come from a distribution in a particular family. They are
then described as *robust*. Robustness is by no means a universal property of
parametric tests. Because they require fewer assumptions for their validity,
nonparametric methods are usually more robust than their parametric counterparts.

Nonparametric methods are often the only ones available for data that
simply specify order (ranks) or counts of the number of events, or of individuals, in various categories.

In most statistical problems, no matter whether parametric or nonparametric methods are appropriate, what we can deduce depends on what
assumptions can validly be made. An example illustrates this.

Example 1.1

Two machines produce metal rods. For each, 2.5 percent of all rods produced have
a diameter exceeding 30 cm.

This condition is met if the first machine produces items having a normal distribution with mean 27 mm and standard deviation 1.53 mm. This is because, for any
normal distribution, 2.5 percent of all items have a diameter at least 1.96 standard
deviations above the mean, so 2.5 percent exceed $27 + 1.96 \times 1.53 \approx 30$.

The condition is also met if the second machine produces items with diameters uniformly distributed between 20.25 and 30.25 mm (i.e., with mean diameter 25.25 mm). Once again the condition that 2.5 percent have diameters exceeding
30 mm is met. This follows because any interval between 20.25 and 30.25 mm of
width 0.25 mm contains a proportion 1/40 (i.e., 2.5 percent) of total production.

This uniform distribution is unlikely to be met in practice in this context, but
the example shows that we may have the same proportion of defectives in two
populations, yet each has a different mean and their distributions do not even belong
to the same family. We consider a more realistic situation involving different means
in Exercise 1.1

If we make an additional assumption that distributions of diameters for each
machine differ, if at all, only in their means, then if we know the proportion over
30 mm in samples of, say, 200 from each machine, we could test whether the means
can reasonably be supposed to be identical. The test would not be efficient. It would
be better to measure the diameter of each item in smaller samples, and then use an
appropriate test.

Means and medians are widely used to indicate where distributions are centred. Both are formally described as *measures of centrality* or *measures of location*. Not all distributions have a mean, but all have a median. If the mean exists, the mean and the median have the same value for a symmetric distribution. Their values differ for asymmetric, or skew, distributions. The Cauchy distribution is a well-known example of a symmetric distribution that has no mean. It has a well-defined median, this being zero for the standard Cauchy distribution.

Tests and estimation procedures for measurement data are often about centrality measures, e.g.,

- Is it reasonable to suppose that a sample comes from a population with a prespecified mean or median?
- Do two samples come from populations whose means differ by at least 10?
- Given a sample, what is an appropriate estimate of the population mean or median? How good is that estimate?

Variation or *spread* or *dispersion* is often of interest also. Buyers of new cars or computers want not only a good average performance but also consistency, i.e., not too much variation in performance from item to item of the same brand. Each buyer expects his or her purchase to perform as well as those of other buyers of that model. Success of a product often depends upon personal recommendations, so mixed endorsements — some glowing, others warning of niggling faults — are not good publicity.

Dispersion is often measured by *variance* or by *standard deviation*, but these may not exist for all distributions. Also, they are not well suited on their own to describe, or compare, the spread of skew distributions. There are parametric and nonparametric methods for assessing spread or variability.

In other situations we want to assess how well data conform to some hypothesized population distribution function, the approriate test being one for *goodness of fit*. Tests of association or correlation are also of considerable interest.

Nonparametric techniques may be the only ones available when we have limited information. We may want to test if it is reasonable to assume that weights of a large batch of items have a prespecified median, 2 mg, say, when all we know is how many items in a sample of n weigh more than 2 mg. If it were difficult, expensive, or impossible to get exact weights, an available nonparametric approach may be cost effective.

Simple nonparametric methods are also useful when data are in some sense incomplete, like those in Example 1.2.

Example 1.2

In medical studies the progress of patients is often monitored for a limited time after treatment; this may be anything from a few weeks to 5 or 6 years. Dinse (1982) gives data for survival times in weeks for 10 patients with symptomatic lymphocytic non-Hodgkin's lymphoma. The precise survival time is not known for one patient who

was alive after 362 weeks. The observation for that patient is said to be *censored*.
Survival times in weeks were

$$49 \quad 58 \quad 75 \quad 110 \quad 112 \quad 132 \quad 151 \quad 276 \quad 281 \quad 362^*$$

The asterisk denotes the censored observation.

Is it reasonable to suppose that these data are consistent with a median survival
time of 200 weeks? Censored observations cause problems in many parametric tests,
but in Example 2.2 we use a simple nonparametric test to show there is no strong
evidence against the hypothesis that the median is 200. For that interpretation to
be meaningful, and useful, we have to assume the data are a random sample from
some population of patients with the disease.

To confirm that the median might well be 200 is not in itself very helpful. It
would be more useful if we could say that the data imply that it is reasonable to
assert that the median survival time is between 80 and 275 weeks, or something
of that sort. This is what *confidence intervals* (Section 1.4.1) are about. In
the original study Dinse was interested, among other things, in whether the
median survival times differed between symptomatic and asymptomatic cases.
He used this sample and another for 28 asymptomatic cases to compare the
survival time distributions in more detail. In this second sample 12 of the 28
observations were censored at values of 300 or more. We show in Example 9.1
that, on the basis of his data, there is strong evidence that the medians for
symptomatic and for asymptomatic cases are different. These data were also
considered by Kimber (1990).

1.1.3 Historical Notes

The first chapter of the Book of Daniel records that on the orders of Nebuchad-
nezzar certain favoured children of Israel were to be specially fed on the king's
meat and wine for 3 years. Reluctant to defile himself with such luxuries,
Daniel pleaded that he and three of his brethren be fed instead on pulse for
10 days. After that time the four were declared "fairer and fatter in flesh
than all of the children which did eat the portion of the king's meat". This
evidence was taken on commonsense grounds to prove the superiority of a diet
of pulse. Throughout this book we illustrate how we test evidence like this
more formally to justify this commonsense conclusion. The biblical analysis
is informal, but it contains the germ of a nonparametric, as opposed to a
parametric, test. Be warned though that the commonsense conclusion may
not be justified here. Daniel and his three brethren may already have been
fairer and fatter before the "experiment" began!

John Arbuthnot (1710) observed that in each of the 82 years from 1629
to 1710 the number of males christened in London exceeded the number of
females. He regarded this as strong evidence against the probability of a male
birth being 1/2. The situation is somewhat akin to observing 82 heads in 82
consecutive tosses of a coin.

Francis Galton (1892) developed a measure — which he termed a *centisimal scale* — to assess agreement between patterns (categorical data) on corresponding fingertips on left and right hands.

Karl Pearson (1900) proposed the well-known, and sometimes misused, chi-squared goodness-of-fit test applicable to any discrete distribution, and C. Spearman (1904) defined a rank correlation coefficient (see Section 10.1.3) that bears his name.

Systematic study of nonparametric inference dates from the 1930s when attempts were made to show that even if an assumption of normality stretched credulity, then at least in some cases making it would not greatly alter conclusions. This stimulated work by R.A. Fisher, E.J.G. Pitman and B.L. Welch on *randomization* or *permutation tests*, which were then too time consuming for general use. That problem has been overcome with appropriate statistical software, but we shall see later that the raw-data permutation tests proposed by these writers have practical limitations, although the concept is of considerable theoretical interest.

Other developments in the 1930s include work by Friedman (1937), Smirnov (1939) and others.

About the same time it was realized that observations consisting simply of preferences or ranks could be used in permutation tests to make some inferences without too much computational effort. A few years later F. Wilcoxon and others showed that, even if we have precise measurements, we sometimes lose little useful information by ranking them in increasing order of magnitude and basing analyses on these ranks. Indeed, when assumptions of normality are not justified, analyses based on ranks, or on some transformation of them, may be the most efficient available. They often enjoy the characteristic we have already referred to as *robustness*, and which we describe more fully in Chapter 14.

From the 1940s nonparametric methods became practical tools either when data were by nature ordinal (ranks or preferences), or as reasonably efficient methods that reduced computation even when measurements were available, providing those measurements could be replaced by ranks. At that time hypothesis testing was usually easy, the more important interval estimation described in Section 1.4 was not. This difficulty was overcome by Hodges and Lehmann (1963).

In parallel with the above advances, techniques relevant to counts were developed. Counts often represent numbers of items in categories that may be either *ordered*, e.g., examination grades, or *nominal* (i.e., unordered), e.g., in psychiatry characteristics like depression, anxiety, psychosis, etc.

Many advanced and flexible nonparametric methods are tedious because they involve repeated performance of simple calculations, something computers do well.

The dramatic postwar development of feasible, but until recently often tedious to carry out, nonparametric procedures was described in Noether (1984), but much has happened since then.

Another line of development has been in the use of the computer intensive procedures of *jack-knifing* introduced by Quenouille (1949), and the even-more widely used *bootstrapping* introduced by Efron (1979). The latter has both parametric and nonparametric versions.

Computers have revolutionized our approach to data analysis and to stat-istical inference. Hopes, often ill-founded, that data would fit a restricted mathematical model with few parameters, and emphasis on simplifying con-cepts such as linearity, have often been replaced by the use of robust methods and by EDA to investigate different potential models. These are areas where nonparametric methods sometimes have a central role. *Generalized linear mod-els* described by McCullagh and Nelder (1989) at the theoretical level, and by Dobson (2001) at the practical level, often blend parametric and nonparamet-ric approaches.

Whereas the initial developments in nonparametric methods were inspired by problems arising in small samples, interest in recent years has turned to their use in extracting information from very large data sets, an aspect we touch upon in Chapter 15.

Nonparametric procedures are in no sense preferred methods of analysis for all situations. A strength is their applicability where there is insufficient theory or data to justify, or to test compatibility with, specific distributional models. At a more sophisticated level they are also useful, for example, in finding or estimating trends in large data sets. Such trends may be difficult to detect due to the presence of disturbances usually referred to as "noise".

Recent important practical developments have been in computer software (Section 2.6) to carry out permutation and other nonparametric tests. Results for these may be compared with those given by asymptotic theory, which, in the past, was often used where its validity was dubious.

Since the 1990s there have been many important advances in the use of nonparametric methods in designed experiments, particularly those involving factorial structures or repeated observations on individuals or other experi-mental units. An introduction to these develpments is given in Chapter 8.

1.2 Populations and Samples

When making statistical inferences a key assumption is often that observations are a random sample from some population. That assumption is essential to the strict validity of many inferential procedures, although properties such as robustness may allow its relaxation in some, but not all, circumstances. Specification of the population may or may not be precise. If we select 20 books at random from 100,000 volumes in a library, and record the number of pages in each book in the sample, then inferences made from the sample about the mean, or median, number of pages per book apply strictly only to the population of books in that library. If the library covers a wide range of fiction, nonfiction and reference works it is reasonable to assume that any inferences apply, at least approximately, to a wider population of books. This might be all books published in the United Kingdom, or the United States, or wherever the library is situated. However, if the books in the library are all English language books, inferences may not apply to books in Chinese or in Russian.

If units are selected without replacement, a random sample of size n from a finite population is one where every possible sample of that size has an equal probability of selection. If sampling is with replacement a random sample is one where each item is independently selected with equal probability. This implies that if we arrange the sampled units in the order in which they are selected each possible ordered sample may be obtained with equal probability.

More often, data form samples that may be expected to have the essential properties of a random sample from a vaguely specified population. For example, if a new diet is tested on pigs and we measure weight gains for 20 pigs at one agricultural experimental station, we might assume that these are something like a random sample from all pigs of that or similar breeds raised under such conditions. This qualification is important. Inferences might apply widely only if the experimental station adopted common farming practices and if responses were fairly uniform for many breeds of pig. This may not be so if the experimental station chose an unusual breed, adopted different husbandry practices from those used on most pig farms, or if the 20 pigs used in the experiment were treated more favourably than is usual in other respects such as being kept in specially heated units during the experiment.

The abstract notion of random sampling from an infinite population (implicit in most inference based upon normal distribution theory) often works well in practice, but is never completely true. At the other extreme there are situations where the sample is essentially the whole population. For example, at the early stages of testing a new drug for treating a rare disease there may be just, say, nine patients available for the test and only four doses of the drug. One might choose at random the four patients from nine to receive the new drug. The remaining five are untreated, or may be treated with a drug already in use. Because of the random selection, if the drug has no effect, or is no better than one currently in use, it is unlikely that a later examination would show that the four patients receiving the new drug had responded better than any of the others. This is possible, but it has a low probability, which we can calculate on the assumption that the new drug is ineffective or is no better than the old one. If the drug is beneficial, or better than that currently in use, the probability of better responses among those treated with it is increased. In Example 2.1 in Section 2.1 we formulate an appropriate nonparametric test for this situation.

Because of the many ways data may be obtained we need to consider carefully the validity and scope of inferences. The ten patients for whom survival times were measured in Example 1.2 came from a study conducted by the Eastern Co-operative Oncology Group in the U.S., and represented all patients afflicted with symptomatic lymphocytic non-Hodgkin's lymphoma available for that study. In making inferences about the median or other characteristics of the survival time distribution, it is reasonable to assume these inferences are valid for all patients receiving similar treatment and who are alike in other relevant characteristics, e.g., with a similar age distribution. The patients in this study were all male, so it would be unwise to infer, without further evidence, that the survival times for females would have the same distribution.

Fortunately, the same nonparametric procedures are often valid whether samples are from an infinite population, a finite population, or when the sample is the entire relevant population. What is different is how far these inferences can be generalized. The implications of generalizing inferences is described for several specific tests by Lehmann (1975, 2006, Chapters 1–4).

In Section 1.1.1 we referred to a set of n independent observations from some normal distribution. In this situation independence implies, though this is not a definition of independence, that the value taken by any one observation tells us nothing about, and does not influence, the values of other observations. More formally, the probability that an observation lies in any small interval $(x, x + \delta x)$ is $f(x)\delta x$ where $f(x)$ is the probability density function of the relevant member of the normal distribution family. This notion extends to samples from any continuous distribution where $f(x)$ is the relevant probability density function.

The concept of observations being independent is an important, and by no means trivial, requirement for the validity of many statistical inference procedures.

1.3 Hypothesis Testing

Estimation, a topic we consider in Section 1.4, is often a key aim of a statistical analysis. Estimation may be explained in terms of testing a range of hypotheses, so we need to understand testing even though it is a technique that, in the view of many statisticians, tends to be overused and is even sometimes misused.

We assume familiarity with simple parametric hypothesis tests such as the t-test and chi-squared test, but we review some fundamentals and discuss changes in emphasis made possible by modern computer software. Until such software became widely available hypothesis testing nearly always required the use of tables.

Given a sample of n independent observations from a population having a normal distribution with unknown mean μ the t-test has a fundamental role in making inferences about μ. In the light of the *central limit theorem* the normality assumption may be somewhat relaxed.

We specify a null hypothesis, H_0, that μ has a specified value μ_0 and an alternative hypothesis, H_1, that it has some other value. Formally, this is stated as:

$$\text{Test } H_0: \mu = \mu_0 \text{ against } H_1: \mu \neq \mu_0. \tag{1.1}$$

The t-test is based on a statistic, t, that is a function of the sample values calculated by a formula given in introductory general statistics textbooks. The classic procedure was to use tables to compare the magnitude, without regard to sign, of the calculated t, often written $|t|$, with a value t_α given in tables, the latter chosen so that when H_0 was true

$$\Pr(|t| \geq t_\alpha) = \alpha. \tag{1.2}$$

In practice α nearly always took one of the values 0.05, 0.01 or 0.001. These probabilities are often expressed as equivalent percentages, i.e., 5, 1 or 0.1, and are widely known as *significance levels*. Use of these particular levels was dictated, at least in part, by available tables. In this traditional approach if one obtained a value of $|t| \geq t_\alpha$, the result was said to be *significant* at probability level α, or at the corresponding 100α percent level. The levels 0.05, 0.01 and 0.001 were often referred to respectively as "significant", "highly significant" and "very highly significant". If significance at a particular level was attained, one spoke of rejecting the hypothesis H_0 at that level. If significance was not attained the result was described as *not significant* and H_0 was said to be accepted.

This is unfortunate terminology giving — especially to nonstatisticians — the misleading impression that nonsignificance implies that H_0 is true, while significance implies it is false.

To illustrate the point that traditional acceptance of a null hypothesis does not imply that H_0 is true, suppose that students over 18 years may study at a certain university. Most of the undergraduate courses run for three years, and the majority of the students go to university straight from school at age 18. However, on the courses there are some mature students aged 25 or more, making the overall mean age of the undergraduate students equal to 23. An external investigator who is unaware of the presence of these mature students specifies a null hypothesis H_0 that the mean undergraduate student age is 21; this null hypothesis is clearly untrue. However, suppose a sample of 50 undergraduates is selected and the mean age of these students is exactly 21 years. Because the sample mean is equal to the null hypothesis mean, the P-value for the test will be $P = 1$, the greatest possible value. It might then be tempting to deduce that H_0 is true, as the evidence is highly suggestive. However, this would be an incorrect conclusion.

The rationale behind the test (1.1) is that if H_0 is true, then values of t near zero are more likely than large values of t, either positive or negative. Large values of $|t|$ are more likely under H_1 than under H_0. It follows from (1.2) that if we perform a large number of such tests on different independent random samples when H_0 is true we shall, in the long run, incorrectly reject H_0 in a proportion α of these. Thus, if $\alpha = 0.05$ we would reject the null hypothesis when it were true in the long run in 1 in 20 tests, i.e., in 5 percent of all tests.

The traditional approach is still common, especially in certain areas of law, medicine and commerce, or to conform with misguided policy requirements of some scientific and professional journals.

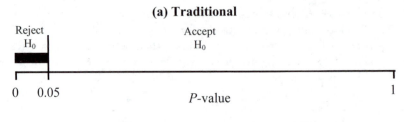

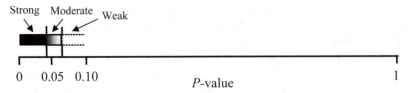

Figure 1.1 *(a) A fixed level approach to significance testing and (b) assessment based on strength of evidence where some flexibility in interpretation may be allowed for P-values close to 0.05.*

Modern statistical software lets us do something more sensible, though by itself still far from satisfactory. The output from any good computer program for a t-test relevant to (1.1) gives the exact probability of obtaining, when H_0 is true, a value of $|t|$ equal to, or greater than, that observed. In statistical jargon this probability is called a P-value. It may be used as a measure of the strength of evidence against H_0 provided by the data — the smaller P is, the stronger is that evidence.

When we decide what P-values are sufficiently small for H_0 to seem implausible, we may speak formally of rejecting H_0 at the exact $100P$ percent significance level. This avoids the difficulty that rigid application of the 5 percent significance level leads to the unsatisfactory situation of H_0 being rejected for a P-value of 0.049, but accepted for the slightly larger P-value of 0.051. Figure 1.1 compares the traditional "fixed level" approach to significance, with the more flexible assessment based on strength of evidence using a P-value.

A serious difficulty still remains. This is that with small data sets one may *never* observe sufficiently small P-values to justify rejection of H_0 even when it is not true. For instance, suppose a coin is tossed 5 times to test the hypothesis

$$H_0: \text{the coin is fair,}$$

against

$$H_1: \text{the coin is biased.}$$

i.e., is either more likely to fall heads, or more likely to fall tails.

In 5 tosses the strongest evidence against H_0 is associated with the outcomes 5 heads or 5 tails. Under H_0 the probability, P, of getting one of these outcomes is given by the sum of the probabilities of $r = 0$ or $r = 5$ "heads" for a binomial

B(5, 0.5) distribution. The probability of each is $(0.5)^5$, so $P = 2 \times (0.5)^5 = 0.0625$. This is the smallest attainable P-value when $n = 5$ and $p = 0.5$, so we never reject H_0 at a conventional $P = 0.05$ level whatever the outcome of the 5 tosses — even if the coin is a double-header. That the experiment is too small is the only useful information given by the P-value in this example.

The situation is different if we increase the experiment to 20 tosses and get 20 heads or 20 tails. This weakness of hypothesis testing, together with the perpetration of myths, such as equating accepting a hypothesis to proof that it is true, has led to justified criticism of what is sometimes called the P-value culture. Krantz (1999) and Nelder (1999) both highlight dangers arising from inappropriate use of, and misunderstandings about, the meaning of a P-value. We draw attention also to an ethical danger in Section 1.5.

Real-world policy decisions are often based on the outcome of statistical analyses. To appreciate the implications of either a formal rejection of, or a decision not to reject H_0, at a given significance level we need further concepts. Suppose we decide to reject H_0 whenever a P-value is less than some fixed value P_0, say. This means that if, in all cases where we do so H_0 is true, we would in the long run reject it in a proportion P_0 of those cases.

Rejection of H_0 when it is true is an *error of the first kind*, or Type I error. A P-value tells us the probability that we are making an error of the first kind by rejecting H_0.

In a t-test, if we reject H_0 whenever we observe a $P \leq P_0$ we do so when k is such that when H_0 holds, $\Pr(|t| \geq k) = P_0$. Such values of t define a *critical*, or rejection, region of size P_0. Using a critical region of size P_0 implies we continue to regard H_0 as plausible if $|t| < k$.

If we follow this rule we shall sometimes (or as we saw above for the 5 coin tosses, in extreme cases, always) continue to regard H_0 as plausible even though, in fact, H_1 may be true. Continuing to accept H_0 when H_1 is true is an *error of the second kind*, or Type II error. Let β denote the probability of a Type II error.

The probability of a Type II error depends in part on the true value of μ, if indeed it is not equal to μ_0. Intuition correctly suggests that the more μ differs from μ_0, the more likely we are to get large values of $|t|$, i.e., values in the critical region, so that β decreases as $|\mu - \mu_0|$ increases.

If we decrease P_0 (say from 0.03 to 0.008) our critical region becomes smaller, so that for a given $\mu \neq \mu_0$ we increase β because the set of values of t for which we accept H_0 is larger. Another factor affecting β is the sample size, n. If we increase n we decrease β for a given μ and P_0. Thus β depends on the true value of μ (over which we have no control) and the value of n and of P_0 determining the size of the critical region. We often have some flexibility in the choice of n and P_0. We have framed our argument mainly in terms of the t-test statistics, but it generalizes to other test statistics.

Despite obvious limitations, P-values used constructively have a basic role in statistical inference. In Section 1.4 we show that a null hypothesis that specifies one single value of a parameter is usually one of many possible hypotheses that are not contradicted by the sample evidence. Donahue (1999) and Sackrowitz and Samuel-Cahn (1999) discuss various distributional properties of the P-value that relate indirectly to uses we discuss here and in Section 1.4.

Fixing the probability of an error of the first kind, whether we denote it by the conventional symbol α, or the alternative P_0, does not determine β. We want β to be small because, in the t-test for example, we want the calculated t-value to be in the critical region when H_0 is not true. The probability $1 - \beta$ is the probability of getting a t-value in the critical region when H_0 is *not* true. It is called the *power* of the test; we want this to be large. For samples from a normal distribution and all choices of n, P and for any μ, the t-test is more powerful than any other hypothesis test of the form specified in (1.1).

The historical choice of significance levels 5, 1 and 0.1 percent as the basis for tables was made on the pragmatic grounds that one does not want to make too many errors of the first kind. It would be silly to choose a significance level of 50 percent, for then we would be equally likely to accept or to reject H_0 when it were true. Even with conventional significance levels, or other small P-values, we may often make errors of the second kind if a test has low power for one or more of these reasons:

- The true μ is close to the value specified in H_0.
- The sample size is small.
- We specify a very small P-value for significance.
- Assumptions required for the test to be valid are violated.

In later chapters we consider for some tests the often nontrivial problem of determining how big a sample is needed to ensure reasonable power to achieve given objectives.

Using small P-values in place of traditional 5, 1 and 0.1 percent significance levels gives more freedom in weighing evidence for or against a null hypothesis. Remembering that $P = 0.05$ corresponds to the traditional 5 percent significance level long used as a reasonable watershed, one should not feel there is strong evidence against a null hypothesis if P is substantially greater than 0.05. However, values of P not greatly exceeding 0.05 often point at least to a case for further studies. In particular, a need for larger experiments.

In this book we shall usually discuss the evidence for or against hypotheses in terms of observed P-values, but in some situations where it is appropriate to consider a hypothetical fixed P-value we use for this the notation α with an implication that we regard any observed P-value less than that α as sufficient evidence to prefer H_1 to H_0. Fixed levels comply with certain long-established conventions, and may be necessary for comparisons of the power of different procedures.

The test in (1.1) is called a two-tail test because the critical region consists both of large positive and large negative values of the statistic t. To be more specific, large positive values of t usually imply $\mu > \mu_0$ and large negative values of t imply $\mu < \mu_0$.

Specification of H_0 and H_1 is determined by the logic of a problem. Two other common choices are

(i)   Test H_0: $\mu = \mu_0$ against H_1: $\mu > \mu_0$.   (1.3)

(ii)   Test H_0: $\mu \leq \mu_0$ against H_1: $\mu > \mu_0$.   (1.4)

Both lead to a one-tail (here right or upper-tail) test, since in each case when the t-test is relevant large positive values of t favour H_1, whereas a small positive value, or any negative value, indicates that H_0 is more likely to hold. The modifications to a one-tail test if the inequalities in (1.3) or (1.4) are reversed are obvious. The critical region then becomes the left, or lower, tail.

For example, if the amount of a specified impurity in 1000 g ingots of zinc produced by a standard process is normally distributed with a mean of 1.75 g and it is hoped that a steam treatment will remove some of this impurity we might steam-treat a sample of 15 ingots and determine the amount of impurity left in each ingot. If the steam is free from the impurity, the treatment cannot increase the level. Either it is ineffective or it reduces the impurity. It is therefore appropriate to test

Test H_0: $\mu = 1.75$ against H_1: $\mu < 1.75$.

If ingots had an unknown mean impurity level, but a batch is acceptable only if $\mu \leq 1.75$, an appropriate test would be

Test H_0: $\mu \leq 1.75$ against H_1: $\mu > 1.75$.

Use of a one-tail test is only justified if it is appropriate to the logic of the problem, as it is in the illustrations just given. It is not appropriate in the situation pertaining in (1.1).

For the t-test some computer packages give a P-value appropriate for a one-tail test, e.g., $\Pr(t \geq t_P) = P$. Because the distribution of t is symmetric, one doubles this probability to obtain P for a two-tail test. The doubling of one-tail probabilities to give the corresponding two-tail test P-value or significance level applies in other parametric tests such as the F-test for equality of variance based on samples from two normal populations, but in these cases the two relevant subregions are not symmetric about the mean. However, in many applications where relevant statistics have a chi-squared or an F-distribution a one-tail (upper-tail) test is appropriate.

A common misconception is that a low P-value indicates a departure from the null hypothesis that is of practical importance. We show why this is not necessarily true in Section 1.4.2.

1.4 Estimation

1.4.1 Confidence Intervals

The sample mean is widely used as a point estimate of the population distribution mean if that mean exists. The sample mean varies between samples, so

we need a measure of the *precision* of this estimate. A confidence interval is
one such measure.

One way to describe a $100(1-\alpha)$ percent confidence interval for a parameter
θ is to define it as the set of all values of θ for which, if any value in that set
were specified in H_0, then the given data would lead to a $P > \alpha$. This implies
that if a confidence interval includes the value of a parameter that is specified
in H_0 there is no strong evidence against H_0. On the other hand a value
specified in H_0 that lies well outside that confidence interval indicates strong
evidence against H_0.

Another common interpretation of a $100(1-\alpha)$ percent confidence interval
is in terms of the property that if we form such intervals for repeated samples,
then in the long run $100(1-\alpha)$ percent of these intervals would contain (or
cover) the true but unknown θ. Confidence intervals are useful because:

- They tell us something about the precision with which we estimate a
  parameter.
- They help us decide (a) whether a significant result is likely to be of practical importance or (b) whether we need more data before we decide whether
  it is.

We elaborate on these points in Section 1.4.2.

A useful way of looking at the distinction between hypothesis testing and
estimation is to regard testing as answering the question:

- Given a hypothesis H_0: $\theta = \theta_0$ about, say, a parameter θ, what is the
  probability (P-value) of getting a sample as or less likely than that obtained
  if θ_0 is indeed the true value of θ?

whereas estimation using a confidence interval answers the question:

- Given a sample, what values of θ are consistent with the sample data in
  the sense that they lie in the confidence interval?

1.4.2 Precision Significance and Practical Importance

Example 1.3

Doctors treating hypertension are often interested in the decrease in systolic blood
pressure after administering a drug. When testing an expensive new drug they might
want to know whether it reduces systolic blood pressure by at least 20 mm Hg. Such
a minimum difference could be of practical importance.

Two clinical trials (I and II) are carried out to test the efficacy of a new drug
(A) for reducing blood pressure. A third trial (III) is carried out with a second
new drug (B). Trial I involves only a small number of patients, but trials II and III
involve larger numbers. The 95 percent confidence intervals for mean blood pressure
reduction (mm Hg) after treatment at each trial are:

Drug A	Trial I	$(3, 35)$
Drug A	Trial II	$(9, 12)$
Drug B	Trial III	$(21, 25)$

In each trial a hypothesis H_0: *drug does not reduce blood pressure* would be rejected at a 5 percent significance level since the confidence intervals do not include zero. This implies strong evidence against H_0. Trial I is imprecise; we would accept in a significance test at the 5 percent level any mean reduction between 3 and 35 units. The former is not of clinical importance; the latter is. This small trial only answers questions about the "significant" mean reduction with *low precision*. The larger Trial II, using the same drug, indicates an average reduction between 9 and 12 units, a result of statistical significance, but not of clinical importance in this context. Compared to Trial I, it has *high precision*. Other relevant factors being unchanged, increasing the size of a trial increases the precision, this being reflected in shorter confidence intervals. Trial III using drug B also has high precision. It tells us the mean reduction is likely to be between 21 and 25 units, a difference of clinical importance. Drug B appears to be superior to Drug A.

For a given test, increasing sample size increases the probability that small departures from H_0 may provide strong evidence against H_0. The art of designing experiments is to take enough observations to ensure a good chance of detecting with reasonable precision departures from H_0 of practical importance, but to avoid wasting resources by taking so many observations that trivial departures from H_0 provide strong evidence against it. An introduction to sample size calculation is given by Kraemer and Thiemann (1987) and it is discussed with examples by Gibbons and Chakraborti (2004), by Hollander and Wolfe (1999), by Desu and Raghavarao (2004) and in many other books and articles. Practical design of experiments is best done with guidance from a trained statistician although many statistical software packages include programs giving recommendations in specific circumstances. In later chapters we show, for some tests, how to find sample sizes needed to meet specified aims.

Our discussion of hypothesis testing and estimation has used the frequentist approach to inference. The Bayesian school adopts a different philosophy, introducing subjective probabilities to reflect prior beliefs about parameters. Some statisticians are firm adherents of one or other of these schools, but a widely accepted view is that each has strengths and weaknesses and that one or the other may be preferred in certain contexts. However, for the procedures we describe sensible use of either approach will usually lead to similar conclusions despite the different logical foundations, so for consistency we use the frequentist approach throughout. For a reasoned argument for and against each approach see Little (2006) and for a comprehensive and realistic review of these and other approaches to inference Cox (2006) is recommended.

1.5 Ethical Issues

Ethical considerations are important both in general working practices (Gillon, 1986) and in the planning and conduct of investigations (Hutton, 1995). The main principles are respect for autonomy, nonmaleficence, beneficence and justice. Many research proposals need to be assessed by ethical committees before being approved. This applies particularly in medicine, but increasing attention is being given to ethical issues in environmentally sensitive fields like biotechnology and the social sciences, where questions of legal rights or civil liberties may arise. The role of statistics and statisticians in what are known as research ethics committees is discussed by Williamson et al (2000). The related issue of the development of guidelines for the design, execution and reporting of clinical trials is described by Day and Talbot (2000).

It is unacceptable to study some issues by allocating individuals to possible groups at random. For instance, in a study of the effects of smoking on health, one could not instruct individuals to smoke or to abstain from smoking. This disregards the autonomy of the study participants. An individual's choice of whether to smoke or not must be respected, and an alternative type of study planned to make this possible.

It is generally good practice to incorporate early stopping rules into a study. If it becomes clear at an early stage of an investigation that a new treatment is much better than the established alternative the study should be closed. A similar decision should be made if the new treatment quickly shows a highly increased risk of harmful side effects. Continuing a study could deprive patients of a more effective treatment or expose participants to unnecessary risks from side effects. Early stopping recommendations are reached through a small number of planned interim analyses that take place throughout the period of the study. Cannistra (2004) discusses ethical issues around the application of early stopping rules.

When planning a new study a comprehensive search of findings from related work is important using, for instance, MEDLINE, a continually updated source of information on articles from medical and biological journals. It is unethical to conduct research that ignores previous work that may be relevant because it is then likely that time, money and scarce resources will not be used to best effect. Nevertheless, results of literature searches need to be interpreted with caution. Studies with interesting findings are more likely to appear in print, leading to publication bias [see Easterbrook et al (1991)].

When there is little relevant prior knowledge it may be prudent to conduct an initial pilot study to highlight potential problems that might arise. Results from a pilot study can also be helpful in choosing an appropriate number of participants for the main study. A sufficiently large number should be involved at the pilot stage to have a reasonable chance of finding the expected difference between the two groups if it really exists. The intended method of statistical analysis also influences the sample size requirement. Small studies often fail to yield useful findings and are thus a poor use of resources. On the other hand, resources can be wasted by recruiting more participants than needed. In medical research, in either situation more patients than necessary are at risk of receiving an inferior treatment. Careful planning should consider the composition of the sample with respect to age, sex, ethnic group, etc., as this will enable problems under investigation to be answered more effectively.

In medical investigations each potential participant should receive a written information sheet outlining the main points about the study. All available information about the possible efficacy and side-effects of treatments involved in the study should be given to the patient. In practice, not all patients will understand, or even wish, to receive details beyond those given in the information sheet, particularly in a sophisticated trial. In this situation, the patient should be given a choice about what information is supplied. Once a trial has been completed, patients who feel that they have received an effective treatment for their health problem may wish to continue with it. Financial constraints and/or the concerns of the patient's general practitioner may prevent long-term use of the treatment; this should be discussed in advance as part of the patient information.

Ethical problems may preclude the use of the same patients to compare two treatments. For instance, in a comparison of two methods of treating oral cancer, both of which involve radiation, the estimated dose of radiation from the combined treatments may be considered unacceptably high for the patient.

The autonomy of the patient should be respected and the patient should only make a decision on whether or not to enter the trial following careful consideration of the information provided. This is particularly important with tests for inherited diseases that only become evident in later life. A positive finding may distress the patient, have serious implications for any children and prejudice life assurance proposals. Patients should give informed written consent to the investigator(s) prior to being entered into a trial and they should be allowed to withdraw from the trial at any time.

Data collected in studies should be kept confidential. In the United Kingdom, for example, computer records should adhere to the principles laid down in the Data Protection Acts. Data used for statistical purposes should not contain patients' names or addresses.

Limited availability of a treatment for experimental use may create ethical problems. Suppose that there were high hopes that a new drug might greatly relieve suffering in severe cases but only enough doses were available to treat six patients. The principles of beneficence and justice suggest that the six patients to receive the drug should be those with the most severe symptoms. In a situation like this, the drug may reduce suffering, but such patients may still, after treatment, be ranked as less well than patients receiving an alternative treatment because, although their condition may have improved, their symptoms may still be more severe than those of patients receiving the alternative treatment. In this situation any statistical analysis should be based on some measure of "degree of improvement" shown by each patient.

At the other extreme, an experimenter might allocate the new drug to the patients with the least severe symptoms. From a research point of view this is misleading, as even if it were ineffective or no better than an existing treatment, these patients may still show less severe symptoms. However, if it is likely that only patients in the early stages of the disease will benefit it is more appropriate from an ethical viewpoint to give the new drug to these patients.

Even when patients are allocated to treatments at random, and we find strong evidence to suggest we should abandon a hypothesis of *no treatment effect*, the statistically significant outcome may be of no practical importance, or there may be ethical reasons for ignoring it. A doctor would be unlikely to feel justified in prescribing the new treatment if it merely prolonged by three days the life expectation of terminally-ill patients suffering considerable distress, but may, from the principle of beneficence, feel bound to prescribe it if it substantially improved survival prospects and quality of life.

Statisticians may be guilty of unethical behaviour. A statistician who performs a number of competing tests — parametric or nonparametric — each producing a different P-value, but only publishes the P-value that is most favourable to the conclusion he or she wants to establish, regardless of whether it is obtained by using an appropriate test, is guilty of unethical suppression of evidence.

Ethical considerations may influence not only how an experiment is carried out (the experimental design) but also what inferences are possible and how these should be made.

1.6 Exercises

Whether the exercises below seem difficult or trivial will depend on the extent of the reader's prior training in statistics. Although not directly concerned with nonparametric methods, they are relevant to the background material covered in this chapter.

Solutions to exercises marked with an asterisk (*) in this and later chapters are discussed briefly in Appendix 2.

1.1   As in Example 1.1, suppose that one machine produces rods with diameters normally distributed with mean 27 mm and standard deviation 1.53 mm, so that 2.5 percent of the rods have diameter 30 mm or more. A second machine is known to produce rods with diameters normally distributed with mean 24 mm and 2.5 percent of rods it produces have diameter 30 mm or more. What is the standard deviation of rods produced by the second machine?

1.2   In a group of 145 patients admitted to hospital with a stroke, weekly alcohol consumption in standard units had a mean of 17 and a standard deviation of 22. Explain why their alcohol consumption does not follow a normal distribution. Is this finding surprising?

1.3   Following a television campaign about the risks of smoking tobacco, the cigarette consumption of a group of 50 smokers decreases by a mean of 5 cigarettes per day with a standard deviation of 8. Explain why the reasoning of Exercise 1.2 cannot be used to show that this distribution is not normal.

*1.4   In Section 1.3 we pointed out that 5 tosses of a coin would never provide evidence against the hypothesis that a coin was fair (equally likely to fall heads or tails) at a conventional 5 percent significance level. What is the least number of tosses needed to provide such evidence using a two-tail test, and what is then the exact P-value?

1.5   A biased coin is such that $\Pr(\text{heads}) = 2/3$. If this coin is tossed the least number of times calculated in Exercise 1.4, what is the probability of an error of the second kind associated with the 5 percent significance level? What is the power of the test? Does the discrete nature of possible P-values cause any problems in calculating the power?

*1.6 If a random variable X_i is distributed $N(\mu, \sigma^2)$ and all X_i are independent it is well known that the variable

$$Y = \sum_{i=1}^{n} X_i$$

is distributed $N(n\mu, n\sigma^2)$. Use this result to answer the following:

The times in minutes a farmer takes to place any fence post are each independently distributed $N(10, 2)$. He starts placing posts at 9 a.m one morning, and immediately one post is placed he proceeds to place another, continuing until he has placed 9 posts. What is the probability that he has placed all 9 posts by (1) 10.25 a.m, (ii) 10.30 a.m and (iii) 10.40 a.m?

*1.7 The following two sample data sets both have sample mean 6.

Set I	13.9	2.7	0.8	11.3	1.3
Set II	2.7	8.3	5.2	7.1	6.7

If μ is the population mean perform for each set t-tests of (i) H_0: $\mu = 8$ against H_0: $\mu \neq 8$ and (ii) H_0: $\mu = 10$ against H_0: $\mu \neq 10$, Do you consider the conclusions of the tests reasonable? Have you any reservations about using a t-test for either of these data sets?

*1.8 Use an available standard statistical software package, or one of the many published tables of binomial probabilities to determine, for samples of 12 from binomial distributions with $p = 0.5$ and with $p = 0.75$, the probabilities of observing each possible number of outcomes for each of these values of p. In a two-tail test of the hypotheses $H_0 : p = 0.5$ against H_1: $p = 0.75$ what is the largest attainable P-value less than 0.05? What is the critical region for a test based on this P-value? What is the power of the test?

CHAPTER 2

FUNDAMENTALS OF NONPARAMETRIC METHODS

2.1 A Permutation Test

Parametric inference assumes observations are samples from populations with distributions belonging to a specified family. We pointed out in the previous chapter that for nonparametric inference we make only weaker assumptions such as one of symmetry, or where two or more populations are involved, that their distributions differ, if at all, only in some measure of location such as their *medians*. This calls for a new approach to hypothesis testing and estimation.

We introduce some basic ideas and illustrate their use primarily by examples. Our first illustration describes the procedure called a *permutation test*.

Example 2.1

Four from nine patients are selected at random to receive a new drug. The remaining five are treated with a standard drug. After three weeks all nine patients are examined by a skilled consultant who, on the basis of various tests and clinical observations ranks the patients' conditions in order from most satisfactory (rank 1) to least satisfactory (rank 9). If there is no beneficial effect of the new drug, what is the probability that the patients who received the new drug are ranked 1, 2, 3, 4?

Selecting four patients "at random" means that any four are equally likely to be given the new drug. If there really is no effect one would expect some of those chosen to end up with low ranks, some with moderate or high ranks, in the post-treatment assessment. It is not impossible, but less likely, that those chosen would be ranked 1, 2, 3, 4 or 6, 7, 8, 9 after treatment.

There are 126 ways of selecting a set of four from nine patients. This may be verified using the well-known mathematical result that the number of ways of selecting r objects from n is $n!/[r!(n-r)!]$. For any integer m, the expression $m!$, called *factorial* m, is the product of all integers between 1 and m. We also define $0! = 1$. Table 2.1 gives all 126 selections. Ignore for the moment the numbers in parentheses after each selection.

If the new drug were ineffective the set of ranks associated with the four patients receiving it are equally likely to be any of the 126 quadruplets listed in Table 2.1. Thus, if there is no treatment effect there is only 1 chance in 126 that the four showing greatest improvement (ranked 1, 2, 3, 4 in order of condition after treatment) are the four patients allocated to the new drug. It is more plausible that such an outcome reflects a beneficial effect of the drug.

Table 2.1 *Possible selections of four individuals from nine labelled 1 to 9 with the sum of the labels (ranks) in parentheses.*

1,2,3,4 (10)	1,2,3,5 (11)	1,2,3,6 (12)	1,2,3,7 (13)	1,2,3,8 (14)
1,2,3,9 (15)	1,2,4,5 (12)	1,2,4,6 (13)	1,2,4,7 (14)	1,2,4,8 (15)
1,2,4,9 (16)	1,2,5,6 (14)	1,2,5,7 (15)	1,2,5,8 (16)	1,2,5,9 (17)
1,2,6,7 (16)	1,2,6,8 (17)	1,2,6,9 (18)	1,2,7,8 (18)	1,2,7,9 (19)
1,2,8.9 (20)	1,3,4,5 (13)	1,3,4,6 (14)	1,3,4,7 (15)	1,3,4,8 (16)
1,3,4,9 (17)	1,3,5,6 (15)	1,3,5,7 (16)	1,3,5,8 (17)	1,3,5,9 (18)
1,3,6,7 (17)	1,3,6,8 (18)	1,3,6,9 (19)	1,3,7,8 (19)	1,3,7,9 (20)
1,3,8,9 (21)	1,4,5,6 (16)	1,4,5,7 (17)	1,4,5,8 (18)	1,4,5,9 (19)
1,4,6,7 (18)	1,4,6,8 (19)	1,4,6,9 (20)	1,4,7,8 (20)	1,4,7,9 (21)
1,4,8,9 (22)	1,5,6,7 (19)	1,5,6,8 (20)	1,5,6,9 (21)	1,5,7,8 (21)
1,5,7,9 (22)	1,5,8,9 (23)	1,6,7,8 (22)	1,6,7,9 (23)	1,6,8,9 (24)
1,7,8,9 (25)	2,3,4,5 (14)	2,3,4,6 (15)	2,3,4,7 (16)	2,3,4,8 (17)
2,3,4,9 (18)	2,3,5,6 (16)	2,3,5,7 (17)	2,3,5,8 (18)	2,3,5,9 (19)
2,3,6,7 (18)	2,3,6,8 (19)	2,3,6,9 (20)	2,3,7,8 (20)	2,3,7,9 (21)
2,3,8,9 (22)	2,4,5,6 (17)	2,4,5,7 (18)	2,4,5,8 (19)	2,4,5,9 (20)
2,4,6,7 (19)	2,4,6,8 (20)	2,4,6,9 (21)	2,4,7,8 (21)	2,4,7,9 (22)
2,4,8,9 (23)	2,5,6,7 (20)	2,5,6,8 (21)	2,5,6,9 (22)	2,5,7,8 (22)
2,5,7,9 (23)	2,5,8,9 (24)	2,6,7,8 (23)	2,6,7,9 (24)	2,6,8,9 (25)
2,7,8,9 (26)	3,4,5,6 (18)	3,4,5,7 (19)	3,4,5,8 (20)	3,4,5,9 (21)
3,4,6,7 (20)	3,4,6,8 (21)	3,4,6,9 (22)	3,4,7,8 (22)	3,4,7,9 (23)
3,4,8,9 (24)	3,5,6,7 (21)	3,5,6,8 (22)	3,5,6,9 (23)	3,5,7,8 (23)
3,5,7,9 (24)	3,5,8,9 (25)	3,6,7,8 (24)	3,6,7,9 (25)	3,6,8,9 (26)
3,7,8,9 (27)	4,5,6,7 (22)	4,5,6,8 (23)	4,5,6,9 (24)	4,5,7,8 (24)
4,5,7,9 (25)	4,5,8,9 (26)	4,6,7,8 (25)	4,6,7,9 (26)	4,6,8,9 (27)
4,7,8,9 (28)	5,6,7,8 (26)	5,6,7,9 (27)	5,6,8,9 (28)	5,7,8,9 (29)
6,7,8,9 (30)				

In a hypothesis testing framework we have a group of four treated with the new drug and a group of five (the remainder) given a standard drug in what is called a *two independent sample experiment*. We discuss such experiments in detail in Chapter 6. The most favourable evidence for the new drug would be that those receiving it are ranked 1, 2, 3, 4; the least favourable that they are ranked 6, 7, 8, 9. Each of these extremes has a probability of 1/126 of occurring when there is no real effect.

If we consider a test of

H_0: *new drug has no effect*

against the two-sided alternative

H_1: *new drug has an effect (beneficial or deleterious)*

the outcomes 1, 2, 3, 4 and 6, 7, 8, 9 are extremes with a total associated probability $P = 2/126 \approx 0.0159$ if H_0 is true.

Table 2.2 *Number of occurences for each sum of ranks of four items from nine.*

Rank sum	10	11	12	13	14	15	16	17	18	19	20
Occurences	1	1	2	3	5	6	8	9	11	11	12

Rank sum	21	22	23	24	25	26	27	28	29	30
Occurrences	11	11	9	8	6	5	3	2	1	1

In classic hypothesis testing terms we speak of rejecting H_0 at an exact 1.59 percent significance level if we observed either of these extreme outcomes. This small P-value provides strong evidence that the new drug has an effect. What if the patients receiving the new drug were ranked 1, 2, 3, 5? Intuitively this evidence looks to favour the new drug. How do we test this?

We seek a statistic, i.e., some function of the four ranks, that has a low value if all ranks are low, a high value if all ranks are high and an intermediate value if there is a mix of ranks for those receiving the new drug. An intuitively reasonable choice is the sum of the four ranks. If we sum the ranks for every quadruplet in Table 2.1, and count how many times each sum occurs we may easily work out the probability of getting any particular sum, and hence the distribution of our test statistic when H_0 is true.

In Table 2.1 the number in parentheses after each quadruplet is the sum of the ranks for that quadruplet, e.g., for 1, 2, 7, 9 the sum is $1 + 2 + 7 + 9 = 19$. The lowest sum is 10 for 1, 2, 3, 4 and the highest is 30 for 6, 7, 8, 9. Table 2.2 gives the numbers of quadruplets having each given sum.

Because there are 126 different, but equally likely, sets of ranks the probability that the rank sum statistic, which we denote by S, takes a particular value is obtained by dividing the number of times that value occurs by 126. For example,

$$\Pr(S = 17) = 9/126 \approx 0.0714.$$

To find what outcomes are consistent with a P-value not exceeding 0.05, we select a region in each tail (since H_1 implies a two-tail test) with a total associated probability not exceeding 0.025. It is easily seen from Table 2.2 that if we select in the lower tail $S = 10$ and $S = 11$, the associated probability is 2/126 and if we add $S = 12$ the associated total probability, i.e., $\Pr(S \leq 12) = 4/126 \approx 0.0317$. This exceeds 0.025, so our lower-tail critical region should be $S \leq 11$ giving $P = 2/126 \approx 0.0159$. By symmetry, the upper-tail region is $S \geq 29$ also with $P \approx 0.0159$. Thus, for a two-tail test the largest symmetric critical region with $P \leq 0.05$ is $S = 10, 11, 29, 30$ and the exact $P = 4/126 \approx 0.0317$.

Some statisticians suggest choosing a critical region with probability as close as possible to a target level such as $P = 0.05$ rather than the more conservative choice of one no larger. In this example, adding $S = 12$ and the symmetric $S = 28$ to our critical region gives a two-tail $P = 8/126 \approx 0.0635$. This is closer to 0.05 than the size (0.0317) of the region chosen above. We reaffirm that ideally it is best to quote the exact P-value obtained, and point out again that the practical argument (though there are further theoretical ones) for quoting nominal sizes such as 0.05

is that many tables give only these, although a few, e.g., Gibbons and Chakraborti (2004) in their Table J and Hollander and Wolfe (1999) in their Table A6, give relevant exact P-values for many sample size combinations and different values of S. Computer programs giving exact P-values overcome any difficulty if the latter type of table is not readily available.

Unless there are strong reasons before the experiment is started to believe that an effect, if any, of the new drug could only be beneficial, a two-tail test is appropriate. We consider a one-tail test scenario in Exercise 2.2.

In Section 1.2 we suggested that in preliminary testing of drugs for treating a rare disease our population may be in a strict sense only the cases we have. However, if these patients are fairly typical of all who might have the disease, it is not unreasonable to assume that findings from our small experiment may hold for any patients with a similar condition providing other factors (nursing attention, supplementary treatments, consistency of diagnosis, etc.) are comparable. When our experiment involves what is effectively the whole population, and the only data are ranks, a permutation test is the best test available. Random allocation of treatments is essential for the test to be valid; this may not always be possible in the light of some ethical considerations that we discussed in Section 1.5.

Tests based on permutation of ranks or on permutation of certain functions of ranks (including the original measurements on a continuous scale when these are available) are central to many nonparametric methods. They are called *permutation* or *randomization* tests. The latter term applies when the permutation process is based on the randomization procedure used to assign treatments to units. That was the situation in Example 2.1, the permutations giving all possible assignments. These tests have an intuitive appeal and comply with well-established theoretical criteria for sound inference. This theoretical basis is summarized by Hettmansperger and McKean (1998) for many different procedures.

Small scale tests of a drug like that in Example 2.1 are often called *pilot studies*. Efficacy of a drug in wider use may depend on factors like severity of disease, treatment being administered sufficiently early, the age and sex of patients, etc. All or none of these may be reflected in a small group of available patients. An encouraging result with the small group may suggest further experiments are desirable. A not very small P-value associated with what looks to be an intuitively encouraging result may indicate that a larger experiment is needed to tell us anything useful.

2.2 Binomial Tests

One observation was censored in the data in Example 1.2. We mentioned that it could be shown that it was not unreasonable, given that data, to accept a hypothesis that the population median was 200. We now consider an appropriate test to justify that conclusion.

Example 2.2

The data for survival times in weeks given in Example 1.2 were

$$49 \quad 58 \quad 75 \quad 110 \quad 112 \quad 132 \quad 151 \quad 276 \quad 281 \quad 362^*$$

The asterisk denotes the censored observation.

We want to test the hypothesis that the median, θ, of survival times for the population from which the sample was obtained is 200 against the alternative of some other value, i.e., to test

$$H_0: \theta = 200 \text{ against } H_1: \theta \neq 200 \tag{2.1}$$

A simple test needs only a count of the number of sample values exceeding 200 (recording each as a "plus"). By the definition of a random sample and that of a population median, if we have a random sample from any continuous distribution with median 200 each sample value is equally likely to be above or below 200. This means that under H_0 the number of plus signs has a binomial $B(10, 0.5)$ distribution.

The probability of observing r plus signs in 10 observations when $p = 0.5$ is given by the binomial formula

$$p_r = \Pr(X = r) = \binom{10}{r}\left(\frac{1}{2}\right)^{10}$$

where

$$\binom{10}{r} = \frac{10!}{r!(10 - r)!}$$

and is called the *binomial coefficient*.

The values of these probabilities, p_r, for each value of r between 0 and 10, correct to 4 decimal places, are

r	0	1	2	3	4	5
p_r	0.0010	0.0098	0.0439	0.1172	0.2051	0.2461

r	6	7	8	9	10
p_r	0.2051	0.1172	0.0439	0.0098	0.0010

In the data 3 observations, including the censored one, exceed 200 so there are 3 plus signs and, from the table above, we see that when H_0 is true the probability of 3 or less plus signs in a sample of 10 is $0.1172 + 0.0439 + 0.0098 + 0.0010 = 0.1719$. There is no strong evidence against H_0, tail probabilities for our observed statistic, the number of plus signs) is $2 \times 0.1719 = 0.3438$. This implies that departures from the expected number of plus signs, 5, as large, or larger, than that observed will occur in slightly more than one-third of all samples when H_0 is true. This simple test, called the *sign test*, is discussed more fully in Section 3.3. The test is *distribution-free* because we have made no assumption about the form of the continuous distribution of the underlying observations. We have only formulated and tested hypotheses concerning possible values of the population median.

When applying a t-test, or most other parametric tests, all values of P between 0 and 1 are possible. For the sign test, however, only certain discrete P-values occur. In this example, for a two-tail test the three smallest are $P = 2 \times (0.0010) = 0.0020$ corresponding to 0 or 10 plus; $P = 2 \times (0.0010 + 0.0098) = 0.0216$ corresponding to 1 or 9 plus; then $P = 2 \times (0.0010 + 0.0098 + 0.0439) = 0.1094$ corresponding to 2 or 8 plus. Next comes the observed $P = 0.3438$. In all cases probabilities have been rounded to four decimal places. For a one-tail test these P-values are all halved. Our statistic — the number of plus signs — has a discrete distribution. This means that, as in Example 2.1, there is no direct way of obtaining a critical region of exact size 0.05 for a two-tail test; we must choose between regions of size 0.0216 or 0.1094.

Once they are recognized, and the consequences appreciated, discontinuities in possible P-values do not cause serious interpretational problems in the analysis of a particular data set. However, these discontinuities do lead to some theoretical difficulties in comparing performance of competing tests.

A device called a *randomized decision rule* has been proposed with the property that in the long run an error of the first kind has, in repeated testing, a probability at a prechosen nominal level, e.g., at 5 percent. In practice, our prime interest is what happens in our one test, so it is better, when we know them, to use exact levels, rather than worry about nominal arbitrary levels. An account of how a randomized decision rule works is given by Gibbons and Chakraborti (2004) (pp. 28–29). They rightly comment that such devices may seem artificial and are "probably seldom employed by experimenters". We suggest they should never be used in real-world applications.

There is, however, when there are discontinuities, a case for forming a tail probability by allocating only one half of the probability that the statistic equals the observed value to the "tail" when determining the size of the "critical" region. This approach has many advocates. We do not use it in this book, but if it is used this should be done consistently.

The sign test provides a basis for forming a confidence interval for the population median.

Example 2.3

We saw in Example 2.2, when using a sign test for a median with a sample of 10, we would, in a two-tail test at the 2.16 percent level, accept H_0 if we got between 2 and 8 plus signs.

Consider again the data in that example, i.e.,

$$49 \quad 58 \quad 75 \quad 110 \quad 112 \quad 132 \quad 151 \quad 276 \quad 281 \quad 362^*$$

where the asterisk represents a censored observation. We have between 2 and 8 plus signs if the median specified in H_0 has any value greater than 58 but less than 281. This implies that the interval (58, 281) is a $100(1 - 0.0216) = 97.84$ percent confidence interval for θ, the population median survival time. Since we would accept any H_0 that specified a value for the median greater than 58 but less than 281, there is considerable doubt about the population median value. It is almost an understatement to say the estimate lacks precision.

Care is needed in interpreting P-values especially in one-tail tests. Most computer programs for nonparametric tests quote the probability that a value greater than or equal to the test statistic will be attained if this probability is less than 0.5, otherwise they give the probability that a value less than or equal to the test statistic is obtained. This is the probability of errors of the first kind in a one-tail test if we decide to reject at a significance level equal to that probability. In practice, the evidence against H_0 is only rated strong if this "tail" probability is sufficiently small, and is in the appropriate tail. In general, we recommend doubling a one-tail probability to obtain the actual significance level for a two-tail test, but see Example 2.4 and the remarks following it. If the test statistic has a symmetric distribution, doubling is equivalent to considering equal deviations from the median value of the statistic in either direction. If the statistic does not have a symmetric distribution, taking tails equidistant from the mean is not equivalent to doubling a one-tail probability.

Example 2.4 exposes another difficulty that sometimes arises due to discontinuities in P-values; namely, that if we only regard the evidence against H_0 as strong enough to reject that hypothesis if $P \leq 0.05$ (or at any rate a value not very much greater than this), we may never get that evidence because no outcome provides it, a problem we have already alluded to with small experiments.

Example 2.4

In a dental practice, experience has shown that 75 percent of adult patients require treatment following a routine inspection. So the number of individuals requiring treatment, S, in a sample of 10 independent patients has a binomial B(10, 0.75) distribution. Here the probabilities for the various values, r, of the statistic S, where r takes integral values between 0 and 10, are given by

$$p_r = \Pr(X = r) = \binom{10}{r} \left(\frac{3}{4}\right)^r \left(\frac{1}{4}\right)^{10-r}.$$

The relevant probabilities are

r	0	1	2	3	4	5
p_r	0.0000	0.0000	0.0004	0.0031	0.0162	0.0584

r	6	7	8	9	10
p_r	0.1460	0.2503	0.2816	0.1877	0.0563

If we had data for another practice and wanted, for that practice, to test H_0: $p = 0.75$ against H_1: $p > 0.75$, the smallest P-value for testing is in the upper tail and is associated with $r = 10$, i.e., $P = 0.0563$. This means that if we only regard $P \leq 0.05$ as sufficiently strong evidence to discredit H_0 such values are never obtained.

There would be no problem here for a one-tail test of H_0: $p = 0.75$ against H_1: $p < 0.75$ since, in the appropriate lower tail, $P = \Pr(S \leq 4) = 0.0162 + 0.0031 + 0.0004 = 0.0197$.

This example also shows a logical difficulty associated with a rule that the appropriate level for a two-tail test is twice that for a one-tail test, for if we get $S = 4$ the two-tail test level based on this rule is $2 \times 0.0197 = 0.0394$. This presents a dilemma, for there is no observable upper tail area corresponding to that in the lower tail. This means that if a two-tail test is appropriate, we shall in fact only be likely to detect departures from the null hypothesis if they are in one direction. There may well be a departure in the other direction, but if so we are highly unlikely to detect it at the conventional level $P \leq 0.05$. Even if we did, it would be for the wrong reason. This is not surprising when, as shown above, the appropriate one-tail test must fail to detect it, for generally a one-tail test at a given significance level is more powerful for detecting departures in the appropriate direction than is a two-tail test at the same level.

An implication is that in this example we need a larger sample to detect departures of the form $H_1 : p > 0.75$. Again, the fairly large P-value associated with the possible critical region for the one-tail test only tells us our sample is too small.

The stipulation that the patients be independent is important. If the sample included three members of the same family it is quite likely that if one of them were more (or less) likely to require treatment than the norm, this may also be the case for other members of that family. We consider situations of this kind in more detail in Section 15.7

There is no universal agreement that one should double a one-tail probability to get the appropriate two-tail significance level — see, for example, Yates (1984) and the discussion thereon. An alternative is that once the exact size of a one-tail region has been determined, we should, for a two-tail test, add the probabilities associated with an opposite tail situated equidistant from the mean value of the test statistic to that associated with our observed statistic value. In the symmetric case, as already pointed out, this is equivalent to doubling the probability, but it seems inappropriate with a nonsymmetric distribution. In Example 2.4 the region $r \leq 4$ is appropriate for a lower-tail test. The mean of the test statistic (the binomial mean np) is here 7.5. Since $7.5 - 4 = 3.5$, the corresponding deviation above the mean is $7.5 + 3.5 = 11$. Because $\Pr(r \geq 11) = 0$, the two-tail test based on equidistance from the mean would have the same exact significance level as the one-tail test.

2.3 Order Statistics and Ranks

Many nonparametric procedures are based on the ordering, or ranking, of detailed observations. In Examples 2.2, we did not use ranks, but ordering was inherent in our procedure. We took order into account in determining the number of survival times that exceeded the hypothesized median.

Ordering data is important in more general statistical contexts, both parametric and nonparametric. We may be interested in the distribution of the largest or smallest observations in a sample to answer questions such as

- On the basis of maximum flood levels recorded in a river over a number of years, what is the probability of the level exceeding, say, 5 m, in future?
- Given a sample of times to first breakdown of a certain brand of computer, what is the probability of a first breakdown being observed within 6 months in one machine in a production run of 1000 machines?

In a parametric context such questions are often answered using families of distributions called *extreme value distributions*. A simple example of the role of order statistics in a parametric context is given in Exercise 2.10.

Greatest and least values in samples are just two examples of *order statistics*. The sample median is also an order statistic.

A detailed account of order statistics and their properties is given by Gibbons and Chakraborti (2004, Chapter 2). Here we only indicate the relevance of these statistics to nonparametric inference, and quote some key results without proof.

Consider a sample of n observations $x_1, x_2, \ldots, x_n$ from a continuous distribution. Continuity implies that there should be no ties and thus observations may be uniquely ordered from smallest to largest. We denote the smallest observation by $x_{(1)}$, the second smallest by $x_{(2)}$ and so on, finally the largest by $x_{(n)}$. It follows that

$$x_{(1)} < x_{(2)} < \cdots < x_{(n)}$$

The $x_{(i)}, i = 1, 2, \ldots, n$ are called the *order statistics*. The minimum order statistic, $x_{(1)}$, is relevant to the study of minimum extremes such as the distribution of shortest times to a machine breakdown, or minimum survival times after some treatment. The largest, $x_{(n)}$, is relevant to the study of floods, or maximum time to failure of a certain type of lightbulb.

The median is widely used as a measure of location in nonparametric inference, and the sample median is defined in terms of order statistics. For a sample of n the median is $x_{[(n+1)/2]}$ if n is odd, and is usually defined as $[x_{(m)} + x_{(m+1)}]/2$ if $n = 2m$ is even. A possible measure of dispersion is the sample range $x_{(n)} - x_{(1)}$. More satisfactory measures are the *interquartile range* or *semi-interquartile range* defined in Section 2.4. One reason for preferring one of the latter is that the extreme order statistics are often strongly influenced by suspect observations associated with the terms *outliers* and *dirty data*.

In nonparametric inference an important concept based on order statistics is the *sample*, or *empirical*, distribution function. For a random sample of size n from a population having cumulative distribution function $F(x)$, the sample, or empirical, distribution function is defined as

$$S_n(x) = \frac{\text{number of sample values } \leq x}{n}$$

For any x the value of $S_n(x)$ is expressible in terms of the order statistics.

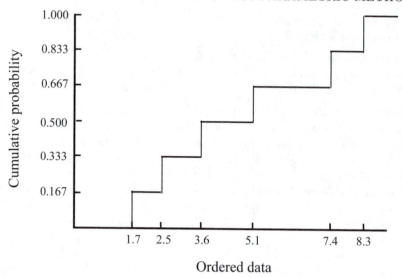

Figure 2.1 *Sample distribution function for a sample of six.*

It is easy to see that

$$
\begin{aligned}
S_n(x) &= 0 && \text{if } x < x_{(1)}, \\
S_n(x) &= i/n && \text{if } x_{(i)} \le x < x_{(i+1)}, \; i = 1, 2, \ldots, n-1, && (2.2) \\
S_n(x) &= 1 && \text{if } x \ge x_{(n)}.
\end{aligned}
$$

The function $S_n(x)$ is a step function with a step of size $1/n$ at $x = x_{(i)}$, $i = 1, 2, \ldots, n$. This is illustrated in Figure 2.1 for six observations

$$1.7, \; 2.5, \; 3.6, \; 5.1, \; 7.4, \; 8.3\,.$$

The sample cumulative distribution function $S_n(x)$ is important because it is closely related to the population cumulative distribution function $F(x)$. This is reflected in the following properties:

- The mean value of $S_n(x)$ is $E[S_n(x)] = F(x)$.
- The variance of $S_n(x)$ is $\mathrm{Var}[S_n(x)] = F(x)[1 - F(x)]/n$.
- $S_n(x)$ is a consistent estimator of $F(x)$ for any fixed x.

The term *consistent estimator* implies that $S_n(x)$ converges in probability to $F(x)$ as n tends to infinity. For a proof of these properties see Gibbons and Chakraborti (2004, Section 2.3).

The assumption that $F(x)$ is continuous rules out, in theory, the possibility of tied observations. In practice tied values in sample data are not uncommon. This may be due to rounding in the recording of data, or to the population distribution being not strictly continuous. We see the practical implications of tied data in specific techniques in later chapters.

2.4 Exploring Data

The addition of nonparametric or distribution-free methods to the procedures for making statistical inferences widens the choice of techniques appreciably, An invaluable first step in selecting an appropriate technique in any given situation is to use *exploratory data analysis* or EDA. Some basic tools of EDA are

- Descriptive statistics.
- Boxplots.
- Histograms and frequency curves.
- Empirical and theoretical cumulative distribution graphs.

Descriptive statistics are commonly presented in lists or tables. The other tools above are by nature graphical. Commonly met descriptive statistics that summarize key features of sample data are the sample mean, median, maximum value, minimum value, standard deviation and quartiles. Slightly less well-known ones, but often of interest when questions of robustness arise, are the *trimmed mean* and the *Winsorized mean*. We introduce the last two in Chapter 14. Statistics such as the mean, median, standard deviation, or other derived quantities, are often referred to as *secondary* data to distinguish them from the original raw or observational data called *primary data*. Most general statistical software packages have a facility for computing a wide range of descriptive statistics.

A study of relevant descriptive statistics may give a quick indication of, for example, whether an assumption of normality appears to be seriously invalidated, or whether it is reasonable to suppose the sample comes from a symmetric or a skew distribution; and if the latter, whether the long tail is to the left or right. Such basic characteristics may be explored more fully by graphical techniques. These are often useful to indicate how well samples reflect population features. We have already indicated in Section 2.3 that the sample, or empirical, distribution function is a consistent estimator of the population cumulative distribution function.

We indicate the use of some basic EDA tools by examples.

Example 2.5

In Appendix 1 we give four small data sets and indicate how they were collected. Table 2.3 gives a set of descriptive statistics useful for summarizing and comparing the data for each of the four sets.

The first row tells us the number of data for each set. The small sample of 13 observations for the *McDelta* clan might be expected to be less informative than the sample of 59 *McAlphas*.

Table 2.3 *Descriptive, or summary, statistics for Badenscallie data given in Appendix 1.*

Clan	McAlpha	McBeta	McGamma	McDelta
Number	59	24	21	13
Mean	61.8	61.1	62.9	48.1
Median	74.0	67.5	77.0	65.0
St. dev	27.52	24.92	26.77	33.45
SE mean	3.58	5.08	5.84	9.28
Minimum	0	0	13	1
Maximum	95	96	88	87
1st quartile	44.0	41.5	33.0	13.0
3rd quartile	81.00	78.75	83.50	80.00
Range	95	96	75	86
IQ range	37.00	37.25	50.50	67.00

The sample mean for *McDelta* is markedly lower than that for the other clans. We may be interested in whether this indicates a shorter average life expectancy for that clan, or whether the difference represents some sampling quirk that might disappear if we had a larger sample.

The medians are all appreciably higher than the means, suggesting that the distributions of ages are asymmetric. This follows because we expect samples to reflect broadly the population characteristics, and for symmetric population distributions the mean and median coincide.

The abbreviation *St. dev* is used in the table for the *standard deviation*, the well-known measure of spread that in the case of a sample from a normal population is an appropriate estimator of the parameter σ. Once again the clan *McDelta* is the odd one out.

SE mean is an abbreviation for *standard error of the mean*. If we denote the sample standard deviation by s, then the standard error of the mean is computed as $s/\sqrt{n}$. Thus, the standard error decreases with sample size for a given standard deviation.

The maximum and minimum ages at death indicate at least one case of infant mortality for each clan except *McGamma*, and at least one nonogenarian survivor for two of the clans.

The *quartiles* divide each ordered sample into four groups of equal size. If we consider the median as dividing the sample into two groups of equal size, the first quartile is in effect the median of the group of lower values and the third quartile is the median of the group of higher values. More formally the first quartile is the median of $x_{(1)}, x_{(2)}, \cdots x_{((n-1)/2)}$ if n is odd and is the median of $x_{(1)}, x_{(2)}, \cdots x_{(n/2)}$ if n is even, with corresponding definitions for the third quartile. The *second quartile* is the sample median. While the third quartiles are similar for all clans, the first

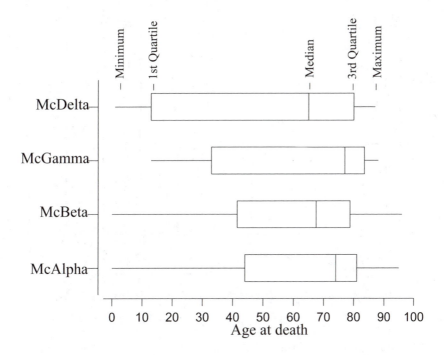

Figure 2.2 *Boxplots for Badenscallie data given in Appendix 1.*

quartile is strikingly low for *McDelta*. In a more formal analysis we may want to know if this can be accounted for by a quirk of the relatively small sample, or if it represents a different age distribution from that of the other clans.

Range and interquartile range, the latter abbreviated in the table to *IQ range*, are respectively the differences *maximum–minimum* and *third quartile–first quartile*. Each is a measure of spread alternative to standard deviation. Of the two, the interquartile range is preferred because range depends only on two observations x_1 and x_n, either of which may represent some unusual, or even a rogue, observation. On the other hand the interquartile range covers an interval containing the central 50 percent of the observations. Intuitively, this may be expected to be a more stable estimate of general variability. As an alternative to the interquartile range the *semi-interquartile range* is often used. As its name implies, it is obtained by dividing the interquartile range by 2. For the clan data the striking differences in interquartile range might be an aspect of the data requiring further analysis.

The five descriptive statistics presented in the order *minimum, 1st quartile, median, 3rd quartile, maximum* consititute a *five number summary*. This is the basis of what is called a *boxplot* or a *box and whisker plot*. Figure 2.2 gives boxplots for each clan for the Badenscallie data based on 5-number summaries easily obtained from Table 2.3.

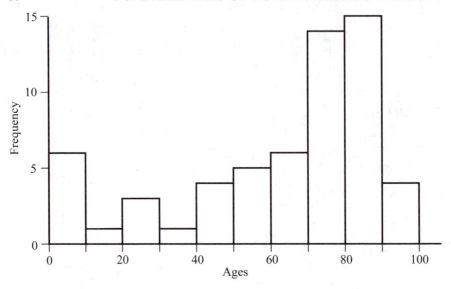

Figure 2.3 *Histogram for clan McAlpha data given in Appendix 1.*

The labels attached to the boxplot for the McDelta clan apply to any box-plot and indicate that the *box* section extends from the first to the third quartile. The vertical line dividing this box into two portions represents the median. The horizontal line outside the boxes extends from the minimum to the maximum.

Including box and whisker plots for all four clans on the one diagram enables useful comparisons of the kind outlined above to be made very easily. In particular, remembering that half the observations lie at or above the median, and half lie at or below the median, we see that for all clans the distribution of ages at death is skewed to the left or lower tail. This is made very clear by the median in all cases being nearer to the third quartile than to the first quartile. Recall that the quartiles are effectively the medians of the lower and upper halves of the data respectively.

Histograms are another widely used graphical device to exhibit key data characteristics. Figure 2.3 is a histogram based on the clan McAlpha data for ages at death with a class interval of 10 years. The long tail to the left is evident. There is also an indication of a mixture of distributions, with a smaller portion of the data indicating *infant mortality* or death before reaching adulthood, while the larger portion represents a more normal (in the physiological but not necessarily in the statistical sense) lifespan peaking at an age close to 80 years.

We pointed out in Section 2.3 that the sample or empirical distribution function was a consistent estimator of the population distribution function. This reflects the fact that as the size of a random sample increases it mirrors

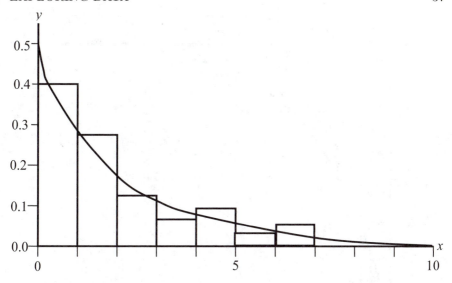

Figure 2.4 *Histogram for a sample of 50 from an exponential distribution with mean 2. The fitted curve is that of the distribution frequency function.*

the population characteristics ever more closely. Modern statistical software packages allow one to draw random samples of any chosen size from a wide range of distributions. For reasonably large samples, i.e., those of at least 50 observations, constructing appropriate histograms and superimposing these on the relevant population distribution frequency function gives a good impression of how effective these matches are.

Example 2.6

A computer-generated sample of 50 observations from an exponential distribution with mean 2 using Minitab gave the histogram in Figure 2.4. All sample values were less than 10, and 20 of them lay in the interval [0, 1), 14 in the interval [1, 2), 6 in the interval [2, 3), and so on. The curve superimposed on the histogram is that of the *frequency function* or *probability density function* of the exponential distribution with mean 2, which has the form:

$$f(x) = \frac{1}{2}e^{-x/2}, \qquad x \geq 0.$$

Statisticians would regard the closeness of the curve to the histogram as an indication that the data might be a sample from this distribution.

For samples smaller than 50 the grouping required to form a histogram may result in a rather poor fit to the population frequency function. However, even for small samples the sample distribution step function usually lies fairly close to the population cumulative distribution function.

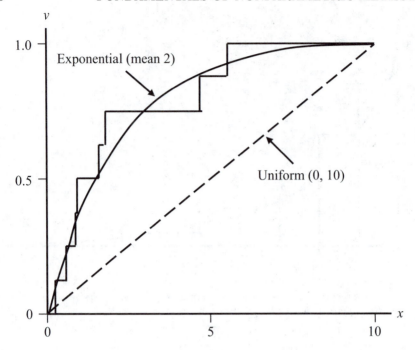

Figure 2.5 *Sample cumulative distribution function (stepped) for a sample of eight from an exponential distribution with mean 2. The curve is the population cumulative distribution function and the straight line is that for a uniform distribution over (0,10).*

Example 2.7

A computer generated sample of eight from an exponential distribution with mean 2 gave the values

$$0.25 \quad 0.53 \quad 0.91 \quad 0.94 \quad 1.56 \quad 1.73 \quad 4.71 \quad 5.50$$

where these have been arranged in ascending order. Figure 2.5 shows the sample cumulative distribution function for these data (stepped function) and the cumulative distribution function for an exponential function with mean 2. This takes the form

$$F(x) = 1 - e^{-x/2}, \qquad x \geq 0.$$

The step function lies close to this population cumulative distribution function. For illustrative purposes the straight line joining the points $(0, 0)$ and $(10, 1)$ on the graph is the cumulative distribution function for a uniform distribution over $(0, 10)$. It is almost self-evident that our sample was not taken from that distribution.

   More sophisticated EDA methods include the so-called P–P and Q–Q plots, abbreviations for plots of probabilities and of quantiles respectively associated with two distributions or with a hypothesized distribution and a sample believed to be from a population having that distribution. A description of how these are used and interpreted is given by Gibbons and Chakraborti (2004, Section 4.7).

The examples in this section only touch on the potential of an EDA approach. Further examples are given throughout this book.

2.5 Efficiency of Nonparametric Procedures

We pointed out in Section 1.3 that the power of a test depends upon (i) the sample size, n, (ii) the choice of the largest P-value to indicate significance (usually denoted in power studies by α), (iii) the magnitude of any departure from H_0 and (iv) whether assumptions that are needed for validity hold.

Most intuitively reasonable tests have good power to detect a true alternative that is far removed from the null hypothesis providing the data set is large enough. We sometimes want tests to have as much power as possible for detecting alternatives close to H_0 even when these are of no practical importance. This is because such tests are usually also good at detecting larger departures, a desirable state of affairs.

If α is the probability of a Type I error, and β is the probability of a Type II error (the power is $1 - \beta$), then the *efficiency* of a test T_2 relative to a test T_1 is the ratio n_1/n_2 of the sample sizes needed to obtain the same power for the two tests with these values of α, β. In practice, we usually fix α at some P-value appropriate to the problem at hand. Then β depends on the particular alternative as well as the sample sizes. Fresh calculations of relative efficiency are required for each particular value of the parameter or parameters of interest in H_1 and for each choice of α, β.

Pitman (1948), in a series of unpublished lecture notes, introduced the concept of *asymptotic relative efficiency* for comparing two tests. He considered sequences of tests T_1, T_2 in which we fix α and then allow the alternative in H_1 to vary in such a way that β remains constant as the sample size n_1 increases. For each n_1 we determine n_2 such that T_2 has the same β for the particular alternative considered.

Increasing sample size usually increases the power for alternatives closer to H_0. Therefore, for large samples, Pitman studied the behaviour of the efficiency, n_1/n_2, for steadily improving tests for detecting small departures from H_0. He showed under very general conditions that in these sequences of tests n_1/n_2 tended to a limit as $n_1 \to \infty$. More importantly, this limit, which he called the *asymptotic relative efficiency* (ARE) was the same for all choices of α, β. A full discussion of asymptotic relative efficiency is given by Gibbons and Chakraborti (2004, Chapter 13).

Bahadur (1967) proposed an alternative definition that is less widely used, so for clarity and brevity we refer to Pitman's concept simply as the *Pitman efficiency*. The concept is useful because, when comparing two tests the small sample relative efficiency is often close to, or even higher, than the Pitman efficiency.

The Pitman efficiency of the sign test relative to the t-test when the latter is appropriate is a rather low $2/\pi \approx 0.64$. Lehmann (1975, 2006) shows that for samples of size 10 and a range of values of the median θ relative to the value θ_0 specified in H_0 with α fixed, the relative efficiency exceeds 0.7. For samples of 20 it is nearer to, but still above, 0.64. Here Pitman efficiency gives a pessimistic picture of the performance of the sign test at small sample sizes.

We have already mentioned that when it is relevant and valid the t-test is the most powerful test for any mean specified in H_0 against any alternative. When the t-test is not appropriate, other tests may have higher efficiency. Indeed, if our sample comes from the double exponential distribution, which has much longer tails than the normal, the Pitman efficiency of the sign test relative to the t-test is 2. That is, a sign test using a sample of n (at least for large samples) is as efficient as a t-test applied to a sample of size $2n$. There are, however, situations where asymptotic relative efficiency may give an unduly optimistic picture of small sample behaviour.

2.6 Computers and Nonparametric Methods

Computer software packages suitable for nonparametric analysis fall into three main categories. The first is specialist menu-driven packages that use exact permutation or related methods for small to medium sized samples and provide Monte Carlo and/or asymptotic tests for larger samples.

The second category are the mainstream menu-driven statistical software packages that allow exact inferences for some, but by no means all, widely used nonparametric tests, or are user-friendly in the sense that they allow the user to write programs to carry out such procedures.

The final category is comprised of versatile interactive statistical packages that have a variety of options, or tools, to perform various data manipulations and statistical operations. These are not menu driven. The user combines relevant tools, often with further options of his or her own creation, to achieve some desired objective. Such programs are by their nature generally less user-friendly than menu-driven packages, but they are often more powerful.

In the first category widely used packages are StatXact 7.0, distributed by Cytel Software Corporation, Cambridge, Ma, and Testimate, distributed by IDV Daten-analyse und Versuchs-planung, Munich, Germany. StatXact gives exact permutation P-values for small samples together with Monte Carlo estimates of these, for a large range of tests. Large sample, or asymptotic, results are also given and there are facilities for computing confidence intervals and also the power of some of the tests for assigned sample sizes and specified alternative hypotheses. Some of the tests in StatXact are also available in SAS. Testimate has considerable overlap with StatXact, but some methods are included in one but not both these packages and there are minor differences between the packages in detail for some procedures. There are also specialized programs dealing with particular aspects of the broad fields of nonparametric and semiparametric inference. These include LogXact, which is especially relevant to logistic regression, a topic only covered briefly in Chapter 15 in this book.

The efficiency of StatXact programs stems from the use of algorithms based on the work of Mehta and his co-authors in a series of papers including Mehta and Patel (1983, 1986), Mehta, Patel and Tsiatis (1984), Mehta, Patel and Gray (1985), Mehta, Patel and Senchaudhuri (1988, 1998). Similar, and other efficient algorithms are used in Testimate, but understanding the algorithms is not needed to use these packages.

General statistical packages such as SAS, Minitab, SPSS, and Stata include some nonparametric procedures. In some of these exact tests are given, but many rely heavily on asymptotic results, sometimes with little warning about when, particularly with small or unbalanced sample sizes, these may be misleading.

In the third category the increasingly popular R, and the closely related S-PLUS are particularly useful for the bootstrap described in Chapter 14, as well as for some of the semiparametric procedures discussed in Chapter 15.

Monte Carlo approximations to exact P-values, or for bootstrap estimation, can often be obtained from standard packages by creating macros that make use of inbuilt facilities for generating many random samples with or without replacement. Packages such as R and SAS have a versatility that makes combining of approaches such as EDA and more formal analyses quick and easy.

Users should test nonparametric procedures in any package programs they use with examples from this book and other sources. In some cases the output will be different, being either more or less extensive than that given in the source of the examples. For instance, output may give nominal (usually 5 or 1 percent) significance levels rather than exact P-values. Sometimes the convention of doubling a one-tail P-value may be used to obtain a two-tail test value, but as indicated in Example 2.4, this may not always be appropriate. Particular care should be taken to check whether exact or asymptotic results are given.

This book is largely about well-established methods, but only modern computing facilities allow us to use them in the way we describe. Solutions to examples, or illustrations in this book using statistical packages, are usually based on StatXacT, Minitab or R, but in many cases it would be equally appropriate to use other well-known packages such as SAS, SPSS, Stata, etc., providing these packages contain relevant programs.

Developments in statistical computer software are rapid and much of what we say about this may be out of date by the time you read it. Readers should check advertisements for statistical software in relevant journals and look for reviews of software in publications such as *The American Statistician* to trace new products.

2.7 Further Reading

Hollander and Wolfe (1999), Conover (1999), Gibbons and Chakraborti (2004), Higgins (2004) and Desu and Raghavarao (2004) give, in some cases, more background for some of the procedures described here. Each book covers a slightly different range of topics, and at varying depths, but all are suitable references for those who want to get a broad picture of the many aspects of basic nonparametrics. Daniel (1990) is a general book on applied nonparametric methods.

A moderately advanced mathematical treatment of the theory behind nonparametric methods is given by Hettmansperger and McKean (1998). Randles and Wolfe (1979) and Maritz (1995) are other recommended books covering the theory at a more advanced mathematical level than that used here. A classic is the book by Lehmann (1975), a revised edition of which appeared in 2006. This book repays careful reading for those who want to pursue the logic of the subject in more depth without too much mathematical detail. Applications in the social sciences are covered by Leach (1979) and by Siegel and Castellan (1988), the latter an update of a book written by Siegel some 30 years earlier.

Noether (1991) uses a nonparametric approach to introduce basic general statistical concepts. Although dealing basically with rank correlation methods, Kendall and Gibbons (1990) give an insight into the relationship between many nonparametric methods. Rayner and Best (2001) give a wide ranging treatment of many standard and a few specialist procedures using methods based largely on partitioning of the chi-squared statistic. Wasserman (2006), despite its title, deals mainly with more advanced modern topics in nonparametric statistics, a few of which we touch upon in Chapters 14 and 15. He gives a lucid introduction to those topics he covers.

Agresti (1984, 1996, 2002) and Everitt (1992) give detailed accounts of various models, parametric and nonparametric, used in categorical data analysis. A sophisticated treatment of randomization tests with emphasis on biological applications is given by Manly (2006). Good (2005) and Edgington (1995) cover randomization and permutation tests. The theory behind rank tests is given by Hájek, Sidák and Sen (1999).

Books dealing with the bootstrap include Efron and Tibshirani (1993), Davison and Hinkley (1997) and Chernick (1999).

2.8 Exercises

2.1 A new type of intensive physiotherapy is developed for individuals who have undergone spinal surgery. Due to limited hospital resources it can only be given to 3 out of 10 patients. The patients are aged:

| 15 | 21 | 26 | 32 | 39 | 45 | 52 | 60 | 70 | 82 |

Explain how a permutation test could be used to investigate whether use of the physiotherapy is related to patient age, (i.e., whether there is a policy to give the treatment to younger as opposed to older groups or *vice versa*). If the patients aged 15, 26 and 32 have the intensive physiotherapy find the *P*-value for a two-tailed test of an appropriate null hypothesis. Comment on your findings.

*2.2 Suppose that the new drug under test in Example 2.1 has all the ingredients of a standard drug at present in use and an additional ingredient that has proved to be of use for a related disease, so that it is reasonable to assume that the new drug will do at least as well as the standard one, but may do better. Formulate the hypotheses leading to an appropriate one-tail test. If the post-treatment ranking of the patients receiving the new drug is 1, 2, 4, 6 assess the strength of the evidence against the relevant H_0.

2.3 An archaeologist numbers some articles 1 to 11 in the order he discovers them. He selects at random a sample of 3 of them. What is the probability that the sum of the numbers on the items he selects is less than or equal to 8? (You do not need to list all combinations of 3 items from 11 to answer this question.)

   If the archaeologist believed that items belonging to the more recent of two civilizations were more likely to be found earlier in his dig and of his 11 items 3 are identified as belonging to that more recent civilization (but the remaining 8 come from an earlier civilization) does a rank sum of 8 for the 3 matching the more recent civilization provide reasonable support for his theory?

2.4 A library has on its shelves 114 books on statistics. I take a random sample of 12 and want to test the hypothesis that the median number of pages, θ, in all 114 books is 225. In the sample of 12, I note that 3 have less than 225 pages. Does this justify retention of the hypothesis that $\theta = 225$? What should I take as an appropriate alternative hypothesis? What is the largest critical region for a test with $P \leq 0.05$ and what is the corresponding exact *P*-level?

*2.5 The numbers of pages in the sample of 12 books in Exercise 2.4 were:

   126   142   156   228   245   246   370   419   433   454   478   503

Find a confidence interval at a level not less than 95 percent for the median θ.

*2.6 In Sect.1.4.1 we associated a confidence interval with a two-tail test. As well as such two-sided confidence intervals, one may define a one-sided confidence interval composed of all parameter values that would not be rejected in a one-tail test. Follow through such an argument to obtain a confidence interval at level not less than 95 percent based on the sign test criteria for the 12 book sample values given in Exercise 2.5 relevant to a test of $H_0 : \theta = \theta_0$ against a one-sided alternative $H_1 : \theta > \theta_0$.

2.7 From 6 consenting patients requiring a medical scan, 3 are chosen at random to undergo a positron emission tomography (PET) scan, the others receiving a magnetic resonance imaging (MRI) scan. Image quality is ranked in order by a hospital consultant from 1 (best) to 6 (worst). Describe how you would test H_0: *scan quality is unrelated to scan method* against (i) H_1: *PET scans are better* (ii) H_1: *the scans differ in quality depending on whether they are from PET or MRI*. Interpret the finding that the consultant rates the three PET scans as the three highest quality images.

2.8 In Example 2.4 we remarked that a situation could arise where we might reject H_0 for the wrong reason. Explain how this is possible in that example.

*2.9   State appropriate null and alternative hypotheses for the example from the book of Daniel about diet in Section 1.1.3. How could you use ranks to calculate the probability that the four receiving the diet of pulses were ranked 1, 2, 3, 4? Calculate this probability assuming that there were 20 young men involved altogether.

*2.10   A sample of 12 is taken from a continuous uniform distribution over the interval $(0, 1)$. What is the probability that the largest sample value exceeds 0.95? (Hint: Determine the probability that any sample value exceeds 0.95. The condition is met if at least one value exceeds 0.95.)

2.11 A sample of 24 is known to come either from a uniform distribution over the interval $(0, 10)$ or else from a symmetric triangular distribution over the same interval $(0, 10)$. The sample values are

4.17	8.42	3.02	2.89	9.77	6.06	2.72	5.12	6.00	4.78	2.62	7.20
1.61	5.92	7.25	8.01	4.76	5.36	5.34	7.59	0.66	7.27	3.39	1.40

Use appropriate graphical or other EDA techniques to get an indication as to which of these distributions is the more likely source of the sample.

CHAPTER 3

LOCATION INFERENCE FOR SINGLE SAMPLES

3.1 Layout of Examples

In the rest of this book we illustrate the logic of many inference methods using examples where, with a few exceptions, we discuss various aspects under the headings:

- *The problem*
- *Formulation and assumptions*
- *Procedure*
- *Conclusion*
- *Comments*
- *Computational aspects*

Summaries and exercises at the ends of nearly all chapters are preceded by indicative, but not exhaustive, lists of fields of application.

The remaining chapters each cover specific fields. These are indicated in broad terms by the chapter titles. However, topics do not fall into water-tight compartments. There are often close links between what at first sight seem unrelated problems. We highlight many such links that are only exposed by looking at a problem from different angles.

3.2 Continuous Data Samples

This chapter is mainly devoted to inferences about centrality measures for a single sample of continuous data, typically measurements. These provide simple illustrative examples of key nonparametric concepts, although in most real problems we compare two or more samples. The methods then used are often easily described as extensions of, or developments from, those for single samples. Single samples are also useful for showing how to deal with certain real-world complications such as the effect of tied data values, censored observations, rogue or outlying values.

Choosing the right inference procedure depends both on the type of data and the information we want. Even when we have complete measurements on some continuous scale, analyses that use only part of the information contained in that data may be more reliable, and more informative, than analyses using the original data. For example, an analysis based on a ranking of the order statistics will in many cases have better power and higher Pitman efficiency than tests based on the original measurement data. These analyses may be more reliable than using a parametric procedure of doubtful validity.

In Examples 2.2 and 2.3 we met a situation where presence of a censored observation precluded direct application of a t-test. We used instead a simple procedure based only on a count of the numbers of observations exceeding the median value hypothesized in the relevant H_0. Detailed consideration of the important topic of censored data in survival analysis is postponed until Chapter 9.

We develop several nonparametric tests in this chapter and compare their performance using simple data sets.

In Sections 3.3–3.5 we show how to compute and interpret some exact P-values and confidence intervals. In practice, even the best currently available software gives exact P-values only for small to medium-sized samples. We present in Section 3.3.3 asymptotic results that often provide good approximations for larger samples. If programs for exact tests are not available, a sensible compromise for small to medium sized data sets is to use Monte Carlo sampling to get good estimates of exact P-values. StatXact provides this facility, but if that package is not available one may still create the relevant samples for Monte Carlo approximation using any software that allows rapid generation of many random samples.

3.3 Inferences about Medians Based on Ranks

We examine in detail a procedure for hypothesis testing and estimation using a permutation test applied to a ranking of the order statistics for the magnitudes of deviation from a hypothesized median. This test, proposed by Wilcoxon (1945), has many attractive features.

3.3.1 The Wilcoxon Signed-Rank Test

Suppose n observations $x_1, x_2, \dots, x_n$ are a sample from some symmetric continuous distribution with unknown median θ. We wish to test the hypothesis H_0: $\theta = \theta_0$ against the two-sided alternative H_1: $\theta \neq \theta_0$. The assumption of symmetry implies that under H_0:

- The sign of each of the n differences $d_i = x_i - \theta_0, i = 1, 2, \dots, n$ is equally likely to be positive or negative.

- A difference of any particular magnitude, $|d|$, is equally likely to be positive or negative.

We have already pointed out that if a symmetric population distribution has a mean this coincides with the median. In those circumstances the test we give may also be formulated in terms of means. Tests against a one-sided alternative are also possible.

To carry out the test we proceed as follows:

- Determine the *magnitude* of the deviation of each data value from the hypothetical median θ_0 specified in H_0.

- Arrange these absolute deviations in increasing order of magnitude and assign ranks to these in ascending order (1 for the smallest, n for the largest).

- Attach a negative sign to each rank that corresponds to a negative deviation (i.e., to ranks assigned to data values below θ_0). The resulting set of ranks with attached signs are referred to as *signed ranks*.

If θ_0 is indeed the population median we expect a near equal scatter of positive and negative ranks. This implies that under the null hypothesis the sum of all positive ranks and the sum of all negative ranks should not differ greatly. A high sum of the positive (negative) ranks relative to that of the negative (positive) ranks implies θ_0 is unlikely to be the population median.

We describe the appropriate permutation test for a sample size $n = 7$, making the assumption that no deviations are tied in magnitude. We denote by S_+ the sum of the ranks associated with positive deviations from the median θ_0 specified in H_0, and by S_- the sum of ranks associated with negative deviations. In the extreme case when all sample values are above θ_0 all ranks will be positive so $S_- = 0$ and $S_+ = 1 + 2 + 3 + 4 + 5 + 6 + 7 = 28$. Similarly, if all sample values were below θ_0 we have $S_+ = 0$ and $S_- = 28$. Either of these results represent the strongest possible evidence against H_0. Intermediate values of S_+ or S_- are more supportive of H_0.

It is easy to see for any allocation of signs to ranks 1 to 7 that $S_+ + S_- = 28$, or equivalently that $S_+ = 28 - S_-$. This implies that we may use any of the statistics S_+, S_- or $S_d = |S_+ - S_-|$ as a test statistic since there is an ordered linear relationship between them.

When $n = 7$ signs may be allocated randomly to ranks in $2^7 = 128$ different and equally likely ways. The distribution of the statistic S_+ now depends only on the sample size, 7, since any data set gives rise to the rank magnitudes $1, 2, 3, 4, 5, 6, 7$.

We may obtain the distribution of S_+ by recording the value of S_+ for all possible sign allocations. It is left as an exercise to establish that the relevant sums and associated probabilities are those in Table 3.1. The probabilities are obtained by dividing the number of times a sum occurs by 128.

Exact P-values for one- or two-tail tests may be obtained using that table and apply to a test for any data without tied values when $n = 7$. For example, in a one-tail test where a small lower-tail value of S_+ is relevant the exact P-value corresponding to an observed $S_+ = 3$ is

$$P = \Pr(S_+ \leq 3) = (1 + 1 + 1 + 2)/128 = 5/128 \approx 0.039.$$

This probability is doubled for a two-tail test.

Table 3.1 *Exact probabilities that S_+ (or S_-) equal a value k in a Wilcoxon signed-rank test for sample size $n = 7$ when H_0 is true.*

k	Probability	k	Probability
0	1/128	15	8/128
1	1/128	16	8/128
2	1/128	17	7/128
3	2/128	18	7/128
4	2/128	19	6/128
5	3/128	20	5/128
6	4/128	21	5/128
7	5/128	22	4/128
8	5/128	23	3/128
9	6/128	24	2/128
10	7/128	25	2/128
11	7/128	26	1/128
12	8/128	27	1/128
13	8/128	28	1/128
14	8/128		

Example 3.1

The problem. The heart rate (beats per minute) when standing was recorded for seven members of a tutorial group. We assume that in the population of such students the distribution of heart rates when standing is symmetric. Given observations of 73, 82, 87, 68, 106, 60 and 97 test the hypothesis H_0: $\theta = 70$ against the alternative H_1: $\theta > 70$ using the Wilcoxon signed-rank test.

Formulation and assumptions. Symmetry implies that under H_0 sample values should have a near-symmetric scatter about $\theta = 70$, so the magnitudes of the sums of positive and of negative deviations from 70 (the d_i) should not differ greatly. The one-sided alternative, H_1, implies that when it holds there are likely to be more positive deviations from 70 than there would be under H_0, and that these positive deviations will tend to be larger than the negative deviations. These properties carry over to the signed ranks of the deviations.

Procedure. For the given data deviations from 70 are $73 - 70 = 3$, $82 - 70 = 12$ and similarly those remaining are 17, -2, 36, -10, 27. The corresponding signed ranks derived from allocating signs to the ranks of the $|d_i|$ are 2, 4, 5, -1, 7, -3, 6. Under H_0 the sum of the positive and negative signed ranks, S_+, S_- should be nearly equal, whereas under H_1 we expect S_+ to be appreciably larger than S_-. If the alternative had been the two-sided H_1: $\theta \neq 70$, small as well as large values of S_+ would support H_1. For these signed ranks $S_+ = 2 + 4 + 5 + 7 + 6 = 24$. From Table 3.1 we find $\Pr(S_+ \geq 24) = 7/128 \approx 0.055$.

Conclusion. When $P = 0.055$ the evidence against H_0 is sufficiently strong to perhaps warrant a similar study with a larger group of students.

Comments. 1. This example shows that, despite our warning in Section 1.3 that small experiments may never produce P-values indicating strong evidence against H_0, an interesting effect may sometimes be suggested even with very small samples. This is encouraging, given the current emphasis on small group work in teaching, where data concerning individuals in the group are often used for illustrative purposes. A teaching advantage of using data from small groups is that tedious arithmetic that may hide the logic of the problem is avoided. However, this example draws attention to a major limitation of a conventional hypothesis test. While it tells us that there is some doubt about the wisdom of retaining H_0 it tells us nothing about what other hypotheses might be plausible. We need a confidence interval to give this information. This is a problem we consider in Section 3.3.2.

2. Modifications for a two-tail test, when that is appropriate, are straight-forward and were mentioned under *Procedure*.

3. We chose S_+ as our test statistic, but we could equally well have chosen S_-, the sum of the negative deviations. This is because, as already pointed out, the sum of the unsigned ranks remains constant for all possible allocation of signs. Computer software sometimes calculates the smaller of S_+, S_-, but each leads to the same relevant P-value, as does $S_d = |S_+ - S_-|$.

4. If we had followed the classic convention requiring $P \leq 0.05$ before asserting significance we would not have rejected H_0 in a formal significance test at the 5 percent level. We see no merit in following such an arbitrary rule in this example, preferring to argue that the case for retaining H_0 is not strongly supported, even though 0.055 slightly exceeds 0.050.

Computational aspects. 1. Using StatXact or Minitab the P-value given above is obtained in a fraction of a second. In passing we note that the normal theory t-test for this example gives $P = 0.051$ for a one-tail test.

Manual calculation of probabilities associated with all values of S_+ or S_- for all but small n is time consuming and error prone. StatXact obtains these values rapidly for sample sizes of practical importance. The number of different values of the rank sums increases as the sample size, n, increases. The differences between the discrete probabilities associated with each sum become smaller, especially in the tails. Also, the distributions of S_+, S_- approach that of a normal distribution as n increases. When $n = 12$ there are $2^{12} = 4096$ possible associations of signs with ranks and the possible signed-rank sums range from 0 to 78. Symmetry means we need only compute afresh the probabilities for each sum between 0 and 39.

Table 3.2 is adapted from StatXact output and gives sufficient information to calculate the complete permutation test distribution of S_+ or S_-. From Table 3.2 we easily find that $\Pr(S_- \leq 13) = 87/4096 \approx 0.021$, and that $\Pr(S_- \leq 14) = 107/4096 \approx 0.026$. For two-tail tests we double these probabilities giving 0.042 and 0.052. Thus, values of the smaller signed-rank sum not exceeding 13 indicate significance at level 4.2 percent or less, i.e., below a conventional 5 percent level. Again, from Table 3.2 we see that

$$\Pr(S_- \leq 9) = 33/4096 \approx 0.008,$$

Table 3.2 *Exact probabilities that S_+ (or S_-) equal a value k in a Wilcoxon signed-rank test for sample size $n = 12$ when H_0 is true. For $k > 39$, by symmetry, $\Pr(S_+ = k) = \Pr(S_+ = 78 - k)$.*

k	Probability	k	Probability	k	Probability
0	1/4096	14	20/4096	28	89/4096
1	1/4096	15	24/4096	29	94/4096
2	1/4096	16	27/4096	30	100/4096
3	2/4096	17	31/4096	31	104/4096
4	2/4096	18	36/4096	32	108/4096
5	3/4096	19	40/4096	33	113/4096
6	4/4096	20	45/4096	34	115/4096
7	5/4096	21	51/4096	35	118/4096
8	6/4096	22	56/4096	36	121/4096
9	8/4096	23	61/4096	37	122/4096
10	10/4096	24	67/4096	38	123/4096
11	12/4096	25	72/4096	39	124/4096
12	15/4096	26	78/4096		
13	17/4096	27	84/4096		

while $\Pr(S_- \le 10) = 43/4096 \approx 0.0105$. It follows that for a one-tail test at a level not exceeding 1 percent that $S_- \le 9$ forms an appropriate critical region with associated $P = 0.008$.

Figure 3.1, based on Table 3.2, shows the frequency function of S_+ or S_- for $n = 12$ in the form of a bar chart. The shape resembles that of the familiar probability density function for the normal distribution. We consider important implications of this approach to normality in Section 3.3.3.

Example 3.2

The problem. For a group of 12 female students, the changes in heart rate (beats per minute) when standing up from lying down are:

$$-2 \quad 4 \quad 8 \quad 25 \quad -5 \quad 16 \quad 3 \quad 1 \quad 12 \quad 17 \quad 20 \quad 9$$

If we assume population symmetry, we may use the Wilcoxon signed-rank test to test H_0: $\theta = 15$ against the alternative H_1: $\theta \ne 15$.

Formulation and assumptions. We arrange the deviations from 15 in ascending order of magnitude and rank these, associating with each rank the sign of the corresponding deviation. We calculate the lesser of the sum of positive and negative signed ranks and, if available, use appropriate computer software to determine P. Alternatively, we may use Table 3.2 although this option is only available because $n = 12$.

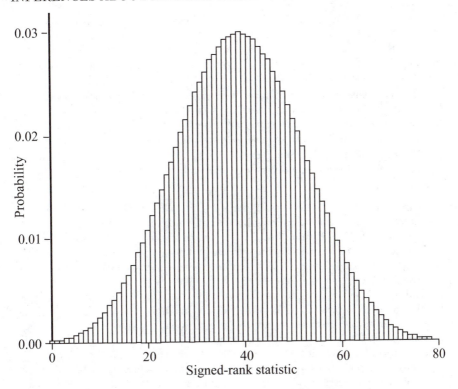

Figure 3.1 *Bar chart illustrating distribution of the Wilcoxon signed-rank statistic when $n = 12$.*

Procedure. Subtracting 15 from each sample value gives the deviations

$$-17 \quad -11 \quad -7 \quad 10 \quad -20 \quad 1 \quad -12 \quad -14 \quad -3 \quad 2 \quad 5 \quad -6.$$

We rearrange these in increasing order of magnitude while retaining signs, i.e.,

$$1 \quad 2 \quad -3 \quad 5 \quad -6 \quad -7 \quad 10 \quad -11 \quad -12 \quad -14 \quad -17 \quad -20$$

whence the signed ranks are

$$1 \quad 2 \quad -3 \quad 4 \quad -5 \quad -6 \quad 7 \quad -8 \quad -9 \quad -10 \quad -11 \quad -12.$$

The sum of the positive ranks, $S_+ = 1+2+4+7 = 14$, is less than the sum, S_-, of the negative ranks. StatXact, Testimate and some general statistical programs give exact P-values in programs designed specifically for the Wilcoxon test, where data may be input either in its original form or as signed ranks. For this example these programs indicate that the two-tail P-value corresponding to $S_+ = 14$ is $P = 0.052$. This value may be confirmed using Table 3.2.

Conclusion. The evidence against H_0 is not very strong, but the P-value is sufficiently small for a further study involving a larger group to be considered.

Comments. 1. Doctors often assume that heart rate measurements are symmetrically distributed. In this example, the median, 8.5, is close to the mean, 9.0, indicating that symmetry is possible. There is, however, some evidence of a departure from normality because there is no particular cluster of values around the mean of 9.0; instead several values are close to zero, and there is another group of values around 17. Given the likely degree of symmetry in the distribution the Wilcoxon test should not be seriously misleading.

2. The two clusters of values may reflect two distinct groups of students: those heavily involved in sports such as hockey or tennis and those who rarely participate. Females who are physically active may not experience a large increase in heart rate on standing compared with those who follow a more sedentary lifestyle.

3. One must be cautious about generalizing the above results to the community at large. For several reasons female students are probably not typical of young adults in general.

Computational aspects. 1. Not all statistical packages give an exact P-value for the Wilcoxon signed-rank test. Some have a program that calculates S_+ or S_- and then may give only a P-value based on asymptotic theory (see Section 3.3.3). If one has no program to give exact P-values, and is not prepared to seek a Monte Carlo approximation for small to medium samples it may be better to refer to published tables even though many give only values required for significance at a nominal 5 or 1 percent level in one- or two-tail tests. Due to discontinuities in the distribution of P-values the exact levels are not precisely 5 or 1 percent. For most published tables the exact values are the closest possible values that do not exceed the nominal significance levels although a few tables give a value as close as possible to the nominal level even if that closest value exceeds the nominal level. Most major published statistical tables [e.g., Neave (1981) or Lindley and Scott (1995)] and some books on nonparametric methods include tables for the Wilcoxon signed-rank test and usually describe which of the above options they use for defining the nominal level. More satisfactory tables in Hollander and Wolfe (1999, Table A4) or Gibbons and Chakraborti (2004, Table H) give exact P-values likely to be of interest for small to medium sample sizes. These tables are not needed if programs like that in StatXact or Testimate are available.

2. The asymptotic P-value given by StatXact for these data for a two-tail test is $P = 0.050$ but this does not include a continuity correction we describe in Section 3.3.3. That correction increases the value to $P = 0.0545$. In the light of the *approximate* normality of the Wilcoxon statistic evident in Figure 3.1 this good agreement with the exact test value is not surprising.

For these data the two-tail t-test value is $P = 0.046$. This does not differ greatly from the exact permutation test value, but gives a timely warning on the dangers of adhering to a rigid significance level such as $\alpha = 0.05$ as a decision-making basis for accepting or rejecting H_0.

There are several tests for symmetry when we are given only the sample values. One proposed by Randles et al. (1980) is described by Hollander and Wolfe (1999, Section 3.9) and also by Siegel and Castellan (1988, Section 4.4), but caution is needed in assessing significance if $n < 20$ in cases where the asymptotic result is on the borderline of significance. However, for these data the test statistic has a value well below any that would suggest asymmetry. The required computations are tedious rather than difficult and the reader should refer to either of the above references for details.

3.3.2 Confidence intervals based on signed ranks

For many nonparametric methods the more important aim of calculating confidence intervals is not so simple as hypothesis testing, but the procedures, once mastered, are often straightforward and not difficult to program by writing a macro if no package program is available.

We outline the theory, but a reader prepared to take this on trust may move directly to Example 3.3 to see an application. We assume a sample of n observations with no ties has been arranged in ascending order $x_1, x_2, \ldots, x_n$. The null hypothesis for the mean or median is specified as H_0: $\theta = \theta_0$. In practice θ_0 will usually lie between x_1 and x_n, but this is not a theoretical requirement.

A deviation $x_i - \theta_0$ will have a negative signed rank for any x_i less than θ_0. Since the x_i are in ascending order, among all negative ranks, if there are any, that of greatest magnitude will be associated with x_1, that of next greatest magnitude with x_2, and so on. Similarly, all x_j greater than θ_0 have positive signed ranks. Of these x_n will have the highest rank, x_{n-1} the next highest and so on.

Consider now the paired averages of x_1 with each of $x_1, x_2, \ldots, x_n$. Suppose the signed rank of the deviation associated with x_1 is $-p$. A moment's reflection shows that each of the averages $(x_1 + x_q)/2$ will be less than θ_0 when the deviation associated with x_q has either a negative rank or a positive rank less than p. Since we are assuming there are no rank ties no observation has positive rank p (since the rank p is already associated with x_1).

If x_q is the smallest observation associated with a positive rank greater than p, this implies that $|x_q - \theta_0| > |x_1 - \theta_0|$. It follows that the average of x_1 and x_q is greater than θ_0. Then, because the data are in ascending order, so is the average of x_1 and any $x_r > x_q$. Thus, the number of averages involving x_1 that are less than θ_0 is equal to the negative rank associated with x_1. Similarly, if we form all the paired averages of any x_i less than θ_0 with each of $x_i, x_{i+1}, \ldots, x_n$ the number of these averages less than θ_0 will equal the (negative) rank associated with x_i. When x_i has a positive signed rank none of the averages with itself or greater sample values will be less than θ_0.

The complete set of averages $(x_i + x_j)/2$ for $i = 1, 2, \ldots, n$ and all $j \geq i$ are called *Walsh averages*, having been proposed by Walsh (1949a, b). It is easily seen that the number of Walsh averages less than θ_0 equals S_- and the number greater than θ_0 equals S_+. We may use Walsh averages to calculate the test statistic S for a given H_0 and to obtain confidence limits for a population mean or median in the way illustrated in the following example.

Example 3.3

The problem. Consider again the data on changes in heart rate for the female students in Example 3.2, viz.,

$$-2 \quad 4 \quad 8 \quad 25 \quad -5 \quad 16 \quad 3 \quad 1 \quad 12 \quad 17 \quad 20 \quad 9.$$

Obtain an estimate of the population median with confidence intervals at levels at least 95 and 99 percent, using Walsh averages.

Formulation and assumptions. We use Walsh averages and the principle enunciated in Section 1.4.1 that the $100(1 - \alpha)$ percent confidence interval contains those values of the parameter that would be accepted in a significance test at the 100α percent significance level.

Procedure. Table 3.3 shows a convenient way to set out Walsh averages. Both the top row and the first column (in italics) of that table give the sample values in ascending order. The triangular matrix forming the body of the table gives the Walsh averages corresponding to the relevant top row and first column values. *Comment 2* below gives tips for calculating these if a program to do so is not available.

An intuitively reasonable point estimator of the population median is the median of the Walsh averages since it follows from the arguments given before this example that for this estimate the sums of both the negative and the positive rank deviations are equal, both being equal to the number of Walsh averages above (or below) this median. The median of the Walsh averages may be determined by inspecting Table 3.3. We see that entries in any row increase from left to right and those in any column increase as we move down, so that the smallest entries are at the top left of the table and the largest are at the bottom right. For a sample size n there are $n(n + 1)/2$ Walsh averages, i.e., 78 in this example. Thus, the median of these lies between the 39th and 40th ordered Walsh average. Inspection of the table establishes that both equal 9.0; so this is the appropriate estimate of the population median (or mean if it exists).

To establish a 95 percent confidence interval we select as end points values of θ that will just be acceptable if $P = 0.05$. In Example 3.3 we found that P is only slightly more than 0.05 if the sum of the lesser of positive or negative ranks is 14.

Table 3.3 *Walsh averages for changes in heart rate from lying to standing.*

	−5	−2	1	3	4	8	9	12	16	17	20	25
−5	−5.0	−3.5	−2.0	−1.0	−0.5	1.5	2.0	3.5	5.5	6.0	7.5	10.0
−2		−2.0	−0.5	0.5	1.0	3.0	3.5	5.0	7.0	7.5	9.0	11.5
1			1.0	2.0	2.5	4.5	5.0	6.5	8.5	9.0	10.5	13.0
3				3.0	3.5	5.5	6.0	7.5	9.5	10.0	11.5	14.0
4					4.0	6.0	6.5	8.0	10.0	10.5	12.0	14.5
8						8.0	8.5	10.0	12.0	12.5	14.0	16.5
9							9.0	10.5	12.5	13.0	14.5	17.0
12								12.0	14.0	14.5	16.0	18.5
16									16.0	16.5	18.0	20.5
17										17.0	18.5	21.0
20											20.0	22.5
25												25.0

Thus, if we choose a value less than the 14th smallest Walsh average we would reject H_0. Similarly, if we choose a value greater than the 14th largest Walsh average we also reject H_0, since in either case the test statistic S would be such that $S \leq 13$.

Thus, for $n = 12$ a confidence interval at not less than a 95 percent confidence level is the interval (14th smallest Walsh average, 14th largest Walsh average). From Table 3.3, by counting from the most extreme value we see that this is the interval $(2.5, 16.0)$. We noted, when discussing Table 3.2 that using $S \leq 13$ as a criterion for significance corresponded to an exact significance level of 4.25 percent in a two-tail test. It is intuitively reasonable to regard the interval $(2.5, 16.0)$ as an actual 95.75 percent confidence interval.

Owing to discontinuities in the distribution of the test statistic, and the fact that the end points of the confidence interval often coincide with observed values, we must regard values outside this interval as ones indicating rejection at a 4.25 percent significance level. We must also, if our exact level is critical, test the end values to see if they would be accepted or rejected at this significance level. Exact tests using StatXact show that we would find evidence against the hypothesized means of 2.5 and 16.0. Thus, in this case all points in the open interval $2.5 < \theta < 16.0$ would be acceptable, but the closed interval $[2.5, 16.0]$ is a conservative 95 percent interval, having a confidence level exceeding 95 percent. In practice it is usual to cite a confidence interval simply as the interval $(2.5, 16.0)$ without being specific as to whether it is open or closed, bearing in mind that discontinuities inherent in the Wilcoxon test statistic distribution may imply (and it will vary from case to case) that the end points may or may not be included as acceptable values at a nominal significance level such as 5 percent. One should not get excited about subtleties due to discontinuity at the end point of intervals. If relevant programs are available it may be worthwhile checking the actual confidence level associated with any quoted interval.

It is left as an exercise to establish that a nominal 99 percent confidence interval (i.e., with at least 99 percent coverage) is the closed interval $[0.5, 18]$.

Conclusion. An appropriate point estimator of the population median is 9.0 and actual 95.75 and 99.08 percent confidence intervals are respectively $(2.5, 16.0)$ and $(0.5, 18)$. This implies that if values of θ outside these intervals are specified in H_0 there will be evidence against these at levels given by the relevant P-values, i.e., $P = 0.0425$ and $P = 0.0092$.

Comments. 1. A possible alternative to the median of the Walsh averages as a point estimator is the mean of all sample values. It is unbiased in the sense that if we sampled repeatedly and took the mean of the sample values as our estimate in all cases, the mean of these estimates would converge to the population mean providing that mean exists. It does not, for example, exist for the Cauchy distribution. However, the median of the Walsh averages is in a technical sense, which we do not discuss here, generally closer to the population median. This estimator is called the *Hodges–Lehmann estimator*, having been proposed by Hodges and Lehmann (1963). A more formal justification for its use when basing the interval on signed ranks is given by Hettmansperger and McKean (1998, Section 1.3).

2. If only confidence limits are required we need write down only a few Walsh averages in the top left and bottom right of the triangular matrix in Table 3.3. If a computer program is available for forming this matrix there is no problem in computing all the averages. If the matrix has to be formed manually when we

have the Walsh averages in the first row, those in the second row may be obtained by adding the same constant, namely half the difference $x_2 - x_1$, to the average immediately above it. To verify this denote by x_{ij} the average of x_i and x_j. This is the entry in row i and column j of the matrix of Walsh averages. Now for any $k > 2$

$$x_{2k} = \frac{1}{2}(x_2 + x_k) = \frac{1}{2}(x_2 - x_1 + x_1 + x_k) = \frac{1}{2}(x_2 - x_1) + x_{1k}.$$

The idea generalizes for all subsequent rows, each entry in row $r + 1$, say, being obtained by adding the same constant $(x_{r+1} - x_r)/2$ to the entry immediately above it.

3. In Example 3.2 we found when testing H_0: $\theta = 15$ that $S_+ = 14$. In Table 3.3 there are 14 Walsh averages exceeding 15.0.

Computational aspects. 1. Minitab includes a program for computing Walsh averages. It gives asymptotic confidence intervals (see Section 3.3.3). These may not be satisfactory for small n. If one has the matrix of Walsh averages generated by some available program it is easy to establish nominal 95 or 99 percent confidence limits using tables of critical values of the statistic S.

2. StatXact includes a program giving the Hodges-Lehmann estimator and confidence intervals at any pre-assigned nominal level and confirms the values obtained above.

3.3.3 Asymptotic Results

Widely available software for exact permutation tests for small to medium-sized samples removes the need to use asymptotic results for such samples. Asymptotic results may not be reliable unless samples are moderately large. How large depends partly upon the procedure being used.

In the case of the Wilcoxon signed-rank test Figure 3.1 suggests that the distribution of the test statistic (S_+ or S_-) approaches normality when $n = 12$. For a sample of n, under the null hypothesis each rank is equally likely to have a positive or negative sign. It follows from the rules for mean and variance of the sum of independent variables that either of these statistics has a mean that is one half of the sum of the magnitudes of the ranks, i.e.,

$$E(S) = \frac{1}{4}n(n + 1),$$

since the sum of the integers from 1 to n is $n(n + 1)/2$. Here S may be either S_+ or S_-

Similarly, the variance of S is one quarter of the sum of squares of the ranks, i.e.,

$$\text{Var}(S) = \frac{1}{24}n(n + 1)(2n + 1),$$

since the sum of the squares of the integers 1 to n is $n(n+1)(2n+1)/6$. These expressions for means and variances are a particular case of a more general result considered in Exercise 3.1.

It follows that for sufficiently large n the distribution of

$$Z = \frac{S - E(S)}{\sqrt{\text{Var}(S)}} \tag{3.1}$$

is approximately standard normal.

The formula (3.1) can be improved by a *continuity correction* that allows for the fact that we are approximating to a continuous distribution with a discrete distribution where S can take only integer values. If S is the smaller of the sums of the positive or negative ranks, the appropriate correction is to replace S by $S + 1/2$. If S is the larger sum it is replaced by $S - 1/2$.

If the statistic $S_W = |S_+ - S_-|$ is used it is easily verified that now (3.1) reduces to

$$Z = \frac{S_W}{\sqrt{\text{Var}(S_W)}},$$

where $\text{Var}(S_W) = n(n + 1)(2n + 1)/6$. The continuity correction is now to subtract 1 from the numerator if $S_+ < S_-$ and to add 1 if $S_+ > S_-$, since changing either S_+ or S_- by 1 changes the value of S_W by 2.

Example 3.4

The problem. For the sample of 12 in Example 3.2 use the asymptotic approximation to test the hypothesis that $\theta = 15$ against the alternative that $\theta \neq 15$.

Formulation and assumptions. We assume a sample of 12 is sufficiently large to justify using (3.1).

Procedure. We know from Example 3.2 that $S_+ = 14$ when $n = 12$ and we have by the formulae given above that $E(S_+) = (12 \times 13)/4 = 39$ and

$$\text{Var}(S_+) = (12 \times 13 \times 25)/24 = 162.5,$$

whence, using (3.1) with a continuity correction gives

$$Z = \frac{14.5 - 39}{\sqrt{162.5}} \approx 1.92.$$

Tables, or appropriate computer software, indicate this is equivalent to $P = 0.055$. Without the continuity correction it is easily shown that $Z \approx 1.96$ equivalent to $P = 0.05$.

Conclusion. Here, as expected from our earlier discusssion based on Figure 3.1, there is close agreement with the permutation test result. The continuity correction here makes little practical difference unless one places undue weight upon a magic cutoff point for significance at $\alpha = 0.05$.

Comment. It is possible to obtain confidence limits for the values of S_+ and S_- that arise with acceptable hypothesized values of θ for some chosen α. One may use these in association with a table of Walsh averages to determine a confidence interval for the true value of θ.

Computational aspects. Programs giving asymptotic results for the Wilcoxon, and for many other standard nonparametric tests are included in most major statistical software packages and as options in specialist programs like StatXact and Testimate. They work well for large samples, but the approximations may or may not be good for medium or small samples. Practice varies about use of a continuity correction. It is usually easy to check whether or not these are used either from software documentation or from given output, which often includes the test statistic value together with its mean and standard deviation. Minitab uses a continuity correction for the Wilcoxon test, but StatXact does not.

3.3.4 Wilcoxon with Ties

In theory, when we sample from a continuous distribution the probability of tied observations, or that of a sample value exactly equalling the population mean or median, is zero. Real observations never have a distribution that is strictly continuous. This may be either because of their nature, or as a result of rounding or limited measurement precision.

We measure lengths to the nearest centimetre or millimetre; weights to the nearest kilogram, gram or milligram; the number of pages in a book in complete pages, although chapter layout often results in part-pages of text. Numbers of words in a sentence must be integers. These realities may produce data with rank ties or zero departures from a hypothesized mean or median. If they do, the exact distribution of S, the lower of the positive or negative rank sums, requires fresh computation for different numbers of ties and ties in different positions in the rank order.

Before the advent of software to compute exact P-values for permutation distributions with ties, a common procedure was, after adjusting the scoring method based on ranks to allow for ties, to use the asymptotic normal approximation we introduced in Section 3.3.3 after certain small adjustments. This usually works well for a few ties if n is reasonably large, but may have bizarre consequences for small n. Another approach for smaller samples was to use tables for the no-tie case assuming that the relevant value of S would be little affected by ties. This need not be so.

When two or more ranks are equal in magnitude, but not necessarily of the same sign, the most widely used procedure, and the one we adopt, is to replace them by their mid-rank, an idea easily explained by an example. If 7 observations are 1, 1, 5, 5, 8, 8, 8 and we want to use a Wilcoxon signed-rank test with $H_0 : \theta = 3$, the relevant signed differences from 3 are $-2, -2, 2, 2, 5, 5, 5$. The magnitude of the smallest difference is 2, and four differences have that magnitude. The mid-rank rule assigns to these ties the mean of the four smallest ranks $1, 2, 3, 4$; i.e., we allocate to these observations the signed ranks $-2.5, -2.5, 2.5, 2.5$. The remaining three differences are all 5, and we give each the mean of the remaining ranks 5, 6 and 7; i.e., each is ranked 6. The exact permutation test is based on the signed mid-ranks

Table 3.4 *Exact probabilities that S_+ (or S_-) $= k$ in a Wilcoxon signed-rank test for sample size $n = 7$ when H_0 is true and mid-ranks are 2.5 (four times) and 6 (three times).*

k	Probability	k	Probability
0.0	1/128	14.5	12/128
2.5	4/128	16.0	3/128
5.0	6/128	17.0	18/128
6.0	3/128	18.0	1/128
7.5	4/128	19.5	12/128
8.5	12/128	20.5	4/128
10.0	1/128	22.0	3/128
11.0	18/128	23.0	6/128
12.0	3/128	25.5	4/128
13.5	12/128	28.0	1/128

$-2.5, -2.5, 2.5, 2.5, 6, 6, 6$. An appropriate test statistic is $S_- = 5$, the sum of the negative mid-ranks.

If we test using the permutation distribution for the no-ties case given in Table 3.1 we find $\Pr(S_- \le 5) = 10/128 \approx 0.078$. However, with ties, if we use the mid-rank procedure the distribution of S_- under H_0 is not that in Table 3.1. The exact distribution with ties depends both on how many ties there are and where they lie in the rank sequence. We can work out exact significance levels with ties for small to medium-sized samples. For any tie pattern StatXact or Testimate give exact P-values for one- and two-tail tests. The former will also compute the complete distribution of S_+ or S_- for samples that are not too large. For the tie pattern above, for a sample of 7, StatXact gives the exact distribution in Table 3.4.

Comparing Tables 3.1 and 3.4 we see that in each case the distribution of the test statistic is symmetric. That is why we may base our test on either S_+ or S_-. However, the distribution with no-ties is unimodal (i.e., it has one maximum) with values of the statistic confined to integral values increasing in unit steps, while that in Table 3.4 is heavily multimodal with S taking unevenly spaced, and not necessarily integer, values. Discontinuities are also more marked.

From Table 3.4 we easily see that $\Pr(S_- \le 5) = 11/128$, compared to the no-tie probability $10/128$ found from Table 3.1. Despite many ties in this example and the differences in the distribution of the test statistic in the two cases, it may appear that ties have little effect on our conclusion. However, this is not always the case.

Example 3.5

Ties often result from rounding. The data 1, 1, 5, 5, 8, 8, 8 considered above could arise from rounding to the nearest integer in either of the sets:

$$0.9 \quad 1.1 \quad 5.2 \quad 5.3 \quad 7.9 \quad 8.0 \quad 8.1$$
$$0.9 \quad 1.0 \quad 4.8 \quad 4.9 \quad 7.9 \quad 8.0 \quad 8.1$$

For a Wilcoxon test with $H_0 : \theta = 3$ the signed differences from 3 are:

$$-2.1 \quad -1.9 \quad 2.2 \quad 2.3 \quad 4.9 \quad 5.0 \quad 5.1$$
$$-2.1 \quad -2.0 \quad 1.8 \quad 1.9 \quad 4.9 \quad 5.0 \quad 5.1$$

with signed ranks:

$$-2 \quad -1 \quad 3 \quad 4 \quad 5 \quad 6 \quad 7$$
$$-4 \quad -3 \quad 1 \quad 2 \quad 5 \quad 6 \quad 7$$

There are now no tied ranks in either set and for the first set $S_- = 3$, while for the second set $S_- = 7$. Table 3.1 gives $\Pr(S_- \leq 3) = 5/128 \approx 0.039$ and $\Pr(S_- \leq 7) = 19/128 \approx 0.148$. Thus a one-tail test at a level not exceeding 5 percent would indicate an implausible hypothesis in the first case, but not in the second. Rounding to the nearest integer may greatly affect the strength of evidence against the null hypothesis.

The above example is a simple illustration of difficulties of interpretation with ties due to rounding, a process that makes small perturbations or changes to most observations. Many statistical analyses are sensitive to such changes. The impact of ties due to rounding in this example is more extreme than one is likely to meet in practice, since it results in only two different magnitudes for ranks. For larger samples and not too many ties, the effect on P-values is less marked.

Example 3.6

The problem. The number of words in the first sentences in 12 randomly selected paragraphs from Fisher (1948) were:

$$12 \quad 18 \quad 24 \quad 26 \quad 37 \quad 40 \quad 42 \quad 47 \quad 49 \quad 49 \quad 78 \quad 108$$

Use the Wilcoxon signed-rank test to test for the population median θ the hypothesis $H_0: \theta = 30$ against the alternative $H_1: \theta > 30$.

Formulation and assumptions. To justify the Wilcoxon test, we assume the population is symmetric. The deviations of observations from 30 are ranked with due regard to sign, using mid-ranks for ties. The lesser of S_+ and S_- is then found.

Procedure. The signed deviations from 30 are

$$-18 \quad -12 \quad -6 \quad -4 \quad 7 \quad 10 \quad 12 \quad 17 \quad 19 \quad 19 \quad 48 \quad 78$$

and the corresponding signed ranks, using mid-ranks for the tied magnitudes at 12 and 19, are

$$-8 \quad -5.5 \quad -2 \quad -1 \quad 3 \quad 4 \quad 5.5 \quad 7 \quad 9.5 \quad 9.5 \quad 11 \quad 12$$

giving $S_- = 16.5$. A one-tail test is appropriate. StatXact gives the exact probability of observing $S_- \leq 16.5$ with this pattern of tied deviations when $n = 12$ as $P = 0.0405$.

Conclusion. The evidence implies H_0 is implausible at a 4.05 percent significance level, so there is fairly strong evidence against H_0.

Comments. 1. Before we had software that gave the exact distribution of the test statistic when there are ties, the advice to resort to asymptotic (large sample) results of a type we describe below had little justification for small samples.

2. The data values 78, 108 throw doubt on the validity of the symmetry assumption. Had these observations been replaced by, say, 51, 53 our Wilcoxon rank-sum test would give the same result, but then there would have been no intuitive reason to doubt the symmetry assumption. In this case, but not in all such situations, the test is said to be *robust* against the outliers.

Computational aspects. With packages like StatXact it is a useful exercise to generate exact distributions for a number of tied situations for small sample sizes to develop a feeling for the effect of ties on standard results. An extreme case arises with the observations 4, 4, 8, 8, 8, 8, 8 for the Wilcoxon test when we set $H_0 = 6$. Then all ranks of deviations are tied apart from signs and it is easy to see that the Wilcoxon test statistic distribution is now the same as that for the sign test given in Section 2.2. This is discussed more fully in Section 3.4. In this sense, a sign test is a special case of the Wilcoxon test where all deviation magnitudes are tied at the same value.

When there are ties the asymptotic test requires modification to the formulae given in Section 3.3.3. This involves the variance and the precise modification depends upon the pattern of ties. Several writers, including Hollander and Wolfe (1999, Section 3.1) give a general formula to cover all possible patterns of ties. An alternative procedure, which has the advantage that it extends to many other situations covered in this book, is to regard each rank or tied rank as a *score*. We denote by s_i the score associated with the sample value x_i. In the Wilcoxon procedure and several others that we shall consider, tests are based on the sum S of the positive (or negative) scores. Under the null hypothesis a key feature is that each score s_i is equally likely to have a positive or negative sign. It follows that any such statistic S has mean value $\sum_{i=1}^{n} |s_i|/2$ and variance $\sum_{i=1}^{n} s_i^2/4$ (see Exercise 3.1). From the central limit theorem we then have that for large n the distribution of

$$Z = \frac{S - \frac{1}{2}\sum |s_i|}{\frac{1}{2}\sqrt{\sum s_i^2}} \tag{3.2}$$

is approximately standard normal. Equation (3.1) is a special case of (3.2). Modern computer software makes it easy to use (3.2) given any set of scores such as those for the midranks in Example 3.6.

Several alternative ways of handling ties in ranked data occur in the literature, but the midrank procedure is the one most commonly used in practice. A critical review of some half-dozen procedures will be found in Gibbons and Chakraborti (2004, Section 5.6).

With ties in signed ranks the method for obtaining confidence intervals needs modification. If computer software for confidence intervals in these circumstances is not available one might use the Walsh average method described in Section 3.3.2, then use an available computer program in a trial and error manner to study appropriate end point adjustments. Such adjustments are usually small unless there is heavy tying.

If one or more sample values equal the mean hypothesized under H_0, the associated deviations are zero. Statisticians are divided on how best to proceed in this case. Some argue that we should omit such observations from our computations because they are uninformative about any alternative hypothesis. The test is then applied to a reduced sample. An alternative we recommend is to rank all deviations temporarily giving any zero the rank 1 because it is the smallest difference (or the appropriate tied rank if there is more than one zero). After all signed ranks are allocated we then change the rank(s) associated with zero difference to 0, leaving the other ranks as allocated. To obtain exact tests in the latter situation a suitable computer program is needed for the exact distribution of the test statistic, else we may resort to asymptotic results using the form (3.2) with the adjusted ranks as scores providing n is reasonably large.

Deleting zero differences is the simplest procedure if no suitable program for exact probabilities of S for the modified rank statistic is available, but the method recommended above has the theoretical advantage of providing a more powerful test.

3.4 The Sign Test

3.4.1 A Closer Look

We look more closely at assumptions behind, and possible complications with the sign test, introduced in Section 2.2. We extend some of the concepts developed in Example 2.3. For the sign test — like most tests — the larger the sample the greater the power and the shorter the confidence interval for a given confidence level.

A symmetry assumption is not needed, nor is an assumption of continuity. Indeed, providing all the x_i are mutually independent they need not all have the same distributions, but they then must be from distributions with identical medians. Remember that for skew distributions the median does not equal the mean.

Example 3.7

The problem. The percentage water content of agricultural land often varies only slightly over an area such as a large field apart from a few local patches of exceptionally high or low percentages associated with especially poor or good drainage. The potential range of percentage water content is exemplified in a set of readings for nearly 400 soil quadrats on 22 June 1992 given by Gumpertz, Graham and Ristiano

(1997). We use two small subsets (with values slightly rounded) from this large data set to illustrate some points about single sample nonparametric analysis. The use we make of these data has no direct relevance to the paper from which our samples were taken. The percentage water contents for our selected quadrats, arranged in ascending order of magnitude, are:

| Sample I | 5.5 | 6.0 | 6.5 | 7.6 | 7.6 | 7.7 | 8.0 | 8.2 | 9.1 | 15.1 |

| Sample II | 5.6 | 6.1 | 6.3 | 6.3 | 6.5 | 6.6 | 7.0 | 7.5 | 7.9 | 8.0 |
| | 8.0 | 8.1 | 8.1 | 8.2 | 8.4 | 8.5 | 8.7 | 9.4 | 14.3 | 26.0 |

We use these data with a sign test for the population median to test in each case $H_0: \theta = 9$ against the alternative $H_1: \theta \neq 9$.

Formulation and assumptions. If the population median is 9, then any observation in a random sample is equally likely to be either above or below 9. Proceeding as in Example 2.3 we associate a plus sign with values greater than 9. If the hypothesis $H_0: \theta = 9$ holds, the number of plus signs has a binomial distribution with $n = 10, p = 0.5$, i.e., a $B(10, 0.5)$ distribution for the first sample, and a $B(20, 0.5)$ distribution for the second sample. For the alternative $H_1: \theta \neq 9$ a two-tail test is appropriate.

Procedure. In the first sample there are 2 values, 9.1 and 15.1, greater than 9, implying 2 plus signs. We gave a table of $B(10, 0.5)$ probabilities in Example 2.3 from which we easily calculate the probability of 2 or less or 8 or more plus signs to be $P = 0.109$, a value confirmed by any computer program for the sign test. In the second sample there are 3 values, 9.4, 14.3 and 26.0 greater than 9. The probability of 3 or less plus signs for a $B(20, 0.5)$ distribution is easily calculated, although there is seldom any need to do this with the ready availability of computer software or tables for a sign test. Any relevant program or table should give $P = 0.003$ for a two-tail test.

Conclusion. With the sample of 10, since $P = 0.109$ there is no convincing evidence against H_0 since P clearly exceeds the conventional 5 percent level. With a sample of 20 the evidence against H_0 is very strong since $P = 0.003$ corresponding to a 0.3 percent significance level.

Comments. 1. We used a two-tail test since there was no *a priori* reason to confine alternatives to only greater or only less than the value specified in H_0.

2. The results are consistent with our remark in Section 1.3 that increasing sample size generally increases the power of a test.

Computational aspects. With adequate tables of binomial probabilities for various n when $p = 0.5$, computer programs might almost be regarded as optional extras for the sign test, but StatXact and Testimate and nearly all major general statistics packages include programs for this or equivalent procedures.

If one or more sample values coincide with the value of θ specified in H_0 we cannot logically assign either a plus or minus to these observations. They may be ignored, and the sample size reduced by 1 for each such value, e.g., if in the sample of 10 in Example 3.7 we specify $H_0 : \theta = 8.0$ we treat our problem as one with 9 observations giving 3 plus signs. Doing this rejects evidence that strongly supports the null hypothesis, but which is uninformative about the direction of possible alternatives. Another proposed approach is to toss a coin and allocate a plus to a value equal to θ if the coin falls heads and a minus if it falls tails: yet another is to assign a plus or minus to such a value in a way that makes rejection of H_0 less likely. The coin-toss approach usually makes only a small difference, and has little to commend it since it only introduces additional randomness, being in a sense more "error". The last approach is ultraconservative. Lehmann (1975, p. 144) discusses pros and cons of these choices in more detail. These and some more sophisticated options for dealing with this problem are discussed by Rayner and Best (2001, Section 2.2).

3.4.2 Confidence Intervals

Example 3.8

The problem. Obtain approximate 95 percent confidence intervals for the median water content using the sign test with the samples in Example 3.7. Determine the exact confidence level in each case.

Formulation and assumptions. We seek all values for the population median θ that would not be rejected in a two-tail test at a 5 percent significance level.

Procedure. We use the argument developed in Section 1.4.1. Consider first the sample of 10. From the table given in Example 2.3 (or any published tables or using a computer program that calculates probabilities for any given n, p) we easily find that a critical region consisting of 0, 1, 9 or 10 plus signs has size $2 \times (0.0010 + 0.0098) = 0.0216$ and that this is the largest exact size less than 0.05. Thus we reject H_0 if it leads to 1 or fewer, or 9 or more, plus signs. We retain H_0 if we get between 2 and 8 plus signs. It is clear from the sample values 5.5 6.0 6.5 7.6 7.6 7.7 8.0 8.2 9.1 15.1 in Example 3.7 that we have between 2 and 8 plus signs if θ lies in the open interval from 6 to 9.1, i.e., if $\theta > 6$ and $\theta < 9.1$. This implies an exact confidence level of $100(1 - 0.0216) = 97.84$ percent for the interval $(6, 9.1)$. For any θ not in this interval we would reject the hypothesis that it is the population median at the $100 - 97.84 = 2.16$ percent significance level.

We leave it as an exercise to show, using appropriate tables or a relevant computer program, that for the sample of 20 in Example 3.7 the interval $(6.6, 8.4)$ is an actual 95.86 percent confidence interval.

Conclusion. For the sample of 10 an exact 97.84 percent confidence interval for the median is $(6, 9.1)$ and for the sample of 20 an exact 95.86 percent interval is $(6.6, 8.4)$.

Comments. 1. The sample median is an appropriate point estimator of the population median when we make no assumption of symmetry. For our samples these are respectively 7.65 and 8.0. The rationale for this choice is that it gives equal numbers of plus and minus signs, the strongest supporting evidence for H_0. Hettsmansperger and McKean (1998, Section 1.3) justify this choice on stronger theoretical grounds.

2. Increasing the sample size shortens the confidence interval.

3. The samples contain values 14.3, 15.1 and 26.0 that all appreciably exceed the median, suggesting a skew distribution of water content with a long upper tail. This is in line with the common experience that after rain, for example, one often finds a few small areas in a field excessively wet due to poor local drainage.

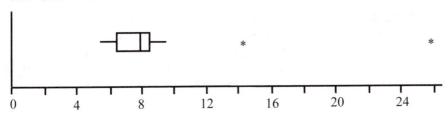

Figure 3.2 *Boxplot highlighting outliers for Sample II in Example 3.7.*

4. Skewness makes normal theory (*t*-distribution) tests or confidence intervals inappropriate. Remember that the mean and median do not coincide for a skew distribution. In passing we note that normal theory *t*-test based estimation for the sample of 10 in this example gives a mean of 8.13 with a 95 percent confidence interval (6.2, 10.0). For the sample of 20 it gives a mean of 8.77 and a 95 percent confidence interval (6.7, 10.9). Both intervals are slightly longer than those obtained using the sign test, and lead to the somewhat bizarre result that the confidence interval for the larger sample is wider than that for the smaller one. Further they apply to the means rather than the medians, and thus are shifted to the right, being symmetric about the sample mean. It is easy to see that this shift is a consequence of the presence of the high right-tail values 14.3, 15.1 and 26.0, i.e., those that suggest skewness.

A useful EDA tool to illustrate this skewness in the larger sample is a boxplot. Many software packages, e.g., Minitab, have a modification to the boxplot as described in Section 2.4 that indicates potential outliers by symbols indicating their precise position. The Minitab output for this example is illustrated in Figure 3.2.

5. In determining confidence intervals the difficulty of observations equal to a hypothesized median does not arise, so there is no question of ignoring some data points and reducing the number of useful observations. Gibbons and Chakraborti (2004, Section 5.4) point out that if there are many values equal to a hypothesized median it may be more appropriate to obtain a confidence interval for the median and accept the hypothesized value if it falls within the confidence interval. This approach protects against the loss of power implicit in the reduced effective sample size when zero differences from a hypothesized median are excluded.

Computational aspects. Given suitable tables there is little need for a computer program to calculate sign-test confidence intervals. Many general statistical packages will generate the relevant binomial probabilities if these are not available in tables. In Comment 4 above we considered *t*-distribution based 95 percent confidence intervals. For direct comparison with the nonparametric intervals, many statistical packages provide parametric (normal theory, *t*-distribution) confidence intervals at any specified level such as 95.86 or 97.84 percent if one wants to make closer comparisons at exact levels.

In Comment 4 above we remarked that *t*-distribution confidence intervals are not appropriate because the data are skew. In Example 3.6, based on the Fisher sentence length data, we drew attention to possible skewness in the data. This raises doubt about the likely validity of confidence intervals based on the t-distribution or the Wilcoxon signed-rank test. We leave it as an exercise (see Exercise 3.7) to show that for the Fisher sentence length data a 95.3 percent Wilcoxon confidence interval based on Walsh differences is (27.5, 62.5). The 95 percent interval based on the t-distribution is not very different, being (27.2, 61.1). It is interesting to consider the sign-test based interval for those data.

Example 3.9

Following arguments similar to those in Example 3.8 and applying them to the sentence length data set given in Example 3.6, i.e.,

$$12 \quad 18 \quad 24 \quad 26 \quad 37 \quad 40 \quad 42 \quad 47 \quad 49 \quad 49 \quad 78 \quad 108$$

it is easily verified (using appropriate binomial tables when $n = 12$, $p = 0.5$, or a suitable computer program) that the interval (24, 49) is an exact 96.14 percent confidence interval. This is appreciably shorter than the interval (27.5, 62.5) obtained in Example 3.6.

The sign test does not assume a symmetric distribution, and the consequence is that when the assumption of symmetry is violated the sign test sometimes (but not always) has higher Pitman efficiency, resulting in greater power in the hypothesis testing context. This often translates to shorter confidence intervals. However, where a symmetry assumption is valid the sign-test approach usually leads to wider confidence intervals than either the normal theory or Wilcoxon test methods. An exception is certain symmetric distributions with very long tails, e.g., the double exponential, where the sign test has higher Pitman efficiency than the signed-rank test, and may give a shorter confidence interval.

For the sign test the asymptotic approximation is a special case of (3.2) obtained by assigning a score zero to each sample value below the hypothesized median and a score 1 to each sample value above (or vice versa). More directly, it follows from the usual normal theory approximation to the binomial when $p = 0.5$ that if S is the number of successes (here the number of values above the hypothesized median), then

$$Z = \frac{S - 0.5n}{0.5\sqrt{n}}. \tag{3.3}$$

The approximation is improved by a continuity correction replacing S by $S + 0.5$ if S is less than its mean and by $S - 0.5$ otherwise.

3.5 Use of Alternative Scores

3.5.1 Raw data permutation tests

The scoring system for the Wilcoxon test is based on ranks, or more specifically signed ranks of deviations. These became the relevant base for a permutation test.

Historically, an earlier test was proposed by Pitman (1937a) who developed ideas put forward by Fisher (1935) for a randomization test for hypotheses about the mean or median of a symmetric distribution when we have a random sample of observations from a population that has that type of distribution. Pitman's prime aim was to show that his test, at least for samples that were not very small, lead to similar inferences to those based on a t-test that requires a normality assumption for strict validity. In effect he verified an earlier assertion by Fisher that using the t-test or other normal theory tests gave a good approximation to an exact randomization test even when the normality assumption was no longer valid. This justified continued use of the then easier-to-compute t-test as the basis for inferences. Without modern computers the calculations for the Pitman test are prohibitive except for trivially small samples. The test is seldom used in practice because, as we shall see in Section 3.6 it lacks robustness.

We describe it briefly mainly because of its historical interest. We call the test a *Pitman test*, but the names *raw data test* and *Fisher-Pitman test* are also used. Modern computer packages such as StatXact or Testimate make computation easy. However, illustrating the test with a simple numerical example helps us understand its main features. Comparing it with other tests highlights some weaknesses.

The assumptions for validity of the Pitman test are the same as those for the Wilcoxon signed-rank test. Instead of signed ranks of deviations from the hypothesized median θ_0, it uses the signed deviations themselves to form a test statistic. The appropriate test statistic is either the sum of the positive deviations or the sum of the negative deviations. A simple example suffices to highlight both the strengths and the weaknesses of the test.

Example 3.10

The problem. We consider the data set for heart rate (beats per minute) used in Example 3.1 with the same assumptions as we made there. The observations were 73, 82, 87, 68, 106, 60 and 97. We test the hypothesis H_0: $\theta = 70$ against the alternative H_1: $\theta > 70$ using the Pitman test.

Formulation and assumptions. Symmetry implies that under H_0 sample values should have a near-symmetric scatter about $\theta = 70$, so the magnitudes of the sums of positive and of negative deviations from 70 should not differ greatly. The one-sided alternative H_1 implies that when it holds there are likely to be more positive deviations from 70 than there would be under H_0 and that these positive deviations will tend to be larger than the negative deviations.

Procedure. We found in Example 3.1 that the deviations from 70 are $73 - 70 = 3$, $82 - 70 = 12$ and similarly those remaining are $17, -2, 36, -10, 27$. The sum of the positive deviations (i.e., differences from 70) is an intuitively reasonable statistic to help assess the strength of the evidence against H_0 in support of the specified one-sided alternative H_1. If the alternative had been the two-sided H_1: $\theta \neq 70$ small as well as large sums of positive deviations would also have supported H_1.

For a randomization test based on S_+, the sum of the positive deviations, we form the appropriate randomization distribution in a manner similar to that in Example 3.1 for the statistic S used there. The key to obtaining the randomization distribution of S_+ when H_0 holds is that under that hypothesis the sign of each deviation is equally likely, on account of the symmetry assumption, to be positive or negative. Thus any combination of plus and minus signs attached to a set of deviations having the observed magnitudes is equally likely. In the current example with 7 observations there are $2^7 = 128$ possible allocations of sign (2 to each of the 7 differences). Each allocation is equally likely under H_0. Recording and ordering all of them without a computer is tedious. We need not do so if we only want to obtain the relevant P-value when that is small, because we then only need to know how many of the equally likely allocations of signs give an S_+ greater than or equal to that for our sample, i.e., $S_+ = 3 + 12 + 17 + 36 + 27 = 95$. Obviously the greatest possible value of S_+ occurs when all the deviations have positive signs and then $S_+ = 3 + 12 + 17 + 2 + 36 + 10 + 27 = 107$. The next highest sum, 105, is attained when only the smallest deviation, 2, has a negative sign. In order of decreasing magnitude it is easily verified that other sums greater than the observed $S_+ = 95$ occur only when negative signs are attached to 3 only (sum 104), 2 and 3 (sum 102) and 10 only (sum 97). If a negative sign is attached to 12 only, or to both 10 and 2, (as above) the sum is 95. Thus, there are seven sums greater than or equal to the observed $S_+ = 95$. Since all are equally likely the relevant P-value for assessing the strength of evidence against H_0 is $P = 7/128 \approx 0.055$.

Conclusion. The evidence against H_0 when $P = 0.055$ is sufficiently strong to perhaps warrant a similar study with a larger group of students.

Comments. 1. Our conclusion is identical with that reached in Example 3.1.

2. Unfortunately, while modern computer programs enable us to perform a Pitman hypothesis test with relative ease it is harder, though not impossible, to obtain a confidence interval for the true θ.

3. Modifications for a two-tail test when that is appropriate are straightforward and were mentioned under Procedure.

4. We chose S_+ as our test statistic but we could equally well have chosen S_-, the sum of the negative deviations. This is because the sum of the magnitudes of the deviations remains constant for all possible allocation of signs (in this example always having the value 107). The statistic has a symmetric distribution because for any of the 128 sign permutations we always have $S_+ + S_- = 107$ or $S_- = 107 - S_+$. Computer software sometimes calculates the smaller of S_+, S_-, but each leads to the same relevant P-value. Another statistic that is sometimes used is $S_d = |S_+ - S_-|$. For this example $S_d = |2S_+ - 107|$ so there is a one-to-one ordered relationship between S_d and S_+. One may even use the one sample t-test statistic instead of S_+ or S_- because it is not difficult to show (Exercise 3.2) that again there is a one-to-one ordered relationship between the values of t and S_+. Remember though that the distribution of this statistic t is not that given by the usual t-tables that assume normality, but is identical with that for S_+.

5. The approach has a strong intellectual appeal because it uses all the information in the data, but it lacks robustness against departures from symmetry. This is explained in Section 3.6.

6. Because the possible values of S_+ depend upon values of the deviations that, in turn, depend upon the data values it follows that the S_+ corresponding to a given

P-value varies between data sets even for samples of the same fixed size. The test is called a conditional test because it is conditional upon the data values. A practical implication of this is that the P-value associated with H_0 must be computed afresh for each data set.

Computational aspects. 1. Using StatXact the P-value obtained above and the complete distribution of S_+ can be calculated for this example in less than one second. That program also gives an asymptotic approximation that is strictly valid only for large samples. In effect it uses the signed deviations as scores in (3.2)

For these data the asymptotic P-value for the one-tail test is $P = 0.051$, which, despite the small sample size, is close to the exact P-value, but such close agreement is not always obtained for small samples. In passing we note that the normal theory t-test for this example also gives $P = 0.051$.

2. To estimate P by Monte Carlo sampling one would randomly allocate, each with probability $p = 0.5$, a plus or a minus sign to every difference for, say, $10\,000$ samples, and record the number, r, of these samples giving values of S_+ greater than or equal to 95. Then $P^* = r/10\,000$ is an estimate of the exact one-tail P.

3.5.2 Normal scores

In essence transformation of a sample of n continuous ordered data to ranks replaces the sample values by something like values we would expect when sampling from a uniform or rectangular distribution over $(0, n)$ — more precisely, the rank r corresponds to the $r/(n+1)$th quantile of that distribution. Might further transformation of these ranks increase the Pitman efficiency or have other advantages?

In many parametric analyses a normality assumption is a key feature, and when it can validly be made it often leads to mathematical and computational simplicity. Exact Pitman tests are only practical for small to medium sample sizes even with the best software. It is not unreasonable to hope that if we start with data that resemble a sample from a normal distribution, that asymptotic or large sample results will approach exact results more rapidly, in the sense that they may be reasonable for not very large samples.

Such hopes inspired a study of transformations to give data more like those from a normal distribution as a basis for exact tests while retaining desirable properties such as the test being unconditional on the actual data. This is the situation for the Wilcoxon signed-rank test (but not for the Pitman test). A possible transformation is suggested by our remark that ranks correspond to quantiles of a uniform distribution. Why not transform ranks to values, i.e., *scores*, corresponding to quantile values for a normal distribution? More specifically, why not choose the *standard normal distribution*, for then it is easy to make the transformation using tables of the standard normal *cumulative distribution function* (which we now abbreviate to cdf). For example, if we have the very nonnormal sample 2, 3, 7, 21, 132, we first replace these values by the ranks 1, 2, 3, 4, 5. We then transform these to normal scores. These are the 1/6th, 2/6th, 3/6th, 4/6th and 5/6th quantiles of the standard normal distribution. These may be read from standard normal cdf tables, but many software programs compute these as one of a set of possible data transformations. The normal score corresponding to the $1/6 = 0.1667$ quantile is the x value such that the standard normal cdf, commonly denoted by $\Phi(x)$, is $\Phi(x) = 0.1667$. From tables or from software we find $x = -0.97$. For the $2/6 = 0.3333$ quantile we find $x = -0.43$ and for the 3/6th (mean or median) $x = 0$. By symmetry the remaining quantiles are $x = 0.43$ and $x = 0.97$. These quantile scores are often called *van der Waerden scores* having been proposed by van der Waerden (1952, 1953).

Alternatives to the above scores are discussed by Conover (1999, Section 5.10) and others. These include expected normal scores where the ith order statistic, $x_{(i)}$ is replaced by the expectation of the ith order statistic for the standard normal distribution. Fisher and Yates (1957, Table XX) give expected normal scores corresponding to ranks for $n \le 50$. In practice, using either van der Waerden or expected normal scores usually lead to similar conclusions. Both need adapting for use in the one sample situation. We consider here only van der Waerden scores.

Unfortunately, direct application of van der Waerden scores in an analogue to the Wilcoxon signed-rank test is not possible because the rationale of the test demands that we allocate signs to magnitudes of ranks, but van der Waerden scores are of equal magnitudes but opposite signs.

This difficulty may be partly overcome by obtaining positive scores to which appropriate signs may be allocated. We may do this by using only some of the positive quantiles of the standard normal distribution as scores. Appropriate signs, i.e., those for Wilcoxon signed ranks, are then attached to these. Recommended quantiles corresponding to the ranks $1, 2, 3, ..., n$ are the $(n+2)/(2n+2)$th, $(n+3)/(2n+2)$th, $(n+4)/(2n+2)$th, $\ldots, (2n+1)/(2n+2)$th quantiles. Effectively, the procedure is making use of the quantiles of the well-known folded normal distribution. One then proceeds using these scores as though they were the original data in a Pitman-type test. The fact that the original quantiles are not symmetrically distributed about their median and the impact of allocating negative signs detracts from the intuitive appeal of the transformation, because these factors distort the normality structure we hoped to achieve. Our experience with a few examples is that the method generally gives results similar to a Wilcoxon signed-rank test. For the problem considered in Example 3.6 this scoring system gives a one-tail $P = 0.035$ (see Exercise 3.21).

3.5.3 Other Scores

Scores other than normal scores are sometimes appropriate. We use some of these in more sophisticated problems. They are needed when data that are essentially nonnormal occur or when there are complications due to censoring of data. An example is given in Section 9.2. In analyzing survival data what are called Savage, or exponential, scores are often appropriate. We introduce these in Section 9.3.

We considered hypothesis tests about the median for a sample with censored data in Examples 2.2 and 2.3. The use of the sign test is only possible if censorship occurs at some value above the median specified in H_0. Even in this situation censorship may cause difficulties in the corresponding method for forming confidence intervals. There is now a vast literature on the use of both parametric and nonparametric methods for analysis of censored data. We discuss some basic elementary methods in Chapter 9. A more detailed treatment of the analysis of censored data, especially in the context of survival data, is given by Desu and Raghavarao (2004).

3.6 Comparing Tests and Robustness

We know that some tests are more sensitive to departures from assumptions than others. We illustrate this with the aid of examples.

Example 3.11

The problem. For each of the data sets used in Example 3.7, viz.,

Sample I	5.5	6.0	6.5	7.6	7.6	7.7	8.0	8.2	9.1	15.1
Sample II	5.6	6.1	6.3	6.3	6.5	6.6	7.0	7.5	7.9	8.0
	8.0	8.1	8.1	8.2	8.4	8.5	8.7	9.4	14.3	26.0

compare and comment upon the performance of the t-test, Wilcoxon signed-rank test, sign test and Pitman test for H_0: $\theta = 9$ against the alternative H_1: $\theta \neq 9$. For the last three compare both the exact and asymptotic test performances.

Formulation and assumptions. We omit details of computation, quoting results obtained using Minitab for the t-test and StatXact for the remaining tests.

Procedure. The relevant exact and asymptotic P-values are

Sample	I	I	II	II
Test	Exact	Asymptotic	Exact	Asymptotic
t-test	0.331	–	0.823	–
Pitman	0.352	0.305	0.916	0.817
Wilcoxon	0.102	0.092	0.015	0.017
Sign	0.109	0.058	0.003	0.002

StatXact does not use the continuity correction for the sign test. That has an appreciable effect in these examples. For instance, for the sample of 20 a continuity correction increases the two-tail asymptotic P to 0.004. This is not closer to the exact value, whereas surprisingly, for the smaller sample it does better, increasing P to 0.114, which is closer to the exact value. This may reflect increasing importance of the continuity correction if the sample size is not large. It indicates a need for caution about using asymptotic values even for moderate n. It is in the tails (i.e., for low values of P) that asymptotic values sometimes show the greatest relative departure from exact probabilities.

When there are ties a continuity correction is not usually used for the Wilcoxon test as it has little theoretical justification, although in practice it sometimes tends to bring results closer to those for the exact permutation test. The t-test $P = 0.331$ and 0.823 are close to the values for the Pitman test.

Conclusion. The data in both samples are, as already pointed out, slightly skewed to the right, especially in the sample of 20. Both the Pitman and Wilcoxon tests indicate lower efficiency relative to the sign test in the larger sample, but the effect of skewness is less for the Wilcoxon test than it is for the Pitman test, because the transformation to ranks does not give the more extreme values of 14.3, 15.1 and 26.0 the influence they have in the Pitman test. Indeed, they would have had the same rankings had they had the values 12.6, 12.7 and 12.8.

Comments. 1. In view of the conclusion that our rankings would not have been altered if the three largest observations had been replaced by 12.6, 12.7 and 12.8 when there would be only a small indication of skewness, it is interesting to see how the Pitman test would perform if these values are used in it. Intuition suggests that the P-values should then be nearer those given by the Wilcoxon test (which, of course, are unaltered by the change). Indeed, with this change the exact P-value for the Pitman test reduces dramatically to $P = 0.123$ for the smaller sample and to $P = 0.036$ for the larger, both much closer to the results for the Wilcoxon test. This should cause no surprise because there is now little indication of skewness and so a key validity requirement for both tests is nearer to being satisfied.

2. The similarity of results for the t-test and the Pitman test for the original data is consistent with the theoretical result that the distributional characteristics of each test is similar. The poor performance of these tests where there is a breakdown in assumptions reflects a shared lack of robustness. The Wilcoxon test on the other hand shows, for reasons indicated above, greater robustness against moderate skewness.

Computational aspects. The wide availability of software to carry out asymptotic tests makes them appealing. Three notes of caution are:

- Large sample sizes are needed to ensure consistently good approximations, especially in the tails, so beware of asymptotic tests when $n < 20$.

- A breakdown of an assumption such as a requirement for symmetry may make asymptotic results of little value (as is also true for exact tests).

- Practice varies regarding use of continuity corrections. It is usually easy to check whether or not these are used either from documentation included with software or from the given output, which often includes the test statistic value together with its mean and standard deviation.

Asymptotic confidence intervals are obtainable using the form of (3.1) relevant to the Wilcoxon test without ties, or for the sign test, using (3.3). For the Wilcoxon test the asymptotic result enables us to obtain an approximate value of S for significance at a suitable level. Walsh differences may then be used to get the approximate confidence interval. This works quite well if there are no ties and it gives a starting point for trial-and-error refinement when there are many ties. For the sign test we can estimate the number of positive or negative signs that determine the end points of our interval. An example shows how this works.

Example 3.12

The problem. For the larger data set in Example 3.11 use the asymptotic formula for the Wilcoxon test and that for the sign test to establish approximate 95 percent confidence intervals for the population median.

Formulation and assumptions. The key to establishing end points of 95 percent confidence intervals is that if the statistic used for computing P-values for some hypothetical value of the median θ is such that $|Z| = 1.96$ then that θ is the end point of the approximate confidence interval.

Procedure. Consider first the Wilcoxon test. For the sample of 20 observations in Example 3.12 either direct computation, or use of statistical software for the asymptotic test, establishes that the statistic S has mean 105 and standard deviation 26.78. Thus, if for some hypothesized mean a calculated S is such that

$$(S - 105)/26.78 = \pm 1.96$$

it follows that such S are appropriate values for calculating the end points of the confidence interval using Walsh averages. Solutions are $S = 52.82$ and 157.18. For practical purposes we round these to 53 and 157. In fact we only need calculate one of these in view of the symmetry in counts of positive and negative ranks; the smaller value 53 indicates that we should reject the 53 largest and 53 smallest Walsh averages. If no program is available to compute these averages it is a matter of tedious arithmetic computing the more extreme averages to establish that the approximate interval is $(7.15, 8.45)$.

For the sign test it follows immediately from (3.3) that when $n = 20$ the mean value of the test statistic (the number of positive signs or of negative signs) is 10 and the standard deviation is $\sqrt{20}/2 \approx 2.236$, whence the lower limit is the value S such that $(S - 10)/2.236 = -1.96$, i.e., $S = 5.62$. Rounding to the nearest integer suggests that we should reject H_0 if there are six or fewer positive or negative signs. For these data this gives the approximate interval $(7, 8.2)$.

Conclusion. The asymptotic results give approximate 95 percent confidence intervals based on the Wilcoxon statistic as $(7.15, 8.45)$ and based on the sign test statistic as $(7, 8.2)$.

Comments. 1. The 95 percent confidence interval based on the t-test statistic is $(6.7, 10.9)$; this is considerably longer than those obtained in this example and reflects the inappropriateness of the t-test procedure when data are clearly skew. Situations do arise where the Wilcoxon based intervals may be similar to the t-test based intervals in the presence of skewness. In practice, how well the Wilcoxon method performs in estimation problems depends upon the number of outlying observations in the sample because these influence only the more extreme Walsh averages. If there are only a few outliers when n is large, these are less influential than the same number of outliers for smaller n.

2. Given programs for exact tests such as that in StatXact it is easy to determine the exact coverage of the intervals and also to adjust the limits by a trial and error method to give better coverage. In this example the Wilcoxon interval $(7.15, 8.45)$

is associated with a lower-tail $P = 0.0235$ and an upper-tail $P = 0.030$. Although one is below, and the other above 0.025, the interval is reasonable for most practical purposes. For the sign-test interval with these end points some observations coincide with the limits, so it is useful to perform an exact test for P-values just above and below these end points. Using StatXact we find the interval (6.95, 8.25) has only an 88.46 percent coverage, but the interval (7.05, 8.15) has an even lower 73.7 percent coverage. Again, there is an indication here that one needs to be cautious about asymptotic results. Use of a continuity correction in determining the asymptotic limits may help. In Example 3.8 we established the 95 percent interval based on the sign test was (6.6, 8.4).

Computational aspects. Although programs giving exact P-values, or good Monte Carlo approximations thereto, are needed to refine asymptotic limits, limits based on asymptotic theory are usually reasonable for moderately large samples and can be computed using any program that deals adequately with the relevant asymptotic nonparametric tests. Some programs (e.g., Minitab) quickly generate tables of Walsh averages, so if the Wilcoxon intervals are required once S corresponding to $Z = \pm 1.96$ has been computed it is easy to get the Walsh averages corresponding to end points. Minitab will itself give asymptotic limits for the Wilcoxon test and exact limits for the sign test.

In Example 3.12 we found an approximate 95 percent confidence interval for the population median using the Wilcoxon signed-rank procedure was (7.15, 8.45) while that based on the sign test was displaced slightly to the left and was slightly shorter being (7, 8.2). However, the confidence levels were not quite identical. The interval (6.7, 10.9) based on the normal theory t-test is appreciably longer, reflecting the inappropriateness of the t-test approach when applied to obviously skew data. On the other hand one commonly meets situations where the sign-test interval is longer than that given by the Wilcoxon test, which in turn is longer than that based on the t-test.

In such examples, although there might be some evidence of departure from normality, if the departure is not severe and there is not much evidence of skewness these results are in general agreement with the finding based on Pitman efficiency that when a sample comes from a normal distribution the efficiency of the t-test is greater than that of any other test. The Pitman efficiency of the Wilcoxon test can be shown to be $3/\pi$ (approximately 95.5 percent) while that of the sign test is $2/\pi$ (approximately 63.7 percent) when a sample is from a normal distribution.

Two kinds of departure from normality that are common are one where the sample may come from some distribution such as the exponential or gamma distribution, which are known to be asymmetric, or one where the sample comes from a mixture of distributions. Mixtures may arise in several ways. For example, data might be results of chemical analyses carried out in two different laboratories, one of these giving less precise results than the other, and perhaps also incorporating a consistent error (called *bias*) due to an equipment fault. Mixtures may also occur in accurate data. We explained why this was so for the data in Example 3.8 (Comment 3). Another way skew or long-tail distributions may occur is by the intrusion of rogue observations often called *outliers*. If we

Table 3.5 *95 and 99 percent confidence intervals for data sets (i) and (ii) in Example 3.13 based on normal theory (t), Wilcoxon signed ranks (W) and the binomial sign test (B). For each set the shortest interval at each confidence level is indicated by an asterisk.*

		95 percent	99 percent
	t	$(2.79, 16.88)^*$	$(-0.12, 19.79)$
Data set (i)	W	$(2.5, 17.0)$	$(0.5, 20.0)^*$
	B	$(1.0, 17.0)$	$(-2.0, 20.0)$
	t	$(0.70, 23.97)$	$(-4.10, 28.77)$
Data set (ii)	W	$(2.5, 18.5)^*$	$(0.5, 34.5)$
	B	$(1.0, 17.0)^*$	$(-2.0, 20.0)^*$

know why they are rogues there may be a good case for rejection, but often we do not know if suspect observations are really rogues (e.g., incorrectly recorded readings, observations on units that are not members of the population of interest, etc.) or if we are sampling from a population where a small proportion of outlying values is the norm. While many parametric tests like the *t*-test are sensitive to certain departures from assumptions, the Wilcoxon test may be less so because it makes fewer assumptions, and the sign-test even less sensitive because it in turn makes even fewer assumptions. These latter tests exhibit the property we have already referred to as *robustness*, which we discuss more fully in Chapter 14. Test and estimation procedures are robust if they are little influenced by fairly blatant departures from assumptions.

Example 3.13 shows that the Wilcoxon test may be more robust than the *t*-test if there is a marked departure from symmetry. We also see that in such circumstances the sign test, not unexpectedly, may do even better since it makes no symmetry assumption.

Example 3.13

Suppose that the data in Example 3.2, i.e.,

$$-2 \quad 4 \quad 8 \quad 25 \quad -5 \quad 16 \quad 3 \quad 1 \quad 12 \quad 17 \quad 20 \quad 9$$

are amended by omitting the observation 25 and replacing it (i) by 35 and (ii) by 65. These changes induce obvious upper-tail skewness. Table 3.5 gives the nominal 95 percent and 99 percent confidence intervals based on the normal theory *t*, the Wilcoxon signed-rank statistic and the sign test in each case.

For the moderately skewed set (i) we have a mixed picture, the *t*-test does best at the 95 percent level and the Wilcoxon test at the 99 percent level, but the other methods do nearly as well in each case. For the more markedly skewed set (ii), the *t* limits are highly unsatisfactory, and those based on Wilcoxon are moderately satisfactory at the 95 percent level but not satisfactory at the 99 percent level.

However, the binomial-based sign-test limits are the same as those for set (i) and are not influenced by the extreme observation at 65. For set (ii) they are the same width as the Wilcoxon limits at the 95 percent level although the exact coverage (96.14 percent) is slightly greater than the exact Wilcoxon coverage (95.5 percent). This is more a reflection of the discontinuities in possible available levels than anything else. At the 99 percent level the sign test limits are the clear winner. This is not surprising since no symmetry assumption is needed.

Are such extreme values as that in data sets (i) and (ii) realistic from a medical point of view? A change in heart rate of 35 beats per minute might just be possible, but the value of 65 in set (ii) is a data error. It is good statistical practice to check data carefully to eliminate errors, but one often has to analyse data where there is no indication that an extreme value, or apparent outlier, is not a correct observation. This is where one may do better with the simple sign test than with a test requiring more stringent assumptions. On the other hand it is generally inappropriate to use the sign test when there is little or no evidence of skewness.

3.7 Fields of Application

Insurance

The median of all motor policy claims paid by a British insurance company in 2005 is £1570. Early in 2006 the company thinks claims are higher. To test this, and to estimate the likely rise in mean or median, a random sample of 25 claims is taken. The distribution of claims will almost certainly be skew, so a sign test would be appropriate.

Medicine

The median systolic blood pressure of a group of boys prior to physical training is known. If the blood pressure is taken for a sample after exercise the sign test could be used to test for a shift in median. Would you consider a one- or a two-tail test appropriate? If it appears reasonable to assume a symmetric distribution a Wilcoxon test or even a normal theory t-test is more appropriate. Even if an assumption of symmetry in systolic blood pressures before exercise is reasonable, this may not be so after exercise. For instance, the increase after exercise might be relatively higher for those above the median blood pressure at rest and this could give rise to skewness. A physiologist asking questions about changes should know if this is likely.

Engineering

Median noise level under the flight path to an airport might be known for aircraft with a certain engine type. The actual level will vary from plane to plane and from day to day depending on weather factors, the precise height each plane flies over the measuring point, etc. If the engine design is modified a sample of measurements under similar conditions to that for the old engine may indicate a noise reduction. A one-tail test would be appropriate if it were clear the modification could not increase noise. We are unlikely to be able to use a true random sample here, but if taken over a wide range of weather conditions, the first, say, 40 approaches using the new engine may broadly reflect characteristics of a random sample.

Biology

Heartbeat rates for female monkeys of one species in locality A may have a symmetric distribution with known mean. Given heartbeat rates for a sample of similar females from locality B, the Wilcoxon test could be used to detect a shift or to obtain confidence limits for the true mean.

Education

A widely used test of numerical skills for 12-year-old boys gives a median mark of 83 when a teacher works through illustrative examples in front of the class. A new method of teaching involves groups of pupils working on problems together under teacher supervision. The asymptotic approximation to the Wilcoxon test could be used to test for a shift in median if symmetry could be assumed. If not, a sign test would be preferred.

Management

Records give the mean and median number of days absent from work for all employees in a large factory for 2005. The number of days absent for a random sample of 20 is noted in 2006. Do you think such data are likely to be symmetric? In the light of the answer to this question one may select the appropriate test for indications of changes in the absentee pattern.

Geography and Environment

Estimates of the amount of cloud cover at the site of a proposed airport are taken at a fixed time each day over a period. The site might be rated unsuitable if the median cover were too high. A confidence interval for the true median would be useful.

Local Differences

A health promotion campaign encourages children to spend at least 20 minutes per day in vigorous physical exercise. A national survey following the campaign shows that this target has been met. The time spent in such exercise by children living in a particular town could be estimated by asking pupils to complete a physical activities questionnaire. One might use a Wilcoxon test to assess whether it is reasonable to assume the mean is 20 minutes for that town.

It is not unusual in a number of contexts to find that national averages applicable to a large area do not apply locally. In the UK, for instance, the average price per litre of petrol in London is very different from that in the Scottish Highlands and both (especially the latter) differ from the nationwide average. The average price in euros of a litre of milk in Italy, France and Germany may each differ from the average for the whole European community.

Industry

The median time people stay in jobs in a large motor assembly plant in Germany is known to be 5.2 years. Employment times for all 55 employees who have left a UK plant in the last 3 months are available. Although this may not be a random sample it may be reasonable to use the data to test whether the median time for UK workers is also 5.2 years. If the distribution appears skew a sign test would be appropriate.

Astronomy

Astronomers frequently estimate quantities such as the mass of a particular star or the diameter of a planet in several observatories each using non-identical equipment. The resulting measurements are often far from normally distributed, forming either a longer tailed or skewed distribution of readings. Providing reasonable assumptions can be made about the nature of measurement errors nonparametric methods of analysis may often be appropriate.

Society and the Law

The number of drivers per 1000 motorists found to be over the legal alcohol limit when stopped at roadside random checks in 2004 is 5.7. In the year 2006 the penalties imposed on drivers committing this offence are doubled. In 2006 the numbers per thousand motorists over the legal limit is recorded in each of 12 different police areas chosen at random. These numbers may be used to test a claim that in the country as a whole the number of offenders per 1000 drivers tested has decreased since 2004. If there is evidence of skewness in the sample data a sign test may be more appropriate than either the Wilcoxon signed rank or a t-test.

Drug Effectiveness

A widely used drug is known to increase the time insomniacs sleep by a median time of 1.4 hours per night. The manufacturer of a new and cheaper drug claims it is even more effective. An experiment is conducted in which increased sleeping times are recorded for each of 25 patients. Depending on the nature of the data a t-test, a Wilcoxon signed-rank test, a normal scores test or a sign test might be used to test the claim. If θ is the population median the appropriate test would be that of H_0: $\theta = 1.4$ against H_1: $\theta \neq 1.4$ despite the manufacturer's claim that his new drug is more effective. Such a claim may not be backed by experimental evidence. In Exercise 3.11 a one-tail test is justified because there is a firm assertion that an effect, if any, will be beneficial.

3.8 Summary

For the **Wilcoxon signed-rank test** a suitable test statistic is S, the sum of the lesser of the positive, S_+, or negatiive, S_-, signed-rank differences from the mean or median specified in H_0 (Section 3.3.1). Tables, or more satisfactorily, appropriate computer software may be used for significance tests. Ties require special treatment (Section 3.3.4) and suitable computer software is then needed to determine exact permutation distributions. **Walsh averages** (Section 3.3.2) provide an alternative hypothesis testing procedure and they are particularly useful for estimation and calculating confidence intervals. An assumption of population symmetry is needed for validity of test and estimation procedures. Asymptotic normal approximations (Section 3.3.3) work well for sample sizes $n \geq 20$ if there are no (or very few) ties. Modification of the denominator in the asymptotic test statistic is required for numerous ties, even for relatively large n.

The **sign-test** statistic is the number of observations above or below a median specified in H_0 (see e.g., Example 3.7). Significance is determined using binomial probability tables for $p = 0.5$ and various n, or else from a suitable computer program. The method is easily used to obtain confidence intervals (Section 3.4.2). For hypothesis testing modifications are required to deal with values equal to the median specified in H_0 (Section 3.4.1). Normal theory approximations (Section 3.4.2) may be used for sample sizes greater than 20. No assumption of symmetry is required for validity of the sign test.

For the **Pitman** or raw data permutation test a suitable statistic (Section 3.5.1) is the sum S_+ of the positive data deviations from a hypothesized mean or median θ assuming the sample is from a symmetric distribution. Appropriate computer software is required for significance tests and even with such software confidence limits can only be obtained by a cumbersome trial-and-error approach. The procedure lacks robustness and often gives similar results to a t-test even when the latter is not appropriate. It is conditional upon each particular data set and is not widely used in practice.

Normal scores (Section 3.5.2) aim to make data more like samples from a normal distribution. Test results are usually not very different from those given by the Wilcoxon test. There appear to be no simple satisfactory methods for obtaining confidence intervals using these transformations.

3.9 Exercises

*3.1   Verify the expressions given for the mean and variance in (3.2) for statistics based on sums S_+ or S_- of signed scores $s_i, 1 = 1, 2, \ldots, n$ where the sign associated with each s_i is equally likely to be plus or minus.

3.2   In Comment 4 on Example 3.10 we asserted that the usual t-statistic could be used in place of S_+ as the test statistic for the Pitman test because there was a one-to-one correspondence between the ordering of the two statistics. Establish that this is so. (Hint: show that the denominator of the t-statistic is invariant under all permutations of the signs of the deviations d_i.)

3.3 In Comment 3 on Example 3.2 we suggested that for a variety of reasons one should be cautious about extending inferences about heartbeat rates for female students to the population at large. What might some of these reasons be?

3.4 Using the data in Table 3.1 for the distribution of the Wilcoxon S when $n = 7$, construct a bar chart like that in Figure 3.1 showing the probability function for S. Discuss the similarity, or lack of similarity, to a normal distribution probability density function.

*3.5 Establish that the permutation distribution of the Wilcoxon signed-rank statistic for testing the hypothesis $H_0 : \theta = 6$, given the observations 4, 4, 8, 8, 8, 8, 8 has a distribution equivalent to that for the sign test of the same hypothesis. Would this equivalence hold if the null hypothesis was changed to $H_0 : \theta = 7$?

3.6 Establish nominal 95 percent confidence intervals for the median based on the Wilcoxon signed-rank test for the following data sets. If an appropriate computer program is available use it to comment on the discontinuities at the end points of your estimated intervals based on Walsh averages.

$$\begin{array}{lccccccccccc} Set\ I & 1 & 1 & 1 & 1 & 1 & 3 & 3 & 5 & 5 & 7 & 7 \\ Set\ II & 1 & 2 & 2 & 4 & 4 & 4 & 4 & 5 & 5 & 5 & 7 \end{array}$$

3.7 Form a table of Walsh averages for the Fisher sentence length data given in Example 3.6, and use it to obtain 95 and 99 percent confidence intervals.

*3.8 The numbers of pages in the sample of 12 books given in Exercise 2.5 were

$$126 \quad 142 \quad 156 \quad 228 \quad 245 \quad 246 \quad 370 \quad 419 \quad 433 \quad 454 \quad 478 \quad 503$$

Use the Wilcoxon signed-rank test to test the hypothesis that the mean number of pages in the statistics books in the library from which the sample was taken is 400. Obtain a 95 percent confidence interval for the mean number of pages based on the Wilcoxon test and compare it with the interval obtained using a t-test under an assumption of normality.

*3.9 Apply the sign test to the data in Example 3.2 for the hypotheses considered there.

3.10 For the sample of 20 in Example 3.7 if θ is the population median test the hypothesis H_0: $\theta = 9$ against the alternative H_1: $\theta \neq 9$ using the sign test by computing any relevant binomial probabilities directly from the binomial probability formula. It is not necessary to determine the complete distribution to obtain the relevant P value. To perform the test for H_0: $\theta = 7.5$ against the alternative H_1: $\theta > 7.5$ additional terms in the distribution will be needed. Either calculate these, or use tables or computer software to carry out the appropriate test.

3.11 Before treatment with a new drug 11 people with sleep problems have a median sleeping time of 2 hours per night. A drug is administered and it is known for good scientific reasons that if it has an effect it will increase sleeping time, but some doctors doubt if it will have any effect. Are their doubts justified if the hours per night slept by these individuals after taking the drug are:

$$3.1 \quad 1.8 \quad 2.7 \quad 2.4 \quad 2.9 \quad 0.2 \quad 3.7 \quad 5.1 \quad 8.3 \quad 2.1 \quad 2.4$$

*3.12 Kimura and Chikuni (1987) give data for lengths of Greenland turbot of various ages sampled from commercial catches in the Bering Sea as aged and measured by the Northwest and Alaska Fisheries Center. For 12-year-old turbot the numbers of each length were:

Length (cm)	64	65	66	67	68	69	70	71	72	73	75	77	78	83
No. of fish	1	2	1	1	4	3	4	5	3	3	1	6	1	1

Would you agree with someone who asserted that, on this evidence, the median length of 12-year-old Greenland turbot was almost certainly between 69 and 72 cm?

3.13 Use the Wilcoxon signed-rank test to test the hypothesis that the median length of 12-year-old turbots is 73.5 using the data in Exercise 3.12.

*3.14 The first application listed in Section 3.7 involved British insurance claims. The 2005 median was £1570. A random sample of 14 claims from a large batch received in the first quarter of 2006 were for the following amounts (in £):

1175 1183 1327 1581 1592 1624 1777 1924 2483 2642 2713 3419 5350 7615

What test do you consider appropriate for a shift in median relative to the 2005 median? Would a one-tail test be appropriate? Obtain a 95 percent confidence interval for the median based upon these data. If all amounts were converted to, say, Euros or to $US, would your conclusions be the same?

*3.15 The weight losses in kilograms for 16 overweight women who have been on a diet for 2 months are as follows:

4 6 3 1 2 5 4 0 3 6 3 1 7 2 5 6

The firm sponsoring the diet advertises "Lose 5 kg in 2 months". In a consumer affairs radio programme they claim this is an "average" weight loss. You may be unclear as to what the sponsors mean by "average", but assuming the sample is effectively random do the data support a median weight loss of 5 kg in the population of dieters? Test this without an assumption of symmetry. What would be a more appropriate test with an assumption of symmetry? Carry out this latter test.

3.16 A pathologist counts the numbers of diseased plants in randomly selected areas each 1 metre square on a large field. For 40 such areas the numbers of diseased plants are:

21	18	42	29	81	12	94	117	88	210
44	39	11	83	42	94	2	11	33	91
141	48	12	50	61	35	111	73	5	44
6	11	35	91	147	83	91	48	22	17

Use histograms and boxplots to decide whether there is evidence of skewness or outliers. Use nonparametric tests to find whether it is reasonable to assume the median number of diseased plants per square metre might be 50 (i) without assuming population symmetry, (ii) assuming population symmetry. For these data and the evidence provided by a boxplot and histograms do you consider the latter assumption reasonable?

*3.17 A parking attendant notes the time cars have been illegally parked after their metered time has expired. For 16 offending cars he records the time in minutes as:

10 42 29 11 63 145 11 8 23 17 5 20 15 36 32 15

Obtain an appropriate 95 percent confidence interval for the median overstay time of offenders prior to detection. What assumptions were you making to justify using the method you did? To what population do you think the confidence interval you obtained might apply?

*3.18 Knapp (1982) gives the percentage of births on each day of the year averaged over 28 years for Monroe County, New York. Ignoring leap years (which make little difference), the median percentage of births per day is 0.2746. Not surprisingly, this is close to the expected percentage on the assumption that births are equally likely to be on any day, that is, $100/365 \approx 0.274$. We give below the average percentage for each day in the month of September. If births are equally likely to be on any day of the year this should resemble a random sample from a population with median 0.2746. Do these data confirm this?

0.277	0.277	0.295	0.286	0.271	0.265	0.274	0.274	0.278	0.290
0.295	0.276	0.273	0.289	0.308	0.301	0.302	0.293	0.296	0.288
0.305	0.283	0.309	0.299	0.287	0.309	0.294	0.288	0.298	0.289

3.19 A standard method for assembling a routine component in a manufacturing process is known to take a trained operative a median time of 14.2 minutes to complete. A new method of assembling that same component is proposed. Twenty operatives are given practice sessions using the new method, after which each claims he or she fully understands the process. They are then each timed for assembling one new component. The times in minutes they take are:

14.2	11.3	12.7	19.2	13.5	14.4	11.8	15.1	12.3	11.7
13.2	13.4	14.0	14.1	13.7	11.9	11.8	10.7	11.3	12.2

Form a boxplot for the data. Follow this by any tests you consider appropriate to indicate whether the median assembly time in mass production is likely to differ from 14.2. Also obtain nominal 95 percent confidence intervals for the median time associated with any test you use. Compare in each case the intervals based on exact procedures with the related asymptotic approximate intervals.

3.20 In a class of 52 students 36 are categorized as being overweight. In an attempt to reduce obesity these 36 all volunteer to go on a diet. Each is weighed at the start of the diet period, and after 3 weeks any gain or loss in weight is recorded. A negative sign indicates a loss. Form a boxplot of the gain or loss in weight and comment on any features that may influence a choice of a test to determine whether or not the diet has induced a change in median weight. Use what you consider the most apropriate method to test whether the diet has affected median weight. The weight changes (in kg) are:

−1.2	1.4	0.2	−0.7	−6.4	−2.7	−8.6	−1.7	−2.2
0.1	−0.4	−4.2	−1.6	1.2	−1.3	−2.4	3.1	−0.2
−4.5	−6.3	−1.7	0.0	0.2	−3.7	1.1	−2.3	−0.1
−7.3	0.2	−1.4	−0.9	−2.0	0.0	1.1	−0.3	−1.1

Obtain an appropriate nominal 95 percent confidence interval for the median change in weight.

3.21 For the data on sentence lengths in Example 3.6 use the modified van der Waerden scores procedure proposed in Section 3.5.2 to test the hypothesis $H_0: \theta = 30$ against $H_1: \theta > 30$.

CHAPTER 4

OTHER SINGLE-SAMPLE INFERENCES

4.1 Other Data Characteristics

Inferences about centrality using a sample from a continuous distribution covered in the previous chapter are a small but important part of single-sample data analysis. Restriction to samples from continuous distributions is a major limitation. Even if we restrict ourselves to such a sample, we may want to test whether it is reasonable to make the further inference that the sample is one from some specific distribution, or at least from some specific family of distributions.

We might also want to test whether it is reasonable to assume that the observations constitute a truely random sample. If the order of the observations is important we may want to test for the presence of a trend.

Data may consist of counts rather than measurements, and we may want to make inferences from these.

We consider some, but not all, of these situations in this chapter.

4.2 Matching Samples to Distributions

We often want to know whether observations are consistent with their being a sample from some specified continuous distribution. Kolmogorov (1933, 1941) devised a test for this purpose.

Given observations $x_1, x_2, \ldots, x_n$ we ask if the data are consistent with their being a sample from some completely specified distribution. This might be a uniform distribution over $(0, 1)$ or over $(20, 30)$, or a normal distribution with mean 20 and standard deviation 2.7.

Kolmogorov's test is distribution-free because the procedure does not depend upon what the distribution under study is, providing it is completely specified under H_0. In Section 4.2.3 we look at modifications to answer questions like "Can we suppose these data are from a normal distribution with unspecified mean and variance?"

4.2.1 Kolmogorov's Test

We pointed out in Section 2.3 that the sample, or empirical, cumulative distribution function $S_n(x)$ is a consistent estimator of the population cumulative distribution function. This provides the motivation for the Kolmogorov test.

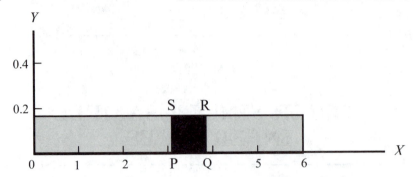

Figure 4.1 *Probability density function of a continuous uniform distribution over* $(0,6)$.

The continuous uniform distribution (sometimes called the rectangular distribution) is a simple continuous distribution that arises when a random variable has the same probability of taking a value in any small interval of length δx lying entirely in a fixed and specified finite interval (a, b).

Suppose pieces of thread each 6 cm long are clamped at each end and a force is applied until they break. If each thread breaks at the weakest point and this is equally likely to be anywhere along its length, then the breaking points will be uniformly distributed over the interval $(0, 6)$, where the distance to the break is measured in centimetres from the left-hand clamp.

The probability density, or frequency, function has the form

$$f(x) = 0,\ x \le 0;\ f(x) = 1/6,\ 0 < x \le 6;\ f(x) = 0,\ x > 6.$$

The rectangular form of $f(x)$ over $(0, 6)$ explains the name rectangular distribution. If the thread always breaks, the probability is 1 that it breaks in the interval $(0, 6)$, so the total area under the density curve must be 1. Only the density function above satisfies both this condition and that of equal probability of a break in any small segment of length δx, no matter where that segment lies within that interval. The function is graphed in Figure 4.1 and is essentially a line running from 0 to 6 at height $1/6$ above the x-axis. The lightly shaded area represents the total probability of 1 associated with the complete distribution. The heavily shaded rectangle PQRS between $x = 3.1$ and $x = 3.8$ represents the probability that the random variable X, the distance to the break, takes a value between 3.1 and 3.8. Since $PQ = 3.8 - 3.1 = 0.7$ and $PS = 1/6$, this area is $0.7/6 \approx 0.1167$. The probability of a break occurring in any segment of length 0.7 lying entirely in the interval $(0, 6)$ is also 0.1167.

The Kolmogorov test uses not the probability density function, but the cumulative distribution function (cdf), a function that gives $\Pr(X \le x)$ for all x in $(0, 6)$.

The cdf is written $F(x)$ and for any x between 0 and 6, $F(x) = x/6$. It has the value 0 at $x = 0$ and 1 at $x = 6$. It is graphed in Figure 4.2. These notions

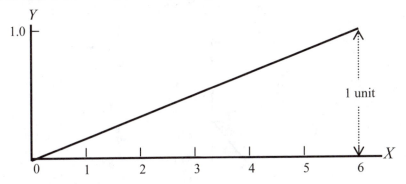

Figure 4.2 *Cumulative distribution function (cdf) for a uniform distribution over* $(0, 6)$.

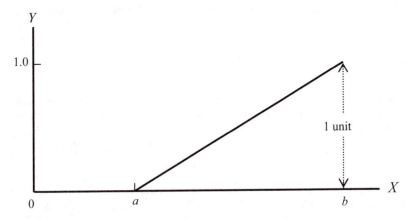

Figure 4.3 *Cumulative distribution function (cdf) for a uniform distribution over* (a, b).

generalize to a uniform distribution over any interval (a, b). The cdf becomes $F(x) = (x - a)/(b - a)$, $a < x \leq b$, specifying a straight line rising from zero at $x = a$ to 1 at $x = b$. Clearly $F(a) = 0$ and $F(b) = 1$. This is illustrated in Figure 4.3. We met another special case of the rectangular distribution cdf where $a = 0$ and $b = 10$ in Figure 2.5.

For any continuous distribution the cdf is a curve starting at zero for some x and increasing as we move from left to right until it attains the value 1. It never decreases as x increases and is said to be *monotonic nondecreasing* (or *monotonic increasing*). Figure 4.4 shows the cdf for the standard normal distribution.

Recognizing that $S(x)$ should not depart violently from $F(x)$ for a sample from a distribution with that specific cdf is basic to Kolmogorov's test. The test statistic is the maximum difference in magnitude between $F(x)$ and $S(x)$.

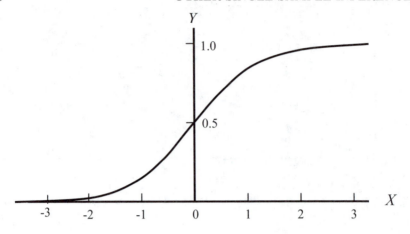

Figure 4.4 *Cumulative distribution function (cdf) for the standard normal distribution.*

Example 4.1

   The problem. The distances from one end at which each of 20 threads 6 cm long break when subjected to strain are given below. Evaluate and graph $S(x)$. For convenience the distances are given in ascending order.

$$
\begin{array}{cccccccccc}
0.6 & 0.8 & 1.1 & 1.2 & 1.4 & 1.7 & 1.8 & 1.9 & 2.2 & 2.4 \\
2.5 & 2.9 & 3.1 & 3.4 & 3.4 & 3.9 & 4.4 & 4.9 & 5.2 & 5.9
\end{array}
$$

   Formulation and assumptions. From (2.2) it is clear that $S(x)$ increases by $1/20$ at each unique x value corresponding to a break, or by $r/20$ if r break distances coincide.

   Procedure. When $x = 0$, $S(x) = 0$. It keeps this value until $x = 0.6$, the first break. Then $S(0.6) = 1/20$, and maintains this value until $x = 0.8$ when it jumps to $2/20$, and so on until $x = 3.4$. At $x = 3.4$ there are 2 breaks so $S(x)$ increases by $2/20$. Its value at each step is given in Table 4.1.

   Conclusion. Figure 4.5 shows the form of $S(x)$ for these data.

   Comment. If a sample comes from a population with cdf $F(x)$, the step function $S(x)$ should not depart markedly from $F(x)$. If breaks are uniformly distributed over $(0, 6)$ one expects, for example, about half of these to be in the interval $(0, 3)$, so that $S(3)$ should not have a value very different from 0.5, and so on. In Figure 4.5 we show also the cdf for a uniform distribution over $(0, 6)$ reproduced from Figure 4.2. Intuition may suggest this is not a good fit.

Example 4.2

   The problem. Given the 20 breaking points and the corresponding $S(x)$ values in Table 4.1, is it reasonable to suppose breaking points are uniformly distributed over $(0, 6)$?

Table 4.1 *Values of S(x) at step points.* * *indicates a repeated value.*

x	0.6	0.8	1.1	1.2	1.4	1.7	1.8	1.9	2.2	2.4
$S(x)$	0.05	0.10	0.15	0.20	0.25	0.30	0.35	0.40	0.45	0.50

x	2.5	2.9	3.1	3.4*	3.9	4.4	4.9	5.2	5.9
$S(x)$	0.55	0.6	0.65	0.75	0.80	0.85	0.90	0.95	1.00

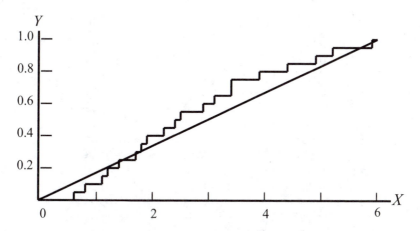

Figure 4.5 *Sample cumulative distribution function (cdf) for thread breaks (step function) together with the cdf for a uniform distribution over (0,6) (straight line).*

Formulation and assumptions. The maximum difference in magnitude between $F(x)$ and $S(x)$ is compared with a tabulated value to assess the goodness of fit. Or preferably, software is used to determine the exact P-value.

Procedure. Table 4.2 shows values of $F(x)$ and $S(x)$ at each break point given in Table 4.1. It also gives at each of these points the difference $F(x_i) - S(x_i)$. In view of the stepwise nature of $S(x)$, the maximum difference may not be in this set. Inspecting Figure 4.5 it is clear that a greater difference may occur immediately before such a step. There $S(x)$ has the value attained at the previous step, so a maximum may occur among the $F(x_i) - S(x_{i-1})$. These differences are also recorded in Table 4.2.

Figure 4.5 shows that $S(x)$ lies above $F(x)$ for much of the interval $(0,6)$. The entry of greatest magnitude in the last two columns of Table 4.2 is -0.18 when $x_i = 3.4$.

Published tables give values of the minimum difference required for significance at nominal 5 and 1 percent levels for various n, and in some cases also for other percentage levels, but modern software makes these somewhat redundant. If required, tables may be found in Gibbons and Chakraborti (2004), in Conover (1999), or in many general statistical tables. The theory for computing exact P-values is more complicated

Table 4.2 *Comparison of $F(x)$ and $S(x)$ for thread breaks.*

x_i	$F(x_i)$	$S(x_i)$	$F(x_i) - S(x_i)$	$F(x_i) - S(x_{i-1})$
0.6	0.10	0.05	0.05	0.10
0.8	0.13	0.10	0.03	0.08
1.1	0.18	0.15	0.03	0.08
1.2	0.20	0.20	0.00	0.05
1.4	0.23	0.25	−0.02	0.03
1.7	0.28	0.30	−0.02	0.03
1.8	0.30	0.35	−0.05	0.00
1.9	0.32	0.40	−0.08	−0.03
2.2	0.37	0.45	−0.08	−0.03
2.4	0.40	0.50	−0.10	−0.05
2.5	0.42	0.55	−0.13	−0.08
2.9	0.48	0.60	−0.12	−0.07
3.1	0.52	0.65	−0.13	−0.08
3.4	0.57	0.75	−0.18	−0.08
3.9	0.65	0.80	−0.15	−0.10
4.4	0.73	0.85	−0.12	−0.07
4.9	0.82	0.90	−0.08	−0.03
5.2	0.87	0.95	−0.08	−0.03
5.9	0.98	1.00	−0.02	0.03

than that for situations we have met so far and Gibbons and Chakraborti (2004, Section 4.3) give a detailed account. Tables indicate that the statistic must have a value of at least 0.294 for significance at the 5 percent level in a two-tail test. The exact test for these data gives $P = 0.458$.

Conclusion. There is no evidence against H_0.

Comments. 1. A two-tail test is appropriate if H_1 specifies that the observations may come from any other unspecified distribution. Sometimes the alternative is that the cdf must lie either wholly at or above, or wholly at or below, that specified in the null hypothesis. In the thread problem we may know that the test equipment places strains on the thread that, if not uniformly distributed, will be greater at the left and decrease as we move to the right. This would increase the tendency for breaks to occur close to the left end of the string. Then the cdf would always lie above the straight line representing $F(x)$ for the uniform distribution. In that case a maximum difference with $S(x)$ greater than the uniform distribution $F(x)$ would, if significant, favour H_1, making a one-tail test appropriate.

2. In commenting on Figure 4.5 we noted that the step function did not appear to match the population cdf very well. The Kolmogorov test does not indicate a meaningful departure, a point we take up again in Section 4.2.2.

Computational aspects. Some general statistical packages include programs for the Kolmogorov test but may use only asymptotic results applied at nominal significance levels. These are not satisfactory even in that context unless sample sizes are at least 20, or sometimes more. The exact test program in StatXact avoids these difficulties and allows a comparison with an asymptotic approximation to the *P*-value, which often shows a surprising difference.

Because it only uses the difference of greatest magnitude the Kolmogorov test appears to waste information. Tests do exist that allow for all differences, but these tend to have few, if any, advantages. This is not so surprising as intuition may suggest, for the value of $S(x)$ at any stage depends on how many observations are less than the current x. Therefore, at each stage we make the comparison on the basis of accumulated evidence.

For a discrete distribution the Kolmogorov test tends to give too few significant results and in this sense is conservative. We consider preferred tests for discrete distributions in Section 12.4.

4.2.2 Comparison of Distributions: Confidence Regions

If, in Figure 4.5 we had drawn step functions everywhere at distances 0.294 units above or below that representing $S(x)$, with adjustments at each end to stop them falling below zero or above one, we would, since 0.294 is the critical difference for significance at the 5 percent level, have a 95 percent *confidence region* for $F(x)$ in the sense that any $F(x)$ lying entirely between these two new step functions would be an acceptable $F(x)$ when testing at a 5 percent significance level in a two-tail test. This interval is not very useful in practice.

A common practical situation is one where it is reasonable to assume that the data may come from one of a (usually small) number of completely specified distributions. We might then use a Kolmogorov test for each. In each test the relevant $S(x_i)$ will be the same for a given data set. As a result of these tests we may reject some of the hypotheses but still find more than one is acceptable. It is then useful to get at least an overall eye comparison of how well the match of $S(x)$ is to each acceptable $F(x)$. Ross (1990, p. 85) gives a useful procedure for such eye comparisons.

Basically this consists of plots of the maximum of the two deviations given for the last two columns of each row in a table like Table 4.2 against the $S(x_i)$ for that row. Plots may easily be done for several different hypothesized population distributions on the same graph.

Example 4.3

The problem. Compare the fit of the thread break data in Example 3.9 to (i) a uniform distribution over $(0, 6)$ and (ii) the distribution with cumulative distribution function over $(0, 6)$ given by

$$F(x) = x/5, \ 0 < x \le 3; \quad F(x) = 0.2 + 4x/30, \ 3 < x \le 6. \tag{4.1}$$

Table 4.3 *Values of* $x_i, S(x_i), d_i = max[F(x_i) - S(x_i), F(x_i) - S(x_{i-1})]$ *for population distributions specified in* (i) *and* (ii) *of Example 4.3.*

x_i	$S(x_i)$	$d_i(\text{i})$	$d_i(\text{ii})$
0.6	0.05	0.10	0.12
0.8	0.10	0.08	0.11
1.1	0.15	0.08	0.12
1.2	0.20	0.05	0.09
1.4	0.25	0.03	0.08
1.7	0.30	0.03	0.09
1.8	0.35	−0.05	0.06
1.9	0.40	−0.08	0.03
2.2	0.45	−0.08	0.04
2.4	0.50	−0.10	0.03
2.5	0.55	−0.13	−0.05
2.9	0.60	−0.12	0.03
3.1	0.65	−0.13	−0.04
3.4	0.75	−0.18	−0.10
3.9	0.80	−0.15	−0.08
4.4	0.85	−0.12	−0.06
4.9	0.90	−0.08	−0.05
5.2	0.95	−0.08	−0.06
5.9	1.00	0.03	0.04

Formulation and assumptions. All information needed for the Kolmogorov test procedure for the distribution (i) was obtained in Example 4.2. We require similar information for the alternative distribution in (ii). We make a graphical comparison using the format suggested by Ross.

Procedure. For (i) the values of $S(x)$ of interest are those in column 3 of Table 4.2. In each case the corresponding deviation is the value of greatest magnitude in columns 4 and 5. For (ii) the values of $S(x)$ are again those in Table 4.2. However, fresh values of $F(x)$ for each x_i must be calculated using (4.1). For example, when $x_i = 1.2$, (4.1) gives $F(1.2) = 1.2/5 = 0.24$. Using this and the values of $S(1.1)$ and $S(1.2)$ given in Table 4.2 we find $F(1.2) - S(1.2) = 0.24 - 0.20 = 0.04$ and $F(1.2) - S(1.1) = 0.24 - 0.15 = 0.09$. Thus, corresponding to $x = 1.2$, the maximum deviation is 0.09 just before the step in $S(x)$ at $x = 1.2$. In Table 4.3 we set out for each x_i the corresponding $S(x_i)$ and maximum deviations for the population distributions specified in (i) and (ii). These deviations are plotted against $S(x)$ in Figure 4.6.

Although it is not essential we have joined consecutive points in each case by straight line segments. The parallel broken lines at $y = \pm 0.294$ represent the critical

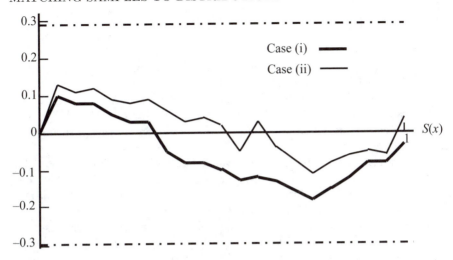

Figure 4.6 *Plot of deviations $F(x) - S(x)$ against $S(x)$ for the two $F(x)$ functions specified in Example 4.3.*

value for significance at the 5 percent level. We reject H_0 at a formal 5 percent significance level only if we observe a deviation outside these lines.

Conclusion. There is no strong evidence against either of the population hypotheses specified in (i) and (ii).

Comment. The deviations, except for the early observations, are rather smaller for (ii), suggesting (but only marginally) that this may be a slightly better fit. The hypothesis (ii) implies a uniform but higher probability of breakages between 0 and 3 cm relative to a uniform distribution, but a lower probability between 3 and 6 cm. If other hypotheses are of interest relevant deviations for these may be entered in Figure 4.6.

4.2.3 Tests for Normality

The Kolmogorov test is valid for any completely specified continuous distribution. If we specify in H_0 only that the sample has a normal distribution and estimate the parameters μ, σ^2 from the data we may still use the Kolmogorov statistic, but the associated P-values differ from those for the fully specified model.

For this important practical situation a test proposed by Lilliefors (1967) is useful. We assume observations are a random sample from some unspecified continuous distribution, and test whether it is reasonable to suppose this is a member of the normal family. To this extent our test is not distribution-free, but it is nonparametric in the sense that we do not specify parameter values.

The basic idea can be extended to test compatibility with other families of distributions such as a gamma distribution with unspecified parameters, but separate tables of critical values or appropriate software to estimate P-values are needed for each family of distributions. Conover (1999, Section 6.2) describes some difficulties in obtaining critical values. One may only get approximate values of the statistic corresponding to precise P-values such as $P = 0.05$ or 0.01. Alternatively Monte Carlo approximations for the exact P-value corresponding to an observed value of the statistic may be used. StatXact provides a program for the latter. Conover (1999, Table A14) gives approximate critical values for sample sizes up to 30 and asymptotic approximations for larger samples.

Another test for normality was given by Shapiro and Wilk (1965). The observed data are standardized to have zero mean and unit standard deviation. The test is based on the correlation between the ordered standardized observed data and the expected values of the order statistics of a standard normal variable. If the data are normal, the correlation should be close to 1. The test requires special tables or appropriate software for implementation. An account of how the test works, without detail of the theory, is given by Conover (1999, Section 6.2) who also provides extensive tables that are required if relevant computer software is not available. This test has good power against a range of alternatives but the rationale is less easy to describe by intuitive arguments. StatXact includes a program for asymptotic estimates of P-values. Except for very small samples, where exact results are available, these usually prove adequate.

Tests for normality may be important in deciding whether to apply nonparametric methods in a particular problem. We describe Lilliefors' test and quote results for analysis of the same data using the Shapiro–Wilk test.

Example 4.4

The problem. Data are given in Appendix I for ages at death of four clans in the Badenscallie burial ground in Wester Ross, Scotland, They have already been introduced in Section 2.4. From all 117 ages recorded a random sample of 30 was taken and the ages for that sample were, in ascending order:

11	13	14	22	29	30	41	41	52	55	56	59	65	65	66
74	74	75	77	81	82	82	82	82	83	85	85	87	87	88

Is it reasonable to suppose the death ages are normally distributed?

Formulation and assumptions. In Lilliefors' test the Kolmogorov statistic is used to compare the standard normal cdf $\Phi(z)$ with the standardized sample cdf $S(z)$ based on the transformation

$$z_i = \frac{x_i - m}{s}$$

where m is the sample mean and s the usual estimate of population standard deviation, i.e.,

$$s = \sqrt{\frac{\sum_i x_i^2 - (\sum_i x_i)^2/n}{n-1}}$$

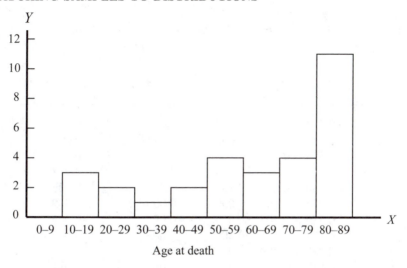

Figure 4.7 *Histogram of death ages at Badenscallie.*

Procedure. We find $m = 61.43$, $s = 25.04$. Successive z_i are calculated: e.g., for $x_1 = 11$, we find $z_1 = (11 - 61.43)/25.04 = -2.014$. We treat this as a sample value from a standard normal distribution and tables (e.g., Neave, 1981, pp. 18–19) show $\Phi(-2.014) = 0.022$. Table 4.4 is set up in analogous manner to Table 4.2. Here $\Phi(z)$, $S(z)$ represent the standard normal and sample cdfs respectively. These computations are done automatically in StatXact. It is not essential, but we have included the x values, an asterisk against a value implying it occurs more than once. The largest difference is 0.192, occurring in the final column when $x = 74$. Tables such as Table A14 in Conover (1999, p. 548) give the 1 percent critical value when $n = 30$ for a one-tail test to be 0.183. StatXact computes a Monte Carlo approximation to the exact P-value for this particular data set as a worked example in the manuals for StatXact version 4 and later. Approximate P-values differ slightly between simulations.

Conclusion. The Monte Carlo approximations to exact P-values provide strong evidence that the sample is not from a normal distribution.

Comments. 1. The StatXact manuals also use these data to illustrate the Shapiro-Wilk test and obtain an asymptotic estimate $P = 0.0007$. Shapiro, Wilk and Chen (1968) carried out studies that show this test has good power against a wide range of departures from normality.

2. A glance at the sample data suggests that a few males died young and that the distribution is skew with a large number of deaths occurring after age 80. Figure 4.7 shows these data on a histogram with class interval of width 10. Readers familiar with histograms for samples from a normal distribution will not be surprised that we reject the hypothesis of normality. Indeed life-span data, whether for man, animal or the time a machine functions without a breakdown, will often by their nature have a nonnormal distribution.

Table 4.4 *The Lilliefors normality test: Badenscallie ages at death. An* * *implies a repeated value.*

x_i	z_i	$\Phi(z_i)$	$S(z_i)$	$\Phi(z_i) - S(z_i)$	$\Phi(z_i) - S(z_{i-1})$
11	-2.014	0.022	0.033	-0.011	0.022
13	-1.934	0.026	0.067	-0.044	-0.007
14	-1.894	0.029	0.100	-0.071	-0.038
22	-1.575	0.058	0.133	-0.075	-0.042
29	-1.295	0.098	0.167	-0.069	-0.035
30	-1.255	0.105	0.200	-0.095	-0.062
41*	-0.816	0.207	0.267	-0.060	-0.007
52	-0.377	0.353	0.300	0.053	0.086
55	-0.257	0.399	0.333	0.066	0.099
56	-0.217	0.414	0.367	0.047	0.081
59	-0.097	0.461	0.400	0.061	0.094
65*	0.142	0.556	0.467	0.089	0.156
66	0.183	0.572	0.500	0.072	0.105
74*	0.502	0.692	0.567	0.125	0.192
75	0.542	0.706	0.600	0.106	0.139
77	0.622	0.733	0.633	0.100	0.133
81	0.781	0.782	0.667	0.115	0.149
82*	0.821	0.794	0.800	-0.006	0.127
83	0.861	0.805	0.833	-0.028	-0.005
85*	0.942	0.827	0.900	-0.073	-0.006
87*	1.021	0.846	0.967	-0.121	-0.054
88	1.061	0.856	1.000	-0.144	-0.111

Computational aspects. Because there is no analytic formula for exact P-values, only a Monte Carlo approximation is available for Lilliefors' test. These are given in StatXact together with a confidence interval for that estimate. This will usually be quite short if the estimate is made on a large number of Monte Carlo samples. In practice it is customary to use anything between 1000 (minimum) and perhaps 1,000,000 samples if high precision is required. For the Shapiro-Wilk test the distribution of the test statistic appears to be known only for sample sizes of six or less and is very nonnormally distributed for larger sample sizes. However it may be transformed to approximate normality and the transformation provides a basis for determining the asymptotic P-value given in StatXact.

4.2.4 Relevant Inference

There are situations where inferences about means or medians are by nature irrelevant to sensible deductions. Tests such as Kolmogorov's or Lilliefors' test may then be more useful. We comment upon one such case.

Example 4.5

Figure 2.3 showed a histogram with class interval 10 for ages at death for the 59 members of the McAlpha clan recorded in Appendix 1.

The mean and median are respectively 61.8 and 74. The age range with most deaths is 80–89. Striking features in Figure 2.3 are the skewness of the data to the left and the bimodal nature with a small peak (secondary mode) of infant deaths in the range 0–9 followed by lower death rates in 10-year intervals until age 60. The death rate then rises sharply to the primary mode at the 80–89 range.

One might go through formal motions of attaching confidence intervals to the mean or median using an approach based on the t-test (obviously irrelevant because of nonnormality – Lilliefors' test indicating $P < 0.001$). The Wilcoxon test is not valid because of asymmetry, while the sign test is valid but not very interesting in the light of the bimodality. Such approaches are at worst wrong, and at best pointless. In this example the histogram is probably the most useful aid in interpreting the data.

What is the relevance of any formal inference procedure in a case like this? From what population is this a random sample? It is not a random sample of all males buried at Badenscallie, for clans and families have differing life-span characteristics. Perhaps the pattern of death ages is a reasonable approximation to that for the McAlpha clan in Wester Ross, Sutherland and the Western Isles, where the clan is well established. The main use we can envisage for these data would be for comparisons with patterns for other clans.

Difficulties in choice of analytic tools are not confined to nonparametric methods. Analyses making various assumptions often draw attention to interpretation problems.

Using modern software it is easy to apply several tests to the same data. Indeed, for centrality tests StatXact allows one to use Wilcoxon and normal scores tests and also to base tests on any scores one chooses. Do not abuse this freedom; a scoring system should be chosen on logical grounds. Remember our ethical warning in the penultimate paragraph of Section 1.5. If different choices are logical under differing assumptions, and for, say, centrality tests these lead to inconsistent results, further analysis (e.g., applying Lilliefors' test) or more simply drawing a histogram, or using some other appropriate EDA tools may suggest the most appropriate choice. Looking at data critically is a habit well worth cultivating.

For the McAlpha data it is clear that a sign-test-type confidence interval for the median is less inappropriate than many others. But this begs the question as to whether the median is a useful summary statistic for bimodal data.

4.3 Inferences for Dichotomous Data

In Section 4.1 we pointed out that sometimes data consist of counts. The simplest situation is when we have counts of numbers in each of two categories.

This is sometimes described as *dichotomous data*. Sometimes dichotomous data arise as secondary data derived from some other type of raw data. This was the situation in Example 3.7, where we applied the sign test using counts of the numbers of continuously distributed observations above or below some hypothesized median. Under the null hypothesis, the independence condition ensured that the count of observations above (or below) the hypothesized median had a binomial distribution with $p = 0.5$. In this section we extend these ideas to other situations where we associate appropriate binomial distributions with dichotomous data.

4.3.1 Binomially Distributed Data

In Example 2.4 we used binomial distribution properties to show how to assess the strength of evidence against a hypothesis that a sample came from a $B(10, 0.75)$ distribution, i.e., to test the hypothesis H_0: $p = 0.75$ against H_1: $p > 0.75$. Our interest was in the parameter p.

In Example 1.2 we applied the sign test to hypotheses about the median of measured survival times (the primary data), basing the test not on these primary data but on the derived or secondary data of numbers of observations above or below a hypothesized median. For the sign test the parameter $p = 0.5$ is relevant to H_0, and although the test was based on that parameter, we were interested not in the parameter itself, but in the median of an underlying continuous distribution of survival times.

In this section we give other examples of inferences about a binomial parameter p. It is convenient to refer to each of the n observations as the outcome of a *trial*. The key assumptions needed for a count to have a binomial distribution are:

- The n trials are mutually independent.
- Only one of two possible outcomes A, B is observed at each trial.
- At each trial there is a fixed probability p associated with the outcome A and a probability $q = 1 - p$ associated with the outcome B.

The outcomes A, B are often labelled *success* and *failure*, but these names are not always literally appropriate, so we designate the outcomes as *event* A and *event* B. The requirement that p be constant is often violated in a strict sense. For example, when UK children are skin-pricked for sensitization to grass, dust mites and cats around 7 percent show a skin reaction to all three allergens. This does not imply that there is a probability $p = 0.07$ that a child given the skin tests will react to all three tests positively. For instance, for children with asthma the probability of three positive reactions is around 0.14, whereas for children without asthma it is about 0.05. However, if a random sample is taken from a large group of UK children that is representative of the population in terms of asthma, it is reasonable to expect approximately 7 percent of those in the random sample to have reactions to all three allergens. For a similar sample of children from a country with a desert climate, and where there are few domestic cats, there is medical evidence to suggest that the proportion of children reacting to the grass and cat allergens may be lower. An appropriate statistical test could be based on the binomial distribution corresponding to a sample of n with the null hypothesis H_0: $p = 0.07$ against the alternative H_1: $p < 0.07$. However, in a comment on Example 4.10 we point out a complication that may invalidate the test.

Tests based on the binomial distribution also arise in quality control problems. Here we may be interested in the proportion of items in large batches having a specific attribute — often a defect. A widely used procedure is to take a random sample of n items from a batch of N items and for a buyer to accept the whole batch if the number of defective items, r, in that sample does not exceed some small fixed number s. Otherwise the batch is rejected. The consequences of such a procedure can be studied using binomial distribution theory. The results are only approximate because the usual sampling method of selecting items without replacing each before we select later items (sampling without replacement) does not lead to a binomial distribution for r. However, if N is very much larger than n, as it often is in practice, the binomial approximation is satisfactory. The scheme we outline below is usually embellished in practice, but we do not consider here the many possible elaborations.

Example 4.6

The problem. A contract for supplying rechargeable batteries specifies that not more than 10 percent should need recharging before 100 hours of use. To test compliance with this requirement a potential purchaser takes a random sample of 20 from a large consignment and finds that 3 need recharging after being used for less than 100 hours. Assess the strength of evidence that the batch is below specification.

Formulation and assumptions. The test is distribution-free in the sense that we make no assumption about the distribution of times before a recharge is needed. However, we do assume that there is a fixed probability p, say, that any one battery will need recharging in less than 100 hours. We also assume that the failure time for each battery is independent of that of any other battery. The null hypothesis is H_0: $p \leq 0.10$ and the alternative is H_1: $p > 0.10$.

Procedure. Tables for the binomial distribution, or standard statistical software, indicate that when $p = 0.10$ the probability that a sample of 20 contains 3 or more batteries that fail to meet specifications is $P = 0.323$.

Conclusion. Since the probability that 3 or more batteries will need recharging in less than 100 hours when $p = 0.10$ is nearly one-third there is little evidence for preferring the alternative that the failure rate in the large consignment exceeds 10 percent.

Comments. 1. This example does little more than highlight serious limitations of hypothesis testing. Three failures in 20 items represents a 15 percent failure rate in the sample, so the data would clearly support a null hypothesis that specified $p = 0.15$ and also many higher values. Confidence intervals for p will be informative. We show in Example 4.7 that a two-sided confidence interval with at least 95 percent coverage for the true p is the interval (0.032, 0.379). This implies that we would only reject a hypothesized value of p falling outside that interval when we use the conventional $P \leq 0.05$ as sufficient evidence against that hypothesis. The sample evidence here suggests only that the batch percentage failing to live up to specification is likely to be between about 3.2 and 37.9 percent.

2. Having obtained a confidence interval, two further questions of interest are

- (i) What is the power of a test for H$_0$: $p \leq 0.10$ against various alternatives?

- (ii) How may we use larger samples to make a test more precise (in the sense of shortening the confidence interval or increasing the power for specific alternatives)?

We address these questions in Examples 4.8 and 4.9.

Computational aspects. In Examples 4.6–4.9 we discuss computational aspects under *Procedure* as they form an integral part of that.

4.3.2 Confidence Intervals for Binomial Proportions

For a binomial distribution, if an event A occurs in r out of n trials the sample estimate of the binomial probability, p, is $\hat{p} = r/n$. We use the notation $X = r$ where $r = 0, 1, 2, \ldots, n$ to indicate that the event A occurs in exactly r among n trials.

To find a two-sided confidence interval for p with at least 95 percent coverage we seek for its upper limit a value p_u such that if the binomial parameters are n, p_u, then the value of $\Pr(X \leq r)$ is as close as possible to, but does not exceed, 0.025. Well-known binomial distribution theory indicates that p_u must satisfy the equation

$$\sum_{i=1}^{r} \binom{n}{i} (p_u)^i (1 - p_u)^{n-i} = 0.025. \qquad (4.2)$$

A similar approach provides an equation for a lower limit, p_l. These equations may be modified to give any required level by replacing 0.025 by another appropriate value, e.g., 0.005 for a 99 percent interval.

The equation (4.2) generally has no simple analytic solution for p_u. The classic way of overcoming this difficulty was recourse to tables or charts obtained by numerical methods, where these limits were presented for several confidence levels (usually 90, 95 and 99 percent) for a range of values of n and r. Table A4 in Conover (1999) is one such table. Many computer packages including Minitab and StatXact compute exact confidence limits at any level chosen by the user.

For reasonably large n if neither exact tables nor relevant computer programs are available an asymptotic approximation may be used. If we observe $X = r$ occurrences of event A in n trials for any p such that $0 < p < 1$, then

$$Z = \frac{p - \hat{p}}{\sqrt{\hat{p}(1 - \hat{p})/n}}$$

has, for large n, approximately a standard normal distribution. The approximation is good for $n > 20$ providing p is close to 0.50, but appreciably larger values of n are needed if p is close to 0 or 1. To obtain an approximate 95 percent interval for p we follow the usual normal distribution confidence interval procedure and find the upper and lower limits as the values of p that satisfy the equation

$$Z = \frac{p - \hat{p}}{\sqrt{\hat{p}(1 - \hat{p})/n}} = \pm 1.96. \tag{4.3}$$

To obtain limits for other confidence levels 1.96 is replaced by the appropriate value, e.g., 1.64 if a 90 percent interval is required.

If the event does not occur at all, i.e., $(r = 0)$, then the estimate of the binomial probability is zero. An upper limit for p can be readily obtained (using a tail probability of 0.05 as the lower tail is not meaningful and (4.2) reduces to

$$(1 - p_u)^n = 0.05.$$

Correspondingly, if $r = n$ a lower limit can be obtained by solving

$$(p_l)^n = 0.05.$$

Sackett et al. (1991) tabulate p_l and p_u when $r = 0$ and $r = n$ for a selection of sample sizes between 1 and 300.

Example 4.7

The problem. For the data in Example 4.6 confirm that an exact 95 percent confidence interval for p is $(0.032, 0.379)$. Determine also 90 and 99 percent intervals. Obtain an asymptotic approximation to the 95 percent interval.

Formulation and assumptions. Assuming that the number of batteries in a sample of 20 that fail to last 100 hours before needing a recharge has a binomial distribution, the exact 95 percent confidence limits are given by (4.2) and the corresponding equation for the lower limit. For the 90 percent and 99 percent limits the 0.025 on the right-hand side is replaced by 0.05 and 0.005 respectively. Because the equations have no analytical solution the required limits are obtained in practice from tables, or by using appropriate software. The relevant asymptotic limits are given by (4.3).

Procedure. From Conover (1999, Table A4) one finds the following exact intervals for this problem where $n = 20$ and $r = 3$.

90 percent	$(0.042, 0.344)$
95 percent	$(0.032, 0.379)$
99 percent	$(0.018, 0.450)$

Both Minitab and StatXact give the same values and these should also be obtained from any other software correctly claiming to give exact intervals.

For the asymptotic limits we use (4.3) with $n = 20$ and $\hat{p} = 3/20 = 0.15$. The solutions give the 95 percent interval $(-0.0065, 0.306)$. Most computer software will give these limits as an alternative to exact limits, but not all include the timely warning in Minitab that these may be unreliable for small samples.

Conclusion. Evidence from this sample of 20 indicates only that it seems likely that the overall proportion in a large batch lies somewhere between about 3–4 percent and about 35–40 percent.

Comments. 1. We have already indicated that the asymptotic approximation requires n to be considerably greater than 20 for small p. The asymptotic 95 percent lower limit, -0.0065 is a statistical nonsense, being outside the permissible range $0 \le p \le 1$. Tables of exact values in Conover cover values of $n \le 30$ but with small values of p the asymptotic result may still be unreliable even when $n > 30$, meaning that unless one has access to a table for larger values of n one needs statistical software for an exact interval. For the case $n = 30$ and $r = 2$ an exact 95 percent interval for p is $(0.008, 0.221)$. The normal approximation gives the nonsense interval $(-0.023, 0.156)$. However, if $n = 30$ and $r = 13$ the exact 95 percent interval is $(0.255, 0.626)$ and the asymptotic approximation is a reasonable $(0.256, 0.611)$.

2. In this example we considered a two-sided confidence interval, but in Example 4.6 we were interested in a one-tail hypothesis test. The lower limit of the one-sided 95 percent confidence interval for the test considered there corresponds to the lower limit for the two-sided 90 percent interval, i.e., the 95 percent one-sided interval is in this example $(0.042, 1)$.

4.3.3 Power and Sample Size

In Examples 4.6 and 4.7 there is considerable doubt about the population value of p. In practical quality control problems a common situation is that where a purchaser would ideally like all batches of items such as the rechargeable batteries that are accepted to contain not more than 10 percent that require recharging before 100 hours of use. However, taking marketing realities such as production cost into account, the purchaser may be prepared to accept a few batches where as many as, say, 20 percent fail to live up to this requirement, but be reluctant to accept batches where more than 20 percent fail. By the same token the supplier does not want the purchaser to reject batches where less than 10 percent will fail.

If we test H_0: $p \le 0.10$ against the alternative H_1: $p > 0.10$ and agree the evidence is strong enough to reject H_0 if the associated P-value is $P \le 0.05$ this means there is a probability not greater than 0.05 that the purchaser will reject a batch that would be acceptable (i.e., one that in reality contains 10 percent or less that are defective). Be careful in your reading to distinguish between the binomial probability (lowercase p) and the test P-value (capital P). In acceptance sampling terms $P = 0.05$ is called the *producer's risk* because in agreeing to accept these conditions it is the maximum long run probability that an acceptable batch will be rejected. In conventional hypothesis testing terms producer's risk is the probability of an error of the first kind. If the consumer is prepared to accept a few batches with more than 10, but less than 20, percent defective the consumer and producer might agree that they should devise a test procedure that ensures a certain preassigned high probability, say $P^* = 0.90$, that one does not accept a batch with 20 percent or more defective. This means there is only a probability $1 - P^* = 0.10$ that a batch with 20 percent or more defective will slip through. This last probability is called the *consumer's risk* because in agreeing to accept these terms it is the maximum probability that an unacceptable batch will be purchased. It is the probability of an error of the second kind, and P^* is the power. A practical problem is to choose the sample size n such that there is both a preassigned small probability P that batches for which $p \leq 0.10$ will be rejected, and also a preassigned high probability P^* that batches for which $p \geq 0.20$ will also be rejected.

Here the concept of power (Section 1.3) is useful. Power depends on the value of P chosen to indicate significance, i.e., as sufficient evidence to make H_0 unacceptable. It also depends upon the sample size n and upon the value of the binomial p specified for an alternative hypothesis.

Two relevant questions are:

- Given a sample of size n what is the power of a test of H_0: $p = 0.1$ against H_1: $p = 0.2$ if we agree to reject H_0 for any P less than some fixed preassigned value in a one-tail test?

- What sample size n will ensure that a test of H_0: $p = 0.1$ against H_1: $p = 0.2$, where we reject H_0 for any P less than some fixed value (e.g., $P = 0.05$), has a given power P^* for that alternative?

The rationale for these choices of H_0, H_1 is that if the true p is less than 0.1 we are less likely to reject a good batch than would be the case if $p = 0.1$ and that if $p > 0.2$ the power of our test will exceed that when $p = 0.2$. These are desirable characteristics for both producer and consumer.

If we observe r occurrences of event A in a sample of size n, and want to test H_0: $p = p_0$ against H_1: $p > p_0$, and agree that there is sufficient evidence to prefer H_1 for all P-values less than or equal to some fixed value α, then the critical region of size α consists of all values of $r \geq r_0$ where r_0 is the smallest value of r such that the binomial distribution $B(n, p_0)$ probabilities for all $r \geq r_0$ have a sum not exceeding α. In practice we either read r_0 from tables, or now more commonly obtain it using computer software.

Having determined r_0 it is easy to find the power of this test against a single-valued alternative H_1: $p = p_1$ because this is simply the $\Pr(r \geq r_0)$ when H_1 is true. This probability may be read from tables or obtained using appropriate computer software.

For large n asymptotic approximations may be used. These are based on the fact that under H_0 the distribution of

$$Z = \frac{\hat{p} - p_0}{\sqrt{p_0(1 - p_0)/n}} \tag{4.4}$$

is approximately standard normal. If, for example, $\alpha = 0.05$ for an upper-tail test we set $Z = 1.64$ in (4.4) and solve for $\hat{p}$ given n and p_0. To get an asymptotic expression for the power when $H_1: p = p_1$ is true, we replace p_0 by p_1 in (4.4) and insert the solution $\hat{p}$ that we have just obtained in this amended expression for Z and calculate the resulting Z, say $Z = z_1$. The asymptotic approximation to the power is $\Pr(Z \geq z_1)$.

Example 4.8

The problem. For a sample of 20 batteries from a large batch we wish to test $H_0: p = 0.1$ against $H_1: p = 0.2$ where p is the population proportion of batteries requiring recharging in less than 100 hours. We will prefer H_1 for any $P \leq 0.05$. What is the minimum number of faulty batteries in the sample for which we prefer H_1? Obtain both exact and asymptotic expressions for the power of this test.

Formulation and assumptions. Let X be the number of occurrences, r, of the event A (battery needs recharging in less than 100 hours). Then under H_0, X has a B(20, 0.1) distribution. We prefer H_1 if our sample gives r_0 or more occurrences of A where r_0 is the least r such that $\Pr(X \geq r_0) \leq 0.05$ when H_0 is true. Once we know r_0 the power is obtained by computing for that r_0 the corresponding probability under H_1. Asymptotic results for large n use the normal approximations given above for these probabilities.

Procedure. If suitable tables are available these may be used, but it is often easier to use software. We outline the procedure for both methods. When $n = 20$, $p = 0.10$, Table A3 in Conover (1999) indicates that $\Pr(X \leq 4) = 0.9568$ whence $\Pr(X \geq 5) = 1 - 0.9568 = 0.0432$. Thus we prefer H_1 if, and only if, $r \geq 5$. To determine the power we require $\Pr(X \geq 5)$ for a B(20, 0.2) distribution. The same tables indicate that now $\Pr(X \geq 5) = 1 - Pr(X \leq 4) = 1 - 0.6296 = 0.3704$.

In the asymptotic approach under H_0 we have, using (4.4)

$$Z = (\hat{p} - 0.1)/\sqrt{0.1 \times 0.9/20} = 1.64$$

whence $\hat{p} = 0.2100$. Thus the relevant power is given by

$$\Pr[Z \geq (0.2100 - 0.2)/\sqrt{0.2 \times 0.8/20}] = \Pr(Z \geq 0.1118).$$

From standard normal distribution tables this probability is 0.4555.

Most computer software easily generates P-values corresponding to any number of successes for one- or two-tail tests. For example, the 1-proportion program in Minitab confirms that $\Pr(X \geq 5) = 0.043$ and the same program using $p = 0.2$ indicates $\Pr(X \geq 5) = 0.370$ confirming our result for the power. Minitab, also gives a separate program relating power to sample size, but uses an asymptotic approximation. This gives an approximate power 0.432, which differs somewhat from our 0.4555. The discrepancy is explained by the estimate being sensitive to round off. If we replace 1.64 by the more precise 1.645 the estimated power is 0.434, close to the Minitab estimate. The discrepancy between the true power and the asymptotic estimate reflects the unsatisfactory nature of the asymptotic result for even moderate n associated with small p. StatXact, gives the exact power directly for any sample size for any pair of values of p specified in H_0 and H_1.

Conclusion. The power of our test against an alternative $p = 0.20$ is only 0.370 if we choose $P \leq 0.05$ (or more precisely because of discontinuities $P \leq 0.043$) as our criterion for not accepting $p = 0.10$.

Comments. 1. In the discussion preceding this example we suggested a practical aim might be to reject a batch with probability (power) 0.90 if $p \geq 0.20$. The power 0.370 is far short of this requirement.

2. The unsatisfactory nature of asymptotic power calculations with the n and p used here is less acute with larger samples.

3. In Section 1.3 we indicated that reducing the P-value chosen as acceptable evidence against H_0 generally results in reduced power against a particular alternative. In this example tables or appropriate software indicates that if we set the requirement that P must not exceed 0.02 the power of the one-sided test against the alternative $p = 0.2$ is reduced from 0.370 to 0.196. However, due to discontinuities in possible P-values the exact P is not 0.02 but is 0.011.

Modern software gives exact P-values for one- or two-tail tests for any n, p and r. This effectively means we can obtain the values of $\Pr(X \geq r)$, $0 \leq r \leq n$ for any $B(n, p)$ distribution. This is useful for developing tests such as those for attribute sampling considered in this section that have preset producer's risk (i.e., P-values for rejection of H_0) and consumer's risk (i.e., power against a specified H_1 that gives the maximum p the consumer considers tolerable).

Example 4.9

The problem. Determine the power of the test H_0: $p = 0.10$ against H_1: $p = 0.20$ when $P \leq 0.05$ when $n = 50, 100, 150, 200, 300$. Compare exact values with the asymptotic approximation. Also determine the minimum n to ensure a power of at least 0.9 for this test.

Formulation and assumptions. For illustrative purposes we use programs in Minitab and in StatXact. Modifications for other packages with binomial distribution programs are generally straightforward.

Procedure. We describe the procedure fully for the case $n = 100$ and only quote results for other n. We urge readers to confirm the results using available software. When $n = 100$ and $p = 0.10$ it is intuitively reasonable to expect a P-value not exceeding 0.05 if r, the number of batteries needing recharging before 100 hours, is close to 15. Minitab and StatXact both give the following values

r	15	16	17
P	0.073	0.040	0.021

establishing $r = 16$ as the critical value. To obtain the power we need to know $\Pr(X \geq 16)$ when $p = 0.20$. StatXact gives this probability, the exact power, to be 0.871 directly without needing to first ascertain, as we just did, that $r = 16$ is the critical cut-off value. The sample size and power program for proportions in Minitab gives an asymptotic approximation which in this case indicates a power of 0.897 for a nominal $P = 0.05$. For our exact $P = 0.040$ the asymptotic power approximation is 0.882. The asymptotic approximation is reasonable for this sample size. For $n = 100$ and the remaining sample sizes of interest with the largest exact P-values meeting the requirement $P \leq 0.05$, the exact power and the asymptotic estimate of power based on the corresponding exact P are:

n	50	100	150	200	300
P	0.025	0.040	0.044	0.043	0.038
Exact power	0.556	0.872	0.963	0.989	0.999
Asymptotic power	0.617	0.882	0.962	0.988	0.999

How best to estimate the sample size n needed to ensure an exact power 0.9 may depend on available software. Clearly the size lies between 100 and 150. Since asymptotic power calculations seem reasonable for samples exceeding 100 one might use an asymptotically based facility in Minitab that gives an estimate of sample size having any required power for the relevant critical P-value (here $P = 0.05$). One may fine tune this using a program for exact binomial probabilities in a way we now describe. For our test the power-sample size program in Minitab indicates the required $n = 102$. We used this as a starting estimate for the exact result. When $n = 100$ we found the critical region was defined by $r \geq 16$. When $n = 102$ the Minitab binomial program relevant to our one-tail test shows that $r = 16$ gives $P = 0.047$, while $r = 15$ gives $P = 0.083$. The exact power for the alternative $p = 0.20$ corresponding to $P = 0.047$ is given by Minitab as 0.890. In practice this may be accepted as a suitable approximation, or one might seek some insurance by choosing a slightly greater sample size such as $n = 105$. With that latter choice Minitab gives the exact $P = 0.032$ with power 0.865! The apparent anomaly of reduced power with increased sample size is simply a reflection of the impact of discontinuities in P-values and possible values of the power (which remember is also a probability with associated discontinuities). Our nearest P-value above $P = 0.032$ would have been $P = 0.058$ and the associated power then becomes 0.914. StatXact also gives exact power for the above, or other sample sizes, directly. These indicate that a sample size of 109 has power 0.901 in the problem considered here.

Conclusions. In general, for a specified single-value alternative hypothesis power increases with sample size for a fixed P-value but there are some irregularities due to discontinuities in possible exact P-values. These irregularities are generally of little practical importance.

Comment. Although one should be aware of the effects of discontinuities in P-values in assessing evidence for abandoning a hypothesis and in calculating power and sample sizes to achieve certain aims the importance of such effects should not be overemphasized. While it is clearly unwise, for example, to use asymptotic results for small samples, the discrepancies due to discontinuities in possible P-values are usually small in larger samples. In this context it is also worth remembering that nearly all mathematical models of real world situations are only approximations. Hopefully these will reflect key features of reality, but there will always be some discrepancies. For instance, we have already pointed out that when we take a sample of n from a large batch of N items by sampling without replacement the binomial distribution is only an approximation, albeit a good one.

4.3.4 A Caution

In Section 4.3.1 we considered a situation where, in the population at large, 7 percent of all children skin-pricked for sensitization to grass, dust mites and cats had a positive skin reaction to all three allergens. However, the presence or absence of asthma meant that each child did not have the same probability of showing three positive skin tests. We indicated there that if a random sample were taken from that larger population it is not unreasonable to expect this to reflect the population proportion of 7 percent showing three positive tests, because statistical theory backs an intuitive hunch that a random sample should reflect fairly closely characteristics of a population. On these grounds we indicated that it may be reasonable to take a random sample of children from a country with a desert climate and few domestic cats to see whether there is a lower proportion of positive skin tests in that population. We formalize a simple test based on the binomial assumption in Example 4.10 but warn in the *Comment* that the analysis may still not be valid.

Example 4.10

The problem. It is known that 7 percent of UK children give positive skin tests to grass, dust mites and cats. It is thought that a desert climate and few domestic cats may reduce the proportion of positive tests. A random sample of 54 children from such a country were given the allergen tests and only one of these had positive reactions to all three allergens. Is there sufficient evidence of a lower incidence compared with the UK?

Formulation and assumptions. We justify a one-tail test of H_0: $p = 0.07$ against H_1: $p < 0.07$ because there are good medical reasons for assuming that the proportion should be lower under such conditions. If the direction of any effect on sensitization were unknown then a two-tail test of H_0: $p = 0.07$ against H_1: $p \neq 0.07$ would be appropriate.

Procedure. As exact tables for $n = 54$ are not widely available and a small p makes asymptotic results unreliable one should resort to appropriate software. For the one-tail test StatXact or Minitab gives an exact $P = 0.101$. The exact two-sided nominal 95 percent confidence interval for p is (0.0005, 0.0989).

Conclusion. The evidence of a reduction is not strong. If it were important to detect even a small reduction to, say, $p = 0.05$ a larger experiment would be needed to give reasonable power.

Comment. In Section 4.3.1 we pointed out possible difficulties if particular groups of patients have markedly different probabilities of showing a reaction. We stressed the need for a random, or effectively random sample, to overcome this problem. Another type of problem may arise in, for instance, the testing of a new drug. For the new drug there may be a quite different set of factors influencing the likelihood of an individual showing side-effects compared with the original drug. For instance, with one antibiotic for the treatment of pneumonia people with asthma may be more likely to exhibit side-effects, whereas for a new antibiotic the reverse may be the case. In these circumstances what could be an overall beneficial effect can sometimes be masked or distorted in our sample. The solution lies in either further experimentation or at least in using more detailed observations and tests. For example, one may require an analysis that takes into account other factors that might influence outcomes for individuals, e.g., factors such as weight, age, gender of patients. The result is a more complex analytic problem and indeed most real life statistical problems are more complicated than the simple ones discussed in this section.

4.4 Tests Related to the Sign Test

4.4.1 An Alternative Approach to the Sign Test

We described the sign test in detail in Section 3.4. An alternative, but equivalent, approach more in the spirit of the work in Section 4.3 is sometimes used and we illustrate this by an example.

Example 4.11

The problem. For the second data set in Example 3.7, i.e.,

$$
\begin{array}{cccccccccc}
5.6 & 6.1 & 6.3 & 6.3 & 6.5 & 6.6 & 7.0 & 7.5 & 7.9 & 8.0 \\
8.7 & 8.1 & 8.1 & 8.2 & 8.4 & 8.5 & 8.7 & 9.4 & 14.3 & 26.0
\end{array}
$$

if θ is the population median, test the hypothesis H_0: $\theta = 9$ against H_1: $\theta \neq 9$ using the sign test. Obtain a 95 percent confidence interval for θ using the methods developed in Section 4.3.

Formulation and assumptions. Wide availability of programs giving exact P-values provides a useful facility for testing and estimation in a sign-test situation.

Procedure. For illustrative purposes we use the basic statistics 1-proportion option in Minitab. Here $n = 20$ and the number of observations below the hypothesized median is $r = 17$. The sign test calls for evaluation of P for a two-tail test with $p = 0.5$ and the Minitab program gives at the same time a confidence interval for the true p at a nominal level, which we set at 95 percent. The program gives $P = 0.003$ and a nominal 95 percent confidence interval for p of $(0.621, 0.967)$. Allowing for rounding, this P-value agrees with that obtained in Example 3.7 and indeed here we did nothing essentially different from what was done there. To obtain a confidence interval for θ we now decrease r by steps of 1 until we get a 95 percent confidence interval for p that just includes the value $p = 0.5$. This straightforward stepwise process indicates that when $r = 15$ the confidence interval is $(0.508, 0.913)$, while for $r = 14$ it is $(0.457, 0.881)$. Thus, when $r = 14$ we would regard $p = 0.5$ as acceptable with $P > 0.05$, but not when $r = 15$. Symmetry of the binomial distribution when $p = 0.5$ implies that we would accept when $r = 6$ but not when $r = 5$. This implies that an appropriate 95 percent confidence interval for the median is from the sixth to the fifteenth largest observations, i.e., the open interval $(6.6, 8.4)$, agreeing with the result in Example 3.8.

Computational aspects. The procedures in Examples 3.7 and 3.8 and in this example are basically equivalent. Which is preferred is largely determined by the availability of appropriate tables or computational software.

4.4.2 Quantile Tests

The sign test is relevant to population medians. We are often interested in other *quantiles* of a population. For example, a quantile such that three-quarters of the population values lie below it and one quarter above is known as the upper (or third) quartile while a quantile such that r tenths of the observations lie below it and $(10 - r)$ tenths of the observations lie above it where $r = 1, 2, \ldots, 9$ is called the rth decile. Many tests and estimation procedures involving quantiles proceed along similar lines to those about medians by simply replacing $p = 0.5$ by a value appropriate to the relevant quantile. For continuous distributions quantiles are unique, but for discrete distributions conventions given in most standard textbooks are needed to give unique quantiles. We have already met the usual convention for the median of a set consisting of an even number, say, $2m$, observations. We arrange them in ascending order and denoting the rth ordered observation by $x_{(r)}$ we define the median as $(x_{(m)} + x_{(m+1)})/2$. Note that the fifth decile is the same as the median.

Example 4.12

The problem. A central examining body publishes the information that "three-quarters of the candidates taking a mathematics paper achieved a mark of 40 or more" (i.e., the first population quartile is 40). One school entered 32 candidates for this paper of whom 13 scored less than 40. The president of the Parents' Association argues that the school's performance is below national standards. The headmaster counters by claiming that in a random sample of 32 candidates it is quite likely that 13 would score less than the lower quartile mark even though 8 out of 32 is the expected proportion. Is his assertion justified?

Formulation and assumptions. Denoting the first quartile by Q_1 a test for the headmaster's assertion is a test for H_0: $Q_1 = 40$ against H_1: $Q_1 \neq 40$.

Procedure. We associate a minus with a mark below 40; thus for the sample of 32 we have 13 minuses (and 19 pluses). If the first quartile is 40 then the probability is $p = 0.25$ that each candidate in a random sample has a mark below 40 (and thus is scored as a minus). The distribution of minuses is therefore binomial B(32, 0.25). Using any program that gives a 95 percent confidence interval for p given $r = 13$ minuses (event A) when $n = 32$ we find the relevant confidence interval is (0.24, 0.59). This includes $p = 0.25$. The associated P-value is $P = 0.0755$.

Conclusion. If we observe 13 minus signs we would not reject at a conventional 5 percent significance level the hypothesis H_0: *first quartile is 40*. Nevertheless, there is some evidence against this hypothesis — enough to worry many parents and some may be reluctant to give the headmaster's claim the benefit of the doubt until further evidence were available.

Comments. 1. Remember that even if we use formal significance levels, non-rejection of a hypothesis does not prove it true. It is only a statement that the evidence to date is not sufficient to reject it. This may simply be because our sample is too small. We may make an error of the second kind by nonrejection. Indeed, since the above confidence interval includes $p = 0.5$ we would not reject the hypothesis that the median is 40.

2. We used a two-tail test. A one-tail test would not be justified unless we had information indicating the school's performance could not be better than the national norm. For example, if most schools devoted three periods per week to the subject but the school in question only devoted two, we might argue that lack of tuition could only depress performance. Indeed, a one-tail test at the conventional 5 percent level would reject H_0 if there are more than 12 minuses. If we observe 13 minuses, and feel a one-tail test with $P = 0.05$ as a significance indicator is appropriate, and want to use the method described in this example, we might determine a 90 percent confidence interval for p and reject H_0 if $p = 0.25$ is below the lower limit for this interval. Think carefully about this to be sure you see why. Have you any reservations about the accuracy of this approach?

3. The headmaster's claim said "if one took a random sample." Pupils from a single school are in no sense a random sample from all examination candidates. Our test only establishes that results for this particular school are not too strongly out of line with national results in the sense that they just might arise if one took a random sample of 32 candidates from all entrants.

4. Tests for a third quartile are symmetric with those for a first quartile if we interchange plus and minus signs.

4.4.3 The Cox–Stuart Test for Trend

We indicated in Section 4.1 that when observations are ordered in time we may be interested in evidence of some trend. Cox and Stuart (1955) proposed a simple test for a monotonic trend, i.e., an increasing or decreasing trend. A monotonic trend need not be linear; it need only express a tendency for observations to increase or decrease subject to some local or random irregularities.

Consider a set of independent observations $x_1, x_2, \ldots, x_n$ ordered in time. If we have an even number of observations, $n = 2m$, say, we take the differences

$$x_{m+1} - x_1, \ x_{m+2} - x_2, \ \ldots, \ x_{2m} - x_m \,.$$

For an odd number of observations, $2m + 1$, we proceed as above omitting the middle value x_{m+1} and calculate $x_{m+2} - x_1$, etc. If there is an increasing trend we expect most of these differences to be positive, whereas if there were no trend, and observations differed only by random fluctuations about some median, these differences (in view of the independence assumption) are equally likely to be positive or negative. A preponderance of negative differences suggests a decreasing trend.

This implies that under the null hypothesis of no trend, the plus (or minus) signs have a $B(m, 0.5)$ distribution and we are in a sign test situation.

Example 4.13

The problem. The U.S. Department of Commerce publishes estimates obtained from independent samples each year of the mean annual mileages covered by various classes of vehicles in the United States. The figures for cars and trucks (in thousands of miles) are given below for each of the years 1970–83. Is there evidence of a monotonic trend in either case?

Cars	9.8	9.9	10.0	9.8	9.2	9.4	9.5
	9.6	9.8	9.3	8.9	8.7	9.2	9.3

Trucks	11.5	11.5	12.2	11.5	10.9	10.6	11.1
	11.1	11.0	10.8	11.4	12.3	11.2	11.2

Formulation and assumptions. The figures for each year are based on independent samples so we may use the Cox–Stuart test for trend. Without further information a two-tail test is appropriate as a trend may be increasing or decreasing.

Procedure. For cars relevant differences are $9.6 - 9.8$, $9.8 - 9.9$, $9.3 - 10.0$, $8.9 - 9.8$, $8.7 - 9.2$, $9.2 - 9.4$, $9.3 - 9.5$ and all are negative. Appropriate tables or computer software immediately establish that seven negative signs or seven positive signs when $p = 0.5$ has an associated $P = 0.016$.

Conclusion. There is strong evidence of a downward monotonic trend.

Comments. 1. For trucks the relevant differences have signs $-, -, -, -, +, +, +$. When we only have seven data values 3 plus and 4 minus (or 3 minus and 4 plus) signs provides the strongest possible evidence to support a hypothesis of no monotonic trend. The fact that the first four differences are all negative and the last three all positive suggests the possibility of a decreasing trend followed by an increasing trend (i.e., a nonmonotonic trend) rather than random fluctuations. The sample is too small to establish this, but in Section 4.5 we describe a *runs test* appropriate when we have larger samples for testing whether, in circumstances like these, fluctuations are random.

2. Periodic trends are common. For example, at many places in the northern hemisphere mean weekly temperature tends to increase from February to July and decrease from August to January. A Cox–Stuart test applied to data of this type might either miss such a trend (because it is not monotonic) or indicate a monotonic trend for records over a limited period (e.g., from February to June). Conover (1999, Examples 4 and 5, pp. 172–175) shows how, in certain circumstances the Cox–Stuart test may be adapted to detect periodic trends by reordering the data.

3. If the same samples of cars and trucks had been used each year the independence assumption would not hold; inference would then only be valid for vehicles in that sample, for anything atypical about the sample would influence observations in all years. With independent samples for each year, anything atypical about the sample in any one year will be incorporated in the random deviations from trend in that year.

4.5 A Runs Test for Randomness

Many statistical inferences are valid only when the data are a sample or samples of independent (random) observations. Methods for obtaining a random sample — especially in simulation or Monte Carlo studies — often depend upon sophisticated mechanisms that purport to be equivalent to repeatedly tossing a fair coin or to repeatedly selecting one of the 10 digits between 0 and 9 inclusive, each having a probability 0.1 of selection.

One characteristic of the coin tossing situation is that in the long run there should be approximately equal numbers of heads and tails and a characteristic of random digit selection is that in a long string of such digits each digit should occur approximately the same number of times. In the coin tossing situation a sign test would be appropriate to see whether this frequency requirement is being met. We give in Section 12.4 a method which may be applied to testing whether there is evidence that digits are not occurring with equal probability.

In an ordered sequence randomness implies more than compliance with frequency criteria. For example, if the outcomes, in order, of a computer process that purports to simulate 20 tosses of a coin were

$$\text{H H H H H T T T T T T T T T T H H H H H} \tag{4.5}$$

we would suspect the process did not achieve its aim. We might be equally surprised if the ordered outcomes were

$$\text{H T H T H T H T H T H T H T H T H T H T} \tag{4.6}$$

but reasonably happy with

$$\text{H H T H T T T T H T H H H T H T H H H T T H} \tag{4.7}$$

A characteristic that reflects our reservations about (4.5) and (4.6) is the number of runs, where a run is a sequence of one or more heads or tails. In (4.5) there are three runs — a run of 5 heads, then 10 tails, then 5 heads. In (4.6) there are 20 runs, each consisting of a single head or a single tail. Intuitively we feel that (4.5) and (4.6) have respectively too few, and too many, runs for a truly random sequence. The sequence (4.7) has an intermediate number of runs, namely 13.

Both numbers of runs and lengths of runs are relevant to tests for randomness. The distribution theory for runs was developed by Whitworth (1886) and a detailed treatment of runs tests is given by Bradley (1968, Chapters 11, 12) and a concise account by Gibbons and Chakraborti (2004, Chapter 3). We consider only a test based on the number of runs, r, in a sequence of N ordered observations of which m are of one kind (e.g., H) and $n = N - m$ are of another kind (e.g., T). We reject the hypothesis that the outcomes are independent or random if we observe too few or too many runs. Computation of the probability of observing any given number of runs under the hypothesis of randomness is a subtle application of combinatorial mathematics and we only quote the outcome.

The random variable R specifies the number of runs. We consider separately the cases r odd and r even. For r odd we set $r = 2s + 1$ and

$$\Pr(R = 2s + 1) = \frac{\binom{m-1}{s-1}\binom{n-1}{s} + \binom{m-1}{s}\binom{n-1}{s-1}}{\binom{N}{m}}.$$

If r is even we set $r = 2s$ and

$$Pr(R = 2s) = 2 \times \frac{\binom{m-1}{s-1}\binom{n-1}{s-1}}{\binom{N}{m}}.$$

The test for randomness is based on the relevant tail probabilities associated with small and large numbers of runs. These were tabulated for $n, m \leq 20$ by Swed and Eisenhart (1943) and tables based on theirs are given by Siegel and Castellan (1988, Table G). StatXact computes exact P-values. Sprent (1998, Example 6.16) shows how to compute the distribution of R for a small sample for given m, n. An asymptotic test is also available and is based on known results for the mean and variance of R. These are shown after some tedious algebra (see, e.g., Gibbons and Chakraborti, 2004, Section 3.2) to be

$$E(R) = 1 + \frac{2nm}{N}$$

and

$$Var(R) = \frac{2nm(2nm - N)}{N^2(N - 1)}.$$

Asymptotically

$$Z = \frac{R - E(R)}{\sqrt{Var(R)}}$$

has a standard normal distribution. The approximation is improved by the usual numerator continuity correction, i.e., adding 0.5 if $R < E(R)$ and subtracting 0.5 if $R > E(R)$.

In practice a two-tail test is most often relevant, a significant result implying nonrandomness, but one-tail tests are meaningful in the sense that few runs imply clustering of like values while many runs imply an alternating pattern.

Example 4.14

The problem. Apply a runs test for randomness to each of the sequences of heads and tails in (4.5), (4.6) and (4.7).

Formulation and assumptions. In each of (4.5) and (4.6) $N = 20$ and $m = n = 10$, while $r = 3$ for the former and $r = 20$ for the latter. In (4.7) $N = 20$, $m = 11$, $n = 9$ and $r = 13$. Small or large numbers of runs indicate nonrandomness.

Procedure. Tables for $N = 20$ and various values of m, n such as those referenced above indicate that in both these cases a nominal $P = 0.05$ is associated with $r \leq 6$ or $r \geq 16$. It is more satisfactory to compute exact P-values for the observed R. For these examples StatXact gives the following exact and asymptotic two-tail P-values:

	Exact	Asymptotic
Data set (4.5)	0.0002	0.0006
Data set (4.6)	< 0.0001	0.0001
Data set (4.7)	0.4538	0.4575

Conclusion. There is strong evidence of nonrandomness in data sets (4.5) and (4.6) but no evidence in set (4.7). The low value of the statistic R for (4.5) reflects clustering while the high value in (4.6) suggests alternation. These are clear data characteristics in the respective cases.

Comment. The above test is not so restrictive as it might seem. For example, in sequences of supposedly random digits between 0 and 9 we may count the numbers of runs above and below the median value of 4.5 and apply the test. It might also be applied in this case to runs of odd and even digits.

4.5.1 A Runs Test for Three or More Categories

If we have three or more categories, and clustering or alternation is suspected, Mood (1940) showed that it is possible to derive the relevant tail probabilities for the numbers of runs. No illustrative example was given. Barton and David (1957) extend Mood's work and apply their theory to falls in share prices on the London Stock Exchange. They give tables for the distribution of the number of runs for samples of up to 12 observations, both for 3 and 4 different outcomes. Shaughnessy (1981) tests for randomness in time-ordered residuals from regression analyses by applying multiple runs distributions. The tables of critical values in the paper are mainly for groups of similar size, and so are of limited application. Schuster and Gu (1997) develop algorithms for calculating exact distributions for the number of runs for multiple outcomes using the software system Mathematica. Smeeton and Cox (2003) describe a method for estimating the distribution of the number of runs by simulating permutations of the sequence of categories using Stata. A recent comprehensive text by Balakrishnan and Koutras (2002) explores the theoretical aspects of a wide range of run problems.

An asymptotic test based on the normal distribution makes use of the mean and variance of the number of runs (Schuster and Gu, 1997). Let the random variable R specify the number of runs. Suppose that the total number of observations is N, with k different outcomes and n_i observations of the ith kind, $i = 1, 2, \ldots, k$. Writing $p_i = n_i/N$ for the proportion of observations of type i we get

$$E(R) = N \left(1 - \sum_{i=1}^{k} p_i^2 \right) + 1$$

and

$$\text{Var}(R) = N \left[\sum_{i=1}^{k} (p_i^2 - 2p_i^3) + (\sum_{i=1}^{k} p_i^2)^2 \right]$$

Barton and David (1957) suggest that the normal approximation is adequate for $N > 12$, whatever the composition of the subsets in terms of number.

Example 4.15

The problem. Smeeton and Cox (2003) give data on the method of delivery for 17 consecutive births in a South London hospital. There are four methods of delivery and there is interest in whether method of delivery is clustered in time. The sequence of births is:

$A \quad A \quad A \quad A \quad B \quad A \quad C \quad C \quad A \quad A \quad A \quad A \quad A \quad D \quad D \quad A \quad A$

The codes are A for normal delivery, B for forceps, C for elective Caesarean, and D for emergency Caesarean. We test the null hypothesis H_0 of randomness for the methods of delivery.

Formulation and assumptions. Using the notation above, $N = 17$, $k = 4$, $n_1 = 12$, $n_2 = 1$, $n_3 = 2$, $n_4 = 2$, and $R = 7$. Since clustering is of interest, a one tail alternative hypothesis is appropriate. Clustering is indicated by a small number of runs, providing evidence against H_0.

Procedure. The distribution of the number of runs is obtained from 100,000 simulations. The estimate for the P-value of $R \leq 7$ is compared with the asymptotic normal approach, and the exact P-value with $N = 17$, and the numbers in the four categories as given. Here $E(R) = 9.0$, $Var(R) = 1.6886$. The results for the P-value are:

Simulation	Asymptotic normal	Asymptotic normal (continuity correction)	Exact
0.0969	0.0619	0.1242	0.0970

Conclusion. There is only weak evidence of clustering for the method of delivery in this sample of births.

Comments. 1. The relatively weak evidence against H_0 is due to the small sample size. A longer sequence from the same data set produced clear evidence of clustering. This was due to elective Caesareans being almost always performed during normal working hours, whereas other types of birth can occur at any time between 0000 and 2359 hr.

2. Although the sample size is 17, it is possible to obtain the exact P-value for this example by extrapolation from the table in Barton and David (1957) using the pattern of differences between the table rows and columns. This approach is only practical if most of the categories have just one or two observations.

3. The P-value obtained by simulation is impressively similar to the exact P-value. However, the P-values from the asymptotic approach are not close, suggesting that this method should only be applied to samples that are at least moderately large.

4. Runs tests lack power to detect clustering partly because interchanging just two observations can increase the number of runs by 2. However, there are few alternatives if, as in this example, the categories cannot be ordered.

Computational aspects. 1. Multiple runs distributions can be simulated using a procedure within Stata (Smeeton and Cox, 2006).

2. The run distribution for this example was obtained in around 15 seconds. Computational times can be significant for longer sequences and a greater number of simulations.

4.6 Angular Data

In some investigations measurements are made on directions, e.g., wind direction at an airport at noon on successive days, or the bearings at which released pigeons disappear over the horizon, or the successive stopping positions of a roulette wheel.

In other studies the data, although not measured directly as angles or directions, can be represented in this way. For instance, the time of day at which babies are born in a large hospital, or the days during the year in which new cases of leukaemia are diagnosed in a certain region are effectively angular measurements. The names *circular* or *directional* measurements are also used, the former because observations of this type are appropriately represented as points on the circumference of a circle. Such data are sometimes analysed as though they are distributed along a line without taking the directional aspect into account. This is not always a sensible course of action. Using a *straight line* approach observations of 1° and 359° would lead to a mean and median of 180°. The values then appear to be very different. If, however, the two points are plotted on the circumference of a circle, the two directions are similar and a more appropriate average would be obtained from the direction bisecting the smaller arc between the two directions, in other words, 0°. If the data consisted of two observations of 31 December and 2 January the "straight-line" approach leads to an average around 1 July, whereas a directional analysis leads to the more reasonable average of 1 January. Measures of centrality (e.g., the median) and spread (e.g., the range) may be adapted for use with angular data.

The book by Mardia (1972) on directional data analysis is written from a mathematical perspective. Fisher (1993) provides an introduction based around real life examples from many fields. A more recent treatment of theory and practice is given by Mardia and Jupp (2000).

One often wants to know whether data support a hypothesis that a sample is from a population uniformly distributed on the circumference of a circle or whether they indicate clustering (Mardia, 1972). The *Hodges-Ajne test* (Hodges, 1955; Ajne, 1968) and the *range test*, the latter developed from work on the circular range by Fisher (1929), address this situation.

Another question of interest is whether points are symmetrical around a particular angular direction such as the North–South line. The sign test and the Wilcoxon signed-rank test are easily adapted for this situation.

4.6.1 The Hodges–Ajne Test

This test is used to investigate whether a sample of n observations on a circle could arise from a uniformly distributed population. The alternative is that in the population observations are more concentrated within a particular arc of the circumference; outliers may nevertheless occur well away from this arc.

To carry out this test, a straight line is drawn through the centre of the circle; this divides the observations into two groups. The line is rotated about the centre to a position at which there is a minimum possible number of points, m, on one side of this line. If the points were regularly spaced they would lie at the vertices of a regular polygon and then either $m = n/2$ if n is even or $m = (n-1)/2$ if n is odd.

For observations around the circle from a uniform distribution there will be some variation in the angles between adjacent points and the value of m will generally be lower. The lowest values of m occur when there is a clustering of points on one side of the line. Under the assumption of uniformity, the probability that m is no more than a value t is shown by Mardia (1972) to be:

$$\Pr(m \leq t) = \frac{\binom{n}{t}(n - 2t)}{2^{n-1}}, \quad t < n/3. \tag{4.8}$$

Mardia gives a table of critical values for this test.

Example 4.16

The problem. A midwife recorded the times of birth for 12 consecutive home deliveries. She was interested in whether births tended to occur at particular times of the day. The times (rearranged in order throughout the day) were 0100, 0300, 0420, 0500, 0540, 0620, 0640, 0700, 0940, 1100, 1200, 1720.

Since, on a 24-hour circular clock, one hour corresponds to $360/24 = 15$ degrees, the successive angles on the circle (assuming midnight corresponds to $0°$) are $15°$, $45°$, $65°$, $75°$, $85°$, $95°$, $100°$, $105°$, $145°$, $165°$, $180°$, $260°$. We test the hypothesis H_0 that the times of birth have a uniform distribution around the circle.

Formulation and assumptions. A line is drawn through the centre of the circular plot of these data and rotated until the number of points on one side of the line takes the minimum value, m. A small value for m provides evidence against the assumption of uniformity made in H_0.

Procedure. A circular plot of the data (Figure 4.8) shows that $m = 1$ (e.g., when the line runs from $10°$ to $190°$). Only a value of zero for m represents an even greater degree of clustering. The appropriate P-value (the probability that m is equal to zero or one, assuming that H_0 is true) given by (4.8) is 0.059.

Conclusion. The evidence against H_0 when $P = 0.059$ casts some doubt on the assertion that births are equally likely at any time throughout the day.

Comments. 1. Inspection of the data shows that most of the babies are born in the morning, particularly between 0300 and 0700. If further evidence for this pattern could be obtained, the information might be useful in planning maternity services and in preparing mothers for the births of their babies.

2. This test is especially useful where some observations may be clustered together but a few may be more distant. For example, injuries from fireworks in the UK are concentrated around 5 November and deaths from drowning are more common during summer months, but such events are not confined exclusively to these periods.

4.6.2 The Range Test

Like the Hodges–Ajne test, the range test is also based on the null hypothesis that a sample of n observations on the circle could be from a uniformly distributed population, but now the alternative is that in the population all

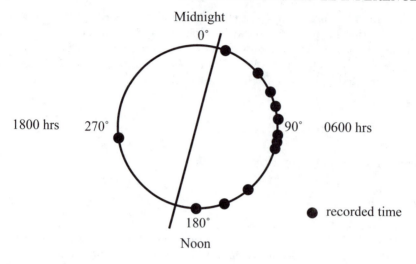

Figure 4.8 *Circular plot of delivery time data in Example 4.15 on a 24-hour clock.*

observations come from a particular arc. This test is therefore not appropriate in the presence of outliers.

To perform this test, the smallest arc that contains all of the points is found. The length of this arc, w, is the circular range. As we have already pointed out, if the points are spaced at regular intervals they will lie at the vertices of a regular polygon and the circular range will then take the maximum possible value. This is $360(n-1)/n°$. The circular range will be small if all points occur close together. Under the assumption of uniformity, the probability that the circular range w is no more than r radians was shown by Mardia (1972) to be:

$$\Pr(w \le r) = \sum_{k=1}^{\nu}(-1)^{k-1}\binom{n}{k}\left[1 - \frac{k(2\pi - r)}{2\pi}\right]^{n-1} \tag{4.9}$$

where ν is the largest value of k such that $1 - k\{(2\pi - r)/2\pi\}$ exceeds zero. Mardia gives a table of critical values for the range test.

Example 4.17

The problem. Smeeton and Wilkinson (1988) give data for a female psychiatric patient who repeatedly attempted to commit suicide. There was evidence to suggest that these attempts occurred during one particular part of the year. Records showed that attempts had occurred on 2 June 1980, 3 June 1980, 8 June 1980, 18 June 1980, 4 July 1980, 5 June 1981, 6 June 1981 and 31 July 1981.

The successive angles on the circle (assuming 0° is the start of the year) are 151°, 152°, 157°, 167°, 182°, 154°, 155°, 209°. We test the null hypothesis H_0 that the dates of the suicide attempts have a uniform distribution.

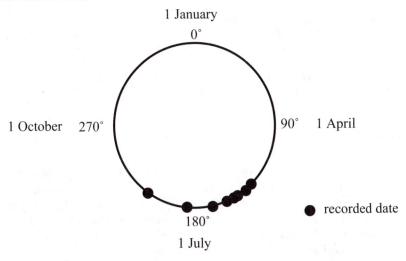

Figure 4.9 *Recorded dates of eight suicide attempts on an annual clock where* $0°$ *represents 1 January.*

Formulation and assumptions. The circular range w is the length of the smallest arc that contains all of the dates. If the suicide attempts are uniformly distributed around the circle then the circular range will be relatively large. By contrast, if the dates form a tight cluster the circular range will be small, providing evidence against H_0.

Procedure. In this example, the circular range is the difference between the largest and smallest angle ($209° - 151° = 58°$). This is illustrated on the circular plot in Figure 4.9. Converting $58°$ to radians, $r = 1.0123$ radians. The appropriate P-value given by (4.9) is $P < 0.0001$ providing overwhelming evidence against H_0.

Conclusion. In view of the strength of evidence against H_0 when $P < 0.0001$, it is reasonable to accept that for this patient clustering of suicide attempts occurs in June and July.

Comments. 1. If the points within a cluster lie on both sides of $0°$ the difference between the largest and smallest values does not give the circular range. It is then most easily obtained from the circular plot.

2. Unlike the Hodges–Ajne test, the range test is only useful when all observations lie in one cluster, since the circular range is highly susceptible to outliers. The occurrence of a further attempt on, say, 5 February, illustrates this.

3. The detection of a single cluster of points containing all the observations is particularly relevant in psychiatry. Extra support can be targeted at the patient during the appropriate period. For the patient in this example, the data suggest that each year the anniversary of a distressing event that happened around the beginning of June could be triggering a series of suicide attempts that diminish in frequency over the subsequent weeks.

4. The circular range increases with sample size so that over subsequent years suicide attempts by this individual may be recorded outside of June and July,

even if the underlying model remains the same. In fact, follow-up data for this patient (to September 1987) subsequently became available. In 1982, three attempts were recorded between 10 April and 10 June. In 1983, four attempts were recorded between 22 August and 14 December. Following this, no further attempts were recorded. Taking all of the evidence into account the most likely explanation is that any anniversary effect gradually dampened out.

4.6.3 Median or Symmetry Tests for Angular Data

In many sports (archery, cricket, target shooting, netball, bowls, etc.) the objective — at least at some phase of the activity and for some participants — is to hit a small target by dispatching an appropriate missile (e.g., throwing, kicking or pitching a ball, firing an arrow or bullet). In practice, even the most skilled exponents will tend sometimes to go left of target and sometimes right of target (or perhaps also above or below target).

It is often of interest to see whether there is a tendency to go more often to the left than to the right or vice versa. If we know the total number of times there is a deviation to the right or to the left in n attempts we might use a sign test to see if there is evidence that the median direction is or is not that of the target. If there were evidence of a tendency to deviate more often to one side than the other the players, or their coaches, may want to take corrective action. In some cases we may also have information on the angular deviations from the target line for each attempt made by a player. We may use those angular deviations with appropriate signs (say plus to the right, and minus to the left) and after ranking absolute values use a Wilcoxon signed-rank test to carry out what is usually called in this context a *test of symmetry*. In doing so we make an assumption that the directional distribution of the attempts is symmetric. The question of interest then is whether it is symmetric about the target direction (zero deviation).

Of course, for the serious player symmetry of shots around the target is not enough; the deviations themselves must be small. Also, in sports involving a two-dimensional target, such as archery or darts, the falling of the missile above or below target has also to be addressed in order to ensure an accurate hit.

These ideas are also relevant in scientific and other applications and extend to testing whether angular data indicate a symmetric distribution about some specified direction. An example of the use of the Wilcoxon test this way in a geological context is given by Fisher (1993, Example 4.17). Where these tests are appropriate they follow the usual procedure for the Wilcoxon test and the sign test given in Chapter 3.

4.7 Fields of Application

Some examples given in Section 3.7 for tests of centrality may be appropriately used for more general tests about distributions such as Kolmogorov's,

Lilliefors' or the Shapiro–Wilk tests. We give here other examples where tests about distributions may be relevant.

Biology

Heart weights are observed for a number of rats used in a drug-testing experiment. The Kolmogorov test could be used to see whether weights are consistent with a normal distribution of mean 11 g and standard deviation 3 g if these were established values for a large batch of untreated rats from the same source. This approach is appropriate if it were uncertain how the drug might affect heart weight, especially if it were felt that it might affect characteristics such as spread or symmetry.

Forestry

The volume of usable timber per tree is obtained for 50 randomly selected trees from a mature forest. If we want to know if it is reasonable to assume volumes are normally distributed with unspecified mean and variance Lilliefors' or the Shapiro–Wilk test would be appropriate.

Time-Dependent Responses

Sometimes tests or observations on individuals have to be performed one at a time and those tested or observed may be able to discuss the test with other subjects due to be tested later. This might influence the performance of later candidates. For example, in medical and other schools, borderline candidates in an examination are sometimes given a *viva voce* examination. Individuals tested earlier may pass hints to those due for later testing that improve their performance. The marks for candidates (taken in the order in which they were tested) may be used in a runs test to detect any trend from independent responses. With a range of scores, the values can be divided into two groups using the median (see Exercise 4.11).

Sport

A football team's results are recorded consecutively throughout a season as W (won), D (drew) or L (lost). A multiple runs test could be used to see whether there is any clustering of particular outcomes.

Pollution Levels

It is widely believed that many pollution levels are increasing with time. Annual observations over many years at a particular point on the earth's surface or in the atmosphere of levels of ozone or other pollutants might be tested for evidence of a trend using a Cox-Stuart test.

Genetics

Certain laws of simple Mendelian inheritance imply that progeny of plant crosses should occur in the ratio 3:1 with respect to some characteristics, e.g., three-quarters may produce crinkled seed and one-quarter smooth seed, or three-quarters may be tall plants and one-quarter short plants. If, in a sample of n progeny, r show the less frequent characteristic, for which in theory $p = 0.25$, this information may be used to test the strength of the evidence that the data are consistent with the hypothesis of a B$(n, 0.25)$ distribution for the characteristic.

Comparison with Standards

When new products, techniques, or services are introduced (for convenience we call these *replacements*) one often wants to compare their performance with current products, techniques, or services (which we call *standards*). A common procedure is to measure one or more characteristics on a sample of the replacements and to test hypotheses about how these characteristics compare with those for the standards. For instance, it may be known that 80 percent of customers categorize a standard toaster as *satisfactory*. For a replacement toaster, 12 out of 18 people rate it satisfactory. Is this sufficient evidence to indicate the replacement has a lower satsifaction rating than the standard? One may also want a confidence interval for the satisfaction rating for the replacement based on this sample result.

The distribution of the time callers telephoning a standard customer service telephone call centre have to wait to be connected to an adviser might be known to be exponentially distributed with a mean of 10 minutes. If a replacement centre is set up and the connection times for a sample of 30 customers is recorded, this data could be used in a Kolmogorov test to determine whether there is evidence of a change in the waiting time distribution. This would be particularly appropriate if it were expected that characteristics other than the mean might have changed, or that the times might not even continue to be exponentially distributed.

Tasting Experiments

To test whether people can detect differences in taste between, say, two different wines, A and B, a set of n tasters are each asked to taste three different samples. Two of these are of one type of wine and the remaining one of the other. Each taster is asked to state which of the three samples is the odd one. If they cannot discriminate on a basis of taste and are just guessing, the number of correct guesses will have a B$(n, 1/3)$ distribution and a test of this hypothesis may be based on a $1/3$ quantile sign test along the lines developed in Section 4.4.2.

Multiple Choice Examinations

In a multiple choice examination each candidate is given n questions each containing 4 statements, one of which is true and the rest of which are false. Candidates must select the statement they believe is true. If a candidate always guesses the probability of correct selection for each question is $p = 0.25$. This hypothesis may be tested using a binomial quartile-based sign test where the distribution of correct choices under the hypothesis of guessing is B$(n, 0.25)$.

Geology

The directions of fractures or fault lines in rock structures may appear to show a preferred orientation. Tests to assess the strength of evidence that this is so may be based on those described in Section 4.6.

Movement of Animals

Birds or animals are often believed to have preferred directions of movement under certain circumstances. For example, there have been many studies of the behaviour of homing pigeons when released to ascertain whether they show an immediate inclination to head towards home or whether they fly at random for some time before aligning to the correct flight path home. Studies have also been made to determine whether birds on a known flight path (e.g., during migration) become disoriented when passing close to electro-magnetic radiation sources and tend then to fly at random. Angular distribution tests are appropriate.

Other One-Sample Problems

Many one-sample problems have not been covered in detail in this or the previous chapter. An interesting one is that where ordered observations taken before a fixed time τ have one distribution, while after that time they have a distribution differing only in the centrality parameter. A problem of interest then is to determine τ. This problem is discussed by Pettit (1979, 1981) and for the case of several change-over times by Lombard (1987).

4.8 Summary

The **Kolmogorov test** (Section 4.2.1) indicates whether data are an acceptable fit to a completely specified continuous distribution. Often several distributions may give an adequate fit. In this case graphical methods (Section 4.2.2) are useful to get an eye comparison of the relative goodness of fit. **Lilliefors' test** for normality with unspecified mean and variance (Section 4.2.3) uses the Kolmogorov statistic, but separate tables or relevant software are needed to test for significance or to estimate P-values. The **Shapiro–Wilk** test is theoretically more complicated than Lilliefors' test, but is often more powerful.

Data consisting of the number of occurrences of one from two possible outcomes in a series of trials often give rise to a **binomial distribution** and inferences are often required about the parameter p associated with that outcome (Section 4.3). Inferences about quantiles of data from continuous or other measurement distributions may also be based on the value of p associated with that quantile. Studies of power and the sample sizes needed to obtain a given power against a single specified alternative are relatively straightforward when the binomial distribution is relevant to primary data. Appropriate software and extensive tables are available for binomial inference. When p is close to 0 or 1 asymptotic results may be reliable only for very large samples.

The **Cox–Stuart test** for trend is a novel application of a sign test.

A basic **runs test** (Section 4.5) applies to numbers of runs for dichotomous outcomes (e.g., heads or tails). In a sequence of N such ordered outcomes the hypothesis of randomness or independence is rejected if there are too few or too many runs. Simple modifications to the test make possible tests for runs above and below the median for more general data. These may provide evidence for detecting certain types of departures from independence.

Runs tests can also be applied to data having several categories. This is generally to investigate possible clustering of outcomes of the same kind, indicated by a lower than expected number of runs.

Angular or **directional** data (Section 4.6) are often best represented by points on the circumference of a circle. Some standard general parametric and nonparametric test procedures are directly applicable to angular data but more commonly modifications are needed. Care is needed when defining concepts such as mean and median for angular data.

4.9 Exercises

4.1 In the data sets in Appendix 1 we give the ages of death for 21 members of the McGamma clan. Perform an appropriate test to determine whether it is reasonable to assume age at death is normally distributed for that clan.

*__4.2__ The negative exponential distribution with mean 20 has the cumulative distribution function $F(x) = 1 - e^{-x/20}$, $0 \le x < \infty$. Use a Kolmogorov test to determine if it is reasonable to assume the excess parking times in Exercise 3.17 are a sample from this distribution.

*__4.3__ Are the insurance claim data in Exercise 3.14 likely to have come from a normal distribution? Test using Lilliefors' test and the Shapiro–Wilk test.

4.4 Test whether the data in Exercise 3.14 may have come from a uniform distribution over the interval (1100, 7700).

*__4.5__ A supermarket indicates in contracts given to suppliers of oranges that not more than 1 percent of the fruit in any consignment should show visible signs of damage. Realism forces it to agree that it will use a sampling scheme with producer's risk set at $P = 0.10$ for rejection of samples with less than 1 percent blemished and consumer's risk set at $1 - P^* = 0.05$ of accepting batches with 3 percent or more blemished. The sampling scheme used is to select a number n from a large batch of N. For each of the sample sizes $n = 100, 200, 300, 400, 500$ determine the maximum number of blemished fruit that would indicate the condition for the producer's risk is being met. In each case find the corresponding consumer's risk. What is the smallest value of n and for what r, the maximum number of permissible defectives in an acceptable batch of that size n, will the above conditions for producer's and consumer's risks be met?

4.6 A commentator on the 1987 Open Golf Championship asserted that on a good day 10 percent of top-class players could be expected to complete a round with a score of 68 or less. On the fourth day of the championship weather conditions were poor and the commentator remarked before play started that one might expect the weather to increase scores by four strokes per round, implying that 10 percent of players might be expected to return scores of 72 or less. In the event 26 of the 77 players competing returned scores of 72 or less. Regarding the players as a sample of 77 top-class players and assuming the commentator's assertion about scores on a good day is correct, do these fourth-day results suggest the commentator's assertion about scores in the poor weather conditions prevailing was (i) perhaps correct, (ii) optimistic or (iii) pessimistic?

*4.7 In a pilot opinion poll 18 voters from one electorate selected at random were asked if they thought the British Prime Minister was doing a good job. Six (one-third) said *Yes* and twelve (two-thirds) said *No*. Is this sufficient evidence to reject the hypothesis that 50 percent of the electorate think the Prime Minister is doing a good job?

The pilot results were checked by taking a larger sample of 225 voters. By coincidence 75 (one-third) answered *Yes* and 150 (two-thirds) answered *No*. Do we draw the same conclusion about the hypothesis that 50 percent of the electorate think the Prime Minister is doing a good job? If not, why not?

4.8 The journal *Biometrics* 1985, **41**, p. 830, gives data on numbers of medical papers published annually in that journal for the period 1971–81. These data are extended below to cover (in order) the period 1969–85. Is there evidence of a monotonic trend in numbers of medical papers published?

   11   6   14   13   18   14   11   22   19   19   25   24   38   19   25   31   19

*4.9 The UK Meteorological Office monthly weather summaries published by HMSO give the following annual rainfalls in mm for 15 stations in the UK during 1978. The stations are listed in order of increasing latitude. Is there evidence of a monotonic trend in rainfall from South to North?

Margate, 443; Kew, 598; Cheltenham, 738; Cambridge, 556; Birmingham, 729; Cromer, 646; York, 654; Carlisle, 739; Newcastle, 742; Edinburgh, 699; Callander, 1596; Dundee, 867; Aberdeen, 877; Nairn, 642; Baltasound, 1142.

4.10 Rogerson (1987) gave the following annual mobility rates (percentage of population living in different houses at the end of the year than at the beginning) for people of all ages in the USA for 28 consecutive postwar years. Is there evidence of a monotonic trend?

18.8	18.7	21.0	19.8	20.1	18.6	19.9	20.5	19.4	19.8
19.9	19.4	20.0	19.1	19.4	19.6	20.1	19.3	18.3	18.8
18.3	18.4	17.9	17.1	16.6	16.6	16.1	16.8		

*4.11 A psychologist is testing 16 applicants for a job one at a time. Each has to perform a series of tests and the psychologist awards an overall point score to each applicant. A high score indicates a good performance. As each applicant may discuss the tests with later applicants before the latter are tested it is suggested that those tested later may have an unfair advantage. Do the applicants' scores (in order of testing) given below support this assertion?

   62   69   55   71   64   68   72   75   49   74   81   83   77   79   89   42

Use an appropriate runs test. Do you consider the Cox-Stuart test (Section 3.2.3) may also be appropriate? Give reasons for your decision.

4.12   The Office for National Statistics (1998) gives the numbers of deaths from railway accidents in England and Wales for 1981 to 1996 as

$$
\begin{array}{cccccccc}
76 & 92 & 105 & 86 & 91 & 81 & 103 & 92 \\
71 & 132 & 71 & 57 & 48 & 63 & 43 & 60
\end{array}
$$

Use an appropriate form (or forms) of the runs test to examine nonrandomness.

4.13   A circular ring road is constructed around a town. A map of the local area shows the town hall is situated at the centre of the circle. The positions of road accidents along the ring road are measured as bearings from the town hall. During a period of road works to the east of the town, the accidents that occur have bearings $10°, 35°, 82°, 87°, 94°, 108°, 125°$. Is there any evidence that the accidents are linked to the road works?

*__4.14__   For a children's television quiz show, a circular board with a stationary vertical pointer attached to the centre is divided into quadrants. The board is spun and contestants are asked questions on sport, popular music, current affairs and science according to the quadrant indicated when the board comes to rest. Angles are marked around the edge of the board. For ten consecutive questions, the angles indicated at rest are $15°, 46°, 114°, 137°, 165°, 183°, 195°, 215°, 271°, 328°$. Do you think that on average all types of questions are equally likely?

*__4.15__   An archer fires arrows at a target that is on a bearing of $145°$ to him. Angles of fire for 10 arrows are $139°, 141°, 146°, 148°, 150°, 152°, 153°, 155°, 158°, 160°$. There is concern that the arrows are tending to fall to one side of target so one wants to test the hypothesis $H_0: \theta = 145$ against the alternative $H_1: \theta \neq 145$. Use the sign test and the Wilcoxon test to do this. Would you question the validity of using the latter test for these data?

4.16   Twenty batteries are selected at random from a large batch and the time, t to complete discharge is recorded for each. Find an approximate 95 percent confidence limits for the lower quartile expressed in terms of the appropriate order statistics, $t_{(i)}$.

4.17   A football team obtains the following results for a series of 12 matches:

$$W \quad W \quad W \quad D \quad D \quad W \quad W \quad L \quad L \quad L \quad L \quad W$$

where W, D and L indicate that they won, drew or lost respectively. Use the normal approximation for the distribution of the number of runs to look for evidence of clustering. If you have suitable software also use a simulation approach. The exact one-tail probability is $P = 0.0261$. Comment on the accuracy of your estimate(s) for P.

CHAPTER 5

METHODS FOR PAIRED SAMPLES

5.1 Comparisons in Pairs

Single-sample methods illustrate basic ideas, but have limited applications. Practical problems usually involve two or more samples that may or may not be independent. We consider paired samples in this chapter, a dependent-sample situation where many, but not all, problems reduce to a single-sample equivalent.

5.1.1 Studies Using Pairing

Many factors must be taken into account in the way paired samples are selected and used. We illustrate the main ones using medical examples. Most of these factors are important in this field, although many of them are also relevant in other applications.

To compare two stimuli if it is practicable it is sometimes appropriate, and advantageous, to apply both to each of, say, n individuals. The main reason for this is that responses to any stimulus often vary markedly between individuals. However, some important issues must be addressed before proceeding this way.

A dentist may wish to compare two mouthwashes for the treatment of ulcerative gingivitis in the mouth. He or she could use the number and size of the mouth ulcers in a patient to assess the severity of the problem. The patients could also indicate their perceived level of pain due to these ulcers.

One way of comparing the two mouthwashes, known as a crossover trial, is to divide the patients into two groups. Each patient in the first group is treated with one mouthwash (A) for a certain period, an assessment of the patient's response is made and then the patient is transferred to the other mouthwash (B). Following the second course of treatment the patient is again assessed. Each patient in the second group is treated and assessed with the mouthwashes in reverse order. At the end of the study the two treatments are compared on the basis of findings from the patient assessments.

An advantage of this design is that patients act as their own control. In an ideal world the only difference between the two time periods would be the type of mouthwash used, which would make a comparison straightforward. Unfortunately, there may be a carry-over effect of the first treatment into the second period if the second mouthwash is administered before the effects of the first have completely worn off. Sometimes an interval or wash out period is allowed between the two treatments in an attempt to reduce this effect. Also, response to either mouthwash may be affected by the prior or subsequent administration of the other (interaction is the technical statistical term). In an extreme case administration of one mouthwash after the other might make a patient's condition worse, whereas if either were given separately, both may be beneficial. In passing, we note that attention should be paid to the patient's diet as particular components might influence the occurrence or healing of mouth ulcers.

We do not consider the statistical analysis of crossover trials at this stage, but describe the procedure for a simple experiment in Section 8.5. A more general account of such designs is given by Senn (2002).

Studies involving ordered application of alternative treatments can only be performed with chronic conditions where the first treatment received is unlikely to lead to complete recovery or death. Ulcerative gingivitis is such a condition (asthma and eczema are others) in which for many patients, improvement under treatment is swift but the condition deteriorates once the treatment is withdrawn. A practical difficulty is that assessment of ulceration by a dentist is notoriously subjective, as is patient assessment of pain.

To allow for subjectivity a double blind procedure should be employed in which neither the patient nor the dentist making the assessments knows the order of presentation (which should be random). Mouthwashes can be made to have identical appearance and taste. The washes are identified by codes available only to research workers or administrators who have no direct contact with the patient. If tablets are used, they should be of the same size, colour and taste. Double blinding can also be used to allow for the placebo effect. This is a psychological response shown by some patients to the knowledge that treatment is being received, even though the solution or tablet (known as a placebo) does not contain an active ingredient. These aspects of a study should be carefully explained in the patient information sheet if applicable.

If it is undesirable for ethical, or practical, reasons to give patients both treatments, we might use pairs of individuals chosen so that the members of each pair are as alike as possible in all relevant characteristics. This process is known as *matching*. For example, a female aged 30 years might be matched with a similarly aged female of the same ethnic group, each showing a similar relevant morbidity status. In each pair, one of the women chosen at random receives the first treatment, the other the second. In this case we look at the assessment differences between members of each pair.

As mentioned in Section 1.5, patients should be free to withdraw from a study at any stage. This means that some patients may not be taking their allocated treatment by the close of the study or may have withdrawn from the study altogether. It is recommended that the results of studies that compare two or more groups of patients be analysed on an *intention to treat* basis in which the groups are compared as they were originally chosen. An attractive alternative might be to analyse the patients by their actual treatment at the end of the study. This may introduce bias however, as those who are unwilling, or unable, to continue in the study are likely to have different characteristics

Table 5.1 *Mean response time (ms) to digital information.*

Subject	1	2	3	4	5	6
LVF(1)	564	521	495	564	560	481
RVF(2)	557	505	465	562	544	448
(1) − (2)	7	16	30	2	16	33

Subject	7	8	9	10	11	12
LVF(1)	545	478	580	484	539	467
RVF(2)	531	458	560	485	520	445
(1) − (2)	14	20	20	−1	19	22

(e.g., greater health problems) compared with the other patients. '*Intention to treat* analysis also gives a fairer assessment of the impact of these strategies in the real world. These comments are relevant to planning an experiment, no matter whether we use a parametric or nonparametric method of analysis.

Returning to the study of mouth ulcer treatment, a patient's assessment of pain may involve no more than indicating which mouthwash, if either, gave the greater relief (e.g., first mouthwash gave most relief, second mouthwash gave most relief, both equally good or both ineffective). One may score mouthwash A gave more relief as a *plus*, mouthwash B gave more relief as a *minus* and no difference as a *zero*. Individual patient scores — *plus, minus,* or *zero* — provide the basis of a sign test of H_0: *drugs are equally effective* against an appropriate one- or two-tail alternative.

5.1.2 Further Examples

We give four specific examples.

I. Geffen, Bradshaw and Nettleton (1973) wanted to know whether certain numbers presented in random order were perceived more rapidly in the right (RVF) or left visual fields (LVF), or whether there was no consistent difference, it being a matter of chance whether an individual responded more quickly in one field or in the other. For each of 12 subjects the mean response times to digital information in each field were measured. Response times varied much more between individuals than they did between fields for any one individual. The data and the differences LVF–RVF for each individual are given in Table 5.1.

This table shows quicker response times in the RVF for all but subject 10. No difference exceeds 33 ms, whereas in either field differences between some individuals exceed 100 ms. For example, it is $580 - 467 = 113$ between subjects

9 and 12 in the LVF. Without matched pairs these differences might swamp the smaller, but relatively consistent, differences between fields for individuals, an aspect we explore further in Exercise 6.6.

II. A course organizer might compare two teaching methods such as lectures and computer assisted learning (CAL) by pairing students so that, if possible, each member of a pair is of the same gender and has the same previous knowledge of the subject. For each pair, one member is allocated to lecture classes and the other to the CAL material. At the end of the course the students take the same tests and the results are interpreted in terms of pairwise differences.

III. Using double blind marking (a method favoured by some institutions) one can compare consistency between two examiners. The two examiners mark the same series of essays without disclosing their assessments to one another. The differences between the marks awarded by each examiner for each essay are then compared to see whether one examiner consistently awards a higher mark, or whether differences have some other pattern, or whether they appear to be purely random.

IV. To compare two animal diets using pairs of twin lambs the diets are fed one to each twin, and growth is measured over a period: attention is focused on growth differences between twins within each pair. Because of genetic similarity each of a pair of twin lambs fed on identical diets tends to grow at a similar rate. When fed different diets any consistent differences in growth may be attributed to the effects of diet.

In summary, the aim of pairing is to make conditions, other than the treatment or factor under investigation, as alike as possible within each pair. The differences within pairs provide a measure of any treatment effect that takes the form of a shift in the distribution.

5.1.3 Single-Sample Analysis of Matched Pairs

Differences between paired observations provide a single sample that can be analysed by methods developed in Chapters 3 and 4. We must, however, consider the assumptions about observations on each member of the pair, and what precautions in experimental procedure are needed to validate analyses. These points are best brought out by examples.

Example 5.1

The problem. Using the LVF, RVF data in Table 5.1, assess the strength of the evidence for a consistent response difference between the two fields for individuals. Obtain a 95 percent confidence interval for that difference.

Formulation and assumptions. We denote the observation on subject i in the RVF by x_i and that in the LVF by y_i. We analyse the differences $d_i = y_i - x_i$, $i = 1, 2, \ldots, n$. The d_i are independent, since each refers to a different individual.

Under H_0: *the median of the differences is zero* the d_i are equally likely to be positive or negative and a sign test is justified. If we assume a symmetric distribution of the d_i under H_0 we may use a Wilcoxon signed-rank (or a normal scores) test. There are several different response patterns in the two visual fields that could result in a symmetric distribution of the differences under H_0.

In particular, if we assume response times are identically and independently (but not necessarily symmetrically) distributed in the two fields for any individual (but this distribution need not be the same for every individual) the difference for each individual will be symmetric about zero. This follows because if X and Y have the same independent distributions, then $X - Y$ and $Y - X$ will each have the same distribution and therefore must be symmetrically distributed about zero.

We also get a symmetric distribution of differences if the LVF and RVF responses for any individual have different symmetric distributions providing each has the same median or mean if the latter exists. It is well to be aware of such subtleties, but a Wilcoxon test is clearly justified when we assume identical distributions for LVF and RVF for any one individual under the null hypothesis. The alternative hypothesis of interest is usually that there is a shift in centrality only, as indicated by a shift in one field or the other in the median of otherwise identical distributions. Often the d_i values themselves indicate whether there is serious asymmetry that suggests this condition may not hold. If we assume also that the differences are approximately normally distributed, a t-test is appropriate. We may test whether normality is a reasonable assumption for the d_i by Lilliefors' test or by the Shapiro–Wilk test, but these tests may have low power for small samples, unless the departure from normality is very marked.

Procedure. We work with the single sample composed of the differences d_i and use these in the way developed in Chapter 3, so we only sketch details. Denoting the centrality parameter or measure of treatment effect (assuming this to be only a centrality or location shift) by θ, the null hypothesis is H_0: $\theta = 0$, so the ordered signed deviations are simply the ordered differences obtainable from Table 5.1, i.e.,

$$-1 \quad 2 \quad 7 \quad 14 \quad 16 \quad 16 \quad 19 \quad 20 \quad 20 \quad 22 \quad 30 \quad 33$$

and the corresponding signed ranks are

$$-1 \quad 2 \quad 3 \quad 4 \quad 5.5 \quad 5.5 \quad 7 \quad 8.5 \quad 8.5 \quad 10 \quad 11 \quad 12.$$

Using the sign test for zero median difference we have 1 minus and 11 plus signs. Appropriate tables or software immediately give $P = 0.0063$, providing very strong evidence that responses differ between fields. The nominal 95 percent confidence interval based on the sign test is $(7, 22)$ since we reject H_0 at a nominal $P = 0.05$ level only for 2 or fewer, or 10 or more, minus signs. For the Wilcoxon test $S_- = 1$, corresponding to a two-tail $P < 0.001$. Indeed, you should only need a pocket calculator to show that $P \approx 0.00098$ in this example.

For a confidence interval with coverage of at least 95 percent, tables indicate we require the 14 greatest and least Walsh averages. These averages may be obtained using Minitab, or StatXact may be used to find the interval directly. Many other software packages include a program giving at least an asymptotic approximation. The interval turns out to be $(9.5, 23.5)$. The Hodges–Lehmann point estimator of the median (i.e., the median of the Walsh averages) is 17.25.

If we assume normality, a two-tail t-test gives $P < 0.001$ and a 95 percent confidence interval $(10.1, 22.9)$ centred about the mean 16.5. Normal theory gives the shortest confidence interval. The fact that the interval is only slightly displaced for the sign test implies reasonable symmetry. Lilliefors' test statistic, 0.152, is well below the value required for significance, and for that test StatXact gave a Monte Carlo estimate for the exact $P = 0.617$. For the Shapiro-Wilk test StatXact gave $P = 0.685$, so there is no evidence of nonnormality.

Conclusion. The sign, Wilcoxon and t-tests all point to strong evidence against H_0 so we conclude that response times in the RVF are faster. Consistency of the confidence intervals given by these approaches suggests the mean difference is between about 10 and 22 ms.

Comments. 1. The set of differences we compare are independent (a necessary condition for validity of our centrality tests) because each difference is calculated for a different individual.

2. If the response rate in the LVF had been measured before that in the RVF for all subjects a difficulty in interpretation would arise. The result might then imply a learning process, people responding more quickly in the RVF because they were learning to react more rapidly. There could be a mixture of a learning effect and an inherent faster response in the RVF. We avoid this difficulty if we decide at random which field — left or right — is to be tested first for each individual. That was done in this experiment. This should balance out, and largely annul, any learning factor. Another approach is to achieve balance by selecting six subjects (preferably at random) to be tested first in the LVF. The remaining six are tested first in the RVF. Such balanced designs provide a basis for separating a learning effect from an inherent difference between field responses, although a somewhat larger experiment would be needed to do this by appropriate parametric or nonparametric methods.

Computational aspects. Most statistical packages that have programs for the procedures discussed in this example allow one to enter either the data for each field for each individual and then computes differences automatically or else to enter the differences themselves as the raw data.

A modification to the above procedure lets us test the hypothesis that a centrality difference has a prespecified value θ. As in Chapters 3 and 4, we simply consider deviations from that value. This effectively shifts the origin to the hypothetical median, so that for our revised data (the deviations) we test the hypothesis that the centrality parameter for that population takes the value zero.

Example 5.2

The problem. Eleven children are given an arithmetic test; after 3 weeks' special tuition they are given a further test of equal difficulty (we say more about this in the *Comments*). Their marks in each test (out of 90) and the individual differences are given in Table 5.2. Do these support a claim that the average improvement due to extra tuition is 10 marks?

Formulation and assumptions. The question essentially is whether, if we assume these children are a random sample from some hypothetical population (perhaps children of the same age trained in the same educational system or studying the same syllabus), it is reasonable to suppose the mean mark difference is 10?

Table 5.2 *Marks (out of 90) in two arithmetic tests.*

Pupil	A	B	C	D	E	F	G	H	I	J	K
First test	45	61	33	29	21	47	53	32	37	25	81
Second test	53	67	47	34	31	49	62	51	48	29	86
(Second − First)	8	6	14	5	10	2	9	19	11	4	5

Procedure. We consider deviations from 10 for the mark differences in the last line of Table 5.2. These deviations are:

$$-2 \quad -4 \quad 4 \quad -5 \quad 0 \quad -8 \quad -1 \quad 9 \quad 1 \quad -6 \quad -5$$

The differences arranged in order of magnitude with appropriate signs are

$$0 \quad -1 \quad 1 \quad -2 \quad -4 \quad 4 \quad -5 \quad -5 \quad -6 \quad -8 \quad 9$$

and the ranks without signs are

$$1 \quad 2.5 \quad 2.5 \quad 4 \quad 5.5 \quad 5.5 \quad 7.5 \quad 7.5 \quad 9 \quad 10 \quad 11.$$

Using the rules given in Section 3.3.4 for mid-ranks with ties, together with the convention mentioned at the end of that section for replacing the rank 1 associated with the zero difference by 0, we get signed ranks:

$$0 \quad -2.5 \quad 2.5 \quad -4 \quad -5.5 \quad 5.5 \quad -7.5 \quad -7.5 \quad -9 \quad -10 \quad 11.$$

Our statistic is the sum of the positive ranks $S_+ = 0 + 2.5 + 5.5 + 11 = 19$. For the exact permutation test StatXact gives $\Pr(S+ \leq 19.0) = 0.124$. Doubling this value for a two-tail test we immediately see there is no strong evidence against H_0, since $P = 0.248$.

While the sample is too small to justify an asymptotic procedure this example gives a two-tail $P = 0.229$, comparing reasonably well with the exact $P = 0.248$.

Conclusion. There is no firm evidence against the hypothesis that the mean improvement may be 10.

Comments. 1. How does one decide if two arithmetic tests are equally difficult? Might not the improved marks in the second test imply it was easier? The statistician should seek the assurance of the educationalist conducting the test that reasonable precautions have been taken to ensure equal difficulty. Sometimes standard tests that have been tried on large groups of students with results that show convincingly that they are for all practical purposes of equal difficulty, are used in such situations.

2. Confidence intervals for the mean or median difference may be obtained in the usual way. The nominal 95 percent confidence interval for these data given by StatXact is (5, 12), in close agreement with the normal theory t-test interval (5.14, 11.76).

3. It may not be realistic to test simply for a centrality shift. Sometimes pupils who perform well initially benefit little from extra tuition; in this example Pupil K cannot possibly improve by more than 9 marks. Likewise very poor pupils may find

the concepts of arithmetic difficult to grasp and gain little from extra tuition. Often only those in the mid-ability range show appreciable benefit. A statistician may find evidence of this simply by looking at the data — and there are indeed tests for such tendencies. Deciding what should be tested or estimated is often a topic for fruitful discussion between statistician and experimenter.

Example 5.3 is based on data for a group of 77 first-year medical students available to one of us (NCS).

Example 5.3

The problem. The data below are differences (*systolic blood pressure after exercise* — *systolic blood pressure before exercise*) measured in mm Hg for a random sample of 24 from the group of 77 students. Obtain 95 percent confidence limits for the population difference based on (i) the sign test, (ii) the Wilcoxon signed-rank test and (iii) the *t*-test, and comment on the appropriateness of each. For convenience we have arranged the differences in ascending order.

−5 −5 0 2 10 15 15 15 18 20 20 20 20 22 30 30 34 40 40 40 41 47 80 85

Formulation and assumptions. We use standard methods developed in Chapter 3 and in this chapter.

Procedure. We omit computational details since examples of similar calculations have already been given. The intervals quoted below were obtained using relevant statistical software.

Conclusion. The nominal 95 percent confidence intervals are:

Sign test	(15, 40)
Wilcoxon	(17.5, 35)
t-test	(16.87, 35.96)

Comments. 1. Not surprisingly, the sign test interval is the longest. There is slight evidence of skewness, but none of the intervals is seriously displaced relative to the others. The Wilcoxon interval being shorter than that based on normal theory may be attributed to a slightly heavier upper tail (especially the values 80, 85) than one would expect with a sample from a normal distribution. Lilliefors' test for normality gives a test statistic value 0.162, and an estimated $P = 0.1071$ based on 10,000 simulations using StatXact. For the Shapiro–Wilk test there is stronger evidence against normality, the statistic 0.8999 corresponding to $P = 0.0205$.

2. The only reason for taking a random sample of 24 from 77 observations was to provide a convenient illustrative example. Almost certainly one would use all observations in any detailed studies of the effects of exercise on blood pressure (and interest would be in the distribution of the differences, and not just in the mean or median difference).

3. The variation in the size of the differences is unsurprising with blood pressure measurements taken by inexperienced first-year medical students. Readings are usually recorded to the nearest mm Hg but in practice they may only be recorded to the nearest 5 mm Hg by those unfamiliar with the technique. The student exercises

would be performed with a range of enthusiasm. A few extremely high differences and a few small negative changes are therefore understandable.

4. Notwithstanding Comment 3, there was clearly an error on the original computer printout we used. For one student the systolic blood pressure measurement after exercise was recorded as 15 mm Hg and gave a difference of $15 - 118 = -103$. A systolic blood pressure reading of less than 80 mm Hg is unlikely (unpublished data from South London). The value of 15 mm Hg for systolic blood pressure after exercise is clearly incorrect. There is a strong suspicion that the final digit has been omitted and that the true reading should be between 150 and 159. It is easy to spot such a discrepancy on a printout and in practice one would then attempt to track down the source of error (and if possible make the needed correction). Possible sources of error here are failure of a printer to reproduce a character, a mistake in entering the original data, or in initial reading or recording of the data by an inattentive student.

The original purpose and manner of the data collection could affect the likely accuracy of the recorded results. With increasing use of computer packages to process data such an error might go undetected. Had this value, -103, been included in our sample of 24 in place of the entry -5 we would have obtained the following 95 percent confidence intervals:

$$
\begin{array}{ll}
\text{Sign test} & (15, 40) \\
\text{Wilcoxon} & (17.5, 34.5) \\
t\text{-test} & (7.83, 36.83)
\end{array}
$$

The reader who has difficulty explaining the differences between these results and those recorded above in terms of the effect of an outlier should refer to Section 3.4.2. In particular it is worth noting the Wilcoxon interval is little changed because the introduction of a value -103 has a tendency to make the sample distribution, if anything, a little more symmetric than the original sample. However, the tails become rather longer than one would expect for a normal distribution and this has the effect of elongating the t-based interval. For these amended data Lilliefors' test gave a Monte Carlo estimated $P = 0.0083$ and a Shapiro–Wilk test gave $P = 0.0001$, both hinting strongly at nonnormality implicit in the tail values.

Practical experience in handling data highlights the crucial importance of detecting data errors. Many computer programs include output to help detect outliers; e.g., a printout of maximum and minimum values often (but not always) highlights a glaring data error.

5.2 A Less Obvious Use of the Sign Test

The way the data are presented in Example 5.4 does not make it obvious that a sign test is relevant.

Example 5.4

The problem. Members of a mountaineering club have long argued about which of two rock climbs is the more difficult. Hoping to settle the argument one member

Table 5.3 *Outcome of two rock climbs.*

		First climb	
		Succeeded	Failed
		---	---
Second climb	Succeeded	73	14
	Failed	9	12

checks the club log book. This records for any climb by a member whether or not it is successfully completed. The log shows that 108 members have attempted both climbs with the outcomes summarized in Table 5.3. Is there evidence that one climb is more difficult?

Formulation and assumptions. A climber succeeding in both climbs, or failing in both, provides no information about relative difficulty. Such cases are ties so far as comparing difficulty is concerned. If we had additional information, for instance about each climber's personal assessment of the difficulty, the situation would be different. As it is, our only meaningful comparators of difficulty are numbers who succeed at one climb, but fail at the other. From Table 5.3 we see that 9 succeed at the first climb but fail at the second. We may think of this as a *plus* for the first climb. Also, 14 fail on the first climb but succeed at the second; we may think of this as a *minus* for the first climb.

This is a sign test situation. If the climbs are of equal difficulty a *plus* or *minus* is equally likely for each climber. Thus under H_0: *climbs are equally difficult* the number of 'plus' signs has a $B(n, 0.5)$ distribution where n is the total number of plus and minus signs.

Procedure. There are 9 plus and 14 minus signs so $n = 23$. StatXact gives an exact test $P = 0.4049$. Alternatively, using the normal approximation to the binomial distribution we find, using (3.3), that $Z = (9.5 - 11.5)/(23/4) = -0.83$ giving a two-tail $P = 0.4065$, so there is no evidence that one climb is easier.

Conclusion. We retain H_0: *climbs are equally difficult.*

Comments. 1. Because $73 + 12 = 85$ from 108 pairs provide no information on the relative difficulty of the climbs this may seem wasted data, but such data give an indication of how big or small any difference might be. In some situations we need many observations because we are looking for a small difference which may not be distinguishable when the only relevant criteria are *success/failure* or *failure/success* categories. In the context of these data it is likely that most of those who failed at both were less experienced or less enthusiastic than those who achieved one success. Those who succeeded at both were likely to be the more experienced, or the more enthusiastic.

2. The continuity correction of replacing 9 by 9.5 in the numerator of the asymptotic approximation has a marked effect in this example. If it is omitted the two-tail P is reduced from 0.4065 to 0.2971. Our conclusion, however, in this case would not be altered by this appreciable numerical change.

3. It would be interesting to know whether club members generally attempted the climbs in a particular order as a learning effect might then be involved.

Computational aspects. StatXact and also many general packages include programs for this simple test that may be used to obtain the relevant P-value for any sign test.

The test is called *McNemar's test*, having been proposed by McNemar (1947). Conover (1999, Section 3.5) presents it more formally, but it is effectively a paired-sample sign test.

5.3 Power and Sample Size

In Section 4.3.3 we explored the power of some tests for binomial probabilities of the form H_0: $p = p_0$ against single-valued alternatives such as H_1: $p = p_1$ where choice of p_0, p_1 was determined by their relevance to producer's and consumer's risks associated with a sampling scheme. Exact power calculations were relatively straightforward and we illustrated their use in finding sample sizes to meet specified producer's and consumer's risks.

We now consider relationships between power and sample size for single-sample centrality tests relevant both to basic single-sample situations or to differences between matched pairs. Exact results are only available in a limited number of cases and in this context the power for a specified alternative is in general not distribution-free even when the test itself is. This restricts the overall value of exact power computations. Nevertheless, providing one exercises care in interpretation, computing approximate power for given sample sizes before beginning an experiment helps one make efficient use of resources.

We first highlight some key features and limitations using simple examples for the sign test, where only relatively straightforward computations are needed. We then consider briefly the Wilcoxon signed-rank test where power and sample size determinations introduce greater difficulties both in theory and practice.

5.3.1 Power and Sample Size for the Sign Test

Examples 5.5 to 5.7 cover three situations where we use a sign test for a median θ of the form H_0: $\theta = \theta_0$ against H_1: $\theta > \theta_0$ where a specific alternative H_1: $\theta = \theta_0 + 1$ reflects a minimal departure from H_0 of practical importance. Without loss of generality (see, e.g., Example 5.2) we set $\theta_0 = 0$. We confine attention to a one-tail test, but modifications for a two-tail test are straightforward if one uses the principle of doubling one-tail P-values for a two-tail equivalent. If we compute the power of a test for the alternative H_1: $\theta = 1$, then if the true shift exceeds one unit the power will in general be greater. From these examples it will emerge that for three different population distributions all having the same variance, the power of the equivalent tests is different for each, confirming that power is not a distribution-free property. This should cause no surprise in the light of the well-known result that the

Pitman efficiency of tests depends upon the underlying population distribution.

Example 5.5

The problem. For a sample of 10 from a normal distribution with unknown mean θ and standard deviation 2, what is the power of the sign test for H_0: $\theta = 0$ against H_1: $\theta = 1$ when the probability of a type I error is $\alpha = 0.05$? What sample size would ensure a test with power 0.9?

Formulation and assumptions. Under H_0 the number of positive sample values (pluses) has a $B(10, 0.5)$ distribution. Taking the number of plus signs as the test statistic, then under H_1 these will still have a binomial distribution, but the value of p is now given by $\Pr(X > 0) = p_1$, say, where X has a $N(1, 4)$ distribution. Thus the test is equivalent to testing H_0: $p = 0.5$ against H_1: $p = p_1$ and once p_1 is calculated the power study when $n = 10$, and that for determining the sample size required for any specified power proceed in the way described in Section 4.3.3.

Procedure. Since X is $N(1, 4)$ if H_1 holds it follows that $Z = (X - 1)/2$ has a standard normal distribution, whence $p_1 = \Pr(X > 0) = \Pr(Z > -0.5) = 0.6915$. This value is obtainable from tables or appropriate software. The power may then be obtained from tables or software in the way described in Example 4.8. Here we test H_0: $p = 0.5$ against H_1: $p = 0.6915$. Using computer software is preferable, for it is unlikely that tables will be readily available for precisely the latter value of p and an approximation such as $p = 0.7$ may then have to be used. With these values of p, StatXact gives the power as 0.1365 for sample size 10. However, discontinuities are influential here, as the smallest possible exact one-tail P-values under H_0 are $P = 0.001$, $P = 0.011$ and $P = 0.055$. Had we replaced $\alpha = 0.05$ by $\alpha = 0.011$ the power would still be 0.1365, whereas if we set $\alpha = 0.055$ the power becomes 0.3604.

Minitab uses an asymptotic approximation for power, giving this as 0.3914 when $\alpha = 0.05$. One has reservations about asymptotic results for so small a sample, but 0.3914 is broadly in line with the exact values obtained for possible type I errors slightly above a nominal 0.05.

The asymptotic result should be more satisfactory for finding the larger n needed to ensure a power of, say, at least 0.90. The Minitab program gives this as $n = 55$. Finer tuning using exact results from StatXact gives $n = 58$ as the minimum required sample size corresponding to an exact $P = 0.0435$ and an exact power 0.9031.

Conclusion. For a sample of $n = 10$ the power is only 0.1365. This low power in part reflects the large gap between the possible P values 0.011 and 0.055. For power at least 0.90 for the one-tail test, the sample size should be at least 58.

Comments. 1. The Pitman efficiency of the sign test relative to the t-test for samples from a normal population is 0.64. This implies that for large n the relative efficiency is less than two-thirds, suggesting that a sample of about two-thirds the size should have the same power if we apply a t-test. Many statistical packages allow power calculations for the t-test when it is optimal, and in this case these indicate that for a $N(\theta, 4)$ distribution the necessary sample size with power 0.90 with our chosen values of θ in H_0 and H_1 is $n = 36$, broadly in line with that suggested by the Pitman efficiency.

2. Our computed $p_1 = 0.6915$ was based on the strong assumption that our sample was from a $N(\theta, 4)$ distribution. We seldom know the population variance but not the mean and if we had such information (or could only assume normality with the variance unknown) one should prefer the normal theory-based inference to that using the sign test. However, the results obtained here are illuminating for comparisons with the situations covered in Examples 5.6 and 5.7 where a nonnormal population, again with variance 4, is assumed.

Example 5.6

The problem. Given a sample of 10 from a double-exponential distribution with unknown mean θ and standard deviation 2 find the power of the sign test for H_0: $\theta = 0$ against H_1: $\theta = 1$ if the probability of a type I error is $\alpha = 0.05$. What sample size would ensure a power at least 0.9?

Formulation and assumptions. The double-exponential distribution, also known as the Laplace distribution, with mean θ and standard deviation 2 (variance 4) has probability density function

$$f(x) = \frac{1}{2\sqrt{2}} \exp[-|(x - \theta)|/\sqrt{2}]. \tag{5.1}$$

The distribution is symmetric about the mean, θ. Probabilities associated with the tails are greater than those for the normal distribution with the same mean and variance and this is a classic example of a long-tail symmetric distribution.

Under H_0 the number of positive sample values (pluses) again has a $B(10, 0.5)$ distribution. If the number of plus signs is used as the test statistic, these will also have a binomial distribution under H_1 but the parameter p is now given by $\Pr(X > 0) = p_1$, say, where X has the double-exponential distribution (5.1) with $\theta = 1$. Thus, the original test is equivalent to testing H_0: $p = 0.5$ against H_1: $p = p_1$ and the power study for the given sample size and for determining the sample size required for any specified power proceed as described in Section 4.3.3.

Procedure. The value of p_1 is given by setting $\theta = 1$ in (5.1), then integrating over the interval $(0, \infty)$. Integration is straightforward (Exercise 5.14) and gives $p_1 = 0.7534$. For testing H_0: $p = 0.5$ against H_1: $p = 0.7534$ when $\alpha = 0.05$ StatXact gives the exact power 0.2518. Discontinuities in possible P-values are the same as those in Example 5.5 and replacing $\alpha = 0.05$ by $\alpha = 0.011$ does not alter the power, whereas replacing it by $\alpha = 0.055$ gives power 0.5358. The asymptotic power computation given by Minitab when $\alpha = 0.05$ is 0.4805. Again, the asymptotic result is likely to be more satisfactory when seeking the larger n needed to ensure a power of at least 0.90. Minitab in this case gives $n = 30$. Finer tuning using exact results confirms this minimum required sample size corresponding to an exact $P = 0.0494$ and gives an exact power 0.9023.

Conclusion. For a sample of $n = 10$ the power is now 0.2518. This low power in part reflects the large gap between the possible P-values 0.011 and 0.055. A sample of 30 gives a power at least 0.90 for the one-tail test.

Comments. 1. The sample size 30 is approximately half the size (58) needed for the same power when using the sign test with a normal distribution with the same variance. This is in line with the Pitman efficiency of the sign test relative to the

t-test, which is 2 when sampling from a double-exponential distribution. We must remember, however, that the t-test itself has lower efficiency for a sample from a double-exponential distribution than it has for a normal distribution with the same mean and variance.

2. The reason the sign test is appreciably more powerful for testing the same basic hypotheses about θ in this test than it was for an equivalent test in the previous example is that although the same values of θ are specified in both examples, and that for H_0 translates to $p = 0.5$ in each case, that for H_1 transforms to $p = 0.6915$ and $p = 0.7634$ respectively, so that for the second example there is a larger departure from the value of p under H_0. That this produces a more powerful test is in line with our notion of a good test being one where its power increases as the difference between the value specified in H_1 increases relative to a fixed value specified in H_0.

Examples 5.5 and 5.6 featured samples from symmetric distributions. We now consider a skew distribution with a similar one-unit shift to the right. The case here is unusual in practice but it provides a direct comparison with Examples 5.5 and 5.6 because we again take a case where we sample from a distribution with a variance of 4.

Example 5.7

The problem. Given a sample of 10 from a population with an exponential distribution having probability density function

$$f(x) = \frac{1}{2}\exp[-\frac{1}{2}(x + 1.3862 - \theta)], \quad x \geq \theta - 1.3862 \qquad (5.2)$$

which has median θ and standard deviation 2 what is the power of the sign test of H_0: $\theta = 0$ against H_1: $\theta = 1$ if the probability of a type I error is $\alpha = 0.05$? Find also the sample size that will ensure a test with power at least 0.9.

Formulation and assumptions. The distribution specified by (5.2) is skew with a long right tail. As in the preceding example under H_0 the number of positive sample values (pluses) has a $B(10, 0.5)$ distribution. Using the number of plus signs as the test statistic this also has a binomial distribution under H_1 but now $p_1 = \Pr(X > 0)$. Thus, the original test is equivalent to testing H_0: $p = 0.5$ against H_1: $p = \Pr(X > 0)$ and the power study for the given sample size, and for determining the sample size required for any specified power again proceed as described in Section 4.3.3.

Procedure. It is easily shown by straightforward integration of the probability density function over $(0, \infty)$ that under H_1 $\Pr(X > 0) = 0.8244$ (Exercise 5.15). For testing H_0: $p = 0.5$ against H_1: $p = 0.8244$ when $\alpha = 0.05$ StatXact gives the exact power 0.4539. Discontinuities in possible P-values described in Example 5.5 are again relevant. If we replace $\alpha = 0.05$ by $\alpha = 0.011$ the test still has power 0.4539, whereas if we replace it by $\alpha = 0.055$ the power becomes 0.7499. The asymptotic power computation given by Minitab when $\alpha = 0.05$ is 0.6970. Again, here the asymptotic result is likely to be more satisfactory for finding the larger n needed to ensure a power of at least 0.90. The Minitab program in this case gives $n = 17$. Finer tuning using exact results in StatXact shows $n = 18$ as the minimum required sample size corresponding to an exact $P = 0.0481$ and giving an exact power 0.9194.

Conclusion. For a sample of $n = 10$ the power is 0.4539. This is again influenced by the large gap between the possible *P*-values 0.011 and 0.055. To give a power at least 0.90 for the one-tail test a sample of size 18 suffices.

Comments. 1. The sign test is particularly useful for skew distributions and in this example where we have a long right tail the test has moderate power for quite small samples for the alternatives considered. The combination of skewness and a long tail reduces the power of a *t*-test appreciably as a consequence of the breakdown of normality assumptions.

2. As indicated above the example is unrealistic from the practical viewpoint because we seldom meet exponential distributions where we are interested in simple shifts of the complete distribution that affect the mean or median alone. We are more often interested in whether a sample is from one or other of a set of exponential distributions where all positive values of X are possible but where the scale parameter λ is unknown. Changes in λ lead to changes in not only the mean and median, but also in the variance and higher moments. This is more complicated than a simple centrality shift. Nevertheless, the sign test is still valid for studying such changes providing we base our test on allocation of signs to observations depending upon whether they are above or below the median specified in H_0. Exercise 5.16 is intended for readers interested in exploring the power and sample-size relationship for this case.

Finding an adequate sample size to give a good chance of detecting median shifts of interest is a useful exercise but it has practical limitations. Examples 5.5 to 5.7 confirm that even if a test is distribution-free the power computations associated with specific hypotheses about a population mean or median are no longer distribution-free. Further, exact power calculations can only be made easily if we assume all observations are from a population with a specific distribution, whereas we have seen that one strength of the sign test is its applicability to situations where each observation may come from a different distribution, providing only that all these distributions have a common median.

There is also a certain irony in that we may choose a nonparametric or distribution-free test because we are uncertain about the distribution that provides our sample, yet computation of a sample size to guarantee a desired power requires knowledge of the distribution. Nevertheless reasonable approximations are often obtainable if we can make a few rational assumptions about symmetry, length of tails, or other prominent characteristics.

A further limitation to the usefulness of power and sample size calculations applies both to parametric and nonparametric inference. Using a test with good power for rejection of H_0 for differences of importance does not imply that if H_0 is rejected then the difference is important. A confidence interval is more informative about this. To illustrate this point, consider the situations in Examples 5.5 to 5.7. In each we assumed a zero median, θ, under H_0 and that only a value $\theta \geq 1$ represented a departure of interest. If we reject H_0, and calculate a 95 percent confidence interval for θ and this turns out to be (0.2, 1.3) there is considerable doubt about whether $\theta \geq 1$. The departure from H_0 may still be unimportant. We emphasise again the distinction made in Section 1.4.2 between statistical significance and practical importance.

Despite their limitations, approximate power computations are a useful first step in many experimental situations. They may indicate that a proposed experiment is too small to be likely to pick up interesting departures from H_0. More rarely, they may suggest we are squandering resources when a smaller experiment would give all the information we need.

Power and sample size studies provide target sizes for experiments likely to be of value, or if we have limited resources, indicate whether using all of these will be enough to achieve an experimenter's aims.

Even though the power of a simple test like the sign test is not distribution-free, power studies are still useful when we can make broad assumptions about the type of population distribution, e.g., is it symmetric and long tailed, or maybe more like a uniform distribution? Unless it is very small, a sample will often give some hint as to whether it comes from a distribution with one or more such characteristics. If there is evidence that the distribution is fairly symmetric, but has longer tails than those associated with the normal distribution, power studies that are optimal for the double-exponential distribution may provide good approximations.

Example 5.7 showed that for a particular skew distribution with one long tail the sign test performs well compared to the t-test. This is broadly true for many skew distributions with a long tail. Very often, a quick eye inspection of sample values aided perhaps by some EDA tool like a box and whisker plot will indicate that a sample appears to have come from, say, a fairly symmetric long-tail distribution not unlike the double exponential. We might estimate its variance by the sample variance and for a null hypothesis about the median, θ, H_0: $\theta = \theta_0$ and a specific alternative H_1: $\theta = \theta_1$ compute the value $p_1 = Pr(X > \theta_0)$ assuming the distribution of X is really a double-exponential with variance equal to the sample variance. We are effectively assuming that the correct p_1 is likely to differ little from that for the double-exponential distribution. Suppose that for the alternative hypothesis of interest $p_1 = 0.82$ for a double-exponential distribution, then we might be conservative and work out the power for this situation, and also that for a slightly lower p_1, say, $p_1 = 0.78$. It is likely that the true power will lie somewhere near the values given by these approximations.

Noether (1987a) gave a good asymptotic approximation for power of a sign test. Denoting the probability of a type I error by α and that of a type II error by β, so that the power is $1 - \beta$, then the sample size required to obtain that power with a sign test is

$$n = \frac{(z_\alpha + z_\beta)^2}{4(p_1 - 0.5)^2} \qquad (5.3)$$

where p_1 has the meaning assigned to it throughout this section and z_α is the value of a standard normal variable that must be exceeded corresponding to a P-value α when H_0 holds. For example, in a one-tail test with $\alpha = 0.05$, $z_\alpha = 1.645$. A corresponding meaning is attached to z_β.

Example 5.8

The problem. Apply the Noether approximate formula to compute the sample size n for the test considered in Example 5.6 to give a power of at least 0.9 when $\alpha = 0.05$ in a one-tail test.

Formulation and assumptions. The appropriate value of p_1 was shown in Example 5.6 to be $p_1 = 0.7534$ and it is easily verified that $z_\alpha = 1.645$ and $z_\beta = 1.282$.

Procedure. Substitution of the above values in the Noether formula gives $n \approx 33.4$.

Conclusion. Conservative rounding up suggests a sample size of 34 is appropriate.

Comments. 1. This asymptotic result is close to the size $n = 30$ found in Example 5.6. Bearing in mind that we may frequently use such calculations when there is uncertainty about the precise population distribution, calculations using the Noether formula will often be adequate in practice.

2. For the problem in Example 5.5 the Noether formula gives $n = 56.4$, close to the exact value $n = 58$, and in Example 5.7 it gives $n = 22.9$ (exact value $n = 18$), suggesting the approximation is reasonable even when observations are from a highly skewed distribution.

5.3.2 Power and Sample Size for the Wilcoxon Signed-Rank Test

We consider this topic in less detail than we did for the sign test. There are additional difficulties in performing exact tests. Approximate results are often all we can use in practice.

A simple property of the sign test under the alternatives H_1 considered in Section 5.3.1 is that no matter from what distribution we are sampling the test statistic, the number of plus signs, has a binomial distribution. For this statistic, for any given value of θ in H_1 all that differs between samples from different distributions is the value of the binomial parameter p_i. This simplicity is a feature of the statistic used, the number of plus signs.

Although the corresponding statistic for the Wilcoxon signed-rank test, the sum of the positive signed ranks, has a relatively simple symmetric distribution under H_0 it has generally got a rather intractable distribution under any H_1 for all but a few simple population distributions. Exact power calculations have only been made for small sample sizes and for a limited number of population distributions such as the normal and a few t-distributions with small numbers of degrees of freedom, see e.g., Klotz (1963) and Arnold (1965). Even approximations are limited in their usefulness. An approximate formula for power for a given sample size against alternatives close to that in H_0 is discussed in detail by Lehmann (1975, Section 4.2) and an example of its use is given by Hollander and Wolfe (1999, Section 3.1). The result is applicable only if all observations are from the same distribution. In addition, one needs to know the value of the population frequency function at the median or mean specified in H_0 (which, without loss of generality, may be set at $\theta = 0$) and also the value at this median of the frequency function for the sum of two independent variables having this same distribution. Except for a few distributions such as the normal, where the sum also has a normal distribution, computation of the latter requires a good understanding of distribution theory and calculus.

The reason the distribution of the sum of two independent observations comes into the calculation is closely allied to the relevance of the Walsh averages in test and estimation procedures associated with signed ranks. This sum for any two sample values has the same sign as the corresponding Walsh average (which is simply that sum divided by 2) and under H_0: $\theta = 0$ the sign is equally likely to be positive or negative, i.e., $\Pr(x_i + x_j > 0) = 0.5$, whereas under H_1: $\theta = \theta_1$, if θ_1 is positive then $Pr(x_i + x_j > 0) = p_1$ where $p_1 > 0.5$. The value of p_1 depends upon the population distribution and is not always easy to calculate.

Although the signed-rank statistic no longer has a binomial distribution, the approximation due to Noether given in (5.3) may still be used to estimate the sample size having a given power. However, for a given θ_1 this may be difficult to calculate except for some simple distributions. It may also be sensitive to an incorrect choice of population distribution, again illustrating the difficulty due to uncertainty about the population distribution when making power calculations for nonparametric methods.

Example 5.9

The problem. Using an approximation based on (5.3) estimate the sample size needed to guarantee a power of 0.80 using a Wilcoxon test when sampling from a normal $N(\theta, 4)$ distribution for a one-tail test for H_0: $\theta = 0$ against H_1: $\theta = 1$ with probability of a type I error not exceeding $\alpha = 0.05$.

Formulation and assumptions. We require $p_1 = \Pr(X_1 + X_2 > 0)$ where X_1, X_2 are independently $N(1, 4)$. This, together with appropriate values of z_α, z_β are substituted in (5.3) to estimate n.

Procedure. Using normal distribution theory we know that $U = X_1 + X_2$ is distributed $N(2, 8)$ under H_1. Thus $Z = (U - 2)/(2\sqrt{2})$ has a standard normal distribution. It follows that $\Pr(U > 0)$ implies

$$p_1 = \Pr(Z > -1/\sqrt{2}) = \Pr(Z > -0.7071) = 0.7601.$$

Clearly $z_\alpha = 1.645$ when $\alpha = 0.05$ and for power 0.8 we have $\beta = 0.2$ and $z_\beta = 0.842$. Substituting these values in (5.3) gives $n \approx 23$.

Conclusion. A sample size of 23 should nearly meet requirements.

Comments. Minitab, like many other packages, provides a program to determine the sample size to give required power when the optimum normal theory t-test is used in these circumstances and indicates a sample size $n = 25$. This suggests the asymptotic test size of 23 for the Wilcoxon test may be an underestimate since the Pitman efficiency of the Wilcoxon test relative to the optimal test in this case is $3/\pi$, which is slightly less than 1. However, the result is of the right order of magnitude and Pitman efficiency is at best a rough guide for small samples.

5.4 Fields of Application

In most applications, if a numerical value of the difference for each matched pair is available and these do not appear markedly skew, the Wilcoxon test (or an analogous test using normal scores) is likely to be appropriate. The matched pairs t-test is appropriate if the differences $d_i = y_i - x_i$ are approximately normally distributed; sometimes this may be the case even when the distributions of X, Y are each far from normal. If there is evidence of skewness in the differences a sign test is preferable.

Laboratory Instrument Calibration

Two different brands of instrument reputedly measure the same thing (e.g., blood pressure, hormone level, sugar content of urine, bacterial content of sputum), but each is subject to some error. Samples from, say, each of 15 patients might be divided into two subsamples, the first being analysed with one brand of instrument, the second with the other. A Wilcoxon test is appropriate to test for any systematic difference between instruments. When purporting to measure the same thing a systematic difference from the true values in means or medians is often described as mean or median *bias*. The term bias alone is usually taken to imply *mean bias*.

Biology

The heartbeat rate of rabbits might be measured before and after they are fed a hormone-rich diet. The Wilcoxon test is appropriate to investigate a shift in mean. "Before" and "after" measurements are common in medical and biological contexts, including experiments on drugs and other stimuli, which may be either physical or biological (e.g., a rabbit's blood pressure may be measured when on its own and again after it has shared a cage for half an hour with a rabbit of the opposite sex). Confidence intervals for the mean difference are useful both as an indication of the precision of the experiment (Section 1.4.2) and to help in reaching a decision as to whether any statistically significant difference is of practical importance.

Occupational Medicine

An instrument called a Vitalograph is used to measure lung capacity. Readings might be taken on workers involved in the production of a washing powder at the beginning and end of a shift to study the effect on lung capacity of any inhalation of powder.

Agriculture

In a pest control experiment each of 10 plots may contain 40 lettuce plants. Each plot is divided into two halves: one half chosen at random is sprayed with one insecticide, the second with another. Differences in numbers of uninfested plants in each plot can be used in a Wilcoxon test to compare effects of insecticides. Incidentally, pest control experiments are a situation where a normality assumption is often suspect.

Psychology

For sets of identical twins, it is known for each pair which was the first-born. For each individual in a pair the times to carry out a manual task appropriate to their age at the time of the test are observed to see if there is any indication that the first-born tends to be quicker. The choice may lie between a Wilcoxon test and a *t*-test. Confidence intervals for the mean difference will indicate the precision with which any difference is measured, and whether it may be of practical importance.

Road Safety

The impact of the use of mobile telephones on drivers' reaction times in dangerous situations (e.g., those requiring an emergency stop) may be studied using equipment that simulates driving conditions on the road. Drivers could be monitored first without a mobile phone, and then holding a conversation using a hands-free telephone throughout the journey. This is a *response to stimulus* situation of the type mentioned above under the heading *Biology*.

Space Research

Potential astronauts may have the enzyme content of their saliva determined before and after they are subjected to a zero gravitational field in a simulator. Such biochemical evidence is important in determining physiological reactions to space travel.

Education

To decide which of two examination questions is perceived by students to be the harder, both questions could be included in a test paper in which candidates are free to choose neither, one or both of the two questions. Records are taken of the numbers who complete both, neither, only the first, only the second. Numbers in the latter two categories can be used in a McNemar sign test for evidence of unequal perceived difficulty.

Social Policy

An association of government employees want evidence to support their case that salaries in the public sector are generally below those paid for equivalent work in the private sector. They obtain data for the average salaries paid in each sector for each of a number, n, of employment categories matched with respect to working conditions, responsibility, security of employment, etc. They record differences and test to assess whether the evidence supports their case. This approach is likely to be too simplistic in a real-world situation because it may be found that for some equivalent jobs (e.g., for the more skilled categories) the private sector pay rates are higher, while for other equivalents (e.g., unskilled work) the opposite is true. In cases like this certain groups of categories might need to be analysed separately. Ethical considerations may come into play here, especially if the grouping is done after the data are available. Both employer's and employee's organizations are often adept at massaging figures to support their favoured, and often opposing, viewpoint.

5.5 Summary

In Section 5.1.1 we discuss some reasons for pairing in a broad context. When there is independence between pairs (a common assumption in most examples in this chapter) the matched pair sample tests for centrality differences reduce to the analogous single-sample tests considered in Chapter 3. See, in particular, the sign test (Section 3.4), the Wilcoxon signed-rank test (Section 3.3), normal, or modified van der Waerden scores (Section 3.5). General tests for the distributions of the paired differences include the Kolmogorov test (Section 4.2.1) and Lilliefors' test and the Shapiro-Wilk test for normality (Section 4.2.3).

McNemar's test (Section 5.2) is relevant to paired observations to assess changes in attitude or for assessments of relative difficulty. It is equivalent to the sign test (Section 3.4).

Power and sample size calculations for the single sample or matched pair differences are in general not distribution-free. Reasonably good approximations depend upon assumptions about the population distribution. Some power computations are relatively easy for the sign test because the statistic still has a binomial distribution under the alternative hypothesis, whereas the distribution of the signed-rank statistic often proves intractable under alternative hypotheses. Asymptotic approximations may be useful in determining approximate sample sizes needed to attain a given power.

5.6 Exercises

5.1 The blood pressures of 11 patients are measured before and after administration of a drug. The differences in systolic blood pressure (*pressure before–pressure after*) for each patient are:

$$7 \quad 5 \quad 12 \quad -3 \quad -5 \quad 2 \quad 14 \quad 18 \quad 19 \quad 21 \quad -1$$

Use an appropriate nonparametric test to see if the sample (assumed random) provides evidence that the drug causes a significant change in blood pressure.

*5.2 Samples of cream from each of 10 dairies (A to J) are each divided into two portions. One portion from each is sent to Laboratory I, the other to Laboratory II, for bacterium counts. The counts (thousands bacteria ml^{-1}) are:

Dairy	A	B	C	D	E	F	G	H	I	J
Lab I	11.7	12.1	13.3	15.1	15.9	15.3	11.9	16.2	15.1	13.6
Lab II	10.9	11.9	13.4	15.4	14.8	14.8	12.3	15.0	14.2	13.1

Use the Wilcoxon signed-rank test to assess the evidence for any consistent difference between laboratories for subsamples from the same dairy. Obtain also nominal 95 and 99 percent confidence intervals for the mean difference and compare these with the intervals using the optimal method when normality is assumed.

5.3 A hormone is added to one of otherwise identical diets given to each of 40 pairs of twin lambs. Growth differences over a 3-week period are recorded for each pair and signed ranks are allocated to the 40 differences. The lower rank sum was $S_1 = 238$. There was only one rank tie. Investigate the evidence that the hormone may be affecting (increasing or decreasing) growth rate.

*5.4 A psychologist interviews both father and mother of each of 17 unrelated children with learning difficulties, asking each individually a series of questions designed to test how well they understand the problems their child is likely to face in adult life. The psychologist records whether the father (F) or mother (M) shows the better understanding of these potential problems. For the 17 families the findings are

$$F \quad M \quad M \quad F \quad F \quad F \quad F \quad F \quad F \quad F \quad M \quad F \quad F \quad F \quad M \quad F \quad F$$

Is the psychologist justified in concluding that there is a tendency for fathers to show better understanding?

5.5 For each of nine matched pairs of students, one student is allocated to a series of lectures and the other to appropriate computer assisted learning (CAL) material. At the end of the course the students sit the same examination paper. The marks achieved (out of 100) are:

Pair	1	2	3	4	5	6	7	8	9
CAL	50	56	51	46	88	79	81	95	73
Lectures	25	58	65	38	91	32	31	13	49

Analyse these results by what you consider the most appropriate parametric or nonparametric methods to determine whether or not they provide acceptable evidence that CAL material leads to better examination results.

*5.6 One hundred general practitioners attend a health promotion workshop. At the start of the workshop they are asked to indicate whether they are in favour of routinely asking patients about alcohol consumption. They are then shown a video on the health and social problems caused by the excessive consumption of alcoholic drinks. The video is followed by discussion in small groups. After the video and discussion they are asked the original question again. Do the results given below indicate a significant change in attitudes as a result of the video and group discussion?

| | | Before video and discussion | |
		In favour	Against
After video	In favour	41	27
and discussion	Against	16	58

5.7 A canned-soup manufacturer is experimenting with a new-formula tomato soup. A tasting panel of 70 each taste samples of the current product and the new one (without being told which is which). Of the 70, 32 prefer the new-formula product, 25 the current product, and the remainder cannot distinguish between the two. Is there enough evidence to reject the hypothesis that consumer preference is equally divided? Do the data support a claim that as many as 75 percent of those who have a preference may prefer the new formula?

5.8 To produce high quality steel one of two hardening agents A, B may be added to the molten metal. Hardness of steel varies from batch to batch, so to test the two agents 40 batches are each subdivided into two portions. For each batch agent A is added to one portion, agent B to the other. To compare hardness, sharpened specimens for each pair are used to make scratches on each other; that making the deeper scratch on the other is the harder specimen. For the 40 pairs, that with agent B is adjudged harder in 24 cases and that with agent A in 16. Is this sufficient evidence to reject the hypothesis of equal hardness?

*5.9 For a subsample of 10 pairs from the steel batches in Exercise 5.8 a more expensive test is used to produce a hardness index. The higher the value of the index, the harder the steel. The indices recorded were:

Batch no.	1	2	3	4	5	6	7	8	9	10
Additive A	22	26	29	22	31	34	31	20	33	34
Additive B	27	25	31	27	29	41	32	27	32	34

Use an appropriate test to determine whether these data support the conclusion reached in Exercise 5.8.

5.10 On the day of the third round of the Open Golf Championship in 1987 before play started a television commentator said that conditions were such that the average scores of players were likely to be at least three higher than those for the second round. For a random sample of 10 of the 77 players participating in both rounds the scores were:

Player	A	B	C	D	E	F	G	H	I	J
Round 2	73	73	74	66	71	73	68	72	73	72
Round 3	72	79	79	77	83	78	70	78	78	77

Do these data support the commentator's claim? Consider carefully whether a one- or two-tail test is appropriate.

*__5.11__   Scott, Smith and Jones (1977) give a table of estimates of the percentages of UK electors predicted to vote Conservative by two opinion polling organizations, A and B, in each month in the years 1965-70. For a random sample of 15 months during that period the paired percentages were:

A   43.5 51.2 46.8 55.5 45.5 42.0 36.0 49.8 42.5 50.8 36.6 47.6 41.9 48.4 53.5

B   45.5 44.5 45.0 54.5 49.5 43.5 41.0 53.0 48.0 52.5 41.0 47.5 42.5 45.0 52.5

Do these results indicate a significant tendency for one of the organizations to return higher percentages than the other? Obtain an appropriate 95 percent confidence interval for any mean or median difference between predictions during the period covered.

*__5.12__   Pearson and Sprent (1968) gave data for hearing loss (in decibels below prescribed norms) at various frequencies. The data below show these losses for 10 individuals aged between 46 and 54 at frequencies of 0.125 and 0.25 kc s^{-1}. A negative loss indicates hearing above the norm. Is there an indication of a different loss at the two frequencies?

Subject	A	B	C	D	E	F	G	H	I	J
0.125 kc s^{-1}	2.5	−7.5	11.25	7.5	10.0	5.0	7.5	2.5	5.0	8.75
0.25 kc s^{-1}	2.5	−5.0	6.25	6.25	7.5	3.75	1.25	0.0	2.5	5.0

5.13 Apply a normal scores test to the data in Example 5.2.

*__5.14__   Confirm the value given in Example 5.6 for $\Pr(X > 0|\theta = 1)$ for the distribution given in (5.1). What is the corresponding probability conditional upon $\theta = 2$?

5.15 Confirm the value given in Example 5.7 for $\Pr(X > 0|\theta = 1)$ for the distribution given in (5.2). What is the corresponding probability conditional upon $\theta = 1.2$?

*5.16 Determine the sample size needed to have power at least 0.80 for the sign test that the median θ is $H_0: \theta = 1$ against the alternative $H_1: \theta = 2$ with the one-tail $P = 0.05$ if the observations are known to be a random sample from an exponential distribution with frequency function $f(x) = \lambda e^{-\lambda x}$, $x \geq 0$. (Hint: You will need to find values of λ that give medians corresponding to those specified in the null and alternative hypotheses.)

5.17 A dentist working in the Accident and Emergency centre of a dental hospital investigates whether boys are more likely than girls to present with accidental damage to their teeth. Starting with the null hypothesis that boys and girls arrive in equal proportions, use the Noether approximation to estimate the sample size required for a study using a one-tail sign test if the alternative hypothesis is that the proportion of boys is 0.8. Assume that a power of 0.9 is required when a significance level $\alpha = 0.05$ is chosen.

A colleague suggests that girls might in fact outnumber boys presenting, making a two-tail test more appropriate. What is the impact of this on the sample size requirement?

CHAPTER 6

METHODS FOR TWO INDEPENDENT SAMPLES

6.1 Centrality Tests and Estimates

6.1.1 Extension from Single Samples

We often have two independent random samples. That is, samples in which members of the first sample, or group, are independent of each other and are independent of those in the second. We want to make inferences about the two populations freom which the samples are drawn. We denote the observations in the samples by $x_1, x_2, \ldots, x_m$ and $y_1, y_2, \ldots, y_n$, where, for convenience and without loss of generality, we assume $n \geq m$, i.e., that the second sample is at least as large as the first.

In Chapters 3 to 5 we saw that several distribution-free tests and estimation procedures differed only in the scores assigned, e.g., ranks, signs, van der Waerden scores. Relevant assumptions and practical computational matters governed the choice of an appropriate procedure. These considerations extend, with modifications and additions, to the two-sample situation.

We consider first centrality tests that broadly correspond to those met in Chapter 3.

6.1.2 Wilcoxon–Mann–Whitney Test

For two independent samples, the analogue of the one sample Wilcoxon signed-rank test is the *Wilcoxon rank-sum test* proposed by Wilcoxon (1945). An equivalent test, widely referred to particularly, but not only, in the medical literature, as the *Mann–Whitney U test* was developed independently by Mann and Whitney (1947). It is convenient, though not universal practice, to refer to the tests jointly as the *Wilcoxon–Mann–Whitney test*, or for brevity, as the *WMW test*.

The data in Example 2.1 can be viewed as two independent samples. Patients who received the new drug form the first sample, and those who did not form the second. Information recorded on the condition of each patient is already in the form of ranks, so the Wilcoxon–Mann–Whitney approach can be applied immediately. This was effectively what we did in that example.

We consider first the Wilcoxon formulation of the test. The two samples are combined and the data are ranked overall. The original samples are then separated, with each rank being attached to the corresponding observation.

The usual null hypothesis is that the two samples are from identical populations. A common alternative hypothesis is that the population distributions differ only in the mean or median. As we indicated in Example 2.1, if both samples come from the same population (which may be of any continuous form and need not be symmetric) we expect a mix of low, medium and high ranks in each sample. Under the alternative hypothesis we expect lower ranks to dominate in one population and higher ranks in the other. Such a shift in centrality is often referred to as an *additive* treatment effect, i.e., there is a constant difference between two treatments.

The sum of the ranks in the first sample, S_m, can be used to determine the strength of evidence against H_0. The sum S_n for the second sample is an equivalent alternative statistic. In Example 2.1 the smaller group consists of the four patients receiving the new treatment. If their ranks are 1, 2, 3 and 5 (say), low ranks predominate and the rank-sum of 11 is small. If the ranks are 6, 7, 8 and 9, high ranks predominate and $S_m = 30$ is the maximum possible. As we indicated in that example, ranks of 2, 3, 6 and 8 imply no marked dominance of high or low ranks in either group, evidence favouring the null hypothesis. The rank sum is then 19. Values of the rank sum close to the minimum possible (10) or maximum possible (30) provide strong evidence against the null hypothesis. Intermediate values of S_m favour H_0.

The test may also be used when samples are from two distributions with identical cumulative distribution functions under H_0, but under H_1, one cumulative distribution curve (see *Comment* 1 on Example 4.2) lies beneath the other apart from some points where the curves touch. A moment's reflection shows that then under H_1 low or high ranks should dominate in one sample, as opposed to a fairly even distribution of ranks under H_0. Such H_1 are sometimes referred to as *dominance* alternatives. Given the permutation distribution of rank sums under H_0, P-values may be determined in the way described for a particular case in Example 2.1.

Example 6.1

The problem. For some models of pocket calculator the trigonometric function values are obtained by entering the number before pressing the function button (Type A models, say). Other models require the function to be selected before the number is entered (Type B models). A mathematics teacher wished to determine whether a particular Type A model allows calculations to be performed with greater speed compared with a certain Type B model. A class of 21 pupils was randomly divided into groups A (using the Type A model), and B (using the Type B model), with 10 and 11 pupils in the respective groups. The pupils were asked to carry out the same set of trigonometric calculations. The total times in minutes for each member of each group to complete the calculations were

| Group A | 23 | 18 | 17 | 25 | 22 | 19 | 31 | 26 | 29 | 33 | |
| Group B | 21 | 28 | 32 | 30 | 41 | 24 | 35 | 34 | 27 | 39 | 36 |

Do the data indicate that one model of calculator is superior (i.e., leads to more rapid computations)?

Formulation and assumptions. A two-tail test is appropriate. We require the sum of ranks associated with the smaller sample, Group A, in a joint ranking of all data. Alternatively the rank sum for Group B could be used. The P-value associated with this sum lets us assess the strength of evidence against H_0: *the population medians are identical*, where the alternative is that one sample is from a population with greater median. We prefer to think in terms of medians rather than means because there is no need to make an assumption of symmetry. The test will also be valid for a dominance alternative, i.e., for a tendency for the computations to be done more quickly with one model of calculator, although that time difference may vary appreciably between pupils.

Procedure. The sample sizes are $m = 10$ and $n = 11$. Most software programs for the Wilcoxon test will calculate the ranks automatically, but to obtain them manually it helps to arrange data in ascending order within each sample and then allocate ranks. We leave it as an exercise for the reader to show that if this is done the ranks assigned to Group A are 1, 2, 3, 5, 6, 8, 9, 12, 14, 16 with a rank sum of $S_m = 76$. The rank sum for Group B is $S_n = 155$.

If software to give exact P-values is not available some relevant values for S_m are given in many tables. For this example we used the program in StatXact, which indicates that for a two-tail test the relevant exact $P = 0.0159$. The output also confirms the value of S_m given above.

Conclusion. The low P-value provides fairly strong evidence against H_0. On average, pupils appear to perform calculations more speedily with the Type A model.

Comments. 1. Tables for the Wilcoxon–Mann–Whitney statistic are given by Gibbons and Chakraborti (2004, Table J) and by Neave (1981, p. 30). Conover (1999, Table A7) and others give various quantiles for S_m, S_n. Actual, rather than nominal, significance levels may be obtained from computer programs giving the exact permutation distribution.

2. This study is a comparison of two particular models of calculator. It would be unreasonable for the teacher to recommend that pupils purchase any Type A calculator on the basis of just this study. One reason for this is that calculators that are particularly useful for operating with trigonometric functions may not perform so well with other problems such as calculating a mean or standard deviation. Also, there may be calculators not included in the study that perform better than Type A for trigonometric calculations.

3. Suppose that this investigation had been conducted in a mathematics lesson of 40 minutes. The fifth pupil in Group B would then have been unable to complete all of the calculations and this observation would have been censored. In the ranking of the data, however, since this observation would be the only one with a time in excess of 40 minutes a rank of 21 would still be given. A t-test could not then have been validly used since an exact value is required for each observation.

Computational aspects. StatXact and Testimate give exact P-values corresponding to observed S providing m, n are not too large. The procedure *wilcox.test* in R also calculates exact P-values. Many general statistical packages compute S_m but leave the user to consult tables or give an asymptotic result which may be unsatisfactory if, for example, one sample is large but the other small, or if there are many tied ranks.

6.1.3 The Mann–Whitney Formulation

A statistic U, which is a function of the rank sum S, can be calculated for either group in order to determine the strength of the evidence against the null hypothesis. For the first of the two samples this statistic is given by $U_m = S_m - m(m+1)/2$. The equivalent statistic for the (perhaps larger) sample is $U_n = S_n - n(n+1)/2$. We only need compute one of S_m or S_n, for the sum of all the ranks from 1 to $m+n$ is $(m+n)(m+n+1)/2 = S_m + S_n$. Using the relations $U_m = S_m - m(m+1)/2$ and $U_n = S_n - n(n+1)/2$ one easily deduces that each has minimum value zero, and that

$$U_m = mn - U_n, \qquad (6.1)$$

so that again only one of U_m, U_n need be computed. Either may be used in a test, although U_m is generally given in tables.

In the Mann–Whitney approach, either U_m or U_n is calculated directly. To obtain either we count the number of observations in one sample exceeding each member of the other sample. Ranks are not needed, and the procedure eases calculation if no computer program is available for the test. It also forms the basis for determining a confidence interval for a centrality difference.

Example 6.2

The problem. Calculate a test statistic for the data in Example 6.1 using the Mann–Whitney approach.

Formulation and assumptions. We inspect the observations in each sample; they need not be ordered, but counting is easier when they are. It is visually easier to count the number of times each observation in Group A is exceeded by an observation in Group B. This gives U_n, from which U_m can be determined using (6.1).

Procedure. The data from Example 6.1, arranged for convenience in ascending order in each group, are:

Group A	17	18	19	22	23	25	26	29	31	33	
Group B	21	24	27	28	30	32	34	35	36	39	41

The first observation, 17, in Group A is exceeded by all 11 observations in Group B. So also are the observations 18, 19. The observation 22 is exceeded by 10 observations in Group B. Proceeding in this way we find the numbers of observations in Group B exceeding each observation in Group A and then add these, viz.,

$$U_n = 11 + 11 + 11 + 10 + 10 + 9 + 9 + 7 + 6 + 5 = 89.$$

Using (6.1) gives $U_m = 110 - 89 = 21$, easily shown to be consistent with the value of S_m found in Example 6.1.

Conclusion. As in Example 6.1.

Comment. Equivalence of the Wilcoxon and Mann–Whitney formulations is general.

Computational aspects. StatXact and Testimate give exact tail probabilities corresponding to the value of U (in this case $P = 0.0159$). As with the statistic S, many general statistical packages compute U but leave the user to consult tables or else give an asymptotic result based on the normal distribution (see Section 6.1.5).

For this example, Stata or StatXact give an asymptotic $P = 0.0167$, which leads to the same conclusion as the exact test. In other situations the asymptotic result may be unsatisfactory. Many tables only give values of the statistic that correspond to nominal conventional significance levels, rather than exact P-values, but most explain what is given and how to use the tables. For instance, Neave (1981) indicates that if either U_m or U_n do not exceed 26, the two-tailed P-value for Example 6.2 is less than 0.05.

6.1.4 Wilcoxon–Mann–Whitney Confidence Intervals

The Mann–Whitney statistic compares all differences $d_{ij} = x_i - y_j$ between sample values. In computing U_m we allocate a score of 1 if d_{ij} is positive and zero if d_{ij} is negative. To calculate U_n we reverse these scores, i.e., score 1 if d_{ij} is negative and zero if d_{ij} is positive. To calculate a confidence interval for the difference in centrality we need the actual values of some or all d_{ij}. If c is the value of the Mann–Whitney statistic U that we would regard as just providing sufficient evidence to indicate significance at the 100α percent significance level, the $100(1 - \alpha)$ percent confidence limits are given by the $c + 1$ smallest and $c + 1$ largest d_{ij}.

The reasoning here is not unlike that used in establishing a confidence interval using Walsh averages in the one-sample situation based on the Wilcoxon signed-rank approach.

Example 6.3

The problem. Obtain nominal 95 percent confidence limits for a population median shift based on the WMW method using the data in Example 6.1.

Formulation and assumptions. If computer software that calculates a confidence interval directly is not available tables such as those in Neave (1981) indicate that for sample sizes of 10, 11 the critical value is $U_m = 26$ for significance at a nominal 5 percent level. Denoting the sample values by $x_1, x_2, \ldots, x_{10}$ and $y_1, y_2, \ldots, y_{11}$ we require the 27 largest and smallest $d_{ij} = x_i - y_j$. It is not essential to compute all d_{ij}, but for completeness we give these in Table 6.1. These need not be computed manually if suitable software is available either to determine the confidence limits directly, or even if only to compute the required differences.

Procedure. It is easier to compute differences manually if the sample values are ordered. We write those for the first sample in the top row and those for the second sample in the first (left) column. The entries in the body of Table 6.1 are the differences between the data entries at the top of that column and at the left of that row, e.g., the first entry, -4, is $17 - 21 = -4$. Note that the difference between each entry in any pair of rows is a constant equal to the difference between the corresponding entries in the left-hand data column; there is an analogous constant difference between entries in pairs of columns. The largest entries appear in the top right of the table and entries decrease and eventually change sign as we move toward the bottom left.

Table 6.1 *Paired differences for times taken to complete calculations.*

	17	18	19	22	23	25	26	29	31	33
21	−4	−3	−2	1	2	4	5	8	10	12
24	−7	−6	5	−2	−1	1	2	5	7	9
27	−10	−9	−8	−5	−4	−2	−1	2	4	6
28	−11	−10	−9	−6	−5	−3	−2	1	3	5
30	−13	−12	−11	−8	−7	−5	−4	−1	1	3
32	−15	−14	−13	−10	−9	−7	−6	−3	−1	1
34	−17	−16	−15	−12	−11	−9	−8	−5	−3	−1
35	−18	−17	−16	−13	−12	−10	−9	−6	−4	−2
36	−19	−18	−17	−14	−13	−11	−10	−7	−5	−3
39	−22	−21	−20	−17	−16	−14	−13	−10	−8	−6
41	−24	−23	−22	−19	−18	−16	−15	−12	−10	−8

A count shows 21 positive and 89 negative values whence $U_n = 89$ and $U_m = 21$ as found in Example 6.2. The critical value is $U_m = 26$ for significance at least at a 5 percent level, so the lower limit for a 95 percent confidence interval for the median difference $\theta_1 - \theta_2$ is obtained by eliminating the 26 largest negative differences; the next largest negative difference is the required lower limit. Using Table 6.1 gives this lower limit as −13. Similarly, elimination of the 26 largest differences gives the upper limit −2. StatXact, SAS and some other packages provides a program to perform these computations given only the original data and quickly gives these limits. Minitab will also compute the relevant differences.

Conclusion. A nominal 95 percent confidence interval for the difference $\theta_1 - \theta_2$ is $(-13, -2)$ or equivalently, for $\theta_2 - \theta_1$ the interval is $(2, 13)$.

Comments. 1. If we assume normality the 95 percent confidence interval based on the relevant t-distribution is $(-12.6, -1.9)$. The close agreement between this and the Wilcoxon interval is heartening because there is little to indicate serious nonnormality or a difference between population variances in these data apart from a small indication that the Group B data might have a slightly greater spread.

2. Confidence intervals at other levels can be obtained. For example, a 99 percent interval is obtained by rejecting the 18 extreme differences and using Table 6.1 is easily seen to be $(-16, 1)$, a result confirmed by StatXact or SAS or Minitab.

3. The computational process is reminiscent of that with Walsh averages given in Section 3.3.2.

Computational aspects. If one wishes to explore further the exact confidence level one may examine results of hypothesis testing situations at or near the end points of the interval, much as we did for the Wilcoxon signed-rank test. This is achieved in the case of a 95 percent confidence interval by an appropriate addition or subtraction to all observations in one sample and then obtaining exact P-values for the corresponding test of zero median difference. While StatXact computes exact intervals some software packages may use asymptotic approximations based on the asymptotic theory we give for hypothesis testing in Section 6.1.5. If both samples are moderate or large in size asymptotic approximations are usually quite good. They may be unreliable if the samples differ greatly in size, especially when one is small.

The appropriate point estimator of the median difference based on the WMW procedure used above, known as the *Hodges–Lehmann estimator* is the median of the differences in Table 6.1. It is easily verified that in the above example this is -7.5.

6.1.5 Asymptotic approximation

For the WMW test it is not difficult to show that the test statistic S_m has mean $m(m+n+1)/2$ and variance $mn(m+n+1)12$, whence it follows that for large sample sizes

$$Z = \frac{S_m - m(m+n+1)/2}{\sqrt{mn(m+n+1)/12}}.$$ (6.2)

has approximately a standard normal distribution. For large m, n a continuity correction may have little effect, but if one is used the appropriate correction is to add 0.5 to S_m if $Z < 0$ and to subtract 0.5 if $Z > 0$.

For the Mann–Whitney formulation if the statistic U_m is used (6.2) is replaced by

$$Z = \frac{U_m - mn/2}{\sqrt{mn(m+n+1)/12}}.$$ (6.3)

Modifications to the denominators in (6.2) and (6.3) are required for tied data. These are discussed in the next section.

6.1.6 Ties

We use mid-ranks for ties as we did for the Wilcoxon signed-rank test (Section 3.3.4). If software to compute exact P-values is not available and there are only a few ties, basing significance tests on the appropriate critical values for the no-tie case is unlikely to be seriously misleading. If m, n are both reasonably large (say, 15 or greater), a normal approximation we develop below may be used with reasonable confidence. Adjustment is essential only when there are a moderate to large number of ties. Example 6.4 illustrates a situation where ties dominate.

In the Mann–Whitney formulation, if an observation in the second sample equals an observation in the first sample it is scored as 1/2 in counting the number of observations in the second sample exceeding that observation in the first.

For small to medium-sized samples ties present no difficulty if a computer program is available to generate exact probabilities for the WMW test based on the appropriate permutation distribution.

Example 6.4

The problem. We consider a set of data from a second experiment similar to that in Examples 6.1 and 6.2 but where there are now some ties in the times taken by different participants. For convenience the data are given in ascending order but this is not essential, especially if suitable software is used.

Group A   16   18   19   22   22   25   28   28   28   31   33
Group B   22   23   25   27   27   28   30   32   33   35   36   38   38

Do the data indicate that one model of calculator is superior (i.e., leads to more rapid computations)?

Formulation and assumptions. We use the WMW test with mid-ranks for ties.

Procedure. We have $m = 11$ and $n = 13$. Computer programs usually assign ranks or mid-ranks automatically, but this is easily done manually since the data are ordered. The reader should verify that these are

Gp A   1   2   3   5   5   8.5   13.5   13.5   13.5   17   19.5
Gp B   5   7   8.5   10.5   10.5   13.5   16   18   19.5   21   22   23.5   23.5

Unless we need S_m specifically and have no software to compute it there is no need to allocate ranks. It is easier to obtain U_n by simply counting for each first sample value the number of observations that exceed it in the second sample (scoring one half for between-sample ties). These are summed. For example, for each value 22 in Group A there is one tied value, scored as $1/2$, and 12 values exceeding 22 in Group B giving a contribution of 12.5 to U_n. Proceeding in this way we find

$$U_n = 13 + 13 + 13 + 12.5 + 12.5 + 10.5 + 7.5 + 7.5 + 7.5 + 6 + 4.5 = 107.5$$

From (6.1), $U_m = 11 \times 13 - 107.5 = 35.5$. StatXact or other software covering the WMW test, confirm this value for U_m, usually requiring only the original data to do so. For a two-tail test the exact $P = 0.0359$.

Conclusion. There is some evidence against H_0 and the results suggest computations are completed more rapidly with the model tested by Group A.

Comments. 1. Tables such as those in Neave (1981) indicate that values of $U_m \leq 37$ imply $P < 0.05$ for a two-tail test. This one example suggests that a few ties do not seriously upset conclusions based on no-tie critical values for moderate sample sizes.

2. The situation using conventional no-tie tables is less satisfactory when there are many ties, or for ties in unbalanced samples. For example, if we had a sample of 3 with values 1, 2, 2, and a sample of 13 with values 1, 1, 4, 5, 5, 5, 7, 8, 9, 9, 9, 9, 10 it is easy to show that $U_m = 5$. In a no-tie situation two-tail test $\Pr(U_m \leq 5) = 0.0571$ whereas the exact permutation test allowing for ties gives $\Pr(U_m \leq 5) = 0.0464$ as the true P-value for this specific tie pattern. Although the strength of evidence against the null hypothesis is not markedly different, adhering to a rigid 5 percent significance level would lead to differing conclusions.

3. As we demonstrated in Example 3.5, if ties result from rounding, and greater accuracy allows us to break ties, the way the ties break may alter appreciably conclusions regarding significance.

Computational aspects. Software to determine exact permutation probabilities associated with the test statistics U or S is particularly valuable when there are ties.

StatXact and Testimate allow this. Some general programs take ties into account by using mid-ranks but provide only an asymptotic test (which may be unreliable for small $m + n$, or when one of m, n, is small). Alternatively, one may resort to tables appropriate to the no-tie situation, but nominal significant levels can no longer be guaranteed.

A common tie situation is one where we are not given precise measurements, but only grouped data. For example, instead of the complete sample values in Example 6.4 we may be given only numbers of participants taking between 10 and 19 minutes, 20 and 29 minutes, 30 and 39 minutes. We may still calculate the U or S statistics making allowances for ties, but it may now be misleading to use tabulated critical values for these statistics.

Example 6.5

The problem. Instead of the data in Example 6.4 suppose we are given only the numbers taking 10–19, 20–29, 30–39 minutes leading to the data given below. Perform the test in Example 6.4 using this reduced information.

No of minutes	10–19	20–29	30–39
Group A	3	6	2
Group B	0	6	7

Formulation and assumptions. We carry out a WMW test based on mid-ranks using, if available, a program giving exact permutation probabilities.

Procedure. If no suitable program is available, calculating the required value of U_n is reasonably straightforward but needs care. For each of the three ties in 10–19 for Group A there are $6 + 7 = 13$ greater values in Group B. Thus, between them these three ties contribute a total of $3 \times 13 = 39$ to U_n. Similarly, each of the six Group A ties in 20–29 corresponds to six ties and seven greater values in Group B. Thus each contributes a score of $(1/2 \times 6) + 7 = 10$, so that together they contribute a score of $6 \times 10 = 60$. Finally, by a similar argument the two group A ties in 30–39 each contribute 3.5, whence it is easily seen that $U_n = 39 + 60 + 7 = 106$ and now from (6.1) $U_m = 143 - 106 = 37$. In Comment 1 on Example 6.4 we noted that $U_m = 37$ is the minimum value that implies $P \leq 0.05$ in a two-tail test if there are no ties. The exact two-tail permutation probability for $U_m = 37$ for these tied data is $P = 0.0442$.

Conclusion. There is reasonably strong evidence against the hypothesis of equal medians.

Comments. 1. In this specific example the use of conventional no-tie tables is not seriously misleading; however this is not always so. In particular, heavy tying often leads to major discontinuities in possible P-values since the tying may reduce appreciably the number of possible values that U_m may take.

2. By pooling the data into groups we use less information in this test than we did in that in Example 6.4 so it is not surprising that there is a reduction in the strength of evidence against H_0 that is reflected by the slightly higher P-value.

The formulae (6.2) and (6.3) need modification in the presence of ties. We denote by T the sum of scores test statistic (typically S_m, or U_m) based on ranks or tied ranks. We denote the *score* for observation i by s_i. Then the general theory of the relevant permutation distribution gives, after straightforward but tedious application of standard results, the mean value, $E(T)$, and variance, $\text{Var}(T)$, of T to be

$$E(T) = \frac{m}{m+n}\left(\sum_{j=1}^{m+n} s_j\right) \tag{6.4}$$

$$\text{Var}(T) = \frac{mn}{(m+n)(m+n-1)}\left[\sum_{j=1}^{m+n} s_j^2 - \frac{1}{m+n}\left(\sum_{j=1}^{m+n} s_j\right)^2\right]. \tag{6.5}$$

For large m, n

$$Z = \frac{T - E(T)}{\sqrt{\text{Var}(T)}} \tag{6.6}$$

has approximately a standard normal distribution.

Example 6.6

The problem. Ages at death for members of two clans (see appendix) in the Baden-scallie burial ground (arranged in ascending order) are:

McGamma	13	13	22	26	33	33	59	72	72	72	77
	78	78	80	81	82	85	85	85	86	88	

McBeta	0	19	22	30	31	37	55	56	66	66	67	67
	68	71	73	75	75	78	79	82	83	83	88	96

Use the WMW large-sample approximation to test H_0: *the population distributions are identical* against H_1: *the population distribution medians differ.*

Formulation and assumptions. With sample sizes 21, 24 a large-sample approximation should suffice. To adjust for ties we require mid-ranks.

Procedure. The appropriate mid-ranks are:

McGamma	2.5	2.5	5.5	7	10.5	10.5	15	23	23	25	28
	30	30	33	34	35.5	40	40	40	42	43.5	

McBeta	1	4	5.5	8	9	12	13	14	16.5	16.5	18.5	18.5
	20	21	25	26.5	26.5	30	32	35.5	37.5	37.5	43.5	45

One may use either S or U, but once the ranks are recorded it is easier to use S. For McGamma addition gives $S_m = 518.5$. We leave it as an exercise to show that using (6.4) and (6.5) to compute the mean and variance that the denominator in (6.6) is 43.92, while the value of $E(T)$ is 483 (the same as that for the no-tie situation), giving $Z = (518.5 - 483)/43.92 \approx 0.81$. We leave it as a further exercise to show that if we had ignored ties and used (6.2) the only change would be that 43.92 in the denominator would have been replaced by 43.95, so in this case the effect of ties is trivial. This is often the situation unless ties dominate.

Conclusion. Since $P = \Pr(|Z| > 0.81) = 0.418$ there is no plausible evidence against the hypothesis of identical population medians.

Comments. 1. For both m, n greater than 20 one expects asymptotic results to agree closely with those for the exact WMW permutation test. Here the exact two-tail probability is $P = 0.426$.

2. We might argue that rather than H_1 specifying a median shift, a hypothesis of dominance of one distribution over the other may be more appropriate on genetic grounds. Inspection of the data gives a slight hint of a greater spread for the McBeta data, indicating a situation analogous to the classic Behrens–Fisher problem of testing for equality of means for two normal distributions with different variances. This more general problem has been considered by Pothoff (1963), by Fligner and Rust (1982) and others.

Computational aspects. With efficient algorithms computation of the exact permutation distribution tail probability in this case takes only seconds on most PCs, but as we have pointed out above, the asymptotic results are nearly as good for samples of this size and there may seem little point in calculating exact probabilities when an asymptotic result suffices. Of particular interest when using an asymptotic result is whether a reasonable fit in the tails is obtained. For the case $m = 21$, $n = 24$, if $S_m = 641$ the exact test one-tail probability is 0.0001 and the asymptotic probability is 0.0002; if $S_m = 557$ the corresponding probabilities are 0.0317 and 0.0311.

6.2 The Median Test

6.2.1 A Warning

The median test comes with a health warning. The price one pays for its simplicity is that it usually has appreciably lower power than other more appropriate alternatives for all but large samples of 100 or more observations. This is true even when its Pitman efficiency is higher than that of its competitors. This is a situation where the asymptotic or large sample property of the Pitman efficiency is not reflected in smaller samples, especially if the sample sizes m, n differ appreciably.

Freidlin and Gastwirth (2000) make an extensive study of power of the median test relative to commonly used alternatives. These include the WMW test and tests for median shift in otherwise identical distributions based on scores that we describe in Section 6.3 and others that we describe elsewhere. If the test is used for a pure median shift (assuming this is the only distributional difference between the sampled populations) the test is also not robust against breakdowns in assumptions such as equality of variance or other measures of spread. We concur with the conclusion of Freidlin and Gastworth that the median test *should not be recommended for routine use*. Nevertheless, because of its simplicity it is useful for illustrating features such as the importance of clarifying assumptions and the basic ideas behind power studies. A thorough account of the theory behind the median test, and a wider but closely related class of tests, is given in Gibbons and Chakraborti (2004, Section 6.4).

6.2.2 The Basic Test

We assume samples of m, n observations come from distributions that differ, if at all, only by a shift in location. That is, if $F(x)$ and $G(y)$ are the population cumulative distribution functions, then $G(y) = F(x - \theta)$. The usual null hypothesis is H_0: $\theta = 0$ implying identical distributions. If the population medians are θ_x, θ_y an equivalent form of H_0 is H_0: $\theta_x = \theta_y$.

The test is usually against one- or two-sided alternatives that the medians differ. Under H_0 the sample number of observations above the common median has a $B(m + n, 0.5)$ distribution and the number of observations, T_m, above the median in the first sample has a $B(m, 0.5)$ distribution.

The *median test* developed here, was proposed by Mood (1954). It does not test whether that common median has a particular value, but as is also the situation in the two-sample t-test or the WMW test, only whether it is reasonable to suppose both populations have the same unknown median. If both populations have the same median then the combined sample median, M, say, provides a point estimate of that median.

The median test may be based on the statistic T_m defined above. As already indicated, if the samples are from populations with the same median the distributions of the numbers in each above M will be approximately $B(m, 0.5)$ and $B(n, 0.5)$, where the relevant probabilities are now conditional upon the numbers above M in each sample adding to the number above M in the combined sample. When $m + n$ is even, and no sample value equals M, in the combined sample the numbers of values above and below M will each be $(m + n)/2$. If $m + n$ is odd at least one sample value equals M. In all cases where some values equal M, we suggest omitting these and proceeding with a reduced sample. Unless many values equal M this is usually satisfactory.

Conditioning on the total numbers above or below M in the combined sample leads to permutation tests that have much in common with some procedures we develop more generally in Chapter 12. In view of the often poor performance of this test our treatment is mainly used to illustrate principles. The procedure is described by an example.

Example 6.7

The problem. The camera ready copy for producing this book used a special program in common use for setting mathematical text. It is called LATEX. A group of 7 males and 21 females, after an initial training period in using LATEX, were each given a test page of mathematical formulae to set up in LATEX. The time in minutes that each took to set up the page correctly is set out below. Do these times indicate a difference between the median times taken by each gender?

Males	15	16	20	24	27	31	34				
Females	8	9	11	11	14	15	16	16	16	17	18
	18	19	21	23	23	25	25	25	27	32	

Formulation and assumptions. To apply the median test, the samples are pooled and the median time, M, for the combined sample is determined. This is easily seen to be 18.5 minutes. None of the observed values is equal to the median, so there are 14 values greater and 14 less than M. For the males there are 5 values greater than M, whereas for the females there are 9 such observations. If the unknown population medians are θ_1 and θ_2 respectively, an appropriate test is H_0: $\theta_1 = \theta_2$ against H_1: $\theta_1 \neq \theta_2$.

Since, for males there are only 2 values below M and 5 above, there is a suggestion that males may tend to take longer than females. That is, that H_0 may not be true.

To explain the rationale behind the test it is convenient to set out the information on numbers above and below M in each sample in a table:

	Above M	Below M	Total
Males	5	2	7
Females	9	11	21
Total	14	14	28

An important feature of this table is that the row and column totals and the grand total are fixed. There are two rows and two columns of data, so such a table is called a 2×2 *contingency table*. The four data positions are often referred to as cells. That at the intersection of the ith row and jth column is called cell (i, j). In the table above cell $(2, 1)$ contains the entry 9.

The row totals are determined by the sample sizes m, n and the column totals by the rule that M is the combined sample median. The test is based on the permutation distribution of all possible cell entries in a table like that above consistent with the marginal totals. Since the marginal totals are fixed, knowledge of one cell value means that all the other cell values can be determined by subtraction. It is convenient to concentrate on the possible values in cell $(1, 1)$ (i.e., the number of values above M in sample 1). Suppose that a value of 3 is considered. To give the correct marginal totals, by subtraction we find the cell entries in the body of the table must then be

$$3 \quad 4$$
$$11 \quad 10$$

Such an outcome is more favourable to H_0 than the one observed in this example, because the proportions of observations greater than M for each gender are closer to equality. Indeed, the only outcomes as or less favourable to H_0 than that observed are those where cell $(1, 1)$ takes the values 0, 1 , 2, 6 or 7.

These allocations, together with that observed, form an appropriate tail critical region for assessing the evidence for H_1 against H_0. We need the probability associated with this tail when H_0 is true. We assume temporarily that if H_0 is true then M is the value of the common median for the two populations. This may not be true, but it is intuitively a best estimate of the common median to be obtained from our data. We see below that this assumption is not critical to our argument. As in the sign test, under H_0 the probability of getting r observation above the median in a sample of 7 is the binomial probability

$$p_r = \binom{7}{r}\left(\frac{1}{2}\right)^7.$$

That of getting s above the median in a sample of 21 is

$$p_s = \binom{21}{s}\left(\frac{1}{2}\right)^{21}.$$

These are independent in the sense that they refer to different samples. They are, however, conditional upon the fixed column total, i.e., $r + s = 14$. Turning to the column totals, in the combined sample of 28 the probability of observing 14 out of 28 observations above M is

$$p_c = \binom{28}{14}\left(\frac{1}{2}\right)^{28}.$$

We require the probability of observing respectively 5 and 9 observations above the median in Samples 1 and 2, conditional upon observing 14 above the median in the combined sample. From the definition of conditional probability, this is $P^* = p_r p_s / p_c$. Thus, for the data in this example,

$$P^* = \frac{\binom{7}{5}(\frac{1}{2})^7 \times \binom{21}{9}(\frac{1}{2})^{21}}{\binom{28}{14}(\frac{1}{2})^{28}}.$$

We consider a generalization of this result to samples of any size in Exercise 6.20. The binomial probability $p = 1/2$ cancels out in the expression for P^*. This will happen if the binomial analogues of p_r, p_s, p_c are written down for any p; i.e., P^* is independent of p. It suffices to assume only that M is the common value of some unspecified quantile. Strictly, this requires an assumption of identical population distributions, but since for our choice of M the unspecified quantile is likely to be close to the median, the test is unlikely to mislead if one assumes only that the distributions differ little in the neighbourhood of the median. The relevant tail probability is the sum of the P^* for the configurations that are as or more extreme than that observed. That is, when the numbers above the median in the first sample (males) are 0, 1 2, 5, 6 or 7.

Procedure. Evaluation of P^* is tedious. It is commonly written in an equivalent form. For a general 2×2 contingency table with cell entries

$$\begin{array}{cc} a & b \\ c & d \end{array}$$

$$P^* = \frac{(a+b)!\,(c+d)!\,(a+c)!\,(b+d)!}{(a+b+c+d)!\,a!\,b!\,c!\,d!} \qquad (6.7)$$

where $a! = a \times (a-1) \times (a-2) \ldots \times 3 \times 2 \times 1$ is referred to as *factorial a*. Also, by definition, $0! = 1$.

The factorials in the numerator of (6.7) are those for the marginal totals. Those in the denominator are for the grand total and the individual cell entries. For the given data $P^* = [7! \times 21! \times 14! \times 14!]/[28! \times 5! \times 2! \times 9! \times 12!]$. There is appreciable cancellation between numerator and denominator, and using a pocket calculator it is easy to verify that $P^* = 0.1539$. For the value 6 in cell $(1, 1)$ a similar computation gives $P^* = 0.0355$, while for the value 7 in cell $(1, 1)$ $P^* = 0.0029$. A moment's

Table 6.2 *Exact permutation distribution for median test in Example 6.7.*

Cell $(1, 1)$ entry	Probability
0	0.0029
1	0.0355
2	0.1539
3	0.3077
4	0.3077
5	0.1539
6	0.0355
7	0.0029

reflection shows that for values 2, 1 or 0 in cell (1,1) these probabilities are repeated. Thus the relevant P-value for the two-tail alternative to H_0 is

$$P = 2 \times (0.1539 + 0.0355 + 0.0029) = 0.3846.$$

Conclusion. There is no case for rejecting H_0, despite 5 out of 7 values for males being above the combined median.

Comments 1. The high P value in this example reflects the poor power performance noted by Freidlin and Gastwirth, especially when one sample size is small. The corresponding values for the t-test, and the WMW test are respectively $p = 0.104$ and $p = 0.143$.

2. There are marked discontinuities in the permutation distribution, despite there being 28 observations. The fact that only 7 of these are in one sample means that the test statistic can take only one of eight possible values.

Computational aspects. StatXact and Testimate and many general software packages have specific programs to carry out this test, which is usually referred to as the Fisher (exact) test, a test we discuss more fully in Section 12.2.1. StatXact generates the complete permutation distribution over all possible cell $(1, 1)$ values. This is given in Table 6.2. This confirms symmetry of the distribution and explains the marked probability discontinuities between successive possible P-values. In Chapters 3 and 4 we saw that for most statistics discontinuity decreased rapidly with increasing sample size. Because we have 28 observations the level of discontinuity here may surprise, but as explained above, this is because one sample is small relative to the other. Asymptotic results may be unreliable in such circumstances.

Markedly different sample sizes (in Example 6.7 one was three times the other) often occur in clinical trials where there may be only a few patients with a rare illness. Then comparison is often between responses for these patients and a much larger available *control* group without that illness.

Given full measurement data we may establish a confidence interval for the difference between population medians based on the median test permutation distribution. Writing the true difference $\delta = \theta_2 - \theta_1$, a nominal 95 percent confidence interval for δ is (d_1, d_2) where the end points d_1, d_2 are chosen so that if we change each observation in sample 2 (or sample 1) by more than these amounts we would reject the hypothesis that the medians are equal for Population 1 and the amended Population 2. This is relevant because increasing or decreasing all sample (or population) values by a constant k increases or decreases the median by k. The process is rather tedious and in view of the general poor power performance of the median test we omit details, outlining the process only in an example. Several statistical software packages give approximate confidence intervals for the median difference.

Example 6.8

Problem. For the data in Example 6.7 determine a nominal 90 percent confidence interval for the difference in population medians.

Formulation and assumptions. The procedure is based on the Fisher exact test developed in Example 6.7.

From Table 6.2 it is evident that we would reject at an exact $P = 0.0029 \times 2 = 0.0058$ in a two-tail test of the hypothesis that the medians are equal for samples of the given sizes (7 and 21) if the numbers above the median [i.e., those in cell (1,1) of the 2×2 tables developed in Example 6.7] were 0 or 7. Similarly, if we observe 1 or 6 the corresponding P-value is $P = (0.0029 + 0.0355) \times 2 = 0.0768$ Having established this our next step is to determine what constant adjustment is needed to each observation in sample 2 (the females) to just avoid rejecting H_0 in a median test based on Sample 1 and the adjusted Sample 2, but that value being such that we would reject H_0 for any greater adjustment. These adjustments give us a nominal 90 (exact 92.32 percent) confidence limits.

Procedure Inspecting the original data, since we reject H_0 if only one Sample 1 observation exceeds the median this implies that the joint sample median must not exceed 31. Since the number above the median in both samples must not exceed 14 this implies that for the adjusted Sample 2 (females) if we are not to reject H_0 it must have 12 values above the adjusted median. If the adjusted median is not to exceed 31 this means the adjustment must ensure there are 12 values at or above 31. That is, the 12 largest values in the adjusted sample must exceed 31. This occurs if we add to all values the difference between 31 and the smallest of the 12 largest observed value in Sample 2. That value is 17, so we add $31 - 17 = 14$ to all second sample values. This implies that if $\theta_2 - \theta_1 > -14$ we would reject H_0. Similarly, since we reject H_0 if 6 or more values in Sample 1 exceed the combined sample median, by an analogous argument we find we should reject H_0 if $\theta_2 - \theta_1 > 5$.

Conclusion. A nominal 90 percent confidence interval for the difference $\theta_2 - \theta_1$ is $(-14, 5)$.

Comments. 1. As usual, at the ends of this interval we have ties at the combined sample medians for the first sample and the adjusted second sample. To check that we would accept H_0 at the end points we would in effect perform a test with allowance for ties at the joint median. Such a tedious exercise is seldom of much interest.

2. The actual confidence level of the above interval is 92.32 percent since the exact $P \approx 0.077$ (see *Formulation and assumptions* above).

3. Assuming normality, the t-based 95 percent confidence interval for the mean difference is $(-10.67, 1.05)$. This is shorter than that based on median test theory and is necessarily symmetric about the sample means difference. Would you have reservations about using the normal theory test with these data? The WMW based interval is (-11, 1), in close agreement with the interval based on the t-test.

Computational aspects. Given a program for the Fisher exact test for a 2×2 contingency table one can easily determine for any given m, n the number above the median in the first sample which just gives significance, and, because there is only one degree of freedom, all other entries in the 2×2 table follow automatically. It is then, as indicated in the above example, relatively simple to determine confidence limits by appropriate additions or subtractions from all second-sample observations.

Asymptotic approximations for the median test are not required if software for the Fisher exact test is available. If such software is not available an asymptotic form of the Fisher test based on the chi-squared test described in Section 12.2.1 may be used.

A modification of the median test proposed by Mathisen (1943) is called the control median test. Given two independent samples of size m, n, one is designated the *control sample*. There is no loss of generality if we label this Sample I. In applications this often consists of observations for an existing standard procedure. If the Sample I median is M_1, the test statistic, V, is the number of Sample II observations that are less than M_1. If m is even and there are no ties and if no Sample I value is equal to M_1, then if we set $m = 2r$ and there are j Sample II values less than M_1, and no Sample II value equals M_1 there will be $n - j$ Sample II values greater than M_1. It follows that in the combined sample there will be $r + j$ values less than M_1 and $r + n - j$ values greater than M_1.

Under the null hypothesis that both samples are from populations with the same common median, with arguments similar to those used in obtaining the expression for P^* given in Exercise 6.20 we find

$$\Pr(V = j) = \frac{\binom{r+j}{j} \times \binom{r+n-j}{n-j}}{\binom{m+n}{n}} \tag{6.8}$$

Small values of j relative to $n - j$ suggest that the Population II median may be greater than the Population I (or control) median. Large values of j relative to $n - j$ suggest the opposite. In a one-tail test in the first case the corresponding P-value may be computed using (6.8) and summing over all j less than or equal to that observed. In the second case there is an equivalent summation over the observed and larger j.

Only slight modification is needed to the above arguments if m is odd or if some Sample II values equal M_1.

Example 6.9

The problem. Apply a control median test to the data below regarding the Sample I as a control group, to test the null hypothesis against the alternative that the Sample II is from a population with a greater median.

Sample I	49	58	75	110	112	132	151	276	281	362
Sample II	50	58	96	139	152	159	189	225	239	242
	257	262	292	294	300	301	306	329	342	346
	349	354	359	360	365	378	381	388		

Formulation and Assumptions. We obtain the median M_1 of the control group, then ascertain how many Sample II values exceed M_1. The relevant one-tail P-value is obtained by calculating the probabilities associated with this or a greater number, j, above the median using (6.8).

Procedure. The control (first sample) median is $M_1 = (112 + 132)/2 = 122$. From the data it is easily seen that there are 25 second sample values exceeding 122. The other more extreme values for which we require to add the relevant probabilities are 26, 27 or all 28 second sample values exceeding 122. Computationally we may write down relevant 2×2 contingency tables correponding to each of these cases and calculate the probabilities associated with each using (6.7). We leave it as an exercise to show that the relevant 2×2 tables are

$$\begin{array}{cc} 5 & 5 \\ 25 & 3 \end{array} \qquad \begin{array}{cc} 5 & 5 \\ 26 & 2 \end{array} \qquad \begin{array}{cc} 5 & 5 \\ 27 & 1 \end{array} \qquad \begin{array}{cc} 5 & 5 \\ 28 & 0 \end{array}$$

with relevant probabilities

$$\Pr(V = 25) = 0.01688, \quad \Pr(V = 26) = 0.00755,$$
$$\Pr(V = 27) = 0.00256, \quad \Pr(V = 28) = 0.00050,$$

whence the required one tail probability is $P = 0.01688 + 0.00755 + 0.00256 + 0.00050 = 0.02749$

Conclusion There is fairly strong evidence that the median survival time for the Sample II population is greater than that for the control sample population.

Comments 1. It is not difficult to show that if we take the median of the second sample as the control median in general the resulting P-values will be different, but usually not alarmingly so. This may be regarded as a weakness of the test. It does however have Pitman efficency 1 relative to the ordinary median test.

2. Asymptotic normal approximations to the control median test are given by Gibbons and Chakraborti (2004, Section 6.5), but these also give different results dependent on choice of the control sample.

3. The concept is most useful when one sample may be logically chosen as a control because it corresponds to an existing treatment, or perhaps to an untreated sample or a set of patients not suffering from the condition under investigation.

Computational aspects. For the median test we indicated that a program for the Fisher exact test could be used to obtain the relevant P-value. This is not so for the control median. This is because in the case of the median test we consider certain relevant 2×2 matrices with fixed marginal totals, whereas in the control median test we fix the entries in row 1, and the row totals. It is evident by inspection of the four tables given above under *Procedure* that here the column totals are not fixed. One therefore either requires a program dedicated to the control median test or one that will compute probabilities equivalent to those in (6.8) for all relevant j.

6.3 Normal Scores

As in the one-sample case, transforming ranks to give scores with something like the characteristics of samples from a normal distribution is intuitively appealing. In Section 3.5.2 we referred briefly to van der Waerden and expected normal scores. For the two independent samples scenario corresponding to the WMW test van der Waerden scores replace the rank r by the $r/(m+n+1)$th quantile of the standard normal distribution ($r = 1, 2, \ldots, m+n$).

Expected normal scores may also be used, but there is usually little difference between results using these and van der Waerden scores, so we consider here only the latter. Symmetry implies that the mean of the $m + n$ van der Waerden scores is zero if there are no ties. The mean of all van der Waerden scores is not exactly zero with ties if we use quantiles based on mid-ranks. For a few ties the effect is negligible, as adjacent quantile scores are replaced by scores close to their mean. For ties, an alternative to the mid-rank quantile (which is the mean of the tied ranks) is to take the mean of the quantiles corresponding to each of the ranks that are tied. In practice the difference is usually slight, except for large numbers of ties, or for ties in certain positions.

If one has an exact permutation test program that covers the Pitman test, or allows arbitrary choice of scores, that program may be used to compute exact P-values for a test using van der Waerden scores by substituting these scores for the original data. StatXact and a few other packages provide a program, the normal scores test, that uses van der Waerden scores, forming these directly from the original data with the convention that for ties the mean of the quantiles corresponding to the ranks that are tied (the second of the options mentioned above) is used.

Example 6.10

The problem. For the data for computation times given in Example 6.1, viz.,

Group A	23	18	17	25	22	19	31	26	29	33	
Group B	21	28	32	30	41	24	35	34	27	39	36

use van der Waerden scores to assess whether the data indicate that one layout is superior (i.e., leads to more rapid computations).

Formulation and assumptions. The data are ranked and van der Waerden scores assigned to each. Using these scores in a permutation test program allowing arbitrary scores an exact P-value may be computed.

Procedure. It is easiest to use any available software that generate the scores automatically. However, they may also be obtained using tables of the standard normal distribution. For example, for these data $m = 10, n = 11$, so the van der Waerden score for the datum ranked 3 (which here is the observation 19 in Group A) is the 3/22nd quantile of the standard normal distribution, i.e., the value of x such that $\Phi(x) = 3/22 \approx 0.1364$. From tables of the standard normal distribution this is found to be $x = -1.097$. StatXact gives a two-tail $P = 0.0118$ for this example.

Conclusion. There is strong evidence against the null hypothesis of equal medians.

Comment. The P-value is less than that obtained using the WMW procedure and almost identical to that for a t-test. This is not surprising. The data suggest only slight departures from normality, and are of a type where one would not have serious reservations about using a t-test. In these circumstances, although one would expect on the basis of Pitman efficiency, that the Wilcoxon test may be not quite as powerful as the t-test it would be reasonable to expect the normal scores test to perform much like the t-test as it moulds the data to a form with the characteristics of a sample from a normal distribution. Indeed, when sampling from normal populations that differ only in mean, the van der Waerden scores test has Pitman efficiency 1.0 relative to the t-test.

Computational aspects. If software for computing exact P-values with van der Waerden scores is not available an asymptotic approximation valid for large samples may be based on (6.4) to (6.6) where the scores used are the van der Waerden scores. In practice this is reasonable for samples of moderate size because of the normalizing influence of the transformation which enhances the rate of convergence toward the asymptotic approximation.

Normal scores are usually not appropriate for the analysis of survival data or for certain other situations where long-tail distributions (often with censored observations as an added complication) are involved. Alternative scoring methods are available. We discuss some of these in Chapter 9.

6.4 Tests for Equality of Variance

Historically, tests about differences between medians or means have dominated inference for two sample problems, but there is increasing interest in differences in variation, or spread, which may or may not be associated with differences in means or medians. In the field of quality control, for example, this reflects consumer demand that not only should all products of a particular kind perform well on average, but that they should also be consistently reliable.

Care is needed in testing these characteristics when we drop an assumption of normality. If we have two independent samples $x_1, x_2, \ldots, x_m$ and $y_1, y_2, \ldots, y_n$ from normal distributions the appropriate statistic to test equality of population variances, irrespective of whether the means are equal, is the well-known

$$F = \frac{\sum(x_1 - \overline{x})^2/(m-1)}{\sum(y_i - \overline{y})^2/(n-1)},$$

which has an F distribution with $m-1, n-1$ degrees of freedom under the hypothesis that the population variances are equal, irrespective of whether the means are equal.

The normal distribution family is the best-known example of what are called *location–scale parameter* families. These are families of distributions of random variables X, Y having location parameters θ_1, θ_2 and scale parameters φ_1, φ_2 such that $U = (X - \theta_1)/\varphi_1$ and $V = (Y - \theta_2)/\varphi_2$ are identically distributed. For the normal distribution in conventional notation we have $\theta_i = \mu_i$, and $\varphi_i = \sigma_i$, $(i = 1, 2)$ and U, V each have a standard normal distribution.

We shall see below that we may be interested in spread differences between members of families of distributions that are not of the location-scale parameter form. Also, that even when we have such families many of our tests for differences in dispersion assume either that $\theta_1 = \theta_2$, or that we know the value of the difference $\delta_1 = \theta_2 - \theta_1$.

The F-statistic relevant to normal distributions is almost unique. It provides a test for differences between the population spreads or dispersions without needing to know the values of the population location parameters.

So far in this chapter we have considered tests basically designed for shifts in means or medians, but we have pointed out that several of the tests are also relevant to tests of dominance, where the alternative hypothesis is of the form $H_1: G(x) \leq F(x)$ or else $H_1: G(x) \geq F(x)$ with strict inequality for at least some x.

This is important in the context of dispersion because, for many distributions, even when they belong to the same family, it is impossible to have a shift in mean or median without altering other distribution characteristics such as variance. A simple example is the exponential distribution with parameter λ. The probability density function is $f(x) = \lambda e^{-\lambda x}, x \geq 0$, and $E(X) = 1/\lambda$, $\mathrm{Var}(X) = 1/\lambda^2$ and $\mathrm{med}(X) = \ln(2/\lambda)$. Thus a change in λ changes the mean, median and variance, reflecting a dominance situation. The parameter λ is often referred to as a scale parameter.

Where both centrality and spread are influenced by parameter changes there are difficulties in developing completely distribution-free tests for differences in variance, or for more general measures of spread. As already indicated, tests for differences in variance require an assumption that the means or medians are equal, or that we know the difference between them. These points are discussed more fully by Hollander and Wolfe (1999, Sections 5.1 and 5.2) and by Gibbons and Chakraborti (2004, Chapter 9). Most tests for a difference in spread either implicitly, or explicitly, assume samples are from populations with identical medians, or that the difference between medians is known, and that an adjustment has been made to align the samples by addition or subtraction of that difference for the data in one of the samples.

If the sample values, or if a test, indicate equality of medians is untenable it is often suggested that one should adjust sample values to align the sample medians. This has intuitive appeal, and is often a sensible thing to do, but examples exist to show that it may sometimes be counterproductive and, in general, it will tend to give rather too many small P-values.

Here we concentrate mainly on tests that assume population means or medians are identical or can be made so by an addition to or subtraction from one set of sample values, but in practice such assumptions are often at best approximations.

6.4.1 The Siegel–Tukey Test

This test proposed by Siegel and Tukey (1960) is easy to carry out, but like many nonparametric tests for spread or dispersion, is not very powerful. The Pitman efficiency relative to the F-test, when samples are from normal distributions, is only 0.61.

The basic idea behind the test is that if two samples come from populations differing only in variance, the sample from the population with greater variance will be more spread out. If there is a known centrality difference as well, we first align the samples by subtracting the median difference from all values in one sample, but in practice this difference is seldom known. If an unknown centrality difference is assumed, a common practice, as suggested above, is to align the populations by shifting the median (or mean) of one sample to coincide with that of the other sample. This requires an appropriate addition to, or subtraction of, a constant for all observations in one sample. The sample variances are unaltered by this change but, as indicated above, the procedure is not optimal unless the difference between sample medians happens to equal that between the population location parameters in a location-scale model situation. However, the change usually works reasonably for the Siegel–Tukey test if there is no strong sample evidence of differences in skewness.

If we arrange the combined samples in order and allocate rank 1 to the smallest observation, rank 2 to the largest, rank 3 to the next largest, ranks 4 and 5 to the next two smallest, ranks 6 and 7 to the next two largest, and so on, the sum of the ranks for the sample from the population having the greater variance should be smaller than if there were no difference in variance.

Example 6.11

The problem. Davis and Lawrance (1989) give (as part of a larger data set collected for different purposes) the time in hours to two different types of tyre failure under similar test conditions. Failure type A is rubber chunking on shoulder. Failure type B is cracking of the side wall. Do the following data suggest the population variances may differ?

Type A	177	200	227	230	232	268	272	297				
Type B	47	105	126	142	158	172	197	220	225	230	262	270

Table 6.3 *Allocation of ranks for Siegel–Tukey test; type A (adjusted) in bold.*

Value	47	105	126	**130.5**	142	**153.5**	158	172	**180.5**	**183.5**
Rank	1	4	5	**8**	9	**12**	13	16	**17**	**20**
Value	**185.5**	197	220	**221.5**	225	**225.5**	230	**250.5**	262	270
Rank	**19**	18	15	**14**	11	**10**	7	**6**	3	2

Formulation and assumptions. We align the medians of the two samples, then allocate ranks as described above. The lesser sum of ranks associated with one sample is calculated and tested for significance as in the WMW test. Our null hypothesis is that the variances are identical. Without further information a two-tail test is appropriate.

Procedure. The first sample median is $(230 + 232)/2 = 231$ and the second sample median is 184.5. We align for location by subtracting $231 - 184.5 = 46.5$ from all first-sample values and work with the adjusted samples values below for Type A, those for type B being unaltered.

Type A	130.5	153.5	180.5	183.5	185.5	221.5	225.5	250.5
Type B	47	105	126	142	158	172		
	197	220	225	230	262	270		

Assigning ranks using the scheme outlined above is easier if we arrange all sample values in ascending order before allocation. The result of this operation is given in Table 6.3, where type A (adjusted) values are indicated in bold. The sum of the ranks for type A faults is $S_m = 106$, giving $U_m = S_m - m(m + 1)/2 = 70$ and $U_n = mn - U_m = 26$. Tables for the WMW test gives $U = 22$ as the critical value for significance at a nominal 5 percent level in a two-tail test for samples of 8 and 12 observations. StatXact provides a program for an exact Siegel–Tukey test. This indicates that the exact two-tail P-value corresponding to $S_m = 106$ is $P = 0.0979$.

Conclusion. There is only weak evidence against the hypothesis of equal variance; not sufficient to reject H_0 at a nominal 5 percent significance level.

Comments. 1. For these data the WMW test two-tail $P = 0.0268$. The Siegel–Tukey program in StatXact does not adjust the sample medians to be equal. This must be done manually or by means of the program editing facility.

2. We "relocated" the type A data by equating sample medians. Since variance is based on squared deviations from the mean, equating means has intuitive appeal. In practice, which alternatives we choose usually makes little difference. The median is easier to calculate — hardly a justification with modern computing methods — but it may also be more robust if there is some skewness. Lehmann (1975) recommends an alignment based on the Hodges–Lehmann estimator of the median difference.

Computational aspects. After ranks are allocated any program that computes exact probabilities associated with the permutation distribution for the WMW test may be used if a dedicated program such as that in StatXact is not available.

Table 6.4 *Ranks and scores for Ansari–Bradley test in Example 6.11; type A (adjusted) in bold.*

Value	47	105	126	**130.5**	142	**153.5**	158	172	**180.5**	**183.5**
Rank	1	2	3	**4**	5	**6**	7	8	**9**	**10**
Score	9.5	8.5	7.5	**6.5**	5.5	**4.5**	3.5	2.5	**1.5**	**0.5**
Value	**185.5**	197	220	**221.5**	225	**225.5**	230	**250.5**	262	270
Rank	**11**	12	13	**14**	15	**16**	17	**18**	19	20
Score	**0.5**	1.5	2.5	**3.5**	4.5	**5.5**	6.5	**7.5**	8.5	9.5

6.4.2 Ansari–Bradley Type Tests

Several almost equivalent tests based on dispersion of ranks in each sample about the combined sample mean rank were formulated about the same time. If there are $N = m + n$ observations and these are ranked 1 to N in ascending order as in the WMW test procedure the mean rank is $(N+1)/2$. The deviation of rank i from this mean is $d_i = i - (N + 1)/2$. The sum of these deviations over all N ranks is zero. If the spread in one sample is greater than that in the other low and high ranks tend to dominate in that sample. Since the d_i corresponding to low and high ranks have opposite signs the magnitudes of the d_i in each sample are relevant and these are reflected by taking either absolute values of the d_i or the squares of the d_i as scores. The sum of these for one sample is an appropriate statistic in a Pitman-type permutation test for a difference in spread under an assumption that the population medians are identical.

Tests that are in essence equivalent and use the sum of the $|d_i|$ were proposed with minor variations by Freund and Ansari (1957), David and Barton (1958) and by Ansari and Bradley (1960). Gibbons and Chakraborti (2004) refer to them collectively as the Freund–Ansari–Bradley–David–Barton tests. While one may use any program for the Pitman test with these scores as data, StatXact has a program that calculates scores given the original data and uses the name Ansari–Bradley test. Mid-ranks are widely used for ties. This does not alter the mean rank, but affects some of the scores.

Example 6.12

The problem. Apply the Ansari–Bradley test to the data in Example 6.11.

Formulation and assumptions. The test assumes the medians of both populations are equal. We align the data as we did in Example 6.11 by subtracting 46.5 from all first sample values. We work with the adjusted sample values. The combined sample data are ranked and the $|d_i|$ are summed over one sample for use in a Pitman-type permutation test or in a dedicated Ansari–Bradley test program.

Procedure. The original data after adjustment and the ranks and the scores $|d_i|$ are given in Table 6.4. The type A data are shown in bold.

The sum of the scores $|d_i|$ for type A is

$$S = 6.5 + 4.5 + 1.5 + 0.5 + 0.5 + 3.5 + 5.5 + 7.5 = 30.$$

Using either a program for a Pitman test with these scores or a dedicated Ansari–Bradley program gives a two-tail $P = 0.1437$.

Conclusion. There is little evidence of a difference between variances.

Comments. 1. The P-value here is slightly greater than that for the Siegel–Tukey test.

2. The effect of lining up the samples on the basis of sample medians implies that the test is not exact in the sense that there is a tendency to give too many small P-values. Simulation studies have confirmed this tendency although the exact P-values are approached for very large samples.

3. For ties it is usual to use mid-ranks.

The alternative to using the $|d_i|$ as scores is to use the squared d_i in a test called the Mood test. It was proposed by Mood (1954) and is otherwise analogous to the Ansari–Bradley test. StatXact provides a program for this test and for the data considered in Examples 6.9 and 6.10 the exact test indicates $P = 0.1293$. Asymptotic approximations valid for large samples are available for the Ansari–Bradley test and for the Mood test.

6.4.3 The Conover Squared-Rank Test for Variance

If the means of X, Y are respectively μ_x, μ_y then equality of variance implies $E[(X - \mu_x)^2] = E[(Y - \mu_y)^2]$, where $E[X]$ is the expectation of X.

Conover (1980) proposed a test for equality of variance based on the joint squared ranks of squared deviations from the means, i.e., from $(x_i - \mu_x)^2$, $(y_i - \mu_y)^2$. The population means are seldom known, so again we assume it is reasonable to replace them by their sample estimates m_x, m_y. We do not need to square the deviations to obtain the required rankings because the same order is achieved by ranking the absolute deviations $|x_i - m_x|$, $|y_i - m_y|$. We rank these deviations and use as scores the squares of the ranks. The test statistic T is the sum of the scores for one of the samples. For large samples Z given by (6.6) has a standard normal distribution. Conover (1999, Table A9) gives quantiles of T for a range of sample sizes in a no-tie situation.

Programs generating the permutation distribution for arbitrary scores may also be used if exact tail probabilities are required and StatXact provides a specific program for the test.

Example 6.13

The problem. For the data in Example 6.11 use the squared-rank test for equality of variance.

Formulation and assumptions. We require deviations of each observation from its sample mean. The absolute deviations are then ranked, the ranks squared, and the statistic T is calculated.

Table 6.5 *Absolute deviations (Adev) and squared ranks (Sqr), Example 6.12: type A in **bold**.*

Adev	6	8	8	11	17	22	30	34	38	38
Sqr	1	**6.25**	6.25	**16**	25	36	**49**	**64**	**90.25**	90.25
Adev	40	45	50	54	**59**	**61**	75	82	90	133
Sqr	121	144	169	196	**225**	**256**	289	324	361	400

Procedure. Denoting type A sample values by x_i and type B sample values by y_i we find $m_x = 238$ and $m_y = 180$. It suffices to express these to the same order of accuracy as the data, here to the nearest integer. We compute the absolute deviations for each sample value, e.g., for the type A observation 227, the deviation is $227 - 238 = -11$, giving an absolute deviation of 11. Table 6.5 gives the ordered absolute deviations for the combined samples together with squared ranks (squared mid-ranks for ties). Type A deviations are in bold. The sum of the bold squared ranks is $T = 706.5$. With squared ranks as scores we find from (6.5) that $\text{Var}(T) = 78\,522.49$. Also (6.4) gives $E(T) = 1147.6$, whence from (6.6), $Z = (706.5 - 1147.6)/280.219 = -1.574$. However, using an exact program such as that in StatXact, if available, is preferable to an asymptotic result. For these data StatXact gives a two-tail $P = 0.1267$.

Conclusion. The evidence against H_0 is not strong.

Comments. 1. Calculating sample means to one decimal place would avoid data ties, but this makes little practical difference to the calculated P-value although the value of T may be somewhat different, especially if high ranks are involved in ties. For example, if ranks 17 and 18 are replaced by a tied value 17.5 and only one of these tied values is included in T it contributes 306.25 whereas if 17 replaces that tied value it contributes only 289. StatXact avoids ties in this example and gives a $T = 720$ (compared to our 706.5). If ties occur StatXact in this, and in some other programs, deals with them slightly differently from the way we do when we use mid-ranks. The method is given in the StatXact manual, but in practice the two methods usually lead to similar results.

2. Whereas the Siegel–Tukey test with median alignment would establish significance at the formal 5 percent level in a one-tail test if that were appropriate, the Conover squared-rank test just fails to do so. While the test appears in this case to be less powerful than the Siegel–Tukey test, it is reasonably robust. The results are broadly in line with those for the Ansari–Bradley test, which in fact uses a scoring system not very different from that for the Siegel–Tukey test.

3. We could have used ranks of absolute deviations rather than squared ranks. In that case we essentially have a WMW test. Conover (1999, Section 5.3) reports that this leads to a less powerful test than the squared-rank test.

Computational aspects. The Conover scores may be used in any permutation test program for a Pitman-type test if a dedicated program is not available.

6.4.4 The Moses Test for Extreme Reactions

Tests so far discussed have considered spread or dispersion as a *variance* type of characteristic. Suppose we have two samples, A and B, and that Sample A has a smaller range of values than Sample B. One might then wish to investigate whether extreme values (e.g., for reaction or response times) are more likely to occur in the population from which Sample B was drawn.

Moses (1952) proposed a test of extreme reactions. The medians of two associated populations are assumed to be equal. The null hypothesis is that extreme values are equally likely to occur in both populations. The alternative is that extreme values are more likely in the population from which the sample with the larger range was drawn. The two samples are combined, ranked overall and the observations put into rank order. Each observation is labelled either A or B according to the sample from which it originally came.

If both samples come from the same population then each is likely to contain some high, some low and some intermediate values. For instance, if there are seven observations in Sample A and eight in Sample B this might be indicated by the ordering:

B B A A B A B A A B A B B B A

Under the alternative hypothesis, values for Sample A will be concentrated in one part of the series, for example:

B B B A A A A A B A A B B B B

Once the values have been ordered, the ranks of the lowest and highest scores from Sample A are noted. The number of cases contained within these values, including the extremes, is defined as the *span* of the A scores. In the first example above, the span is relatively large (13) whereas in the second case it is quite small (8). The smaller the value of the span, the stronger is the evidence against the null hypothesis. The Moses test determines the probability of observing a span of no more than that recorded, assuming that the null hypothesis is true.

If samples A and B have m and n observations respectively, then the span of A must be between m and $m+n$ inclusive. Suppose that the observed span is $m+k$. Under the null hypothesis, the probability that the span s is no more than this is given by

$$\Pr(s \leq m+k) = \frac{\sum_{i=0}^{k} \binom{i+m-2}{i}\binom{n+1-i}{n-i}}{\binom{m+n}{m}} \tag{6.9}$$

Probabilities for the upper tail of the scan length distribution can be calculated in a similar manner. Exact P-values for the one-tail Moses test can be found using SPSS, although tail probabilities are simple (if tedious) to compute using tabulated values of binomial coefficients.

Table 6.6 *Allocation of ranks for Moses test, Example 6.14; Group A in bold.*

Value	1	3	5	6	**8**	9	10	11	**12**	13
Rank	1	2	3	4	**5**	6	7	8	**9**	10
Value	14	**15**	**18**	19	**20**	**21**	22	23	25	
Rank	11	**12**	**13**	14	**15**	**16**	17	18	19	

Example 6.14

The problem. Nineteen women who had received results from cervical cancer screening completed a questionnaire about their attitude toward receiving detailed information about cervical cancer, a high score indicating a great degree of interest. Those women who had received a negative screening result (Group A) were compared with those with a positive result (Group B). The questionnaire scores (out of 25) were as follows:

$$Group\ A: \quad 8 \quad 9 \quad 10 \quad 12 \quad 13 \quad 15 \quad 18 \quad 20 \quad 21$$
$$Group\ B: \quad 1 \quad 3 \quad 5 \quad 6 \quad 11 \quad 14 \quad 19 \quad 22 \quad 23 \quad 25$$

We test the null hypothesis that extreme values are equally likely to occur in the positive test and negative test populations of women.

Formulation and assumptions. The two groups are combined and the observations ranked. The ranks are written in ascending order. The span of Group A is then obtained by recording the number of cases, including the extremes, contained within the lowest and highest ranks relating to this group. A one-tail test is appropriate, as there are grounds for believing that the scores for women with a positive test result might be more variable (see *Comment* 3).

Procedure. The ranks for the combined observations are obtained. These are shown in Table 6.6. Ranks relating to Group A are given in bold. We have $m = 9, n = 10$. The lowest rank for a Group A observation is 5, the highest is 16, giving a span of 12. The minimum possible value for the span is $m = 9$, thus $k = 3$. Using (6.9) we find $P = 0.0149$. The SPSS program confirms this.

Conclusion. There is considerable evidence against the null hypothesis. One may reasonably conclude that extreme questionnaire scores occur more frequently for patients with a positive screening result.

Comments. 1. The span of the Group A observations is often referred to as the range, and it is a measure highly susceptible to outliers. One way of taking this into account is to decide before the investigation to remove (or trim) one or more Group A observations from the upper and lower extremes. The span is then based on the values that remain. For the above data, SPSS gives a P-value of 0.128 when one observation is trimmed from each end. There is little reason to use a trimmed span in this example, as there are no obvious outliers. Trimming when there are outliers in other contexts is discussed briefly in Section 14.4.

2. If ties occur between two or more members of the same group the tied values are adjacent in the ordering and the span is not affected. In addition, if ties occur

between members of the two different groups the span will generally not be affected. In the absence of trimming, the only exception is where a tie involves either the highest or lowest observation from Group A; the value for the span is then unclear. In SPSS the span is calculated by assuming that the Group A value in the tie is actually the larger. This can overestimate the value of the span and hence the P-value. In the above example, if the highest value in Group A is changed to 22, the resulting tie leads to a P-value 0.04.

3. The findings in this example are in line with the belief of many psychologists that individuals who receive a positive diagnosis for a serious illness tend to be either *blunters* whose reaction to the bad news is denial, or *monitors* who wish to be informed about every detail regarding their illness. Individuals who do not have to face up to such bad news are assumed to have an intermediate degree of interest in that illness. Thus the blunters would have low scores on the cervical cancer questionnaire with the monitors having relatively high scores. This could explain why Group B has more extreme values than Group A.

Computational aspects. Since $n = m + 1$, there is considerable simplification in the numerator of (6.9) for this example.

6.4.5 Some Related Procedures

Klotz (1962) proposed using van der Waerden scores in a test analogous to the Mood test, while Capon (1961) suggested a similar test using expected normal scores. Sukhatme (1957) proposed a test having analogies with the Mann–Whitney formulation of the WMW test, but the scoring is more complicated and there are practical limitations to its use. It is described by Gibbons and Chakraborti (2004, Section 9.7) who point out that it can be adapted to construct confidence intervals for a scale parameter.

Gastwirth (1965) proposed statistics for detecting location and scale differences based on simple scores assigned to top and bottom nonoverlapping fractions of the combined sample ordered data, with zero scores given to any data with intermediate ranks. The motivation for his approach was that tests for centrality or location based on transformation of ranks, such as those using van der Waerden scores, that have high Pitman efficiency relative to optimal normal theory tests when the latter are appropriate give greater weights to scores associated with extreme ranks. Gastwirth was especially interested in tests that were more efficient for dispersion than the tests we have discussed here. A general account of his approach is given in Sprent (1998, Section 6.6).

6.5 Tests for a Common Distribution

In Section 4.2 we developed tests of the hypothesis that a single sample was drawn from some specified distribution. For two independent samples we may want to know if it is reasonable to suppose each comes from the same unspecified distribution. The *Smirnov test* (Smirnov, 1939, 1948) has similarities to the Kolmogorov test developed in Section 4.2.1.

6.5.1 Smirnov Test for a Common Distribution

The hypothesis H_0: *two samples come from the same distribution* may be tested against H_1: *the distributions have different cumulative distribution functions (cdfs)*. We do not specify further the nature of any differences. The distributions may have the same mean but different variances; one may be skew, the other symmetric, etc. We compare the sample cdfs; the test statistic is the difference of greatest magnitude between these two functions.

Example 6.15

The problem. The following data are rates of saliva production (ml/min) for 7 female and 21 male adults, arranged for convenience in ascending order. Is it reasonable to conclude that the samples are from populations with the same distribution?

Females	0.45	0.60	0.80	0.85	0.95	1.00	1.75				
Males	0.40	0.50	0.55	0.65	0.70	0.75	0.90	1.05	1.15	1.25	1.30
	1.35	1.45	1.50	1.85	1.90	2.30	2.55	2.70	2.85	3.85	

Formulation and assumptions. We compute the sample cumulative distribution functions $S_1(x)$, $S_2(y)$ at each sample value. We also compute and record the difference $S_1(x) - S_2(y)$. For samples of size m, n respectively $S_1(x)$, $S_2(y)$ are step functions with steps $1/m$, $1/n$ respectively at each sample value (with multiple steps if there are ties).

Procedure. Table 6.7 gives in successive columns the sample values and corresponding values of $S_1(x)$, $S_2(y)$ and $S_1(x) - S_2(y)$ at each sample point. The difference of greatest magnitude (final column) is 0.524. The difference is positive suggesting that at the saliva production rate at which it occurs, namely 1 ml/min, that the cdf for females is greater than that for males. The preponderance of positive values for the difference $S_1(x) - S_2(y)$ suggests that the corresponding population cdf for females increases more rapidly than that for males. This reflects the greater spread of the male data, suggesting both a shift in location and an increasing spread in the form of an elongated right tail. If there were prior reasons to believe males salivate more rapidly a one-tail test might be justified. However, a two-tail test is appropriate for unspecified general alternatives. An exact test is preferable if a suitable program is available. One is provided in StatXact and in this example gives an exact one-tail $P = 0.0465$ and an exact two-tail $P = 0.0930$.

Conclusion There is a slight indication that the cdf for females may be consistently above that for males for at least many identical values of x, y.

Comments. 1. Sometimes a test for location or variance establishes a difference when a Smirnov test indicates no overall distributional difference. This is because Kolmogorov–Smirnov type tests are often less powerful than tests for specific characteristic differences such as those for location.

2. Like the Kolmogorov test, the Smirnov test may appear not to be making full use of the data because it uses only the maximum difference. As we pointed out in Section 4.2.1, the statistic uses cumulative information. In Section 6.5.2 we indicate that power may be increased by considering all differences.

Table 6.7 *Calculation of the Smirnov test statistic.*

Female(x_i)	Male(y_i)	$S_1(x)$	$S_2(x)$	$S_1(x) - S_2(y)$
	0.40	0.000	0.048	−0.048
0.45		0.143	0.048	0.095
	0.50	0.143	0.095	0.048
	0.55	0.143	0.143	0.000
0.60		0.286	0.143	0.143
	0.65	0.286	0.190	0.096
	0.70	0.286	0.238	0.048
	0.75	0.286	0.286	0.000
0.80		0.429	0.286	0.143
0.85		0.571	0.286	0.285
	0.90	0.571	0.333	0.238
0.95		0.714	0.333	0.238
1.00		0.857	0.333	0.524
	1.05	0.857	0.381	0.478
	1.15	0.857	0.429	0.428
	1.25	0.857	0.476	0.381
	1.30	0.857	0.523	0.332
	1.35	0.857	0.571	0.286
	1.45	0.857	0.619	0.238
	1.50	0.857	0.667	0.190
1.75		1.000	0.667	0.333
	1.85	1.000	0.714	0.286
	1.90	1.000	0.762	0.238
	2.30	1.000	0.809	0.191
	2.55	1.000	0.857	0.143
	2.70	1.000	0.904	0.096
	2.85	1.000	0.952	0.048
	3.85	1.000	1.000	0.000

3. Tables of nominal significance levels may be used if no program for an exact test is available. Asymptotic approximations are available for large samples. Many published tables for the Smirnov test give not critical values, but quantiles which must be exceeded for significance at a given level. Neave (1981, p. 31) gives values for the equivalent $mn[\max|S1(x) - S2(y)|]$ for significance for a wide range of sample sizes.

4. The rationale behind this test is sketched by Sprent (1998, Example 6.14).

Computational aspects. Several standard packages give programs for the Smirnov test (often referred to as the Kolmogorov–Smirnov test), but care should be taken to see whether each test uses only an asymptotic approximation, or a more appropriate approach for small samples.

Chandra, Singpurwalla and Stephens (1981) developed a test of this type for the Weibull distribution, often met with survival data. Modifications of the Smirnov test for discrete data are discussed by Eplett (1982), while Saunders and Laud (1980) devised a multivariate Kolmogorov goodness-of-fit test.

6.5.2 The Cramér–von Mises Test for Identical Populations

The differences $S_1(x) - S_2(y)$ at each sample value are not independent. This makes it difficult to work out the distribution for statistics that take account of all differences. However, for one such test, the *Cramér–von Mises test* (Cramér, 1928; von Mises, 1931), an approximate theory gives simple significance tests that, except for very small samples, are virtually independent of sample size, and these are useful if one only wants broad measures of evidence for or against identity rather than more precise P-values.

The test statistic is a function of the sum of squares of the differences $S_1(x) - S_2(y)$ at all sample points. Denoting this sum of squares by S_d^2 the test statistic is $T = mnS_d^2/(m + n)^2$. In a two-tail test for significance at the 5 percent level T must exceed 0.461; for significance at the 1 percent level T must exceed 0.743.

Example 6.16

The problem. Perform the Cramér–von Mises test for the data in Example 6.15.

Formulation and assumptions. We square and add the differences in the last column of Table 6.7, then form the statistic T.

Procedure. From the last column of Table 6.7 we find the sum of squares of the differences is 1.6945, whence $T = 7 \times 21 \times 1.6945/(28 \times 28) = 0.318$.

Conclusion. Since $T < 0.461$ we cannot conclude that the population cumulative distribution functions differ for at least some x.

Comment The Cramér–von Mises test is often more powerful than the Smirnov test and is easy to use because of the approximation. Accurate tables exist for some values of m, n. The only labour additional to that for the Smirnov test is calculation of the sums of squares of differences. In this example the result is not inconsistent with that for the two-tail Smirnov test.

6.5.3 The Wald–Wolfowitz Runs Test

Wald and Wolfowitz (1940) proposed a test based on runs for a population difference when we are given two independent samples. Observations in the first sample are coded as *zero* and those in the second sample are coded as *one*. They are then combined, ranked and placed in order of size. Each observation is replaced by its code to form runs of zeros and ones. These runs are analysed for clustering or alternation by the method described in Section 4.5. An example is given in Gibbons and Chakraborti (2004, Section 6.2) and in Sprent (1998, Section 6.9). StatXact provides a dedicated program for an exact test. Difficulties arise in assigning runs if there are tied values in different samples. For the data in Example 6.1 the test leads to an exact one-sided P = 0.680 compared to an exact two-sided P = 0.0159 for the Wilcoxon test, reflecting the often low power of this run test. However, as the Wilcoxon test is specifically designed to look at the possibility of differences in location whereas the Wald–Wolfowitz test is designed to pick up unspecified differences it would be fairer to compare performance with that of the Smirnov test or the Cramér–von Mises test.

6.5.4 The Two Sample Runs Test on a Circle

Suppose we have two independent samples of angular data, A and B. In this test, the points are plotted on a circle and labelled A and B according to the sample from which they came. The number of runs around the perimeter of the circle is noted. In proceeding around the circle once, the final run and initial run must be from different groups. The number of runs has therefore to be even (if the extreme points are from the same group when arranged on a line, the number of runs is reduced by one when the points are arranged on a circle).

Where there is a tendency for points from the same group to cluster, the number of runs is relatively small; for alternation the number of runs is relatively large (see Section 4.5).

If R and R_C represent the number of runs when the data are arranged on a line and on a circle respectively, then using the notation of Section 4.5 the lower tail probabilities for the number of runs on the circle, $2s$, are given by:

$$\Pr(R_C \leq 2s) = \Pr(R \leq 2s + 1)$$

with upper tail probabilities calculated in a similar manner.

Example 6.17

The problem. The midwife in Example 4.16 also recorded the gender of each of the 12 home deliveries. She was interested in whether or not the times of births for boys and girls occurred in clusters. The data were as follows (M = male, F = female)

Time	0100	0300	0420	0500	0540	0620
Sex	M	F	F	F	M	M

Time	0640	0700	0940	1100	1200	1720
Sex	F	F	F	F	M	M

Formulation and assumptions. We test the null hypothesis H_0 that for the populations of male and female births the times of delivery are randomly ordered with respect to gender. Since the midwife is interested only in clustering and not in alternation, a one-tail test is appropriate. A small value for the number of runs, $2s$, indicates possible clustering.

Procedure. The data are plotted on a circle. The numbers of boys and girls are $m = 5$ and $n = 7$ respectively. The count of runs starts from the beginning of a particular run. We therefore count runs from the female birth that occurred at 0300. The number of runs, $2s$, is four. As reasoned above, we could not have had five runs on a circle for these two groups. The relevant P-value for the angular case is therefore given by $\Pr(R \le 5)$ where R is the number of runs for such data arranged along a line. This P-value is 0.197.

Conclusion. There is no strong evidence against H_0. On the basis of these data, the times of delivery for boys and girls do not seem to cluster with respect to gender.

Comment. Recall that if the number of runs for the data on a line is odd, the number of runs is one fewer when the data are plotted on a circle. This implies that low P-values are only obtained from small studies in the most extreme situations. For instance, unless both m and n are at least eight, there will only be strong evidence against H_0 (formally $P < 0.05$) if the number of runs is two.

6.6 Power and Sample Size

In the light of the discussion in Section 5.3 for single samples it is not surprising that power and sample size results for the median test and the WMW test are not distribution-free. Not unexpectedly, the situation for the WMW test is more complicated than that for the median test because the latter is an extension from the sign test and therefore essentially only involves binomial distributions whereas the WMW test involves the distribution of ranks which may be complicated under H_1.

6.6.1 Power and Sample Size for the Median Test

The hypotheses only involve possible differences between unspecified population medians θ_1 and θ_2. If we denote this difference by $\delta = \theta_2 - \theta_1$ the null-hypothesis is H_0: $\delta = 0$ and the median test is equivalent to testing whether the samples support a hypothesis that the probability that a sample value is above the common median is $p = 0.5$ for each sample. If we use the combined sample median as our estimate of the common median then under H_0 the probability is only approximately $p = 0.5$ for reasons indicated in Section 6.2. Under an alternative hypothesis H_1 specifying some value $\delta \ne 0$ the probabilities p_1, p_2 for sample values to lie above the combined sample median will be unequal, but estimation of p_1, p_2 and the interpretation of the relationship between the combined sample median and the two population medians depends on both the nature of those populations and the relative sample sizes.

The combined sample median is only easy to interpret if samples are from populations that differ, if at all, only by a median shift. The interpretation is

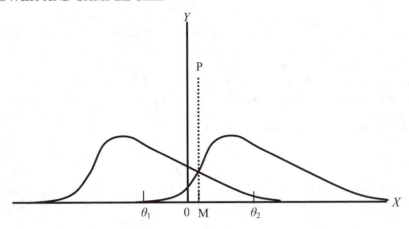

Figure 6.1 *Probability density functions for two distributions that differ only by a shift in median or mean. The point M lies midway between the population medians θ_1, θ_2 and is the median of a population consisting of an equal mixture of the two populations. The combined sample median provides an estimate of this point if both samples are of equal size $m = n$.*

simplest when $m = n$. Then, if H_1 specifies a fixed difference δ between the medians, i.e., $H_1 \colon \theta_2 - \theta_1 = \delta$ it is easily seen that the combined sample median is an intuitively reasonable estimator of a constant midway between θ_2 and θ_1, i.e., an estimator of $\theta_1 + 0.5\delta = \theta_2 - 0.5\delta$. Figure 6.1 shows probability density functions for two arbitrary distributions that differ only in their medians θ_1, θ_2. The dotted vertical line PM meets the x-axis at the point M where $\theta_1 + 0.5\delta = \theta_2 - 0.5\delta$.

If we assume that under H_0 the common value of the median is θ_1 and that the common probability density function $f(x)$ is completely defined, we may calculate the approximate power of the median test under these restrictive assumptions by computing a probability p_1 that under H_1 a sample value from the specified distribution with median θ_1 takes a value greater than $\theta_1 + 0.5\delta$. Similarly, p_2 is computed for a similar distribution with the median shifted to $\theta_1 + \delta$. Programs are now fairly readily available for power calculations for this situation when we have assigned median shifts for completely specified distributions. We illustrate the method in Examples 6.16 and 6.17. However, this approach is unrealistic in many practical situations where we are likely to use the median test, for these are conditions where a WMW test may in general be more powerful.

In more complicated situations where a median shift is the hypothesized difference an approach based on order statistics may be used. However, here again some knowledge of the precise form of the distribution is required. The method is outlined by Gibbons and Chakraborti (2004, Section 6.4) who also give an asymptotic approximation available for unequal sample sizes.

An alternative approach to obtaining the sample size required to obtain a given power that is probably a reasonable approximation if $m = n$ is to use any available program to work out the sample size for a corresponding t-test with the required power and to adjust on the basis of the Pitman efficiency relative to the t-test where this is known. This is discussed briefly in the comments on Examples 6.16 and 6.17. This approach is only appropriate for m, n reasonably large in the light of the findings of Freidlin and Gastwirth (2000) referred to in Section 6.2.1.

Example 6.18

The problem. For samples of size 20 from each of two normal distributions with standard deviation 2 and unknown medians θ_1, θ_2 find the power of the median test of H$_0$: $\theta_2 = \theta_1$ against H$_1$: $\theta_2 = \theta_1 + 1$ when the probability of a type I error is $\alpha = 0.05$. What equal sample sizes would be needed to ensure power of at least 0.80?

Formulation and assumptions. Under H$_0$ the rows in the 2×2 table for the Fisher exact test in Section 6.2 may be considered as the number of positive and negative outcomes in independent samples each of size 20 when sampling from binomial distributions for which the p are equal with approximately the value $p = 0.5$. Under H$_1$ the probability that any value in the sample from the first population has a value greater than $\theta_1 + 0.5$ is given by $p_1 = \Pr(X > \theta_1 + 0.5)$ where X has a N$(\theta_1, 4)$ distribution. This is approximately the probability that a sample value from this distribution takes a value greater than the combined sample median.

Similarly, under H$_1$ the probability that any value in the sample from the second population has a value greater than $\theta_1 + 0.5$ is given by $p_2 = \Pr(X > \theta_1 + 0.5)$ where X has a N$(\theta_1 + 1, 4)$ distribution. This is approximately the probability that a sample value from this distribution takes a value greater than the combined sample median. As a consequence of the symmetry assumption, it is easily verified that $p_2 = 1 - p_1$. Approximate power and sample size determinations now reduce to considering those for the problem of testing H$_0$: $p_1 = p_2 = 0.5$ against H$_1$: p_1, p_2 *take the values computed when* H$_1$: $\theta_2 = \theta_1 + 1$ holds. Computer software exists for the latter test giving either exact or asymptotic approximations.

Procedure. For the first population $Z = (X - \theta_1)/2$ has a standard normal distribution, since X is N$(\theta_1, 4)$. Thus

$$p_1 = \Pr(X > \theta_1 + 0.5) = \Pr(Z > 0.25) = 0.4013.$$

Since $p_2 = 1 - p_1$ this implies $p_2 = 0.5987$, which may be verified directly. We used the program in StatXact for power calculations for the Fisher exact test for comparing two binomials. This provides both an asymptotic estimate of power or an option for exact computation. For $n = m = 20$ for the test considered here the program gives an exact power 0.22.

Larger samples are required to achieve power 0.80. If software is not available to estimate the sample size a trial and error approach is needed. For $m = n = 100$ StatXact gives a power 0.85. For $n = m = 90$ the required power 0.80 is achieved.

Conclusion. For $m = n = 20$ the power is 0.22. Samples of size $m = n = 90$ are required for power 0.80.

Comments. 1. Since the samples are from normal distributions differing only in mean the optimal test is a *t*-test and for the situation in this example the relevant computations (available in many statistical packages) indicate $m = n = 51$ are the sample sizes needed to attain a power 0.8. This is broadly in line with the expected size based on the Pitman efficiency of 0.64 for the median test when the *t*-test is optimal. If we assume the Pitman efficiency is reasonable for the test under consideration this would suggest the sample size $51/0.64 = 80$ approximately. However it should be kept in mind that this is only a limiting result concerning the power for small departures independent of the choice of α, β and that appreciable variations from this estimate are possible for finite samples, specific alternatives, and particular choices of α, β. However, using a sample size based on the relevant Pitman efficiency will usually give an indication of the size needed to give reasonable power in a practical context providing samples are not small or unbalanced.

2. The theory behind power and sample size computations for the Fisher exact test on which the results obtainable from StatXact are based is described in the StatXact 7 manual (Chapter 31). StatXact computes an unconditional power that ignores the restriction implicit in the median test that numbers of observations above and below the combined sample median are fixed. In practice this unconditionality may tend to overestimate the power of this particular test but in our experience this overestimation is only slight in the context here.

3. The restriction to $m = n$ is not essential to the binomial comparison problem, but it is needed to induce the property that the sample median is a reasonable estimate of $\theta_1 + 0.5\delta$. This in turn depends on the assumption that samples come from identical distributions apart from possible differences between medians. Relaxation of these assumptions changes the function of θ_1 and θ_2 that is estimated by M.

4. In the light of common uncertainty about the exact nature of the populations being sampled asymptotic power calculations for comparing two binomials available in many software packages will often suffice for obtaining sample sizes needed to give substantial power since these sample sizes are often fairly large, and an approximate sample size is often all that is required.

Example 6.19

The problem. For samples of size 20 from each of two double exponential distributions with standard deviation 2 and unknown medians θ_1, θ_2 find the power of the median test of H$_0$: $\theta_2 = \theta_1$ against H$_1$: $\theta_2 = \theta_1 + 1$ when the probability of a type I error is $\alpha = 0.05$. What equal sample sizes would be needed to ensure power of at least 0.80?

Formulation and assumptions. As in Example 5.6 a double exponential distribution with mean θ and standard deviation 2 has frequency function

$$f(x) = \frac{1}{2\sqrt{2}}\exp[-|(x - \theta)|/\sqrt{2}]. \tag{6.10}$$

Using arguments similar to those in Example 6.16 under H$_0$ the rows in the 2×2 table for the Fisher exact test in Section 6.2 may be considered as the number of positive and negative outcomes in independent samples each of size 20 when sampling from binomial distributions for which the p are equal with approximately the value $p = 0.5$. Under H$_1$ the probability that any value in the sample from the first population has a value greater than $\theta_1 + 0.5$ is given by $p_1 = \Pr(X > \theta_1 + 0.5)$

where X has a double exponential distribution given by (6.10) with mean θ_1 and this is approximately the probability that a sample value from this distribution takes a value greater than the combined sample median. Similarly, under H_1 the probability that any value in the sample from the second population has a value greater than $\theta_1 + 0.5$ is given by $p_2 = \Pr(X > \theta_1 + 0.5)$ where X has a double exponential distribution with mean $\theta_1 + 1$. This is approximately the probability that a sample value from this distribution takes a value greater than the combined sample median. Because of symmetry it is easily verified that, as in the previous example, $p_2 = 1 - p_1$. Approximate power and sample size determinations now reduce to considering those for the problem of testing H_0: $p_1 = p_2 = 0.5$ against H_1: p_1, p_2 *take the values computed when* $\theta_2 = \theta_1 + 1$.

Procedure. For the first population sampled p_1 under H_1 is easily computed by integrating (6.10) over the interval $(\theta + 0.5, \infty)$ where $\theta = \theta_1$ (see Exercise 6.18). This gives $p_1 = 0.3511$. Since $p_2 = 1 - p_1$ this implies $p_2 = 0.6489$. This may be verified directly by integrating (6.10) over the interval $(\theta + 0.5, \infty)$, where now $\theta = \theta_1 + 1$. The StatXact program used in Example 6.16 gives for $n = m = 20$ for the test considered here an exact power 0.45. Larger samples will be needed to achieve power 0.80. If no software is available to estimate the sample size directly a trial and error approach is needed. For $m = n = 40$, StatXact gives a power 0.79. For $n = m = 41$ the required power 0.80 is achieved.

Conclusion. For $m = n = 20$ the power is 0.45. Samples of size $m = n = 41$ are required for power 0.80.

Comments. 1. The t-test is no longer optimal so a direct comparison is inappropriate. However, in view of its non-optimality the t-test would require a larger sample size to achieve the same power as it would have for the corresponding normal test, where, as indicated in *Comment* 1 on Example 6.16 we require $m = n = 51$ to attain a power 0.8, indicating, in line with the Pitman efficiency of 2.0 relative to the t-test, that the median test is preferable here.

2. Comments 2, 3 and 4 on Example 6.16 are also relevant here.

3. The median test has higher Pitman efficiency than the WMW test for double exponential distributions, that differ if at all, only in medians. However, several authors have indicated that for small or unbalanced samples (i.e., samples of very different sizes) the median test may not have as high a power as the WMW test for some alternatives specified in H_1.

6.6.2 Power and Sample Size for the WMW Test

Added complexities of power and sample size computations for the Wilcoxon signed-rank test relative to the sign test carry over to the WMW test relative to the median test. We consider here only an approximation due to Noether (1987a) similar to that given for the signed-rank test in (5.3) for determining the equal sample sizes to ensure a given power when testing for a median shift only in otherwise identical distributions (which need not be symmetric).

The analogous formula to (5.3) is

$$n \approx \frac{(z_\alpha + z_\beta)^2}{6(p_1 - 0.5)^2} \qquad (6.11)$$

where $p_1 = \Pr[(Y - X) > 0]$. The samples are from populations of random variables X, Y that differ only in medians and $n = m$ is the common sample size. The other terms have the same meaning as in (5.3). The distribution of $Y - X$ depends on the nature of the distributions of X and Y and is in general obtained by standard distribution theory for change of variables. In simple situations like that when X, Y are both normally distributed p_1 is easily computed as we see in the following example.

Example 6.20

The problem. Samples of equal size n are to be taken from two normal distributions each with standard deviation 2 and unknown medians θ_1, θ_2. Use (6.11) to estimate n so that the power of the WMW test of H_0: $\theta_2 = \theta_1$ against H_1: $\theta_2 = \theta_1 + 1$ is at least 0.80 when the probability of a type I error is $\alpha = 0.05$.

Formulation and assumptions. Relevant values of z_α, z_β are obtained from tables or using appropriate software. Normal distribution theory tells us that $Y - X$ is in this case distributed $N(1, 8)$ whence p_1 is easily obtained. Relevant values are substituted in (6.11).

Procedure. We leave it to the reader (Exercise 6.19) to show that $z_\alpha = 1.645$, $z_\beta = 0.842$ and $p_1 = 0.6381$, whence substitution in (6.11) gives $n \approx 54.05$.

Conclusion. Rounding up the result suggests samples of size 55 are needed to attain the required power.

Comment. These results for the normal distribution are consistent with the known Pitman efficiency of $3/\pi \approx 0.95$ relative to the t-test in this case because we noted in Example 6.17 that the t-test sample sizes needed for this power are $n = 51$, giving a relative efficiency for the WMW test of $51/55 \approx 0.93$, in close agreement with the Pitman efficiency $3/\pi$.

Hollander and Wolfe (1999, Section 4.1) generalize (6.11) to unequal sample sizes and discuss some further results for asymptotic approximate power calculations.

6.7 Fields of Application

Little imagination is needed to think of realistic situations where one might wish to compare medians or means of two populations (i.e., look for centrality or treatment differences) on the basis of two independent samples. Here are a few relevant situations.

Medicine

For comparing the efficacy of two drugs for reducing hypertension, blood cholesterol levels, relief of headaches or other conditions, independent samples are often needed because of *interaction* or *hangover* effects if each drug is given to the same patients (Section 5.1). For this, or ethical reasons, it may be inappropriate to give both drugs to any one person even after a considerable time lapse.

Sociology

To explore the possibility that town and country children may attain different levels of physical fitness, samples of each might be scored on an appropriate scale in a fitness test and the results compared nonparametrically.

Mineral Exploration

A mining company has options on two sites but only wishes to develop one. Sample test borings are taken at each and the percentage of the mineral of interest in each boring is determined. These provide the basis for a test for population differences in mean or median levels. If there is evidence that one site is richer the company may want to estimate the difference because development costs, and other factors, may well differ between sites. This would call for confidence intervals and power considerations may come into play to determine whether a larger experiment is need to provide a sound basis for a decision. Results may need to be interpreted with caution because observations from borings close together may not be independent.

Manufacturing

New and cheaper production methods are often tried. Manufacturers may compare products using a new process or raw material with an existing one to assess quality and durability. Interest here is often not only in average quality, but also in differences in variability.

Psychology

Children with learning difficulties may be given a treatment that it is hoped will encourage them to respond to commands. Sixty commands are given to a sample of 10 treated children and to a further sample of 12 untreated children. The number of favourable responses is recorded for each child. Interest will lie in whether there is a response level difference and a confidence interval for any response shift is likely to be informative.

6.8 Summary

The **Wilcoxon–Mann–Whitney test** (Sections 6.1.2 and 6.1.3) is a rank test for differences in centrality or domination of one distribution over the other. The test statistic is the rank sum associated with either sample. An alternative formulation was given by Mann and Whitney. Confidence intervals are given in Section 6.1.4. Asymptotic results are given Section 6.1.5. Adjustments for ties are given in Section 6.1.6.

The **median test** is a test for centrality based on numbers in each of two samples above and below the combined sample median. Confidence intervals

(Section 6.2.2) are available when we have full measurement data. The test is not recommended if assumptions allow use of generally more powerful tests such as the WMW test. A modified test is the **control median test**.

Among tests based on transformation of ranks that using **van der Waerden scores** (Section 6.3) is easy to use, but results are often similar to those given by the WMW test.

For Spread or dispersion the **Siegel–Tukey** test (Section 6.4.1) or **Ansari–Bradley** type tests (Section 6.4.2) are often less powerful and less robust than the **Conover squared-rank** test (Section 6.4.3). The **Moses** test (Section 6.4.4) for extreme reactions is relatively simple, but being based on a range of ranks it is susceptible to outliers. Many dispersion tests require for strict validity that both populations medians or means are the same or that the difference, if any, between them is known.

For comparison of continuous distributions the **Smirnov test** (Section 6.5.1) has analogies with the one-sample Kolmogorov test (Section 4.2.1). The **Cramér–Von Mises test** (Section 6.5.2) is easy to use as conventional significance levels are almost independent of sample size except for very small samples. The **Wald–Wolfowitz runs test** is briefly described in Section 6.5.3. A two-sample runs test can be applied to angular data (Section 6.5.4).

Power and sample size. Studies of **power** and *sample size* (Section 6.6) are reasonably straightforward only for the median test using samples of the same size from distributions that differ only in mean or median, a situation where that test is often not the most appropriate! In similar situations, and some that are slightly more general, reasonable asymptotic approximations are available for the WMW test.

6.9 Exercises

6.1 Use (6.5) to compute the exact variance using mid-ranks in Example 6.6 and compare it with the result given in that example.

*6.2 An alloy is composed of zinc, copper and tin. It may be made at one of two temperatures H (higher) or L (lower). We wish to know if one temperature produces a harder alloy. A sample is taken from each of 9 batches at L and 7 at H. To arrange them in ascending order of hardness, all specimens are scraped against one another to see which makes a deeper scratch (a deeper scratch indicates a softer specimen). On this basis the specimens are ranked 1 (softest) to 16 (hardest) with the results given below. Should we reject the hypothesis that hardness is unaffected by temperature? State any assumptions needed for validity of the test you use.

Temp	H	L	H	H	H	L	H	L	L	H	H	L	L	L	L	L
Rank	1	2	3	4	5	6	7	8	9	10	11	12	13	14	15	16

6.3 If we have samples of m, n where $m + n = 20$ and the only tie is one between ranks 2 and 3 what would be the differences between the van der Waerden scores if the scores for these ties are based on the mid-rank 2.5 and the van der Waerden scores if these are based on the mean of the van der Waerden scores corresponding to rank 2 and to rank 3. Carry out a similar comparison if there were ties for ranks 10, 11 and 12.

*6.4 Hotpot stoves use a standard oven insulation. To test its effectiveness they take random samples from the production line and heat the ovens selected to 400°C, noting the time taken to cool to 350°C after switching off. For a sample of 8 ovens the times in minutes are:

$$15.7 \quad 14.8 \quad 14.2 \quad 16.1 \quad 15.3 \quad 13.9 \quad 17.2 \quad 14.9$$

They decide to explore a cheaper insulation, and using this on a sample of 9 the times taken for the same temperature drop are:

$$13.7 \quad 14.1 \quad 14.7 \quad 15.4 \quad 15.6 \quad 14.4 \quad 12.9 \quad 15.1 \quad 14.0$$

Are the firm justified in asserting there is no real evidence of a different rate of heat loss? Obtain a 95 percent confidence limit for the difference in median heat loss (a) with and (b) without a normality assumption. Comment critically on any differences between your conclusions.

6.5 A psychologist wants to know whether men or women are more upset by delays in being admitted to hospital for routine surgery. He devises an anxiety index measured on patients 1 week before scheduled admission and records it for 17 men and 23 women. These are ranked 1 to 40 on a scale of increasing anxiety. The sum of the ranks for the 17 men is 428. Is there evidence that anxiety is gender-dependent? If there is, which gender appears to show the greater anxiety?

*6.6 Suppose we are given the data for response times in LVF and RVF in Table 5.1, but the information that they are paired is omitted. In these circumstances we might analyse them as independent samples. Would we then conclude that responses in the two fields differed? Does your conclusion agree with that found in Example 5.1? If not, why not?

6.7 Hill and Padmanabhan (1984) give body weights (g) of diabetic and normal mice. Is there evidence of a significant difference in mean body weight? Obtain the Hodges–Lehmann estimate of the difference together with a 95 percent confidence interval. Compare this interval with that based on the t-distribution.

Diabetic	42	44	38	52	48	46	34	44	38				
Normal	34	43	35	33	34	26	30	31	27	28	27	30	37
	38	32	32	36	32	32	38	42	36	44	33	38	

*6.8 The journal *Biometrics* published data on the numbers of completed months between receipt of a manuscript for publication and the first reply to the authors for each of the years 1979 and 1983. The data are summarized below. Is there evidence of a difference in average waiting times between 1979 and 1983?

Completed months	0	1	2	3	4	5	6
1979	26	28	34	48	21	22	34
1983	28	27	42	44	17	6	16

*6.9 The data below are numbers of words with various numbers of letters in 200-word sample passages from the presidential addresses to the Royal Statistical Society by W.F. Bodmer (1985) and J. Durbin (1987). Is there acceptable evidence of a difference between the average lengths of words used by the two presidents?

Number of letters	1–3	4–6	7–9	10 or more
Bodmer	91	61	24	24
Durbin	87	49	32	32

6.10 Carter and Hubert (1985) give data for percentage variation in blood sugar over 1–hour periods for each of two groups of nine rabbits given different dose levels of a drug. Is there evidence of a response difference between levels?

Dose I	0.21	−16.20	−10.10	−8.67	−11.13
	1.96	−10.19	−15.87	−12.81	
Dose II	1.59	2.66	−6.27	−2.32	−10.87
	7.23	−3.76	3.02	15.01	

6.11 The following data are DMF scores for 34 male(M) and 54 female(F) first-year dental students. The DMF score is the total of the numbers of decayed + missing + filled teeth.

M	8	6	4	2	10	5	6	6	19	4	10	4	10	12	7	2	5
	1	8	2	0	7	6	4	4	11	2	16	8	7	8	4	0	2
F	4	7	13	4	8	8	4	14	5	6	4	12	9	9	9	8	12
	4	8	8	4	11	6	15	9	8	14	9	8	9	7	12	11	7
	4	10	7	8	8	7	9	10	16	14	15	10	4	6	3	9	3
	10	3	8														

Use an asymptotic WMW test to determine whether the DMF score differs significantly between males and females. Do tied ranks have much influence on the appropriate test statistic?

6.12 Lindsey, Herzberg and Watts (1987) give data for widths of first joint of the second tarsus for two species of the insect *Chaetocnema*. Do these indicate population differences between the width distributions for the two species?

Species A	131	134	137	127	128	118	134	129	131	115
Species B	107	122	144	131	108	118	122	127	125	124

*6.13 Perform a Siegel–Tukey test on the data in Example 6.11 after shifting one set of sample values to align the sample means.

6.14 Using the data in Example 6.11 carry out a test analogous to the Conover squared rank test but using absolute ranks instead of squared ranks.

6.15 A psychologist notes total time (in seconds) needed to perform a series of simple manual tasks for each of eight children with learning difficulties and seven children without learning difficulties. The times are:

Without difficulties	204	218	197	183	227	233	191
With difficulties	243	228	261	202	343	242	220	239

Use a Smirnov test to find whether the psychologist is justified in asserting these samples are likely to be from different populations. Do you consider a one- or a two-tail test appropriate? If you think the psychologist should have tested more specific aspects of any difference, perform the appropriate tests.

6.16 The numbers of words in the first complete sentence on each of 10 pages selected at random is counted in each of the books by Conover (1980) and Bradley (1968). The results were:

Conover	21	20	17	25	29	21	32	18	32	31
Bradley	45	14	13	31	35	20	58	41	64	25

Perform tests to determine whether there is evidence that in these books (i) sentence lengths show a difference in centrality; (ii) the variances of sentence lengths differ between authors; (iii) the distributions of sentence lengths differ in an unspecified way; (iv) the sentence lengths for either author are not normally distributed.

*__*6.17__* Records from a maternity hospital show that during a particular year, identical and nonidentical twins were born on the following days:

Identical	Jan 3	Feb 28	Apr 14	May 31	Aug 6
	Oct 2	Dec 7			
Nonidentical	Jan 30	Feb 28	Mar 5	Apr 13	Jun 2
	Jun 29	Jul 8	Jul 17	Aug 4	Sep 17
	Oct 9	Oct 26	Nov 17	Dec 11	

Is there evidence of possible clustering of births of identical and nonidentical twins?

6.18 Confirm the value of p_1 under H_1 for the double exponential distribution introduced in Example 6.19 that is quoted in the *Procedure* section of that example.

6.19 Verify the correctness of the required sample sizes given in Example 6.20 for the WMW test in the circumstances given there.

6.20 Show that if we have independent samples of size m, n and there are r values above the combined sample median in the first sample and no sample values are equal to the combined sample median, then the expression for P^* given in Example 6.7 generalizes to

$$P^* = \frac{\binom{m}{r} \times \binom{n}{k-r}}{\binom{m+n}{k}},$$

where $k = (m+n)/2$.

CHAPTER 7

BASIC TESTS FOR THREE OR MORE SAMPLES

7.1 Comparisons with Parametric Methods

Parametric one- or two-sample analyses of continuous data using test and estimation procedures based on normality are modified for three or more samples. Emphasis shifts from the t-distribution and related test and estimation procedures to the analysis of variance (ANOVA) and tests based on the F-distribution. The t-distribution, however, is related to a particular case of the F-distribution. The concept of experimental design becomes more important. Readers familiar with the analysis of variance in a parametric context, or who are not particularly concerned with that aspect, need only skim through this section briefly, or may prefer to proceed directly to Section 7.2.

In this chapter we are concerned largely, but not exclusively, with nonparametric analogues of situations arising in the one-way classification ANOVA and the simplest two-way classifications. Multiway, especially more general two-way classifications including factorial structures and some more advanced related topics are covered in Chapter 8.

Given k independent samples from normal distributions all with the same, but usually unknown, variance and means $\mu_1, \mu_2, \ldots \mu_k$, the basic overall parametric significance test is that of H_0: $\mu_1 = \mu_2 = \ldots, = \mu_k$ against H_1: *not all μ_i are equal*. When the relevant assumptions hold, in ANOVA terminology large values of the statistic

$$F = \frac{between\ samples\ mean\ square}{within\ samples\ mean\ square}$$

indicate evidence against H_0. Under H_0, F has an F-distribution with $k-1$ and $N-k$ degrees of freedom, where N is the total number of observations. When appropriate, H_0 is sometimes expressed as the hypothesis of *no difference between treatments*.

The above F-test is usually a preliminary to more specific tests and estimation procedures concerning possible differences within chosen subsets. Such differences are referred to as *contrasts*, The simplest of these is between pairs of the $\mu_i, i = 1, 2, \ldots, k$. New concepts, such as least significant differences or multiple comparison tests, are then sometimes introduced.

The two-dependent-samples situation in Chapter 5 generalizes to designed experiments that at their most basic are two-way classifications. Then the form of the overall parametric test of H_0: *no difference between treatments* against H_1:*some treatments differ in location* and related estimation procedures depend on the experimental design. A well-known design is that of randomized blocks. The nature of the treatment structure also influences the analysis; factorial experiments not only allow us to study the basic effect of two or more factors, but also whether factors interact with one another. We consider nonparametric analogues in Chapter 8.

In normal theory inference linear models provide a framework that links analysis of variance and linear regression analysis. Nonparametric analyses parallel some aspects of the normal theory linear model, but relaxing assumptions means that some linear model techniques have no direct nonparametric analogue.

We develop first overall tests for treatment differences analogous to those used in the analysis of variance. These generalize some of the methods in Chapters 5 and 6.

Having more than two samples often means there are sufficiently many observations for asymptotic results to be reasonable, but this may not be so if some samples are very small even when the total number of data is large. Overreliance on asymptotic results when some samples are small has in the past reflected a paucity of tables giving even nominal significance levels for permutation tests based on ranks, etc. Programs like StatXact and Testimate and increasing facilities for exact test procedures in general packages such as SAS and SPSS have eased this problem. Experienced users of some of the procedures in R may also be able to apply exact tests.

In this and the following six chapters we shall find that the same nonparametric technique may often be applied to problems that appear at first sight to be different. We have already met examples of such equivalences; e.g., the sign test and McNemar's test (Section 5.2), also that between Wilcoxon's signed-rank test in a specific highly-tied situation and the sign test (Exercise 3.5). We also introduced the median test (Section 6.2) as a special case of the widely applicable Fisher exact test (Section 12.2.1).

7.2 Centrality Tests for Independent Samples

7.2.1 Raw Data Permutation Tests

A permutation test based on raw data developed by Pitman (1938) is seldom used in practice, primarily because of its lack of robustness, but also because exact P-values for a test corresponding to the overall F-test in the one-way classification ANOVA are only obtainable for fairly small samples. Thus, one has to resort to Monte Carlo approximations or to asymptotic results if samples are not all small. StatXact includes a program under the title *One-way ANOVA with general scores* that may be used for the test by taking the raw data as scores.

7.2.2 The Kruskal–Wallis Test

Kruskal and Wallis (1952) extended the WMW test using the Wilcoxon formulation to three or more samples. As well as being applicable to data consisting only of ranks, it is relevant as an overall test for equality of population means or medians when samples are from otherwise identical and continuous distributions. It also has reasonable power for testing for identical population cumulative distribution functions against an alternative that one or more of these cumulative distributions are distinct in the sense of dominating the others. As in the two-sample case, a shift in mean or median is often referred to as an additive treatment effect. This term is widely used in parametric (normal theory) linear models.

Suppose we have k random samples, the ith sample $(i = 1, 2, \ldots, k)$ consisting of n_i observations. We write x_{ij} for a typical element of the ith sample where $j = 1, 2, \ldots, n_i$. The total number of observations is $N = \sum_{i=1}^{k} n_i$. We assume there are no ties, and rank the N observations from smallest (rank 1) to largest (rank N). Let r_{ij} be the rank allotted to x_{ij} and $s_i = \sum_{j=1}^{n_i} r_{ij}$ be the sum of the ranks for the ith sample. We compute $S_k = \sum_{i=1}^{k}(s_i^2/n_i)$. The test statistic is then

$$T = \frac{12S_k}{N(N+1)} - 3(N + 1). \tag{7.1}$$

If all the samples are from the same population we expect a mixture of small, medium and high ranks in each sample, whereas under the alternative hypothesis high (or low) ranks may dominate in one or more samples. Consequently, under the alternative hypothesis S_k will contain the square of at least one relatively large rank sum, leading to larger values of T.

Computing exact permutation probabilities under H_0 by hand is impractical except for small samples. For N moderate or large, T has a chi-squared distribution with $k - 1$ degrees of freedom under H_0.

Example 7.1

The problem. A British estate agency sells properties in villages A, B and C. The homes for sale in these villages are at the following prices $(\pounds K)$:

$$\begin{array}{lcccc}
A & 139 & 145 & 171 & \\
B & 151 & 163 & 188 & 197 \\
C & 199 & 250 & 360 &
\end{array}$$

Use the Kruskal-Wallis test to examine the validity of the hypothesis that the house prices in these samples come from identical populations.

Formulation and assumptions. A suitable null hypothesis is H_0: *the populations from which the samples were drawn have identical medians* while the alternative is H_1: *the population medians are not all equal.*

Table 7.1 *House prices £K and ranks in three villages.*

Village A	price	139	145	171	
	rank	1	2	5	
Village B	price	151	163	188	197
	rank	3	4	6	7
Village C	price	199	250	360	
	rank	8	9	10	

Procedure. The $N = 10$ house prices are ranked overall and the ranks assigned to the appropriate groups. These are shown in Table 7.1.

The sum of ranks is $1+2+5 = 8$ for the houses in Village A, $3+4+6+7 = 20$ for Village B and $8+9+10 = 27$ for Village C, whence $S_k = (8)^2/3+(20)^2/4+(27)^2/3 = 364.33$. Hence (7.1) gives

$$T = \frac{12 \times 364.33}{10 \times 11} - 3 \times 11 = 6.745.$$

This corresponds to an exact $P = 0.010$.

Conclusion. There is strong evidence that house prices differ between villages.

Comments. 1. Properties in Village C seem to be more expensive. This may be because the village has more executive-style homes. One could only make an informed comparison of house prices in the three villages by looking at similar types of property. Remember too that properties offered for sale may not be representative of the local housing stock. We have a small non-random sample which may not be representative of the "population" of village properties.

2. We could use S_k as a test statistic rather than T since all other quantities in T are invariant in the exact permutation distribution (i.e., are the same for any permutation). The use of T is preferred for practical reasons (e.g., availability of tables and for asymptotic approximations).

3. Calculating the P-value by hand is tedious. StatXact computes an exact P directly from the above data. Tables giving critical values of T for small sample sizes are given by Neave (1981, pp. 32–34) while Hollander and Wolfe (1999, Table A12) give upper-tail P-values for large T for a range of sample sizes. These are strictly relevant only to a no-tie situation. Good estimates of exact probabilities are also available using the Monte Carlo facility in StatXact. StatXact also gives the asymptotic chi-squared approximation, as does Stata, and many other general statistical packages. With $T = 6.745$ this gives $P = 0.0343$, considerably different from the exact value indicating, not surprisingly, that the asymptotic result can be misleading for small samples.

4. The test is scale invariant and exactly the same ranks and T value would be obtained if the prices were converted to US\$, euros, or any other currency, at the ruling exchange rate.

7.2.3 The Kruskal–Wallis Test with Ties

As in the WMW test we give mid-rank values to tied data. With ties a more general formula for T is needed. In addition to the terms calculated above, we also calculate $S_r = \sum_{i,j} r_{ij}^2$, where some of the r_{ij} will now be mid-ranks. Readers familiar with the analysis of variance will recognize S_k and S_r as uncorrected treatment and total sums of squares for ranks. An appropriate correction for the mean is subtracted from each, namely $C = N(N+1)^2/4$. The test statistic is

$$T = \frac{(N-1)(S_k - C)}{S_r - C}. \tag{7.2}$$

If there are no ties, this can be shown to be equivalent to (7.1), since with no ties $S_r = N(N+1)(2N+1)/6$.

Example 7.2

The problem. Uniform editions by each of three writers of detective fiction are selected. The numbers of sentences per page, on randomly selected pages in a work by each are

C.E. Vulliamy	13	27	26	22	26		
Ellery Queen	43	35	47	32	31	37	
Helen McCloy	33	37	33	26	44	33	54

Use the Kruskal-Wallis test to examine the validity of the hypothesis that these may be samples from identical populations.

Formulation and assumptions. The null hypothesis is that the populations are identical. We consider the alternative hypothesis that the samples are from populations that are not identically located.

Procedure. Manual calculation of the required statistic is outlined, but this is only needed if no program is available to compute an exact P-value. Ranks are shown in Table 7.2. From that table we find that $s_1 = 1 + 2 + 4 + 4 + 6 = 17$, $s_2 = 72.5$ and $s_3 = 81.5$, whence

$$S_k = \frac{17^2}{5} + \frac{72.5^2}{6} + \frac{81.5^2}{7} = 1882.73.$$

The sum of squares of allocated ranks is $S_r = 2104.5$, and since $N = 18$, $C = 18 \times (19)^2/4 = 1624.5$. We compute

$$T = \frac{17(1882.73 - 1624.5)}{2104.5 - 1624.5} = 9.146.$$

Tables are not readily available for this mix of sample sizes but they show that $T \geq 8.157$ implies strong evidence against H_0 (significant at the 1 percent level). StatXact confirms the above value of T and gives an exact $P = 0.0047$.

Conclusion. There is strong evidence to support a difference in sentence length. Inspection of the data indicates that Vulliamy uses longer sentences (fewer sentences per page).

Table 7.2 *Sentences per page and ranks in samples for three authors.*

Vulliamy	Number	13	22	26	26	27		
	rank	1	2	4	4	6		
Queen	Number	31	32	35	37	43	47	
	rank	7	8	12	13.5	15	17	
McCloy	Number	26	33	33	33	37	44	54
	rank	4	10	10	10	13.5	16	18

Comments. 1. Using uniform editions avoids difficulties that may arise with different page sizes or type fonts.

2. The total number of observations is small ($N = 18$). However, if we perform an asymptotic test $T = 9.146$ is just short of the chi-squared value 9.21 required for significance at the 1 percent level. Indeed $P = 0.0103$ for this chi-squared approximation.

Computational aspects. Monte Carlo estimates are provided by StatXact. These may be useful for sample sizes a little larger than those in this example. The computational burden for exact P-values rapidly increases as N increases. Most standard packages such as Minitab and Stata include the Kruskal–Wallis test, usually computing T but giving only an asymptotic P-value. The procedure *kruskal.test* in R also computes exact P-values.

For the above example most statisticians would consider a parametric analysis of variance to be suitable. This leads to basically similar conclusions (Exercise 7.1).

Boos (1986) developed more comprehensive tests for k samples using linear rank statistics that test specifically for location, scale, skewness and kurtosis. Willemain (1980) tackles the interesting problem of estimating the mean of a population given only the largest observation in each of n samples. That scenario may arise if each of n consultants is given one dose of a new or expensive drug and asked to give this to the patient of greatest need on his or her list. A record may be taken of the change, say, in some level of a blood constituent and the aim may be to estimate the mean, or median, change that could be expected if the drug were used for all patients suffering from the relevant disease.

Shirley (1977) proposed a nonparametric test to determine the lowest dose level at which there is a detectable treatment effect compared with the level for untreated control subjects. An improved version of this test was given by Williams (1986) and House (1986) extended the idea to a randomized block design.

7.2.4 Tests Based on Transformation of Ranks

A modification of the Kruskal–Wallis test is to use van der Waerden scores in place of ranks. The rank or mid-rank r is replaced by the $r/(N+1)$th quantile of a standard normal distribution. Ignoring a slight discrepancy in the case of ties, the mean of these scores is zero. If we evaluate quantities corresponding to S_k and S_r in Section 7.2.2 with ranks replaced by van der Waerden scores and set $C = 0$, we may compute a T analogous to (7.2) for the van der Waerden scores. Again, asymptotically T has a chi-squared distribution with $k-1$ degrees of freedom. The asymptotic result is usually reasonable for all but very small samples. Some packages, including StatXact, now include a program giving exact P-values for samples that are not too large and provide Monte Carlo estimates for larger samples. The exact P-value given by StatXact using van der Waerden scores for the data in Example 7.2 is again 0.047. Expected normal scores may also be used.

7.2.5 The Jonckheere–Terpstra Test

The Kruskal–Wallis test is an omnibus test for differences in centrality or for differences of a dominance nature. If treatments represent, for example, steadily increasing doses of a stimulant we may want to test hypotheses about means or medians, θ_i, of the form H_0: *all θ_i are equal* against

$$H_1 : \theta_1 \leq \theta_2 \leq \theta_3 \leq \ldots \leq \theta_k$$

where at least one of the inequalities is strict, or against an ordered alternative,

$$H_1 : \theta_1 \geq \theta_2 \geq \theta_3 \geq \ldots \geq \theta_k,$$

where again at least one inequality is strict. These are one-tail tests.

A test for such ordered alternatives was given by Jonckheere (1954), but it had been conceived independently by Terpstra (1952). In essence, it extends the WMW test using the Mann–Whitney formulation. The samples must be ordered in the sequence specified in H_1. Exact permutation tests are more readily available for small samples than they are for the Kruskal–Wallis test. An asymptotic test is also available. We discuss this after an example. The asymptotic test is widely available in standard statistical software packages.

Example 7.3

The problem. Hinkley (1989) gives braking distances taken by motorists to stop when travelling at various speeds. A subset of his data is:

Speed (mph)	Breaking distances (feet)			
20	48			
25	33	59	48	56
30	60	101	67	
35	85	107		

Use the Jonckheere–Terpstra test to assess the evidence for a tendency for braking distance to increase as speed increases.

Formulation and assumptions. We test H$_0$: *braking distance is independent of initial speed* against H$_1$: *braking distance increases with speed.* The samples are already arranged in order of increasing speed implicit as the natural order under H$_1$.

Procedure. If there are k samples we calculate the sum, U, of all Mann–Whitney statistics U_{rs} relevant to the rth sample ($r = 1, 2, \ldots, k - 1$) and any sample s for which $s > r$. Thus, for the four samples above we calculate $U_{12}, U_{13}, U_{14}, U_{23}, U_{24}$ and U_{34}. For example, U_{12} is the number of sample 2 values that exceeds each sample 1 value. Here $U_{12} = 2.5$, since there is only one sample 1 value (48), and this is equal to one value in sample 2, and is exceeded by two others (56, 59). Ties are scored as $1/2$ as in the Mann–Whitney statistic. Similarly $U_{13} = 3$, $U_{14} = 2$, $U_{23} = 12$, $U_{24} = 8$ and $U_{34} = 5$. Adding all U_{rs} gives a total $U = 32.5$. Obtaining the exact permutation distribution of ranks is tedious without a suitable computer program. However, the exact test in StatXact indicates $P = \Pr(U \leq 32.5) = 0.0011$, for the one-tail test relevant here.

Conclusion. There is thus clear evidence (not surprising!) that increasing speed increases braking distance.

Comments. 1. When it is relevant, the Jonckheere–Terpstra test is generally more powerful than Kruskal–Wallis. If a Kruskal–Wallis test is applied to these data, the exact $P = 0.0133$ while the asymptotic test gives $P = 0.0641$. It is not surprising that the asymptotic result is unreliable for such small samples. A parametric analysis of variance using these data gives a variance ratio $F = 4.45$ with $3, 6$ degrees of freedom. This again would not indicate significance at the 5 percent level, since $P = 0.057$ for this F-value.

2. Older editions of the UK Highway Code used the formula $d = v + v^2/20$ for minimum stopping distances, d, feet, when travelling at v mph. The mean values in the above data for each speed are fairly close to those obtained from this formula, but there is considerable variation between drivers at any one speed.

Computational aspects. StatXact computes exact P-values for somewhat larger samples than it does for the Kruskal–Wallis test. It is illuminating to compare these with tail probabilities for the asymptotic test given below. In this example the asymptotic result gives $P = 0.0022$. Although twice as large as the exact test level, the discrepancy is not so great as that indicated above for the Kruskal–Wallis test.

The asymptotic Jonckheere–Terpstra test is based on the fact that U as defined in Example 7.3 has a mean $\mathrm{E}(U) = (N^2 - \sum_{i=1}^{k} n_i^2)/4$ and variance $\mathrm{Var}(U) = \{N^2(2N + 3) - \sum_{i=1}^{k}[n_i^2(2n_i + 3)]\}/72$. For large N, and the individual n_i not too small, the distribution of

$$Z = \frac{U - \mathrm{E}(U)}{\sqrt{\mathrm{Var}(U)}}$$

is approximately standard normal.

The sample sizes in Example 7.3 are too small for an asymptotic result to inspire confidence, but for this example, $Z = 2.85$, corresponding, as indicated above, to a one-tail $P = 0.0022$.

The expression for $\mathrm{Var}(U)$ needs adjustment for ties. The adjustment is trivial for relatively few ties but in Section 12.3.2 we show that the existence of many ties may have a dramatic effect. A formula for adjusting $\mathrm{Var}(U)$ for ties is given by Lehmann (1975, p. 235) and a modified version is given by (12.11) in Section 12.3.2.

Tests involving ordered treatment differences that are sometimes of interest include that where we wish to test H_0: *all θ_i are equal* against

$$H_1 : \theta_1 \leq \theta_2 \leq \ldots \leq \theta_{q-1} \leq \theta_q \geq \theta_{q+1} \geq \ldots \geq \theta_k$$

where at least one inequality is strict. These tests are called *umbrella tests*. They were proposed by Mack and Wolfe (1981), and are discussed in some detail by Hollander and Wolfe (1999, Section 6.3). An umbrella test would be relevant if it were believed that a treatment applied at increasing doses was beneficial (e.g., increased yield) up to a certain dose level, while at higher doses a toxic effect made the treatment harmful (e.g., reduced yield).

7.2.6 The Median Test for Several Samples

The median test in Section 6.2 generalizes easily to three or more samples. The alternative hypothesis is now one of a difference between population medians without specifying which populations differ in location, how many differences there are, or their direction. As in the two-sample case, each sample may come from any unspecified population (they need not all have the same distribution). Given k samples where the unknown population medians are $\theta_1, \theta_2, \ldots, \theta_k$ we test H_0: $\theta_1 = \theta_2 = \ldots = \theta_k$ against H_1: *not all θ_i are equal*.

The procedure generalizes that in Section 6.2.1. If M is the combined sample median for all observations, in each sample we count the numbers of observations above and below M. As in Section 6.2.1, we reject sample values equal to M and then work with reduced samples. Assuming sample values that equal M have already been dropped, suppose that, among the n_i observations in sample i there are $a_i > M$ and $b_i < M, i = 1, 2, ..., k$. We record these numbers above and below M in a $k \times 2$ contingency table with k rows and 2 columns. The values a_i, b_i are constrained for each sample so that $a_i + b_i = n_i$, the number in that sample. When no observations equal M, the column totals A and B both equal $N/2$. Table 7.3 illustrates this for a general contingency table with k rows and 2 columns.

The probability P^* of the above cell values when the row and column totals are fixed is given by a generalization of (6.7):

$$P^* = \frac{A!B! \prod_{i=1}^{k}(n_i!)}{N! \prod_{i=1}^{k}(a_i!) \prod_{i=1}^{k}(b_i!)} \tag{7.3}$$

Table 7.3 *A general contingency table for the median test.*

Above M	Below M	Total
a_1	b_1	n_1
a_2	b_2	n_2
.	.	.
.	.	.
a_k	b_k	n_k
---	---	---
A	B	N

where $\prod_{i=1}^{k}(x_i!)$ is the product $(x_1!) \times (x_2!) \times \ldots \times (x_k!)$. As illustrated in the following example, other $k \times 2$ contingency tables with the same marginal totals must be considered to obtain the P-value.

Example 7.4

The problem. Six dental surgeons (I, II, III, IV, V and VI) perform the removal of just one third molar (or wisdom) tooth from adult patients. To investigate the assertion that some surgeons extract these teeth more quickly, the following data were obtained. These relate to 28 patients, each of whom had just one third molar teeth removed. The times (in minutes) taken for the six dental surgeons to operate were:

I	23	25	34	45		
II	7	11	14	15	17	24	40
III	5	8	16	20	26	
IV	12	21	31	38		
V	30	43				
VI	4	9	10	18	19	35

The median extraction time, M, for all 28 patients is 19.5 minutes, so no sample value is equal to the median. Table 7.4 shows the numbers of extraction times above and below M for each surgeon. Use a median test to assess the strength of any evidence against H_0: *all six samples come from populations with the same median extraction time.*

Formulation and assumptions. The test is an extension of the Fisher exact test developed in Section 6.2.2 for the two-sample situation to the $k \times 2$ table having the form in Table 7.3. We omit details but the formula (7.3) is obtained by extending arguments similar to those leading to (6.7). For the data in Table 7.4 a P-value is calculated based on a critical region consisting of all 6×2 contingency tables having the given marginal totals for which, under H_0, P^* does not exceed that observed. For the 2×2 tables in Example 6.7, the more extreme configurations were obvious. For larger contingency tables this is not the case, nor can we attribute these extreme

Table 7.4 *Numbers of observations above and below the combined median M in Example 7.4.*

Dental surgeon	Above M	Below M	Total
I	4	0	4
II	2	5	7
III	2	3	5
IV	3	1	4
V	2	0	2
VI	1	5	6
Total	14	14	28

configurations to a particular tail; i.e., our alternative hypothesis H_1: *not all medians are equal* is essentially two-tail, not unlike the situation in the Kruskal-Wallis test (Section 7.2.2), or indeed the *F*-test in parametric ANOVA.

Procedure. For Table 7.4 the calculation using (7.3) reduces to

$$P^* = \frac{(14!)(14!)(4!)(7!)(5!)(4!)(2!)(6!)}{(28!)(4!)(2!)(2!)(3!)(2!)(1!)(0!)(5!)(3!)(1!)(0!)(5!)}$$

where, by definition, $0! = 1$. Despite cancellations between numerator and denominator, it is tedious to verify that $P^* = 0.00013$ using only a pocket calculator.

The P^* for all other contingency tables with the same marginal totals that have the same or lower values of P^* can be found similarly. Realistically, a computer program or some suitable approximation is required. An appropriate program gives $P = 0.046$ for the comparison of third molar extraction times.

Conclusion. There is moderately strong evidence against H_0, indicating some variation in median extraction times between the dental surgeons.

Comments. 1. Individual sample sizes here are all small (particularly for dental surgeon V). The main differences in Table 7.4 seem to be that surgeon I probably has a median above M and surgeon VI (possibly also surgeon II) has a median below M. If we were to perform separate sign tests of H_0: $\theta_i = M$ for each individual sample we would not be able to demonstrate strong evidence against H_0 (even in a one-tail test) since, for any particular surgeon, the number of patients operated on is very small. In practice much larger samples of patients would be selected for such a study.

2. The results from a comparison of dental surgeons need to be interpreted carefully. Most extractions are completed in around 15 to 20 minutes, but complicated cases can take 1 hour or more. These can be identified in advance from a radiograph, so consultants may deal with serious cases, with less experienced staff operating on the straightforward patients. Higher extraction times for surgeon I could indicate greater responsibility, rather than inexperience.

3. The difficulty in carrying out an exact test without a suitable computer program makes an asymptotic test of interest. We discuss one below.

4. A situation where we may want to test for a common median when samples come from distributions differing in other respects (e.g., spread, skewness, etc.) arises if we compare alternative components that may be used in an industrial process. We may measure a characteristic such as time to failure for a number of replicates of each type of component. A preliminary step may then be to test whether it is reasonable to suppose all may have the same median time to failure. If there is clear evidence they do not, one might reject from further consideration those that appear to have a lower median time. Further choice may depend on preference for a component with little variability in its failure time distribution, or on nonstatistical factors, such as cost.

Computational aspects. The StatXact, Testimate, Stata or other programs for the Fisher exact test (Section 12.2.1) may be used for the median test.

In the notation used in Table 7.3 an asymptotic test makes use of the statistic

$$T = \sum_{i=1}^{k} \frac{(a_i - n_i A/N)^2}{n_i A/N} + \sum_{i=1}^{k} \frac{(b_i - n_i B/N)^2}{n_i B/N} \qquad (7.4)$$

The term $n_i A/N$ is equal to the expected number of cases above M for the ith surgeon, assuming that the null hypothesis is true. Readers familiar with the chi-squared test for independence in contingency tables will recognize (7.4) as the sum of

$$\frac{(observed\ value\ -\ expected\ value)^2}{expected\ value}$$

for all cells, which we discuss in Section 12.2.2. There an alternative form easier to compute manually is given by (12.8). For a $k \times 2$ table T has asymptotically a chi-squared distribution with $k - 1$ degrees of freedom. Widely available chi-squared tables giving minimal critical values for significance at conventional levels and nearly every statistical software package has a program for this test.

Example 7.5

The problem. For the data in Example 7.4 test for equality of population medians using the asymptotic test based on (7.4).

Formulation and assumptions. We calculate all quantities in (7.4) from Table 7.4 and compare the resulting T with the tabulated critical value.

Procedure. The expected numbers ($n_1 A/N$, etc.) are calculated for each cell in Table 7.4. For instance, with surgeon I the expected number of cases above M is

$4 \times 14/28 = 2$. The expected numbers, in the order given in Table 7.4 are:

$$
\begin{array}{cc}
2 & 2 \\
3.5 & 3.5 \\
2.5 & 2.5 \\
2 & 2 \\
1 & 1 \\
3 & 3
\end{array}
$$

Substituting these and the observed cell values from Table 7.4 in (7.4) gives

$$T = 2 \times (2^2/2 + 1.5^2/3.5 + 0.5^2/2.5 + 1^2/2 + 1^2/1 + 2^2/3) = 11.15.$$

Tables indicate $T \geq 11.07$ is the critical value of chi-squared with $k - 1 = 5$ degrees of freedom for significance at the 5 percent level.

Conclusion. There is moderately strong evidence against H_0.

Comments. 1. All samples are small, but exact and asymptotic results agree well. For the chi-squared distribution with 5 degrees of freedom $\Pr(T \geq 11.15) = 0.048$, close to the exact $P = 0.046$ obtained in Example 7.4.

2. The alternative hypothesis that not all medians are equal is essentially two-tailed, but the asymptotic test uses a single tail of the chi-squared distribution. This is because we use the squares of the discrepancies $observed - expected$ and these are necessarily positive, no matter what is the sign of the actual discrepancy.

Computational aspects. If StatXact is used for the Fisher exact test it gives an asymptotic test statistic that differs from T given by (7.4). This is because it uses a statistic, called the *Fisher statistic*, which also has an asymptotic chi-squared distribution. For this particular example the Fisher statistic takes the value 10.23 and $\Pr(\chi^2 \geq 10.23) = 0.069$. The reason for this discrepancy is explored further in Section 12.2. For larger samples the two asymptotic statistics are usually in reasonable agreement.

In practice, we may be interested in whether there are differences in medians for some subset of the k populations. One might take pairs of samples and compute, say, 95 percent confidence limits for the median differences for each pair using the method given in Section 6.2.2. Since these confidence intervals will generally be determined with varying precision for different pairs (due to differences in sample size and perhaps differences in the population distributions) direct comparisons involving these intervals sometimes lead to what at first sight appear bizarre consequences. For example, if zero is included in the 95 percent confidence interval for the difference $\theta_2 - \theta_1$ and also in the 95 percent confidence interval for the difference $\theta_3 - \theta_1$ there is no strong evidence against the null hypotheses that $\theta_2 = \theta_1$ or that $\theta_3 = \theta_1$.

It is tempting to conclude that this implies a lack of evidence against the hypothesis $\theta_3 = \theta_2$. This does not follow. For instance, if most of the confidence interval for $\theta_2 - \theta_1$ is positive and most of that for $\theta_3 - \theta_1$ is negative, the likely differences with respect to θ_1 are in opposite directions. As a consequence, differences between possible values for θ_2 and θ_3 may be quite large, and the 95 percent confidence interval for $\theta_2 - \theta_3$ might not include zero. Similar considerations apply for pairwise comparisons in parametric analyses and the practice is not recommended.

If assumptions needed for, say, a Kruskal–Wallis test appear to hold that test should be preferred to the median test, the latter then usually being appreciably less powerful. For the data in Example 7.4, however, one may have reservations about the Kruskal–Wallis test because the sample values suggest appreciable differences in the spread of times taken by different surgeons.

7.3 The Friedman, Quade, and Page Tests

7.3.1 The Friedman Test

We indicated at the beginning of this chapter that we would mainly consider situations closely analogous to those met in one-way classification ANOVA. The tests described in this section apply basically to situations like those met in randomized block experiments where a two-way classification ANOVA involving blocks and treatments is used in parametric analyses. However, as examples show, the tests described here effectively eliminate any block component from the analysis, thus making them effectively an analysis of a one-way classification.

Friedman (1937) extended the sign test to the case of several related samples. In the sign test the two scores (− or +) can be thought of as corresponding to ranks 1 and 2. In the Friedman test, the scores within each block (often a block is an individual) are ranked. For example, if measurements are made at five successive times the five measurements will be ranked from 1 to 5 in order of increasing (or decreasing) magnitude and each reading is then replaced by the corresponding rank.

The null hypothesis is that the population distributions at each time point are the same so that, for instance, the population medians are equal. If the null hypothesis is true, inspection of the readings for a particular time point will reveal a mixture of ranks. If the population medians differ, then for at least one of the time points the ranks should be mainly high or mainly low.

In the randomized block design, instead of several discrete times, there are an equivalent number of treatments. The number of units in each block equals the number of treatments, each treatment being applied to exactly one unit in each block.

In parametric analysis of variance for continuously distributed observations such as measurements we remove differences between blocks as a source of variability before making time point or treatment comparisons. Do not worry about how this is done if you are unfamiliar with the analysis of variance.

For Friedman's test differences between blocks are removed automatically by replacing observations by ranks, the ranking being done separately within each block. The above step is irrelevant if the basic data are already ranks within each block (see Exercise 7.7). Friedman's test in the context of that and similar examples examines the consistency of ranks rather than acting as

Table 7.5 *Pulse rates for students before and after exercise.*

Time Student	I	II	III
A	72	120	76
B	96	120	95
C	88	132	104
D	92	120	96
E	74	101	84
F	76	96	72
G	82	112	76

a test for centrality. This approach, which we discuss further in Section 10.2, was developed independently by M.G. Kendall.

The test may be applied, for example, to a study where some characteristic of each of b patients is measured at each of t time points. It may also be applied to a randomized block design with t treatments applied each to one of t units in each of b blocks. Here there is an equivalence between patients/blocks and between times/treatments. For simplicity in what follows we use the terms *blocks* and *treatments* to cover both situations.

Ranking is carried out separately within each block. We assume that there are no tied ranks. The sum over all blocks of the ranks allocated to treatment i is denoted by $s_i, i = 1, 2, \ldots, t$.

The Friedman statistic is

$$T = \frac{12 \sum_{i=1}^{t} s_i^2}{bt(t + 1)} - 3b(t + 1). \tag{7.5}$$

If b and t are not too small, T has approximately a chi-squared distribution with $t-1$ degrees of freedom. For this no-tie situation, some tables are available for conventional critical values of T, e.g., Neave (1981, p. 34), Hollander and Wolfe (1999, Table A22), Gibbons and Chakraborti (2004, Table N). Modern statistical software provides many programs for exact or asymptotic tests.

Example 7.6

The problem. For a group of seven students, the pulse rate (per minute) was measured before exercise (I), immediately after exercise (II), and 5 minutes after exercise (III). The data are given in Table 7.5. Use the Friedman statistic to test for differences between pulse rates on the three occasions.

Formulation and assumptions. Separately for each student (students corresponding to blocks) we replace each observation by its rank and calculate the Friedman

Table 7.6 *Pulse rates ranked within students.*

Time Student	I	II	III
A	1	3	2
B	2	3	1
C	1	3	2
D	1	3	2
E	1	3	2
F	2	3	1
G	2	3	1
Total	10	21	11

statistic for these ranks to test H_0: *population distributions of pulse rates are identical* against H_1: *at least one time point has a different distribution of pulse rates reflected by a median shift or dominance.* There are no ties within students, so (7.5) may be used.

Procedure. Relevant ranks are given in Table 7.6 along with the rank total for each time point. To use (7.5), we calculate $\sum_{i=1}^{t}(s_i^2) = 10^2 + 21^2 + 11^2 = 662$. We have $b = 7$ students and $t = 3$ time points, whence

$$T = \frac{12 \times 662}{7 \times 3 \times (3+1)} - 3 \times 7 \times (3+1) = 10.57.$$

This gives an exact $P = 0.0027$. By contrast, the chi-squared approximation with $T = 10.57$ (2 degrees of freedom) gives $P = 0.0051$.

Conclusion. There is strong evidence of a difference in pulse rates between the three times.

Comments. 1. The difference between the P-values given by the exact and approximate methods is a cause for concern. For such small data sets it is unwise to rely on the chi-squared approximation.

2. Inspection of the raw data shows that the strength of evidence against the null hypothesis is not surprising. From a physiological point of view, the pulse rate should increase on exercise and fall off to the resting value once the exercise has been completed. Some individuals (e.g., C) appear to take longer than others to return to the resting pulse rate.

3. While the physiological explanation of the differences is sensible for this example care must be taken with measurements repeated in time because, unlike the randomized block situation, time order is fixed, not randomized. This means that evidence against H_0 may be due to some factor other than that under study. In the present study, if all individuals recorded their pulse rates using a clock in the exercise room and because of some unnoticed mechanical defect the clock mechanism slowed only during the time of the second reading, and this was not detected, that could explain

higher readings on occasion II. Alternatively, if for some reason the air temperature at the time of the third reading had dropped dramatically from that at the time of the second reading, this might explain the relatively lower rates at occasion III.

7.3.2 The Friedman Test with Ties

Tied ranks within individuals (blocks) are given mid-rank values. With ties we use a more general formula for T. In addition to the above terms in (7.5) we need $S_r = \sum_{i,j} r_{ij}^2$ where r_{ij} is the rank allocated to treatment i in block j. With no ties $S_r = bt(t + 1)(2t + 1)/6$. It is also convenient to define $S_t = \sum_{i=1}^{t}(s_i^2)/b$.

As with the Kruskal–Wallis test, a correction factor is required. This is $C = bt(t + 1)^2/4$. The Friedman statistic can be written

$$T = \frac{b(t - 1)(S_t - C)}{S_r - C}. \tag{7.6}$$

If there are no ties this is equivalent to (7.5). As with the no-ties case, if b, t are not too small T has an approximate chi-squared distribution with $t-1$ degrees of freedom. Iman and Davenport (1980) suggest that a better approximation than (7.6) is

$$T_1 = \frac{(b - 1)(S_t - C)}{S_r - S_t}. \tag{7.7}$$

which, under a null hypothesis of no treatment difference, has approximately an F-distribution with $t-1$ and $(b-1)(t-1)$ degrees of freedom. If the ranking is identical in all blocks the denominator of T_1 is zero. Iman and Davenport show that in formal significance testing terms this may be interpreted as a result significant at a $P = (1/t)^{b-1}$ significance level. T_1 is relevant whether or not there are ties.

Example 7.7

The problem. Pearce (1965, p. 37) quoted results of a greenhouse experiment carried out by J. I. Sprent (unpublished). The data, given in Table 7.7, are the numbers of nodes to first initiated flower summed over four plants in each experimental unit (pot) for the pea variety Greenfeast subjected to six treatments — one an untreated control, the others named growth substances. There were four blocks allowing for differences in light intensities and temperature gradients depending on proximity to greenhouse glass. The blocks were arranged to make these conditions as like as possible for all units (pots of four plants) in any one block. Use the Friedman statistic to test for differences between treatments in node of flower initiation.

Formulation and assumptions. Within each block we replace each observation by its rank and calculate the Friedman statistic for these ranks to test H$_0$: *no*

Table 7.7 *Nodes to first flower, total for four plants.*

Block	I	II	III	IV
Treatment				
Control	60	62	61	60
Gibberellic acid	65	65	68	65
Kinetin	63	61	61	60
Indole acetic acid	64	67	63	61
Adenine sulphate	62	65	62	64
Maleic hydrazide	61	62	62	65

Table 7.8 *Nodes to first flower, ranks within blocks.*

Block	I	II	III	IV	Total
Treatment					
Control	1	2.5	1.5	1.5	6.5
Gibberellic acid	6	4.5	6	5.5	22
Kinetin	4	1	1.5	1.5	8
Indole acetic acid	5	6	5	3	19
Adenine sulphate	3	4.5	3.5	4	15
Maleic hydrazide	2	2.5	3.5	5.5	13.5

difference between treatments against H_1: *at least one treatment has a different centrality parameter from the others.* We may test using (7.6) or (7.7).

Procedure. Relevant ranks are given in Table 7.8 where we add a column giving rank totals s_i for each treatment. Squaring each rank, and adding, we get $S_r = 361$. The uncorrected treatment sum of squares is

$$S_t = (6.5^2 + 22^2 + 8^2 + 19^2 + 15^2 + 13.5^2)/4 = 339.625$$

and $C = 4 \times 6 \times 7^2/4 = 294$. Using (7.6) $T = 4 \times 5(339.625 - 294)/(361 - 294) = 13.62$, and (7.7) gives $T_1 = 3(339.625 - 294)/(361 - 339.625) = 6.40$.

Using T_1, as recommended by Iman and Davenport, we find that with 5, 15 degrees of freedom, critical values for significance at the 5, 1 and 0.1 percent significance levels are respectively 2.90, 4.56 and 7.57 so we would judge our result significant at the 1 percent level. If instead, we compare T with the critical values of the chi-squared distribution with 5 degrees of freedom the result is significant at the 5 percent, but not the 1 percent, level. For the chi-squared distribution with 5 degrees of freedom $\Pr(T \geq 13.620) = 0.0182$. Comparing T with the critical values given in the tables referenced above for six treatments and four blocks we find significance is indicated at the 1 percent level if $T \geq 12.71$, so again significance is indicated at that level, but remember that Neave's table apply strictly to the no-tie situation. The Friedman test program in StatXact confirms and refines the above findings based on T. The exact P-value based on the relevant permutation test is $P = 0.0045$, while the asymptotic P-value based on the chi-squared distribution is $P = 0.0182$ as already stated.

Conclusion. There is strong evidence of a difference between treatments.

Comments. 1. While for practical reasons we use T or T_1 as our test statistic, we could use S_t because both S_r and C remain unaltered in the permutation distribution, which involves only permuting ranks within blocks.

2. A parametric randomized blocks analysis of variance of the original data gives $F = 4.56$, which corresponds almost exactly to $P = 0.01$.

3. Whereas analysis of variance introduces a sum of squares reflecting difference between block totals, there is no such term in the Friedman analysis because the sum of ranks in all blocks is the same, namely $t(t + 1)/2$.

Computational aspects. For larger samples exact P-value computations like those in StatXact may be slow or even impossible. The Monte Carlo option will then give good estimates of P. The test statistic is calculated in many statistical software packages, but often only asymptotic tail probabilities are quoted. The Friedman test is also available in R using the facility *friedman.test*.

Ranks within blocks might be replaced by normal scores. Experience suggests that such rescoring has few advantages. Indeed, if there are ties, the block differences removed by ranking may be reintroduced, though usually not dramatically.

Ranking within blocks is robust against many forms of heterogeneity of variance, in that it removes any inequalities of variance between blocks (see Exercise 7.10).

Durbin (1951) gives a rank-based Friedman-type test for data in incomplete blocks.

7.3.3 The Quade Test

Quade (1979) proposed a test that is often more powerful than the Friedman test. It also eliminates block differences, but weights the rank scores to give greater weight in those blocks where the raw data indicate possibly more marked treatment effects. Details, together with a numerical example, are given by Conover (1999, Section 5.8). StatXact includes a program for this test. Whereas the Friedman test is basically an extension of the sign test, the Quade test is effectively an extension of the Wilcoxon signed-rank test and is equivalent to it when $t = 2$.

If x_{ij} denotes the observed value for treatment i in block j the range, D_j of values in block j is the difference between the greatest and least x_{ij}, i.e.,

$$D_j = \max_i(x_{ij}) - \min_i(x_{ij}).$$

For the Quade test the D_j are ranked in ascending order from 1 to b, using mid-ranks for ties if needed. If we denote the rank of D_j for block j by q_j

and the Friedman rank of x_{ij} by r_{ij}, then the Quade scores are defined as $s_{ij} = q_j[r_{ij} - (t+1)/2]$. These scores are used in StatXact for an exact permutation test, but if that program is not available the Iman and Davenport (1980) analogue of (7.7) may be used in an asymptotic test. It takes the same form with the Quade scores s_{ij} replacing the r_{ij}. As with the Friedman test, if the denominator is zero this may be interpreted as a result significant at the $P = (1/t)^{b-1}$ significance level.

For the data in Example 7.7 for the Quade test, StatXact gives an exact $P = 0.0019$. In R there is a facility *quade.test* for this test.

7.3.4 An Alternative Extension of Wilcoxon-Type Tests

Hora and Conover (1984) proposed ranking all observations simultaneously without regard to treatments or blocks and carrying out an analysis of variance on the ranks (or normal scores derived from these). The procedure for ranks is described by Iman, Hora and Conover (1984).

When applied to the data in Example 7.7 the relevant F-statistic for treatment difference is $F = 4.96$ (compared with 4.56 for a parametric test and 6.40 using (7.7)). The reader familiar with standard analysis of variance may wish to verify these results (see Exercises 7.4 and 7.9). Ranking data prior to a formal analysis of variance may in some circumstances lead to unfortunate distortions of data characteristics such as interactions — a matter we take up in Section 8.2.

7.3.5 The Page Test for Ordered Alternatives

Page (1963) proposed an analogue to the Jonckheere–Terpstra test for ordered alternatives applicable to blocked data. If the treatments are arranged in the order specified in the alternative hypothesis and s_i is the sum of ranks for treatment i (as in the Friedman test) the Page statistic is

$$P = s_1 + 2s_2 + 3s_3 + \ldots + ts_t.$$

Asymptotically, if H_0: *no treatment difference* holds, then P has a normal distribution with mean $tb(t+1)^2/4$ and variance $b(t^3-t)^2/[144(t-1)]$. Tables (Daniel, 1990, Table A17) give critical values of P for small b, t. The test is discussed in more detail by Daniel (1990, Section 7.3), by Marascuilo and McSweeney (1977, Section 14.12), by Hollander and Wolfe (1999, Section 7.2) and by Gibbons and Chakraborti (2004, Section 12.3).

StatXact includes a program for calculating exact P-values for this test with the usual provision for a Monte Carlo approximation if the sample size is too large. An asymptotic approximation is also given. Hollander and Wolfe (1999, Table A23) give tail probabilities for the Page statistic P given above for a range of treatment and block sizes for which the asymptotic approximation may prove unreliable.

7.4 Binary Responses

In many situations each experimental unit may show one of two responses —
win or lose, succeed or fail, live or die, a newborn child is male or female. For
analytic purposes we may conveniently score such responses as 0 or 1. For
example, generalizing from Example 5.4, suppose five members A, B, C, D,
E of a mountaineering club each attempt three rock climbs at each of which
they either succeed or fail. If a success is recorded as 1 and a failure as 0, the
outcomes may be summarized as follows:

Member	A	B	C	D	E
Climb 1	1	1	0	0	1
Climb 2	1	0	0	1	0
Climb 3	0	1	1	1	1

Cochran (1950) proposed a method applicable to such situations to test the
hypothesis H_0: *all climbs are equally difficult* against H_1: *the climbs vary in
difficulty*. In conventional statistical terms climbs may be regarded as *treatments* and climbers as *blocks*. If we have t treatments in b blocks and binary
(i.e., 0, 1) responses the appropriate test statistic is

$$Q = \frac{t(t-1)\sum_{i=1}^{t} T_i^2 - (t-1)N^2}{tN - \sum_{j=1}^{b} B_j^2} \qquad (7.8)$$

where T_i is the total (of 1s and 0s) for treatment i, B_j is the total for block
j and N is the grand total. The exact permutation distribution of Q is not
easily obtainable, but for large samples Q has approximately a chi-squared
distribution with $t - 1$ degrees of freedom.

   Although not immediately obvious, for two treatments the test reduces to
McNemar's test and in Section 12.5 we give an alternative form of the McNemar test, which is exactly equivalent to Q given above.

   Cochran's test is discussed more fully by Conover (1999, Section 4.6).

7.5 Tests for Heterogeneity of Variance

Most tests for heterogeneity of variance discussed in Section 6.4 extend to
several samples. We illustrate this for the squared-rank test given in Section
6.4.3. The procedure parallels that for the Kruskal–Wallis test with squared
ranks of absolute deviations replacing data ranks as scores.

   The absolute deviations of all sample values from their sample mean are
ranked over all samples and these ranks are squared. The statistic T is identical
to (7.2) except that now s_i is the sum of the squared rank deviations for
sample i and C is the square of the sum of these squared ranks divided by N.

Symbolically, denoting such a squared rank by r_{ij}^2 we find $C = [\sum_{i,j} r_{ij}^2]^2/N$. S_r is here the sum of squares of the squared ranks, i.e., $S_r = \sum_{i,j} r_{ij}^4$.

If all population variances are equal, then for a reasonably large sample, T has a chi-squared distribution with $t - 1$ degrees of freedom. If there are no rank ties the sum of squared ranks is $N(N+1)(2N+1)/6$. Also, for no ties the denominator of T reduces to

$$S_r - C = (N-1)N(N+1)(2N+1)(8N+11)/180.$$

Example 7.8

The problem. As well as increasing with speed it is thought that braking distance may vary more between drivers as speed increases. Use the squared rank test to test for heterogeneity of population variance on the basis of the following subsample from Hinkley (1989) of initial speeds (mph) and braking distance (feet).

Speed	Braking distance			
5	2	8	8	4
10	8	7	14	
25	33	59	48	56
30	60	101	67	

Formulation and assumptions. We obtain the absolute deviations $|x_{ij} - m_i|$ of each observation x_{ij} from its sample mean m_i. These deviations are then ranked over combined samples and the statistic T is calculated and compared to the relevant chi-squared critical value.

Procedure. The means at each speed are $m_1 = 5.5$, $m_2 = 9.67$, $m_3 = 49$, $m_4 = 76$. The absolute deviations together with their ranks (1 for least, 14 for greatest, with mid-ranks for ties) are as follows:

5 mph	Absolute deviation	3.5	2.5	2.5	1.5
	Overall rank	7	4.5	4.5	2
10 mph	Absolute deviation	1.67	2.67	4.33	
	Overall rank	3	6	8	
25 mph	Absolute deviation	16	10	1	7
	Overall rank	12.5	11	1	9
30 mph	Absolute deviation	16	25	9	
	Overall rank	12.5	14	10	

The sum of squared ranks at 5 mph is $s_1 = 7^2 + 4.5^2 + 4.5^2 + 2^2 = 93.5$. Similarly, $s_2 = 109$, $s_3 = 359.25$, $s_4 = 452.25$, so the sum of all squared ranks is 1014 and $S_p = (93.5)^2/4 + (109)^2/3 + (359.25)^2/4 + (452.25)^2/3 = 106\ 587.724$. The reader should confirm that $C = 73\ 442.571$ and that

$$S_r = 7^4 + 4.5^4 + ... + 10^4 = 127\ 157.25$$

whence $T = 8.02$. Tables of the chi-squared distribution indicate that the critical value for significance at the 5 percent level with 3 degrees of freedom is $T = 7.815$.

Conclusion. There is evidence of heterogeneity of variance in the light of significance at a nominal 5 percent level.

Comments. 1. One has reservations about an asymptotic result with such small samples, but we can do little better except in extremely small samples where we might calculate exact permutation distributions (see *Computational aspects* below). For the original data there may of course be differences in centrality whether or not we accept H$_0$: *no variance difference in a squared rank test*.

2. An alternative would be a Kruskal–Wallis test using overall ranks of absolute deviations.

Computational aspects. StatXact provides a program for an exact permutation test. The squares of the rank deviations are submitted as data to the program for one-way ANOVA with general scores and this gives $P = 0.0240$ broadly in line with the asymptotic result based on T. Indeed, if $T = 8.02$ the exact chi-squared $P = 0.046$, a discrepancy that is not surprising with such small samples.

7.6 Some Miscellaneous Considerations

7.6.1 Power Sample Size and Efficiency

In the light of the problems hinted at in our discussions about power calculations in the simpler one- and two-sample situations in earlier chapters it will hardly surprise readers to learn that exact power calculations for most of the procedures considered in this chapter are only available for a few very restricted, and often unrealistic, situations. Exploration of the circumstances in which a specific method is likely to be beneficial and any indication of sample sizes needed to meet specific objectives may be aided by information about Pitman efficiency. Pitman efficiencies for some of the methods developed in this chapter as well as in other situations discussed by them are considered by Hollander and Wolfe (1999, Sections 6.10 and 7.16).

Bünning and Kössler (1999) give some asymptotic power results for the Jonckheere–Terpstra test and some closely related tests and report that simulation studies indicate that their asymptotic results provide good approximations even for moderate sample sizes.

7.6.2 Runs Test for Three or More Samples

The Wald–Wolfowitz runs test may be extended to three or more independent samples. Within each sample observations are allocated the same code, but different codes are assigned to each sample. As in the Wald–Wolfowitz test, runs are created by placing the original observations from the combined groups in order of size. Runs in the codes representing these observations are then investigated for clustering or alternation using the method described in Section 4.5.1.

This technique may prove useful in detecting the presence of repeated measurements that have been pooled together to create a group of seemingly independent observations. Statistical consultants need to be aware that inexperienced researchers involved in small-scale studies are sometimes tempted to combine repeated observations in order to produce samples of a "respectable" size (Bland and Altman, 1994). Suppose that several doctors are each timed performing a specific task on a number of occasions. If the variation in the times recorded for each doctor is much less than that between the doctors, the application of this test may highlight clusters of similar values relating to particular doctors.

Example 7.9

The problem. A new surgical manoeuvre is performed on five occasions by each of two consultants, two medical registrars and two clinical medical students. The times (in seconds) taken for each performance of the manoeuvre were:

Consultant A	97	157	181	215	216
Consultant B	520	527	540	566	580
Registrar C	305	320	345	350	356
Registrar D	454	456	471	485	489
Student E	414	422	423	451	467
Student F	544	550	599	661	665

A member of the surgical team presents the data to a medical statistician not as above but arranged by seniority as three groups of seemingly independent observations:

Consultants	97	157	181	215	216	520	527	540	566	580
Registrars	305	320	345	350	356	454	456	471	485	489
Students	414	422	423	451	467	544	550	599	661	665

without mentioning that the data contain repeated measurements. He asks for assistance in looking for group differences in the times taken. Use the multi-sample form of the Wald–Wolfowitz test to investigate whether these observations might not be truly independent.

Formulation and assumptions. There are $N = 30$ observations, $k = 3$ groups, $n_1 = n_2 = n_3 = 10$ observations in each group. We test the null hypothesis H_0 of randomness in the ordered times within level of seniority. Similar values from a particular individual may result in a cluster of observations with the same seniority code. Clustering being of interest, a one tail alternative hypothesis is appropriate. A small number of runs indicates clustering, providing evidence against H_0.

Procedure. Consultant observations are coded as 0, those for registrars as 1 and student times as 2. The data are combined and ranked overall, each observation being labelled by its code as shown in Table 7.9.

Inspection of Table 7.9 indicates $r = 10$ runs. The distribution of the number of runs generated by 1,000,000 simulations gives an estimate for the P-value for $R \leq 10$ to be $P = 0.000026$.

Table 7.9 *Ranking and coding of the times to complete a surgical manoeuvre.*

Time	97	157	181	215	216	305	320	345	350	356
Rank	1	2	3	4	5	6	7	8	9	10
Code	0	0	0	0	0	1	1	1	1	1
Time	414	422	423	451	454	456	467	471	485	489
Rank	11	12	13	14	15	16	17	18	19	20
Code	2	2	2	2	1	1	2	1	1	1
Time	520	527	540	544	550	566	580	599	661	665
Rank	21	22	23	24	25	26	27	28	29	30
Code	0	0	0	2	2	0	0	2	2	2

Conclusion. There is very strong evidence for clustering of values having the same seniority code. This should raise concern about the independence of these observations. Looking at the data as originally collected shows that these concerns are well founded.

Comments. 1. The small variation in the times recorded for each individual, and the much greater variation between individuals, allows the similar observations to be easily detected as clusters of the same code. This becomes much more difficult if individuals times have greater variability and overall differences between individuals are smaller.

2. Although overall there is a tendency for consultants to be the speediest and students the slowest at this task this relationship is not exact. This should not be a problem as long as repeated values for specific individuals are similar.

3. Tied values within an individual's observations may be dealt with by the usual method of averaging ranks. Ties between observations from different individuals will cause problems in the assignment of clusters if those individuals have different codes. In such a situation a conservative approach is to order the tied values to maximise the number of runs. With such an approach this test becomes difficult to apply, and is likely to have little power with a large number of ties between codes.

Computational aspects. Multiple runs distributions can be simulated using the procedure within Stata mentioned in Example 4.15.

7.6.3 A Quantile-Based Procedure for Minimum Dispersion

Sometimes samples come from several populations for each of which the kth quantile is known to be the same. We may then be interested in selecting the population with the smallest variance, or some other measure of spread. For example, if several different methods of producing an item all lead to the same proportion of defectives (e.g., falling below a minimum permitted weight) we may want to select the production method with the smallest variability of weight among the items produced. A nonparametric approach to this problem is given by Gill and Mehta (1989, 1991).

7.7 Fields of Application

Parametric analysis of variance of designed experiments had historical origins in agriculture but soon spread to the life and natural sciences, medicine, industry and, more recently, to business and the social sciences. Development of nonparametric analogues was stimulated by a realization that data often clearly violated normal theory assumptions and, more importantly, to provide a tool for analysis of ordinal data expressed only as ranks or preferences. Ranking, or preference scores, often combine assessment, either deliberately or subconsciously, of a number of factors that are given different weights by individuals. It is thus of practical interest to see if there is still consistency between the way individuals rank the same objects despite the fact that they may not give the same weight to each factor. Our first example in this section illustrates this point.

Preferences for Washing Machines

Consumers' preferences for washing machines are influenced by their assessment of several factors, e.g., price, reliability, power consumption, washing time, load capacity, ease of operation and clarity of instructions. Individuals weigh such factors differently. A farmer's wife offering bed and breakfast to tourists will rate ability to wash bed linen highly. A parent with a large young family will appreciate ability to remove sundry stains from children's clothing. For many, a low price and running economy may be key factors. No machine is likely to get top ratings on all points, but manufacturers will be keen to achieve a high overall rating for their product from a wide range of consumers.

A number of people may be asked to state preferences (e.g., ranks) for each manufacturer's machine and for those of competitors. A manufacturer wants to know if there is consistency in rankings — whether most people give a particular machine a preferred rating — or whether there is inconsistency or general dislike for some machine. Each consumer is a block, each machine a treatment, in the context of a Friedman test. The hypothesis under test is H_0: *no consistency in rankings* against H_1: *some consistency in rankings*. We discuss this type of situation further in Section 10.2.

Literary Discrimination

A literary expert asserts that short stories by one writer (A) are excellent, those by a second writer (B) are good, and those by a third (C) are of a disappointing standard. To test his claim and judgement he is given a series of short stories to read on typescripts that do not identify the authors and asked to rank them using a scale from 1 (for worst) to 20 (for best). A Kruskal–Wallis test could be used to test the hypothesis that the scores indicate the samples are all from the same population (i.e., no discriminatory ability), but as the samples have a natural ordering — excellent, good, inferior — a Jonckheere–Terpstra test is more appropriate. See Exercise 7.6.

Assimilation and Recall

A list of 12 names is read out to students. It contains in random order, 4 names of well-known sporting personalities, 4 of national and international political figures, and 4 of well-known scientists. The students are later asked which names they can recall and a record is made of how many names each student recalls in each of the three categories. By ranking the results we may test whether recall ability differs systematically between categories, e.g., do people recall names of sporting personalities more easily than those of politicans or scientists? See Exercise 7.5.

Tasting Tests

A panel of tasters may be asked to rank different varieties of raspberry in order of preference. A Friedman test is useful to detect any pattern in taste preference. See Exercise 7.7.

Paint Testing

A paint manufacturer may be interested in testing several new chemical constituents to see whether they improve the waterproofing qualities of paints designed for use on exterior walls. Tests are carried out in a laboratory in which extreme weather conditions are simulated. Replicated measurements are made of the volumes of water in a fixed time period penetrating a single coating of paints each containing one of the constituents. If conditions for a parametric ANOVA are violated a Kruskal–Wallis test might be used to detect evidence of different waterproofing abilities.

Quantal Responses

Four doses of a drug are given to batches of rats, groups of four rats from the same litter forming a block. The only observation is whether each rat is dead or alive 24 hours later. Cochran's test is appropriate to test for different survival rates at the four doses.

7.8 Summary

The **Kruskal–Wallis test** (Section 7.2.2) is a rank analogue of the one-way classification analysis of variance. The test statistics commonly used are (7.1) or, if there are ties, (7.2). If k is the number of samples (7.1) and (7.2) have asymptotically a chi-squared distribution with $k - 1$ degrees of freedom. An alternative is to transform ranks to van der Waerden scores.

The **Jonckheere–Terpstra test** (Section 7.2.5) is appropriate if H_1 orders the treatments. The test statistic is a sum of certain Mann–Whitney statistics. An asymptotic approximation is widely used, but care is needed with ties.

The **median test** (Section 7.2.6) is applicable to k samples from any populations (these may differ from one another in aspects other than the median). The test is essentially a special case of the Fisher exact test, or asymptotically a Pearson chi-squared test applied to a $k \times 2$ contingency table calculating the statistic using (7.4). The test often has low power relative to many of its competitors.

For related samples the **Friedman test** (Section 7.3.1) is applicable to block designs where ranks are allocated within blocks. One test statistic is given by (7.5) or (7.6) Asymptotically these have a chi-squared distribution. Another statistic is given by (7.7). An alternative test is the **Quade test** (Section 7.3.3).

The **Page test** (Section 7.3.5) is an analogue of the Jonckheere–Terpstra test applicable in a randomized block context if H_1 orders treatments.

The **Cochran test** (Section 7.4) is applicable to blocked binary response data and the test statistic (7.8) has an asymptotic chi-squared distribution.

The **Conover squared-rank test** and some other tests for heterogeneity of variance (Section 7.5) extend from those for two samples.

A multiple sample extension of the **Wald–Wolfowitz** runs test (Section 7.6.2) may be useful in checking for possible dependence of observations within a group.

7.9 Exercises

*7.1 Perform a parametric analysis of variance on the data in Example 7.2, comparing your result with those for the Kruskal–Wallis test.

7.2 Reanalyse the data in Example 7.2 using van der Waerden scores.

*7.3 Carry out the asymptotic Jonckheere–Terpstra test for the data in Example 7.3.

7.4 Analyse the data in Example 7.7 using a parametric analysis of variance.

7.5 At the beginning of a tutorial 12 names are read out in random order to 10 medical students. Four are names of prominent sporting personalities (Group A), four are national and international politicians (Group B) and four are scientists who regularly feature in the media (Group C). At the end of the session students are asked to recall as many of the names as possible. The numbers recalled were:

Student	1	2	3	4	5	6	7	8	9	10
Group A	3	1	2	4	3	1	3	3	2	4
Group B	2	1	3	3	2	0	2	2	2	3
Group C	0	0	1	2	2	0	4	1	0	2

Rank the data within each block (student) and use a Friedman test to assess evidence of a difference between recall rates for the three groups. Carry out an ANOVA on the given data. Do the conclusions agree with the Friedman test? If not, why not?

*7.6 The ranks given by an expert in the literary discrimination example in Section 7.7 are

Excellent author	7	9	10	13	14	19	20
Good author	3	5	6	12	16	17	18
Disappointing author	1	2	4	8	11	15	

The work adjudged best is ranked 20. Assess the validity of the expert's claim to discriminate between the authors.

*7.7 Five tasters rank four varieties of raspberry in order of preference. Do the results indicate a consistent taste preference? Mid-ranks are given for ties.

Taster	1	2	3	4	5
Variety					
Malling Enterprise	3	3	1	3	4
Malling Jewel	2	1.5	4	2	2
Glen Clova	1	1.5	2	1	2
Norfolk Giant	4	4	3	4	2

*7.8 Four share tipsters are each asked to predict on 10 randomly selected days whether the London FTSE Index (commonly known as Footsie) will rise or fall on the following day. If they predict correctly this is scored as 1, if incorrectly as 0. Do the scores below indicate differences in tipsters' ability to predict accurately?

Day	1	2	3	4	5	6	7	8	9	10
Tipster A	1	0	0	1	1	1	1	0	1	1
Tipster B	1	1	1	1	0	1	1	0	0	0
Tipster C	1	1	0	1	1	1	1	1	0	1
Tipster D	1	1	0	0	0	1	1	1	0	1

7.9 Replace the data in Table 7.7 by ranks 1 to 24 (using mid-ranks for ties where appropriate) and carry out an ordinary randomized block analysis of variance of these ranks to confirm the F-value quoted in Section 7.3.4.

7.10 Berry (1987) gives the following data for numbers of premature ventricular contractions per hour for 12 patients with cardiac arrhythmias when each is treated with 3 drugs A, B, C.

Patient	1	2	3	4	5	6	7	8	9	10	11	12
A	170	19	187	10	216	49	7	474	0.4	1.4	27	29
B	7	1.4	205	0.3	0.2	33	37	9	0.6	63	145	0
C	0	6	18	1	22	30	3	5	0	36	26	0

Use a Friedman test to investigate differences in response between drugs. Note the obvious heterogeneity of variance between drugs. Carry out an ordinary randomized block analysis of variance on these data. Do you consider it to be valid? Is the Friedman analysis to be preferred? Why?

*7.11 Cohen (1983) gives data for numbers of births in Israel for each day in 1975. We give below data for numbers of births on each day in the 10th, 20th, 30th and 40th weeks of the year.

Day	Mon	Tue	Wed	Thu	Fri	Sat	Sun
Week							
10	108	106	100	85	85	92	96
20	82	99	89	125	74	85	100
30	96	101	108	103	108	96	110
40	124	106	111	115	99	96	111

Perform Friedman analyses to determine whether the data indicate (i) a difference in birth rate between days of the week that shows consistency over the four selected weeks and (ii) any differences between rates in the 10th, 20th, 30th and 40th weeks.

7.12 Snee (1985) gives data on average liver weights per bird for chicks given three levels of growth promoter (none, low, high). Blocks correspond to different bird houses. Use a Friedman test to see if there is evidence of an effect of growth promoter.

Block	1	2	3	4	5	6	7	8
None	3.93	3.78	3.88	3.93	3.84	3.75	3.98	3.84
Low dose	3.99	3.96	3.96	4.03	4.10	4.02	4.06	3.92
High dose	4.08	3.94	4.02	4.06	3.94	4.09	4.17	4.12

Since dose levels are ordered, a Page test is appropriate. Try this also.

*7.13 Lubischew (1962) gives measurements of maximum head width in units of 0.01 mm for three species of *Chaetocnema*. Part of his data is given below. Use a Kruskal–Wallis test to see if there is a species difference in head widths.

Species 1	53	50	52	50	49	47	54	51	52	57	
Species 2	49	49	47	54	43	51	49	51	50	46	49
Species 3	58	51	45	53	49	51	50	51			

7.14 Biggins, Loynes and Walker (1987) considered various ways of combining examination marks where all candidates sat the same number of papers but different candidates selected different options from all those available. The data below are the marks awarded by four different methods of combining results for each of 12 candidates. Do the schemes give consistent ranking of the candidates? Is there any evidence that any one scheme treats some candidates strikingly differently than the way they are treated by other schemes so far as rank order is concerned? Is there any evidence of a consistent difference between the marks awarded by the various schemes?

Cand.	1	2	3	4	5	6	7	8	9	10	11	12
A	54.3	30.7	36.0	55.7	36.7	52.0	54.3	46.3	40.7	43.7	46.0	48.3
B	60.6	35.1	34.1	55.1	38.0	47.8	51.5	44.8	39.8	43.2	44.9	47.6
C	59.5	33.7	34.3	55.8	37.0	49.0	51.6	45.6	40.3	43.7	45.5	48.2
D	61.6	35.7	34.0	55.1	38.3	46.9	51.3	44.8	39.7	43.2	44.8	47.5

*7.15 Chris Theobald supplied the following data from a study of 40 patients suffering from a form of cirrhosis of the liver. One purpose was to examine whether there was evidence of association between spleen size and blood platelet count. Blood platelets form in bone marrow and are destroyed in the spleen, so it was thought that an enlarged spleen might lead to more platelets being eliminated and hence to a lower platelet count. The spleen size of each patient was found using a scan and scored from 0 to 3 on an arbitrary scale, 0 representing a normal spleen and 3 a grossly enlarged spleen. The platelet count was determined as the number in a fixed volume of blood. Do these data indicate an association between spleen size and platelet count in the direction anticipated by the experimenter?

Size													
0	156	181	220	238	295	334	342	359	365	374	391	395	481
1	65	105	121	150	158	170	214	235	238	255	265	390	
2	33	70	87	109	114	132	150	179	184	241	323		
3	79	84	94	259									

CHAPTER 8

ANALYSIS OF STRUCTURED DATA

8.1 Factorial Treatment Structures

In this chapter we assume some familiarity with basic concepts of experimental design; in particular, with factorial treatment structures at least for the case for two factors each at two or more levels.

We describe some developments over the last 10 to 15 years that do not yet appear in many introductory textbooks. Specifically, the concept of *interactions* is extended beyond that usually met in conventional courses on experimental design and analysis in a parametric context. Section 8.4.2 may prove difficult to a reader not familiar with basic experimental design topics and mathematical modelling. In such circumstances that section might be omitted at first reading.

We comment briefly on some aspects of factorial structures that have a particular relevance to nonparametric ot distribution-free alternatives to the classic parametric procedures.

The concept of interaction applies more widely, but we consider first its interpretation in linear models, where the model may include an interaction term as well as *additive effects*. We expect some, but not all readers, will be familiar with these basics.

The simplest case is that of two factors each at two levels. For instance, we may allow a chemical reaction to proceed for two different times — 2 hours and 3 hours. These are the two levels of a first factor, *time*. We may also carry out the experiment at two different temperatures — 75°C and 80°C. These are the two levels of a second factor, *temperature*. The responses, or yields, are the amounts of some chemical produced by a fixed amount of input material for each of the factor combinations

- 2 hours at 75°C;
- 2 hours at 80°C;
- 3 hours at 75°C;
- 3 hours at 80°C.

The experiment is repeated, or replicated, a specified number of times. The average, or the total output, for the same number of replicates for each factor combination is calculated. Apart from random variation in the form of an additive error, which in the classic parametric model is assumed to be distributed $N(0, \sigma^2)$, the output y_{ij} for the first factor at level i and the second at level j $(i, j = 1, 2)$ is specified by a linear model we describe in the next

Table 8.1 *Expected yields in a no-interaction model.*

	Temperature (°C)	
	75	80
Process time (hrs)		
2	X	$X+3$
3	$X+8$	$X+3+8$

section. If each factor has purely additive effects we have a *no-interaction* model.

To clarify these concepts assume temporarily an unrealistic state of perfection where there is no random variation to create error. Suppose that operating for 2 hours at 75°C gives an output X and that this increases to $X+3$ if we leave time unaltered, but increase temperature to 80°C. Also, suppose that output increases to $X+8$ if we leave temperature at 75°C but increase time to 3 hours. We say there is *no interaction* if the result of increasing both time from 2 to 3 hours and temperature from 75 to 80°C is to increase yield from X to $X+3+8 = X+11$. These results for expected yield are summarized in Table 8.1.

In general, for a combination of level i of the first, or row, factor with level j of the second, or column, factor we may denote the true yield (including now an additive error) by y_{ij} and the corresponding expected yield (output apart from error) by x_{ij}. Thus, in Table 8.1

$$x_{11} = X, \ x_{12} = X+3, \ x_{21} = X+8, \ x_{22} = X+3+8 = X+11,$$

and $x_{11} + x_{22} = 2X+11 = x_{12} + x_{21}$, i.e., the diagonal or cross sums in Table 8.1 are equal. This is a fundamental characteristic of an additive (or no interaction) model for an experiment with two factors each at two levels.

In many real life situations the effect of increasing both time and temperature is either to boost or diminish the effect of changing only one factor. This would happen if, for example, x_{22} in Table 8.1 took the value $X+17$ or $X+2$ instead of $X+11$, the other x_{ij} being unaltered. The cross sums are then no longer equal. For the 2×2 factorial model with no interaction the requirement that the cross sums are equal implies

$$x_{11} + x_{22} - x_{12} - x_{21} = 0. \tag{8.1}$$

Because of random error this equality will not hold for the observed yields in each of r replicates, y_{ijk}, where $k = 1, 2 \ldots r$, or for the corresponding total or mean yields over all replicates for each factor combination.

The standard analysis of variance test for interaction is essentially one to assess whether the observed means, which we temporarily denote by y_{ij}. are such that $y_{11.} + y_{22.} - y_{12.} - y_{21.}$ departs from zero by an amount which is too great to be attributed to random variation. If the departure is too great this implies an interaction.

More formally, asserting there is an interaction implies acceptance of a hypothesis

$$H_1 : x_{11} + x_{22} - x_{12} - x_{21} = I, \text{ where } I \neq 0 \qquad (8.2)$$

rather than H_0: $I = 0$.

In the absence of additive errors, if there is no interaction, it is easily seen from (8.1) that at *either* temperature the difference in yield at 3 hours minus that at 2 hours is the same i.e.,

$$x_{21} - x_{11} = x_{22} - x_{12}.$$

If there are additive errors, and no interaction, for the observed mean yields, $y_{ij.}$, in general

$$y_{21.} - y_{11.} \neq y_{22.} - y_{12.},$$

but their mean

$$(y_{21.} - y_{11.} + y_{22.} - y_{12.})/2,$$

is a sensible estimate of the effect of time on yield. This is usually referred to as the *main effect* of time. Analogous arguments can be applied to estimation of a main effect of temperature.

If there is interaction in general this implies

$$x_{21} - x_{11} \neq x_{22} - x_{12},$$

and the concept of using

$$(y_{21.} - y_{11.} + y_{22.} - y_{12.})/2,$$

as an estimate of main effect becomes somewhat meaningless. The concept of main effect is inappropriate when there is interaction. It is then more profitable to examine separately the effect of any one factor at each level of the other factor. Readers familiar with the analysis of factorial treatment structures will know that two-way tables of means corresponding to each factor combination, together with appropriate standard errors, are valuable tools to aid interpretation of factorial treatment data.

The above arguments extend to any set up where we have two factors at each of two levels. In the next section we use numerical examples to indicate the analysis and interpretation of 2 × 2 factorial structures using parametric methods, and then discuss when we may prefer nonparametric methods, and later consider which might be approriate ones. We find that this may depend upon whether or not we retain the linear model.

8.2 Balanced 2 × 2 Factorial Structures

Examples 8.1, 8.2 and 8.3 cover situations where statisticians would happily use the standard parametric analysis of variance and we indicate how they might interpret the results in each case.

Example 8.1

The problem. Pots each with five plants of the same variety of pea are grown in greenhouses at each of the four combinations of two levels of growth hormone (factor A) and with two levels of undersoil heating (factor B). There are six replicates (pots) of each treatment combination and after nine weeks all plants are harvested and weighed. The total weights for the five plants in each plot are recorded.

Conventional notation which we now use differs from that in Section 8.1. We denote the levels of factor A by a_1, a_2 and those of factor B by b_1, b_2. The four possible combination of levels are a_1b_1, a_1b_2, a_2b_1, a_2b_2. The weights in grams totalled for the five plants in each pot are:

a_1b_1	16.2	16.6	17.1	17.3	17.5	17.7
a_2b_1	17.2	17.4	17.5	17.9	18.1	18.3
a_1b_2	18.2	18.5	18.6	19.1	19.2	19.5
a_2b_2	18.1	18.9	19.6	19.9	20.1	20.3

Formulation and assumptions. We assume the usual linear model allowing for an interaction term and specify the usual homogeneous normal error structure.

Procedure. Most standard software programs will produce an analysis of variance table similar to this:

Source	DF	SS	MS	F	P
Main A	1	2.535	2.535	6.99	0.016
Main B	1	18.727	18.727	51.66	0.000
Interaction	1	0.002	0.002	0.00	0.947
Error	20	7.250	0.363		
Total	23	28.513			

and a table of means

	a_1	a_2	Row means
b_1	17.07	17.73	17.40
b_2	18.85	19.48	19.17
Col means	17.96	18.61	18.29

Conclusion. The small P-values for the main effects of A and B indicate both are significant. There is no indication of an interaction.

In the table of means the greater difference between the row means relative to the difference between the column means reflects the more significant main effect of B. The small mean difference between the diagonally opposite means in the body of the table, i.e.,

$$(17.07 + 19.48)/2 - (18.85 + 17.73)/2 = -0.01$$

is clear confirmation of a lack of interaction.

Comment. In the absence of interaction the main effects of factors A and B are sensible measures of the additive effects of each factor.

Example 8.2

The problem. Suppose that in the experiment in Example 8.1 the data were as given there except that for the factor combination a_2b_2 the observations were all increased by 4 units, i.e., they were

$$a_2b_2 \quad 22.1 \quad 22.9 \quad 23.6 \quad 23.9 \quad 24.1 \quad 24.3$$

How does this effect the analysis of variance and the interpretation of results?

Formulation and assumptions. As for Example 8.1.

Procedure. The analysis of variance table is:

Source	DF	SS	MS	F	P
Main A	1	42.135	421'35	116.23	0.000
Main B	1	85.127	85.127	234.83	0.000
Interaction	1	23.602	23.602	65.11	0.000
Error	20	7.250	0.363		
Total	23	158.113			

and the table of means is

	a_1	a_2	*Row means*
b_1	17.07	17.73	17.40
b_2	18.85	23.48	21.17
Col means	17.96	20.61	19.29

Conclusion. In the presence of interaction, a study of main effects is useless, or even misleading. For, example the main effect of A is epitomized by the difference in the marginal column means, i.e., $20.61 - 17.96 = 2.65$. However, at the lower level of B the mean difference is only $17.73 - 17.07 = 0.66$ whereas at the higher level it is $23.48 - 18.85 = 4.63$. These last two differences are often referred to as *particular effects*. They are usually of more interest to experimenters than main effects when there are interactions.

Comments 1. In this example if the experimenter were interested in which factor level produced the greatest mass of plant material it is important to recognise that the effect of factors are not purely additive. Having both factors at the higher level increases yield above that expected with a purely additive model.

2. Situations where factors interact in a manner like that described in this example are common in many fields involving experimental or observational data. For example, there is strong evidence that smoking increases the risk of lung cancer as does exposure to certain forms of asbestos dust. Further, the effects are not merely additive. The lung cancer incidence rate for smokers exposed to asbestos dust are appreciably higher than they would be if the effects were simply additive.

3. Interactions may sometimes mask main effects if the effect of increasing the level of one factor reverses the direction of response to the other factor. This is illustrated in Example 8.3.

Example 8.3

The problem. Suppose that in the experiment in Example 8.1 the data were as given there except that for the factor combination a_2b_2 the observations were all decreased so that they were

$$a_2b_2 \quad 17.9 \quad 18.3 \quad 18.4 \quad 18.6 \quad 18.7 \quad 18.8$$

How does this effect the analysis of variance and the interpretation of results?

Formulation and assumptions. As for Example 8.1.

Procedure. The analysis of variance table is:

Source	DF	SS	MS	F	P
Main A	1	0.107	0.107	0.50	0.489
Main B	1	9.375	9.375	43.64	0.000
Interaction	1	1.707	1.707	7.94	0.011
Error	20	4.297	0.215		
Total	23	15.485			

and the table of means is

	a_1	a_2	Row means
b_1	17.07	17.73	17.40
b_2	18.85	18.45	18.65
Col means	17.96	18.09	18.02

Conclusion. Here with the presence of interaction, a study of main effects is useless. The main effect of A is epitomized by the difference in marginal column means, i.e., $18.09 - 17.96 = 0.13$. However, at the lower level of A the mean difference is $17.73 - 17.07 = 0.66$, while at the higher level it is $18.45 - 18.85 = -0.40$. These last two differences are the *particular effects*. Here the indication is that each particular effect of A is in opposite directions at each level of B.

The above examples cover situations where standard normal theory analysis of variance is appropriate. Standard errors for the individual means and for the marginal means provide a tool for determining confidence intervals for main effects and particular effects, but we do not pursue this standard aspect of the classic parametric analysis here.

Nonparametric procedures become appropriate when assumptions needed for parametric methods may break down. We have seen in earlier chapters that the use of statistics based on ranks are then sometimes appropriate in the sense that they often give similar results to a parametric analysis when that is appropriate, but at the same time overcome difficulties associated with breakdown of assumptions needed in a parametric approach.

A simple idea already mentioned in Section 7.3.4 and discussed in this context by Conover and Iman (1976, 1981) is to rank all observations in ascending order and to perform an analysis of variance on the ranks instead of the original data. This has been shown by many writers to produce unfortunate anomalies. On a more positive note, it provided a stimulus for new approaches to the modelling and study of interactions.

Example 8.4

We rank the data in (i) Example 8.1 and (ii) in Example 8.2 and (iii) in Example 8.3 in ascending order and perform an analysis of variance on these ranks.

For ranks based on the data in Example 8.1 we obtain the following analysis of variance:

Source	DF	SS	MS	F	P
Main A	1	77.0	77.0	5.83	0.025
Main B	1	805.0	805.0	60.93	0.000
Interaction	1	2.7	2.7	0.20	0.658
Error	20	264.3	13.2		
Total	23	1149.0			

leading, as in Example 8.1, to the conclusion that there is no interaction.

For ranks based on the data in Example 8.2 we obtain the following analysis of variance:

Source	DF	SS	MS	F	P
Main A	1	155.04	155.04	21.06	0.000
Main B	1	840.17	840.17	114.11	0.000
Interaction	1	7.04	7.04	0.96	0.340
Error	20	147.25	7.36		
Total	23	1149.50			

Unlike the analysis in Example 8.2, evidence of an interaction has disappeared. Can you see why? It may help if you think about other possible yields at the higher level of each factor that would result in the same rankings as those in this example. Since absence of interaction makes interpretation of main effects meaningful this might seem desirable. However, as indicated in Example 8.2, the presence of interaction in the original data may be important in practice, indicating that effects are not strictly additive. The practical implication in that example is that increased dose of growth hormone together with the higher level of undersoil heating enhance growth appreciably more than increasing only one or the other level. This may have important economic or other implications.

For ranks based on the data in Example 8.3 we obtain the following analysis of variance:

Source	DF	SS	MS	F	P
Main A	1	4.2	4.2	0.28	0.599
Main B	1	770.7	770.7	52.70	0.000
Interaction	1	80.7	80.7	5.52	0.029
Error	20	292.5	14.6		
Total	23	1149.5			

Unlike the situation arising with the data in Example 8.2, tranformation to ranks has not removed evidence for an interaction. Can you suggest a reason for this? Again it may help to think about what other yields for both factors at the higher level would give the same rankings.

Classical parametric methods are appropriate for each of the data sets considered in this section. Example 8.4 illustrates a possible nonparametric approach that is unsatisfactory in the sense that it gives answers incompatible with those given by the parametric method when the latter is appropriate.

Can we develop nonparametric approaches that may be valuable when assumptions such as that of homoscedastic normally distributed errors breaks down, or one that will not be unduly disturbed by a few outliers?

Some answers can be provided only after a more detailed study of the concept and meaning of interactions in a wider context than that of 2×2 factorial structures.

8.3 The Nature of Interactions

Many students first and main encounter with interactions relates to interactions associated with factorial treatment structures in designed experiments. A wide-ranging discussion of these and other important types of interactions and their implications is given in Cox (1984).

It is well-known that in a parametric context interactions may sometimes be reduced, or even eliminated, by a data transformation. Whether it is desirable to do so depends upon the nature of the interaction, and the physical nature of the data involved and the questions being asked.

With factorial treatment structures and continuous data we are often interested in the linear model. If there is no interaction, in classic analysis of variance for two factors A and B at each of a, b levels respectively and r replications of each factor combination the yields, or responses, y_{ijk} may be written

$$y_{ijk} = \mu + \alpha_i + \beta_j + \varepsilon_{ijk},$$

where $i = 1, 2, \ldots, a$, $j = 1, 2, \ldots, b$, $k = 1, 2, \ldots, r$ and

$$\sum_{i=1}^{a} \alpha_i = \sum_{j=1}^{b} \beta_j = 0$$

In the classic parametric model the ε_{ijk} are independently and identically distributed (iid) $N(0, \sigma^2)$. Generally σ is unknown.

The model with interaction is

$$y_{ijk} = \mu + \alpha_i + \beta_j + \gamma_{ij} + \varepsilon_{ijk}$$

with additional constraints $\sum_i \gamma_{ij} = \sum_j \gamma_{ij} = 0$.

We use the notation $y_{ij.}$ for the mean over the r replicates for yields corresponding to factor level combination (ij), $y_{i..}$ for the mean over all units with factor A at level i, $y_{...}$ for the mean of all observations, and so on. Some writers use these notations for the corresponding totals, writing the means as $\bar{y}_{ij.}$, etc. The estimates of the parameters are

$$\hat{\mu} = y_{...}, \quad \hat{\alpha}_i = y_{i..} - y_{...}, \quad \hat{\beta}_j = y_{.j.} - y_{...}, \quad \hat{\gamma}_{ij} = y_{ij.} - y_{i..} - y_{.j.} + y_{...}.$$

If the assumption of a homoscedastic normally distributed error is relaxed, and the error is merely assumed to have the same unknown distribution with zero median and the remaining features of the above linear models are retained, interest centres on finding appropriate distribution-free procedures for testing for interaction, and perhaps for estimating the parameters in the model. Here we use the term distribution-free rather than nonparametric as we are still interested in hypotheses about, or estimation of, the parameters. $\mu, \alpha_i, \beta_j, \gamma_{ij}$. Distribution-free refers to the error structure. Some writers call this approach either *semiparametric* or *robust parametric*.

Example 8.4 showed that a rank ordering of all observations followed by a formal analysis of variance of ranks may be unsatisfactory in the context of the linear model since it may miss a very evident interaction.

The transformation to ranks, although a monotonic transformation of the data, effectively destroys the linearity.

The concept of interaction is often relevant in a broader context than linear models and is used to cover virtually any departure from an additive linear model. It also includes cases where we may not even assume a linear model with additive error for the original response variable.

It is well-known that even in the one-way classification model, homogeneity of error variance can often be obtained by some appropriate transformation of the data. In a broader context a model such as a multiplicative model may sometimes be converted to a linear model (with or without interaction terms) by a logarithmic transformation.

While there are many situations — such as those considered in Examples 8.1, 8.2 and 8.3 — where retention of the linear model is sensible if we are interested in responses in the measured units, there are situations where retaining the linear model is either of less interest, or where that model may not even be appropriate. A linear model is irrelevant where responses, for example, are of a qualitative nature or measured only on an ordinal scale. In these circumstances only certain types of interaction may be of interest.

One consequence of these generalities has been the development of two distinct approaches to nonparametric methods for dealing with interactions. The first retains the *parametric* element of the linear model, e.g., the parametric structure in the two factor situation with the form $\mu + \alpha_i + \beta_j + \gamma_{ij}$. The other approach replaces this structure by an entirely different model that is both completely nonparametric and distribution-free. In this latter situation an analysis may then often be appropriately performed using a rank transformation. This is because without the linear model, a new conception of interaction is relevant.

Before we decide which model is appropriate in a particular case it helps to consider different types of interaction along the lines given in Cox (1984).

Cox categorizes interactions associated with different aspects of data. He suggests that in essence an observation y may be explained in terms of factors of three types.

The first type are *treatment factors* which are under the control of the experimenter (e.g., temperature, pressure, the time a reaction is allowed to proceed in a chemical procees). Treatment factors may be quantitiative, ordinal, nominal or binary.

The second type of factor Cox calls *intrinsic factors*. These include variables such as age or gender of animals in an experiment. The experimenter has no control over such factors in the sense that he or she cannot alter them, although they might be taken into account in the choice of units, and in the allocation of treatments to these units. Variables outside the experimenter's control, such as meteorological conditions may also be regarded as intrinsic factors. Again, intrinsic factors may be quantitative, ordinal, nominal or binary. In many experiments intrinsic factors and evaluation of their effects is of interest, as is any interaction between them and treatment factors.

The third type of factor consists of *nonspecific* or not uniquely characterized factors, such as block characteristics in a randomized block experiment.

Cox points out that three types of *two-factor* interaction are commonly of interest, viz,

- treatment × treatment;
- treatment × intrinsic;
- treatment × nonspecific.

Some of these may not be estimable. For example, in a randomized block experiment where each treatment occurs once and only once in each block it is not possible to test for a *blocks × treatment* interaction. In effect, this is assumed to be zero. If it is, in fact, not zero it is incorporated in, and will increase, the error mean square in an analysis of variance.

If, in a randomized block experiment each treatment is replicated r times within each block we have a situation which is technically equivalent to a two factor experiment with factor A (treatments) at, say, t levels and factor B (blocks) at b levels. Any program for analysis of variance of a two-factor design will provide a test for *blocks × treatment* interaction. In Cox's terminology this is an interaction between between a *treatment* and a *non-specific* factor.

Under the usual parametric linear model assumptions a *blocks × treatment* interaction implies that treatment and block effects are not strictly additive. In practice this is an indication that in one or more blocks treatments are responding differently from the way they respond in the remaining blocks. With the usual normal homogeneous error assumptions this approach is essentially a parametric one, so we do not discuss it further here.

Our main interest is usually in the effect of treatment factors on responses. Intrinsic and nonspecific factors must be taken into account only in the way they may bear on the treatment factors, or if they are in themselves of interest (e.g., we may be interested in whether the response to a drug differs with the

intrinsic factor *gender*). If we are interested in the additional hours sleep
achieved by insomniacs given two different sophorifics, A and B, and obtained
the following results for four patients of each gender

$$
\begin{array}{llcccc}
Male & \text{A} & 1 & 2 & 2 & 2 \\
     & \text{B} & 4 & 4 & 4 & 4 \\
Female & \text{A} & 4 & 4 & 4 & 5 \\
       & \text{B} & 0 & 0 & 0 & 1
\end{array}
$$

it does not need a formal analysis to show that there is an interaction between
the intrinsic factor *gender* and the treatment factor *drug*, for it is evident that
drug B is of greater benefit than drug A for males, but that the opposite
applies for females.

Cox draws attention to the fact that certain types of interaction are order
dependent in the sense that they are not removed by replacing the original data
by monotonic transformations. In particular, this applies to transformation to
ranks, as this is a monotonic trnasformation (i.e., one that does not change
the order of the data). If we replace the data given above on the increased
hours of sleep induced by two drugs by their ranks in increasing order (using
mid-ranks for ties) the data become

$$
\begin{array}{llcccc}
Male & \text{A} & 4.5 & 7 & 7 & 7 \\
     & \text{B} & 12 & 12 & 12 & 12 \\
Female & \text{A} & 12 & 12 & 12 & 16 \\
       & \text{B} & 2 & 2 & 2 & 4.5
\end{array}
$$

and interaction in the sense that the drugs show an opposite ordering in
responses between genders is still clear without any formal analysis.

In this example a linear response model has little justification and certainly the error strcture for the original data could hardly be described as
homoscedastic normal by any stretch of the imagination.

The simple examples considered so far in this chapter have indicated that
the rank transformation may remove interactions of interest when the additive
linear model is appropriate (because linearity will not in general be preserved
under this monotone transformation).

On the other hand when order-based interactions are of interest there may
be no particular reason to consider a linear additive model. A key feature of
order-based interactions is that they are preserved under any monotonic, or
order preserving, transformation, which includes transformation to ranks.

The distinction betweeen these two situations has led to two different approaches to nonparametric or distribution-free analyses involving interactions.
We discuss these in Sections 8.4.1 and 8.4.2 respectively.

8.4 Alternative Approaches to Interactions

In Section 8.4.1 we consider a method proposed by Hettmansperger and
Elmore (2002) in which we assume a linear model that may include an additive

interaction term is appropriate, and we want to detect any interaction and study its nature.

In Section 8.4.2 we consider a different approach where a rank transformation is appropriate, This uses an alternative to the additive linear model. The basic notions stem from work by Akritas and Arnold (1994).

8.4.1 Interaction in an Additive Linear Model

Hettmansperger and Elmore considered the traditional model

$$y_{ijk} = \mu + \alpha_i + \beta_j + \gamma_{ij} + \varepsilon_{ijk},$$

but assuming only that the errors are identically and independently distributed continuously with zero median. They considered the balanced case with r replicates of each factor combination, the factor A having a levels and the factor B having b levels.

Their approach is motivated by the form of the usual test statistic in the analysis of variance for testing the hypothesis H_0: $\gamma_{ij} = 0$ for all (i, j). The test statistic has an F distribution with $(a-1)(b-1)$ degrees of freedom under H_0 with the usual normal error assumption.

They point out that two relevant features of the test statistic for interaction are (i) that the numerator is a sum of squares of cell mean residuals formed by removing estimates of the A-factor and B-factor and overall (mean) effects, and (ii) the denominator is an estimator of error variance. They propose as the first step obtaining residuals after fitting estimates of μ. α, β appropriate to the model

$$y_{ijk} = \mu + \alpha_i + \beta_j + \varepsilon_{ijk}.$$

That is, the additive linear model without an interaction term. These estimates might be, but need not necessarily be, the least squares estimates. If there is interaction, the residuals from fitting this model will reflect the fact that nonzero γ_{ij} have effectively been combined with the error term. This will inflate the error.

Hettmansperger and Elmore base a test for interaction on the *ranks* of the residuals formed above, i.e., $r_{ijk} = rank(y_{ijk} - \hat{\mu} - \hat{\alpha}_i - \hat{\beta}_j)$. The test statistic they propose is a modified analogue of the usual analysis of variance test statistic for interaction. However, the usual F-test is no longer appropriate if the normality assumption for errors is dropped. This, of course, is likely to be the situation when a nonparametric approach is appropriate.

They propose a statistic which is, in effect, the interaction sum of squares of the ranks divided by the total mean square for ranks which can be shown to have, for large values of r, approximately a chi-squared distribution with $(a-1)(b-1)$ degrees of freedom under the hypothesis of no interaction. This approximation is based on theory developed in Chapters 3 and 4 of Hettmansperger and McKean (1998). The statistic is somewhat analogous to that used in the Kruskal–Wallis test.

Some studies based on approximate permutation distributions given by Hettsmanperger and Elmore suggest that this asymptotic result is reasonable for all but very small data sets. Although we have not investigated the approximation in detail, we believe it may well be a conservative test in that it uses the total mean square in the denominator as a scaling factor rather than the error mean square, as in the analysis of variance.

Example 8.5

The problem. Use the method described above to test for interaction in the data in Example 8.2 using least squares estimates for the parameters in the additive model without interaction, then analyzing the ranked residuals.

Formulation and assumptions. If least squares estimators are used the first step is to fit an additive model without an interaction term. The analysis of variance program in most packages provides this facility and will usually give residuals. The residuals are then ranked and an analysis of variance program fitting a model with an interaction term to these residual ranks provides the necessary information to calculate the Hettmansperger and Elmore statistic.

Procedure. We used the Minitab two-way analysis of variance program to fit the pure additive model without interactions. The residuals were

a_1b_1	0.125	0.525	1.025	1.225	1.425	1.625
a_2b_1	-1.525	-1.325	-1.225	-0.825	-0.625	-0.425
a_1b_2	-1.642	-1.342	-1.242	-0.742	-0.642	-0.342
a_2b_2	-0.392	0.408	1.108	1.408	1.608	1.808

These were then ranked and the ranks were:

a_1b_1	14	16	17	19	21	23
a_2b_1	2	4	6	7	10	11
a_1b_2	1	3	5	8	9	13
a_2b_2	12	15	18	20	22	24

These ranks were subjected to an analysis of variance fitting the model allowing for interaction. The Minitab output for this analysis was:

```
Analysis of Variance for residual ranks
Source        DF      SS       MS       F       P
A              1      0.2      0.2     0.01   0.918
B              1      0.0      0.0     0.00   1.000
Interaction    1     840.2    840.2   54.26  0.000
Error         20     309.7    15.5
Total         23    1150.0
```

Not surprisingly for the main effects, here labelled A, B, both sums of squares are near zero because we are analysing the residuals after having fitted the additive constants representing these factors. The P-values given in the above analysis are not relevant as they are based on an analysis of ranks and not all assumptions needed for the usual variance-ratio based F-test apply.

The information from this analysis for determining the Hettmansperger and Elmore statistic Q is the interaction sum of squares 840.2 and the total mean square which is $1150/23 = 50$. The latter may be calculated independently. The reader should verify that for ranked data this quantity is given by $rab(rab + 1)/12$ where, in this example $r = 6$, $a = b = 2$. Thus, for this example, $Q = 840.2/50 = 16.80$. Comparing this with the chi-squared distribution value with 1 degree of freedom we see that $P < 0.001$.

Conclusion. There is strong evidence of an interaction.

Comments. 1. We already found such evidence in Example 8.2 and, as there need be no concern about the normality assumption made in that example, it is somewhat artificial to apply the nonparametric procedure here. However the procedure here may prove more efficient with a more general error assumption. If we use a modification outlined below using an alternative to least squares estimators of the α_i, β_j the procedure may be more robust against outliers.

2. While the above is a test for interaction, it is not directly relevant to estimation of effects. Its importance lies in the fact that strong evidence of interaction indicates that it may be more appropriate to study particular effects than it is to study main effects. We consider below how this might be done.

3. We have only given an asymptotic result. Hettmansperger and Elmore point out technical difficulties in developing an exact permutation test since mean ranks for each factor combination are not invariant under transformation. However, they indicate that for an example they considered that Monte Carlo sampling of permutations yielded a P-value that closely approximated their asymptotic result.

It is possible to use alternatives to the least squares estimators of the mean and main effects parameters. One suggested by Hettmansperger and Elmore is to use *median polish estimators* as described and studied in detail by Emerson and Hoaglin (1983). When this method is applied to the original data, the residuals have row and column medians zero rather than means zero. It is equivalent to fitting by least absolute deviations. A program for median polish is available in Minitab. This calculates the residuals, which may then be ranked and the ranks analysed as in Example 8.5. Because outliers have little influence on the fit using median polish this may result in one or two outliers having very large residuals which may in some cases make the test less sensitive to detection of interactions, although giving a model more appropriate for the reliable data (i.e., those excluding the outliers). We comment further on this point in the next example.

Example 8.6

The problem. Consider again the data in Example 8.2 with the modification of replacing the final entry of 24.3 for both factors at the higher level by a blatant outlier taking the value 16. Use both least squares and median polish to fit a no-interaction model and use the Hettmansperger and Elmore rank of residuals method to test for interaction in each case.

Formulation and assumptions. As in Example 8.5 with additional use of median polish as an alternative to least squares.

Procedure. Proceeding as in Example 8.5, least squares estimation gives $Q = 11.21$, indicating $P < 0.001$. Using median polish and then ranking the residuals using these estimators of μ, α_i, β_j leads to $Q = 6.75$ indicating $P \approx 0.01$

Conclusion. There is strong evidence of interaction.

Comments. 1. That median polish gives a greater *P*-value than least squares is perhaps surprising because for each procedure the residual for the outlying 16 is ranked 1, since it is the negative residual of greatest magnitude. In each case other residuals for that same factor combination are all positive and ranked 20, 21, 22, 23, and 24. However, there is greater variability among ranks of residuals for other treatment combinations using median polish and this tends to dilute the evidence for interaction. This may reflect the slightly lower efficiency of median based methods when the classic normality assumptions apply, although a further study of comparisons between the methods may provide a useful student exercise.

2. A classic least squares analysis of these data allowing for interaction gives an interaction $P = 0.062$, meaning that interaction would then be ignored using conventional significance levels.

As we have indicated, in the presence of interaction it is usually inappropriate to make comparisons on the basis of main effects, but more important to look at the particular effects of one factor at each fixed level of the other. In the case of a 2×2 structure considered in examples so far in this chapter, if there are interactions one might compare the response at each level of factor A at the lower level of factor B using the WMW test, and then do a similar comparison at each level of factor A at the higher level of factor B. Similar inspections might be made for particular effects of Factor B.

In the more general situation of more than two levels of one or both factors an obvious extension would be to perform Kruskal–Wallis tests across all levels of factor A separately at each level of factor B. An example for a 3×2 experiment is given by Hettmansperger and Elmore.

We have already mentioned the common use of tables of means, together with appropriate standard errors, as aids to interpretation of data with factorial structures. A useful EDA extension of these, particularly with larger data sets, is to present boxplots for all treatment combinations. Hettmansperger and Elmore include an excellent illustration of this for the data set they use for illustrative purposes. Those data are reproduced in Exercise 8.1 at the end of this chapter.

A similar approach to testing for interaction is described briefly by Higgins (2004, Chapter 9) who describes the procedure as an *aligned-rank transformation*. He also indicates related methods for testing for main effects, but it would seem desirable to use these only if no significant interaction is detected.

8.4.2 The Akritas and Arnold Distribution-Free Response Model

We have indicated that a linear model involving additive factor effects and a possible interaction term may not be appropriate. Also, even if it is appropriate for the initial data, the linear model will in general no longer hold for overall rank transformations of the data.

Akritas and Arnold (1994) proposed a remodelling of the response variable in terms of distributions rather than parameters. The models they proposed are very general and wide ranging, covering not only factorial structures, but extending to experiments in which repeated measurements on the same units are made. In their initial presentation they also relaxed the restriction we have made so far in this chapter of equal replication of all factor combinations. There are parameteric models and analytic procedures available for many of the situations they cover, but their approach allows an appreciable relaxation of assumptions.

The generality of their model means that in practice fairly complicated algebra is needed to cover all possible scenarios. Their original paper has spawned a vast literature, but perhaps because of the generality of most of the procedures, the basic concepts do not seem to have received much coverage in nonparametric textbooks at the elementary level.

Here we discuss some of the fundamentals of the approach. For those who wish to apply the method to specific problems we recommend as a starting point the comprehensive review by Brunner and Puri (2001), perhaps after first reading at least the introduction to Akritas and Arnold (1994). Other key papers in the area include Akritas, Arnold and Brunner (1997) and Brunner and Puri (1996). All contain numerous references to related and earlier work in this field, much of the latter being of a somewhat *ad hoc* nature, either assuming interactions did not exist, or if they did, that they were of some particular kind.

Inspiration for the Akritas and Arnold model comes from the simplest one-way linear model involving t groups or treatments, which in the classic parametric set-up may be written

$$y_{ij} = \mu + \alpha_i + \varepsilon_{ij}, \quad i = 1, 2, \ldots, t, \quad j = 1, 2, \ldots n_i. \tag{8.3}$$

where $\sum_i \alpha_i = 0$ to ensure uniqueness of the α_i, and the ε_{ij} are iid $N(0, \sigma^2)$. The usual nonparametric approach is to drop the normality assumption on the ε_{ij} requiring only that they have zero median.

In the case $t = 2$ the standard nonparametric procedure is then WMW. The model (8.3) is then a model for location shift. An alternative way of writing this is to specify the distribution of the y_{1j} by a cumulative distribution function $F_1(x)$ and that of the y_{2j} by $F_2(x) = F_1(x - \theta)$. If $\theta \neq 0$ the implication is that the distributions differ only in location. We pointed out in Section 6.1.2 that the WMW test was also appropriate if the alternative to equality of distribution functions was that one distribution *dominated*. That is, that for all x, we had either $F_1(x) \geq F_2(x)$ or $F_1(x) \leq F_2(x)$, with strict inequality for at least some x.

In this latter form the test is completely nonparametric, whereas in the linear shift form there is a parameter θ, even if the test is distribution-free in the sense that the cdfs are not confined to any particular family.

Akritas and Arnold generalized this idea and instead of the model (8.3), assumed only that the y_{ij} had a distribution $F_i(x)$ such that

$$F_i(x) = M(x) + A_i(x), \quad \sum_{i=1}^{t} A_i(x) = 0. \tag{8.4}$$

That is, F_i is the distribution function of all $y_{ij}, j = 1, 2, \ldots, n_i$ and it is expressed as the sum of two distributions.

The null hypothesis of no treatment difference is H_0: *all* $A_i(x) = 0$, or equivalently, the $F_i(x)$ are the same for all i. We denote by $\overline{F}_.(x)$ the average over all i of the $F_i(x)$. We may set $M(x) = \overline{F}_.(x)$. Then $A_i(x) = F_i(x) - \overline{F}_.(x)$. Under the null hypothesis, which implies all $F_i(x)$ are equal, the models in (8.3) with only the normality assumption dropped and that in (8.4) are equivalent. However, under the alternative hypothesis the latter is more general as it does not assume that the Y_{ij} are from families differing only in location. Akritas and Arnold refer to the model using (8.3) with the normality assumption relaxed as the *robust parametric model* and that in (8.4) as the *nonparametric model*.

The nonparametric model may be extended to a two-way factorial structure and this extension shows appreciable differences from the corresponding robust parametric model considered in the preceding sections of this chapter, i.e.,

$$y_{ijk} = \mu + \alpha_i + \beta_j + \gamma_{ij} + \varepsilon_{ijk}$$

with the usual constraints and the errors iid with zero median.

For each given i, j we now assume that the independent Y_{ijk} have distribution function F_{ij} which can be expressed in the form

$$F_{ij}(x) = M(x) + A_i(x) + B_j(x) + C_{ij}(x),$$

subject to constraints

$$\sum_i A_i(x) = \sum_j B_j(x) = \sum_i C_{ij}(x) = \sum_j C_{ij}(x) = 0,$$

where, extending the notation developed for the one way classification model, we express the components in terms of appropriate averages of certain of the F_{ij}. Specifically

$$M(x) = \overline{F}_{..}(x), \quad A_i(x) = \overline{F}_{i.}(x) - \overline{F}_{..}(x),$$
$$B_j(x) = \overline{F}_{.j}(x) - \overline{F}_{..}(x), \quad C_{ij}(x) = F_{ij}(x) - \overline{F}_{i.}(x) - \overline{F}_{.j}(x) + \overline{F}_{..}(x).$$

For the nonparametric model the hypothesis of no interaction implies that $C_{ij}(x) = 0$.

Careful reflection shows that this hypothesis is unchanged by monotonic transformations of the observed y_{ij}. This includes overall ranking of the data. An immediate implication is that an appropriate test will only pick up interactions that are preserved under transformation to ranks in the sense considered by Cox (1984).

Table 8.2 *Numbers of rats exhibiting a given side effect in groups of 50.*

	Males					Females				
Drug I	7	9	9	5	6	6	11	9	10	10
Drug II	8	11	10	10	9	3	4	7	2	4
Drug III	7	9	16	4	8	18	32	19	14	59

Applied to conventional factorial designs, the usual assumption that the responses y_{ijk} are independent is appropriate. It is not essential to assume independence in other situations, and indeed when applied to repeated measurements on the same units such an assumption is inappropriate. Unlike the linear additive model, there is no specified error distribution and thus no assumptions about the variance of the y_{ijk}. The variance need not even exist, as would be the case for a Cauchy distribution.

Because hypotheses that can be tested will still hold for monotone transformations of the original data all the analyses may be based on an overall ranking of the data. As Brunner and Puri (2001) point out this means the procedure may be applied not only to designed experiments giving continuous responses, but to data consisting of counts and indeed to any observational data that may be ranked. In their most general form the relevant test statistics allow for unequal replication in factorial structures and ties may be dealt with by using mid-ranks.

Methods based on this model are especially apposite when linear additive models are inappropriate or irrelevant. Situations of this kind are common in drug testing and other clinical trials where, for example, data may consist of numbers of animals per group of, say, 50 or 100, responding to one of several different drugs, each drug being in this sense a treatment.

If three drugs are assigned each to several groups of 50 rats of each gender, and counts made of the number of animals in each group exhibiting side effects typical counts might be those indicated in Table 8.2

An example of count data arising in a different context is given by Brunner and Puri (2001).

We confine our attention here to the Akritas and Arnold model applied to two factor experiments with the first factor A at a levels and the second factor B at b levels. The simplest case is the 2×2 structure with $a = b = 2$. We shall not give the derivation of the results we use. These may be found in a much broader context in Brunner and Puri (2001). The key features are

- (i) determination of an appropriate measure of the *relative* treatment effect of each factor combination,

- (ii) forming a relevant statistic to test for interactions and when these are not significant equivalent statistics to test for main effects.

We must then formulate appropriate asymptotic tests, or exact small sample tests if feasible. We only consider here asymptotic tests for what Brunner and Puri describe as ANOVA-type statistics. All tests are carried out on the overall ranked data and effects are estimated on the basis of the sample, or empirical, cdfs for these ranked data.

For a 2×2 design with r replicates when using ranks the relative treatment effects are analogous to treatment effects in the classic model where the parameters α_i are estimated by the mean $y_{i..} - y_{...}$, i.e., the mean of all units receiving the ith level of factor A minus the overall mean. The corresponding relative treatment effect in the Akritas and Arnold model is $r_{i..} - r_{...}$.

In the 2×2 situation the main effects and interactions are estimated by the mean rank analogues of the data means in the classic analysis.

However, the analogues of the ANOVA test statistics for mean and interaction differ in that, unlike the linear model with an additive homoscedastically distributed error term, the assumption of constant error variance is no longer relevant.

In the light of remarks on the inappropriateness of considering main effects in the presence of interaction, we recommend that the first test be one for interaction.

The ANOVA type test statistic for interaction proposed by Brunner and Puri takes, for equal replication in an $a \times b$ design the form

$$F_{AB} = \frac{abr(r-1)}{(a-1)(b-1)\sum_{i=1}^{a}\sum_{j=1}^{b} d_{ij}^2} \sum_{i=1}^{a}\sum_{j=1}^{b}(r_{ij.} - r_{i..} - r_{.j.} + r_{...})^2, \quad (8.5)$$

where

$$d_{ij}^2 = \sum_{k=1}^{r}(r_{ijk} - r_{ij.})^2.$$

This is numerically equal to the standard ANOVA statistic for interaction computed for overall ranks. Inspection of the individual d_{ij} in typical examples show that these differ markedly. This has a considerable effect on the relevant degrees of freedom for the asymptotic F-test. Approximate degrees of freedom for general cases are best expressed in matrix notation. Their evaluation is not difficult with appropriate computer software for handling matrix operations. We do not discuss these general forms, but they can be expressed in simple algebraic closed formulae for the 2×2 and one or two other simple structures.

Under the hypothesis of no interaction for the 2×2 structure asymptotically F_{AB} given by (8.5) has an F-distribution with degrees of freedom ν_1, ν_2 given by

$$\nu_1 = \frac{\left(\sum_{i=1}^{2}\sum_{j=1}^{2} d_{ij}^2\right)^2}{\sum_{i=1}^{2}\sum_{j=1}^{2} d_{ij}^4 + \sum_{i=1}^{2}\sum_{j=1}^{2} d_{ij}^2 d_{kl}^2}, \quad (8.6)$$

where the suffix $ij \neq kl$, and

$$\nu_2 = (r-1)\frac{\left(\sum_{i=1}^2 \sum_{j=1}^2 d_{ij}^2\right)^2}{\sum_{i=1}^2 \sum_{j=1}^2 d_{ij}^4}. \tag{8.7}$$

It is easily verified (see Exercise 8.2) that if all d_{ij}^2 are equal then $\nu_1 = 1$ and $\nu_2 = 4(r-1)$, the degrees of freedom associated with the test for interaction in classic ANOVA.

The complexity of the form of the degrees of freedom may seem surprising. However, the form is similar to that proposed by Satterthwaite (1946) for dealing with the classic Behrens–Fisher problem of extending the t-test for equality of means to samples from populations with different variances.

The formula for ν_2 extends to $a \times b$ factorials by taking summation over i from 1 to a and that over j from 1 to b. Although the matrix extension is straightforward, closed expressions for ν_1 become very complicated for $a, b > 2$. We give just one example for $a = 3$, $b = 2$ when

$$\nu_1 = \frac{2\left(\sum_{i=1}^3 \sum_{j=1}^2 d_{ij}^2\right)^2}{2\sum_{i=1}^3 \sum_{j=1}^2 d_{ij}^4 + 4\sum_{i=1}^3 d_{i1}^2 d_{i2}^2 + \sum_{\text{rem}} d_{ij}^2 d_{kl}^2}, \tag{8.8}$$

where the final denominator sum $\sum_{\text{rem}}$ is over all pairs with suffix $ij \neq kl$ not occurring in the previous term, i.e., not in $\sum_{i=1}^3 d_{i1}^2 d_{i2}^2$. If all d_{ij}^2 are equal this reduces to 2, the degree of freedom for interaction in classic ANOVA (Exercise 8.3).

Example 8.7

The problem. Apply the Akritas and Arnold method to the data in Table 8.2 to determine if there is evidence of a significant interaction between the factors A (drug) and B (gender).

Formulation and assumptions. The statistic for testing for interaction is given by (8.5), which has asymptotically an F-distribution with degrees of freedom given by (8.7) and (8.8).

Procedure. The interaction statistic is computed either by extracting the interaction variance ratio from a standard ANOVA analysis of the ranked data or by direct substitution of relevant values in (8.5). The latter is easily done once the d_{ij}^2 are obtained and we need these to evaluate (8.7) and (8.8). With more advanced models discussed in Brunner and Puri it is best to use programs that do matrix operations, but calculations here are easily done with Minitab by writing straightforward macros. For purposes of discussion we give the values of d_{ij}^2. They are

$$d_{11}^2 = 88.2, d_{12}^2 = 156.2, d_{21}^2 = 75.2, d_{22}^2 = 48.8, d_{31}^2 = 265.8, d_{32}^2 = 14.8,$$

whence (8.7) and (8.8) give $\nu_1 \approx 2.3$ and $\nu_2 \approx 15.2$. The value of the test statistic for interaction is $F_{AB} = 20.14$. With 2, 15 degrees of freedom this gives $p < 0.001$.

Conclusion. There is strong evidence of an interaction, which the reader should confirm, and deduce the nature of, from the means of the data given in Table 8.2, or from the table of rank means given in *Comment* 1 below.

The presence of interaction suggests that particular effects rather than main effects should be examined in any further analysis. That is, effects should be compared within each gender separately.

Comments. 1. The means for the original data in Table 8.2 are

	M	F
Drug I	7.2	9.2
Drug II	9.6	4.0
Drug III	8.8	28.4

A clear indication of a likely interaction is that while Drug II shows for males the highest mean side-effect rate by a small margin, in the case of females Drug II has a markedly lower rate than that for either alternative drug. Table 8.2 also indicates immediately that the marked variation in numbers of females showing side effects with drug III suggests considerable variability between the five groups in this factor combination, implying that the error is heteroscedastic if a linear model is considered. That the interaction effect is still exhibited by the rank transformation, is indicated by the following table of rank means:

	M	F
Drug I	11.1	17.6
Drug II	18.6	4.2
Drug III	13.7	27.8

The transformation to ranks does not result in the homogeneity of error variance required in the classic analysis. This is evident from the marked differences between the d_{ij} displayed above under *Procedure*. This is why the classic analysis applied to ranks would be inappropriate. Also, even with the classic model remember that the monotone transformation to ranks may destroy linearity. The Akritas and Arnold model, where it is appropriate, compensates for these difficulties. But we emphasise once again that it may not be appropriate if linearity is an important physically meaningful concept.

2. If the interaction test indicates no significant interaction, then main effects can be tested These follow broadly similar lines to the test for interaction described in this example for $a \times b$ factorial structures with equal replication. We omit details even for the simple cases, but these may be found in Brunner and Puri (2001).

3. In this example it may be appropriate to apply Kruskal–Wallis or some other test to data for each gender separately in view of the strong evidence of interaction.

Readers interested should consult Brunner and Puri (2001) to learn precisely what hypotheses can be tested and how the methods extend to designs with unequal replication and to repeated measurements where the independence assumption is no longer valid and correlations betweeen observations on the same unit at different times have to be taken into account. It is interesting to note that the WMW, the Kruskal–Wallis, Page and several other well-known tests are special cases of the Akritas and Arnold one-way model. Details are given in Brunner and Puri (2001).

8.5 Crossover Experiments

When several treatments are involved, experimenters often wish to give each treatment to each of a number of individuals. Response to any treatment may be influenced by treatments previously received. The situation may occur in clinical trials involving several drugs where the response to one drug may be influenced by residual effects of a drug previously administered, something we mentioned in Section 5.1.1.

A similar situation may arise if several different tools are used to perform a specific task. The performance with tool A might be more satisfactory if it is used before tool B rather than after. This could happen, for example, if tool B created considerable vibration that left the operator's hands shaky when he or she started to use tool A afterwards.

In educational or psychological experiments involving different tests, each being given to all participants, possible learning effects should be considered. Procedures tested early in the sequence may in that way influence performance in procedures tested later in the sequence.

It is sometimes possible to eliminate residual or carryover effects by allowing appreciable time lapses between treatments. Otherwise a design should be used that allows us to take account of such effects.

One device useful in such circumstances is a *crossover* design, mentioned briefly in Section 5.1.1. We describe only the simplest such design for comparing two treatments in what is called a *two-period crossover* design. The method may be extended to any finite number of treatments over more than two periods.

Crossover designs are widely used in pharmaceutical studies. A comprehensive account of their use, together with descriptions of both parametric and nonparametric methods of analysis, is given by Senn (2002). Putt and Chinchilli (2004) review a number of nonparametric approaches to analysis of crossover trials, and include references to more detailed studies.

Koch (1972) gives an example of a nonparametric analysis of one such experiment, some features of which we consider in Example 8.8. Two treatments are applied to each of an even number, $2N$, patients. N of these selected at random are given treatment A first, followed by treatment B. The treatment order is reversed for the remaining N patients. The design enables us to test hypotheses about

- direct effect of treatments;

- any residual effects of treatments;

- period effects.

where *period effects* represent a genuine difference of responses in each period that is independent of the actual treatment applied in that period. This is only meaningful if there are no residual effects of treatments.

Example 8.8

The problem. Koch (1972) analysed the data below from an experiment at the Dental Research Center, University of North Carolina. Ten children were randomly assigned to two groups of five. Two treatments were administered to each child. These were

A   Drink 100 ml grapefruit juice followed by an elixir of Pentobarbital;
B   Drink 100 ml water followed by an elixir of Pentobarbital.

The first group of five received A in period I and B in period II. We denote this sequence by AB. The second group received B then A, described as BA. Data are the amount of the elixir in a 10 ml blood sample taken 15 minutes after administration expressed in units of μg. ml^{-1}. The data were

Child	1	2	3	4	5	6	7	8	9	10
Sequence	AB	AB	AB	AB	AB	BA	BA	BA	BA	BA
Period I	1.75	0.30	0.35	0.20	0.30	7.20	7.10	0.75	2.15	3.35
Period II	0.55	1.05	0.63	1.55	8.20	0.35	1.55	0.25	0.35	1.50

Formulation and assumptions. The large values for treatment B for subjects 5, 6 and 7 relative to other readings suggest heterogeneity, so a nonparamteric test seems appropriate. Koch based his analysis on the WMW test.

Procedure. We first examine the data for each child summed over Period I and Period II. If there is a different carry-over effect (i.e., if the joint response summed over the two periods are different) we would expect the WMW test to indicate a significant difference between the sums for childs 1–5 (AB sequence) and 6–10 (BA sequence). The sums for each child in the order given in the data are

$$Sequence\ AB\quad 2.30\quad 1.35\quad 0.98\quad 1.75\quad 8.50$$
$$Sequence\ BA\quad 7.55\quad 8.65\quad 1.00\quad 2.50\quad 4.85$$

StatXact gives a two tail $p = 0.3095$ for the WMW test. A possible alternative would be a normal scores test which gives $p = 0.2778$, so there is no evidence of a differential residual effect. This does not mean that there is no residual effect from a previous treatment, but if there is any then it applies equally to either sequence.

When there is no evidence of differential residual effects the next step, and the most important one, is to test for a direct effect of treatments. This is based on the *differences* Period I–Period II for all individuals. If there is a difference between treatments we expect the differences for the first five children to exhibit a different location pattern to those same differences for the remaining children because one is measuring the effect A−B, the other B−A, Under H_0: *no treatment difference* both sets of differences should have the same distribution.

The relevant differences are

$$Sequence\ AB\quad 1.20\quad -0.75\quad -0.28\quad -1.35\quad -7.90$$
$$Sequence\ BA\quad 6.85\quad 5.55\quad 0.50\quad 1.80\quad 1.85$$

For these data the WMW test gives a two-tail $p = 0.0159$, a fairly strong indication of a real treatment effect.

Conclusion. There is no evidence of a differential residual effect, but evidence of a location difference between A and B.

Comment. A further test proposed by Koch is only relevant if the hypothesis of no differential residual effect is accepted. This is a test for a period effect, i.e., that the responses A−B may be different for each sequence order. The A−B difference for those in the sequence AB is that given above, while for those given the sequence BA it is the difference given above after reversal of sign. For the above data the WMW test gives a two-tail $p = 0.309$, so there is no evidence for a period effect.

In practice most crossover experiments used in the pharmaceutical industry or in psychological or educational tests, for example, are more complicated than that described in Example 8.8.

8.6 Specific and Multiple Comparison Tests

In parametric analysis of variance after a preliminary overall test interest usually shifts to specific treatment comparisons. In factorial experiments, for example, interest may centre on main effects and interactions. If interactions are indicated, then interest shifts to particular effects of each factor. In other contexts one may be interested in pairwise comparisons for a selected sub-group of treatments or in comparing one specified treatment, often called a *control*, with each of the remaining treatments. One may also want to make comparisons between groups of treatments; for example, if three of a set of seven fertilizer treatments include a nitrate component while the remaining four do not, one may want to make inferences about differences between the overall means or medians for treatments with the nitrate fertilizers and the corresponding measure for nitrate-free fertilizers.

There are nonparametric procedures for assessing such contrasts covering both hypothesis tests and estimation (e.g., obtaining confidence intervals for differences). A detailed treatment is beyond the scope of this elementary text, but an account of the basic theory illustrated with a range of examples is given by Hettmansperger and McKean (1998, Chapter 4). Many contrasts of practical interest are considered by Hollander and Wolfe (1999, Chapters 6 and 7) and by Conover (1999, Chapter 5).

Careful thought should be given to which of many possible contrasts are relevant and to the implications of the fact that not all contrasts are independent of one other. In an experiment in which one treatment may reasonably be regarded as a control, it is often both relevant and sensible to compare the mean or median for this treatment with the corresponding measure for each remaining treatment. Such contrasts may be of interest, for example, when comparing the efficacy of three different drugs in treating a disease and the experiment includes also as a control a group consisting of individuals not receiving any drug.

If an experiment is conducted where the different treatments consist of administring the same drug at a sequence of increasing dose levels, the main contrasts of interest will be in how response changes with dose level. Where each dose is given to a different group of patients, overall tests for increasing or decreasing responses with dose are provided by the Jonckheere–Terpstra or the Page test. In some circumstances umbrella tests of the type mentioned briefly at the end of Section 7.2.5 may also be appropriate. Further tests using specific contrasts may also be relevant to find, for example, if there is a threshold dose which must be exceeded before there is evidence of a real treatment effect.

For both parametric and nonparametric analyses many more-or-less automatic procedures known as *multiple comparison tests* have been devised. They are designed to protect against making unwarranted inferences when many contrasts that are not necessarily independent are examined. The aim is to restrict the number of unwarranted conclusions of real differences that may be reached as a result of carrying out all possible tests. Such conclusions are often referred to as *false positives* although that terminology is also used in a wider sense in diagnostic studies.

If 20 treatments are involved in an experiment an overall parametric test such as the F-test in the analysis of variance, or a Kruskal–Wallis test, may indicate no acceptable evidence of treatment differences. In that situation, a comparison of the largest observed treatment mean with the smallest observed treatment mean, ignoring all others, might well produce a P-value sufficiently small to indicate strong evidence that these means differ, contrary to the conclusion in our overall test that treatments do not differ. This is an ever-present danger if one decides to carry out a test because the data look as though that test might tell us something interesting.

Most statisticians advise strongly against using multiple comparison procedures blindly, especially when there is no obvious treatment structure such as that implied in factorial experiments or a structure where comparisons with a control are of prime interest.

We shall not go into the sometimes deep and often contentious logical arguments for and against multiple comparison methods here. We stress that they should be used with caution. A constructive approach is one where an experimenter, often in association with a statistician, elects (or nominates) comparisons that are of interest before the data are obtained, and makes relevant assessments of evidence about these on the basis of what information is later provided by the data. This is in the spirit of our overall attitude in this book towards interpretation of P-values as a tool for weighing evidence, rather than as a decision tool in a way that is implicit in the assignment of pre-fixed formal significance levels.

One approach widely used for nominated pairwise comparisons is to compute what are known as *least significant differences*. Logically, these can only be justified if applied to prechosen comparisons, i.e., those chosen before the data are collected, or at latest before the data are examined. Even then, they are applied only if an overall Kruskal–Wallis, Friedman or other relevant test indicates strong evidence for differences. The critical P-value or significance level used in multiple comparison tests should be no less stringent than that in the overall test.

When using, for example, the Kruskal–Wallis test the criteria for accepting a location difference between the ith and jth sample are that

1. the overall test indicates significance, and
2. if $r_i.$ and $r_j.$ are the mean ranks for these samples of size n_i, n_j, then

$$|r_i. - r_j.| > t_{N-t,\alpha} \sqrt{(S_r - C)(N - 1 - T)(n_i + n_j)/[n_i n_j (N - t)(N - 1)]} \quad (8.9)$$

where T is given by (7.2) and $t_{N-t,\alpha}$ is the t-value required for significance at level α percent in a t-test with $N - t$ degrees of freedom, other quantities being defined as in Section 7.2.2.

An analogous result holds for van der Waerden scores with sample rank means replaced by sample score means and C now zero and T being the form of the statistic appropriate for these scores.

Nominated comparisons for the Friedman test may be based on either treatment rank means or totals since we have equal replication. The algebraic expression in terms of rank totals is simpler. An analogue of (8.9) gives the requirement for a least significant difference between the treatment rank totals $R_i., R_j.$:

$$|R_i. - R_j.| > t_{(b-1)(t-1),\alpha} \sqrt{2b(S_r - S_t)/[(b - 1)(t - 1)]} \quad (8.10)$$

where $S_r = \sum_{i,j} r_{ij}^2$ and $S_t = \sum_i R_i.^2$.

Many statisticians would agree with Pearce (1965, pp. 21–2) that it is reasonable to compare a subset of all samples (or treatments) if that subset is selected before the data are obtained (or at least before they are inspected) using formulae like (8.9) or (8.10) if one could sensibly anticipate that the specified pair are likely to show a treatment difference that may be of practical interest or importance.

Example 8.9

The problem. In Example 7.2 special interest attaches to differences in sentence length between Vulliamy and Queen. The former is an English author with an academic background and limited fictional output. Queen is a popular and prolific American writer. Use a least significant difference test to determine whether the difference is significant.

Formulation and assumptions. Relevant quantities are mostly available from the solution to Example 7.2, and these are substituted in (8.9).

Procedure. Relevant rank means are $r_1. = 17/5 = 3.4$ (Vulliamy), $r_2. = 72.5/6 = 12.1$ (Queen). Also $S_r - C = 2104.5 - 1624.5 = 480.0$, $T = 9.146$, $N = 18$, $n_1 = 5$, $n_2 = 6$, $t = 3$. Tables give the critical t-value at the 1 percent significance level with 15 degrees of freedom to be 2.95. The left-hand side of (8.9) is $12.1 - 3.4 = 8.7$ and the right-hand side reduces to

$$2.95 \times \sqrt{[480 \times (17 - 9.164) \times 11/(30 \times 17 \times 15)]} = 6.86.$$

Conclusion. Since $8.7 > 6.86$ the difference is formally significant at the 1 percent level, i.e., there is strong evidence of a real difference.

Table 8.3 *Preferences of Blue-tailed Damsel Fly larvae for various prey.*

Larva Prey	1	2	3	4	5	6	Total
Anopheles	1	2	1	1	3	1	9
Cyclops	7	7	7	6.5	6.5	7	41
Ostracod	5.5	5.5	5	5	5	2.5	28.5
Simocephalus (wild)	3	4	4	6.5	1.5	4	23
Simocephalus (domestic)	2	1	6	4	6.5	6	25.5
Daphnia magna	4	3	2	2	1.5	2.5	15
Daphnia longspina	5.5	5.5	3	3	4	5	26

Comment. We use the 1 percent level since the overall test indicated significance at this level.

The description of the experiment and the data in the next example were kindly provided by Chris Theobald.

Example 8.10

The problem. As part of a study of the feeding habits of the larvae of the Blue-tailed Damsel Fly, 6 of the larvae were collected along with six members of each of seven species of prey on which they usually fed. Each larva was placed on a cocktail stick in a glass beaker that contained water and one each of the seven types of prey. Records were kept of the order in which each larva ate the prey. The results are shown in Table 8.3.

Some ties denoted by mid-ranks arose due to failure to observe the order when a larva ate two prey in quick succession. Is there evidence that the larvae prefer to eat some species of prey before others?

The cyclops was usually the last to be eaten. If the experimenter had prior reason to anticipate such an outcome, how might we appropriately examine whether the tendency was statistically significant?

Formulation and assumptions. Evidence of preference requires a Friedman test analogous to that in Example 7.6. A multiple comparison test using (8.10) is one approach to examining the situation regarding cyclops.

Procedure. The Friedman test follows the lines in Example 7.6. If not using StatXact or some other program for a Friedman test the reader should confirm that $t = 7$, $b = 6$, $S_r = 837$, $S_t = 775.75$, $C = 672$, $T = 22.64$ and that $T_1 = 8.47$, indicating a P-value less than 0.01, or probably less than 0.001. Indeed, the Monte Carlo estimate of the exact P-value in StatXact indicates almost certainly that $P < 0.0005$. Setting $\alpha = 0.01$ we find the appropriate t-value with $(b-1)(t-1) = 30$ degrees of freedom is 2.75, so in (8.10) the least significant difference between treatment rank totals given in the last column of Table 8.3 is $2.75 \times \sqrt{2 \times 6(837 - 775.75)/30} = 13.61$.

Conclusion. The difference 13.61 is exceeded between totals for *cyclops* and all other prey except *ostracod*, so there is strong evidence that there is a low preference for *cyclops*.

Comment. We have used the only relevant test given in this book to clarify the situation regarding *cyclops* but it is possible to devise a test comparing *cyclops* with the average rank response for all other prey which would confirm the conclusion that *cyclops* was less favoured. A more sophisticated approach to multiple comparisons is given by Leach (1979, Section 6.2). A useful paper on this topic is that by Shirley (1987), and an important earlier paper is Rosenthal and Ferguson (1965).

We have pointed out that for an overall test involving k treatments there are $k(k-1)/2$ paired comparisions between treatments, If k is large there is a good chance that at least one of these comparisons will indicate a significant difference even when an overall test shows no significant difference between treatments. In particular, if the test is for equality of some location parameter there is a high probability that the treatments corresponding to samples with highest and lowest sample means will indicate a significant difference if those two treatments are compared ignoring all others.

If all pairwise comparisons are of interest one suggestion to overcome this difficulty is to require an appreciably smaller P-value for each of the pairwise comparisons. This is to ensure that what is often called the *experiment-wise* error rate is no greater than some preassigned α such as $\alpha = 0.05$. We achieve this if we only accept pairwise comparisons as significant if the associated

$$p \leq \frac{\alpha}{k(k-1)/2}.$$

Thus, if $k = 6$ and $\alpha = 0.05$ each pairwise test would require $p \leq 0.05/15 \approx 0.0033$.

Depending on what procedure is relevant, each pairwise test may be carried out by an appropriate parametric procedure such as the t-test or a nonparametric one such as the WMW test with $p \leq 0.0033$ as the criterion for significance.

This adjustment is often referred to as a *Bonferroni adjustment*. For large values of k the calculated value of p tends to be conservative as it is based on approximating $(1 - \alpha/n)^n$ by $1 - \alpha$.

8.7 Fields of Application

Experiments involving factorial treatment structures are common in many applications. They were originally used in agricultural field experiments, especially those involving fertilizers with several chemical constituents with several levels of application, or in application to different crop varieties. Factorial structures involving both treatment and intrinsic factors occur in many disciplines. We give a few examples.

Manufacturing

A confectionary manufacturer wishes to test consumer preference for three different chocolate fillings — *strawberry, cherry* and *vanilla*. The company expect that preferences may vary between genders and between younger and older purchasers. It sets up panels of six tasters in each of the four categories *males aged under 30, females aged under 30, males aged over 45, females aged over 45*. Each member of each panel is asked to award a score between 0 and 10 to each of the fillings. The different types of filling may be regarded as a treatment factor and the gender/age combinations as intrinsic factors. A linear additive model might be of doubtful validity, so a test for interaction using the Akritas and Arnold model may be more appropriate. If there is evidence of interaction, particular effects, rather than main effects, are likely to be of interest.

Drug Comparisons

A medical research team may wish to compare the performance of three drugs for reducing blood pressure. They might decide to administer each drug at four different dose levels each to different sets of 10 patients, using both a group of males and a group of females for each drug and dose level. Here there are three factors, drugs (3 levels) doses (4 levels) and gender (2 levels) in a three factor experiment often referred to as a $3 \times 4 \times 2$ structure. If reduction in systolic blood pressure is the observed measurement an additive linear model may be appropriate and the data might be analysed by an extension from 2 to 3 factors of the parametric ANOVA if there were no evidence of outliers or heteroscedastic error structure. If the parametric analysis were inappropriate the method of Hettmanspergert and Elmore might be preferred to that of Akritas and Arnold, as interactions that may be lost in a transformation to ranks may be of clinical importance.

Routine Tasks

In attempts to improve performance at routine tasks such as assembing a set of components an experiment might be conducted where groups of workers carried out the task with each of three different types of tools, in two different seating positions. If there was a suspicion that using factor combinations in different order might induce residual or learning effects that influenced performance at later factor combinations, an extension of crossover designs from that described in a simple case in this chapter might be used.

Comparison with Control

In many applications if there is an overall difference between treatments interest often centres on whether there is a difference between one particular treatment, often that representing the status quo, and referred to as a *control* and each of the other applied treatments. In testing the efficacy of different drugs, or different dose levels of a particular drug, comparisons are often made with responses for units receiving no treatment, or only a placebo.

If a manufacturer is considering obtaining a raw material from one of several alternative sources he is interested in whether the finished product will compare favourably in one (or more) characteristics with those characteristics for a product made from his current source of raw material (the control).

In these applications the situation is similar to that considered in Example 8.10.

8.8 Summary

The appropriate analysis of $a \times b$ factorial designs is heavily dependent on whether or not there are **interactions** (Sections 8.1 to 8.4). Interactions of interest may be between treatment factors or between treatment factors and either intrinsic or nonspecific factors. Some interactions are preserved under monotonic transformations but others are not. Historically, interactions were first defined in association with linear models and the parametric ANOVA requires errors to be iid $N(0, \sigma^2)$.

While the classic additive linear model may be relevant for much continuous data, it is generally not relevant to data such as counts or qualitative ordered data. Even when a linear model is appropriate for raw data, linearity is often not preserved under data transformations.

If the linear model is of practical importance, and there is heteroscedasity in the error variances, or outliers are present, a method proposed by Hettmansperger and Elmore (2002) provides a relevant test for interaction.

If the additive linear model is irrelevant and only interactions that are preserved under monotonic transformations are of interest, a model proposed by Akitaras and Arnold (1994) that defines interaction in terms of distribution functions is often appropriate. This takes a simple form for two-way factorial structures and extends very generally to cover cases of unequal replication as well as to repeated measurement structures.

Crossover designs (Section 8.5) in which each subject or unit receives each treatment, but groups of subjects receive treatments in a different order are valuable to assess residual effects as well as the direct effects of treatments.

Multiple treatment comparisons and **specific treatment comparisons** (Section 8.6) are often based on least significant differences, but these should not be used unless the overall parametric or nonparametric tests indicate significance. Comparisons are best confined to contrasts nominated prior to carrying out an experiment, and certainly before the data are examined, and should be selected because prior beliefs (usually nonstatistical in nature) make them of interest. Of special interest are comparisons between various treatments and a control.

8.9 Exercises

8.1 The data used as an example in Hettsmansperger and Elmore (2002) was originally given in Hahn, Haber and Fuller (1973) and is also analysed in Scheirer, Ray and Hare (1976). The data are measures of time of tail rattling while fighting for 42 pairs of mice selectively bred for brain weight (factor A) — *large, medium, small* in two different environments (factor B) — I, II. Seven pairs were observed in each category. The data are given below. Apply the Hettmansberger and Elmore analysis to these data and consider how the results should be interpreted if there is evidence of an interaction. We recommend that the advice of Hettmansperger and Elmore be followed, and that as a preliminary to more formal analysis, suitable EDA techniques such as boxplots for each factor combination be obtained and studied to highlight important features of the data.

I	*Small*	9.60	8.00	5.14	3.50	3.23	2.66	1.44
I	*Medium*	1.98	1.81	1.37	0.98	0.27	0.00	0.00
I	*Large*	0.77	0.66	0.37	0.40	0.19	0.13	0.00
II	*Small*	4.36	1.49	1.45	1.35	0.33	0.20	0.00
II	*Medium*	2.61	2.09	1.76	1.27	0.84	0.64	0.00
II	*Large*	2.95	2.73	0.91	0.82	0.69	0.11	0.00

8.2 Show that if in (8.6) and (8.7) all d_{ij} are equal then the expressions reduce to the degrees of freedom for interaction and for error in classic parametric ANOVA.

8.3 Show that if in (8.8) all d_{ij} are equal then the expression reduces to the degrees of freedom for interaction in classic parametric ANOVA.

8.4 A crossover design is used to test reading skills of young children. Fourteen children of the same age and taught in a similar manner each read the same preselected passages. One is from a novel by Charles Dickens (A) and the other from a novel by Iris Murdoch (B). Half the children read A in Period I and B in Period II (sequence AB), and other other half read B in Period I and A in Period II (sequence BA). The number of mistakes are recorded as follows

Child	1	2	3	4	5	6	7
Seq	*AB*	*AB*	*AB*	*AB*	*AB*	*AB*	*AB*
Per I	7	4	11	9	0	8	2
Per II	5	3	8	7	2	6	3
Child	8	9	10	11	12	13	14
Seq	*BA*	*BA*	*BA*	*BA*	*BA*	*BA*	*BA*
Per I	5	8	6	8	7	3	4
Per II	6	5	7	8	8	6	5

Do these data indicate differences in the number of mistakes between A and B? Is there evidence that the number of mistakes depend upon which author's work is read first?

8.5 A sportswear manufacturer produces a low-priced shoe (A) for athletes which it recommends for training runs and a higher-priced shoe (B) which it recommends for competitive events. The coach at an athletics club who takes a cynical view of advertising suggests that performance may be no better with a change to the more expensive type for competitions, indeed that the reverse may be true. He recommends a crossover trial in which 16 members of the club of varying ability test them in 100 metre sprints. Eight use shoe A in training runs then use shoe B in a competitive run (combination AB). The remaining 8 use shoe B in training and shoe A in the competitive run (combination BA). Do the experimental results below justify the coach's cynicism? What recommendation should he make to members about which type of shoe to use?

Athlete	1	2	3	4	5	6	7	8
Seq	AB	AB	AB	AB	AB	AB	AB	AB
Per I	10.7	11.4	13.2	12.2	12.9	11.1	15.1	13.7
Per II	11.1	11.3	14.9	12.1	14.2	16.1	14.9	14.3
Athlete	9	10	11	12	13	14	15	16
Seq	BA	BA	BA	BA	BA	BA	BA	BA
Per I	11.3	13.2	13.5	14.4	15.8	11.3	14.1	15.1
Per II	11.1	13.3	12.7	13.9	10.3	11.0	13.1	14.9

*8.6 Shirley (1977) gives the following data for mice in four groups, the first, or control, group being given zero dose of a drug and the remaining groups doses of the drug at increasing levels. The data are reaction times in *seconds* × 10 of mice to stimuli applied to their tails. Interest centred on whether there was evidence of toxicity that would cause increased reaction time as the dose increased, and if so at which level toxicity became apparent. The method of analysis used by Shirley is not described here, but details will be found in the original paper.

Apply methods you consider appropriate from those given in this and the preceding chapter to answer the questions of interest.

Level 0	24	30	20	22	22	22	22	28	30	30
Level I	28	22	38	94	84	30	32	44	32	74
Level II	98	32	58	78	26	22	62	94	78	34
Level III	70	98	94	88	88	34	90	84	24	78

Comment on any special features that you discern in the data and indicate which of these influenced your choice of method of analysis.

8.7 Hand et al. (1994, set 304) give the following data from Dolkart, Halperin and Perlman (1971) on amounts of nitrogen-bound bovine serum albumen in mice. A person interested in these data asks you to use nonparametric methods of analysis to determine whether there are treatment (location) differences between groups and to carry out any specific or multiple comparison tests you consider appropriate, would you be happy to proceed without additional information? If not, what sort of additional information would you require? The three groups are *normal mice*, those that are *alloxan diabetic untreated* and those that are *alloxan diabetic and treated with insulin*.

Normal	156	282	197	297	116	127	119	29	253	172
	349	110	143	64	26	86	122	455	655	14
Alloxan	291	49	469	86	174	133	13	499	168	62
	127	276	176	146	108	276	50	73		
Allox. + Ins	82	106	98	150	243	68	228	131	73	18
	20	100	72	133	465	40	46	34	44	

*8.8 An archaelogist obtains a number of coins of comparable nominal value from each of four dynasties of an ancient civilization. For each he obtains the percentage silver content. This varies appreciably from coin to coin in any one dynasty, but is there an indication that percentage silver content is declining between dynasties? Is it reasonable to conclude that there is a difference between content in the first and fourth dynasties?

Dynasty I	6.2	9.1	9.7	8.2	11.1	5.6		
Dynasty II	7.2	6.9	5.3	8.2	7.7	7.0		
Dynasty III	8.2	7.7	9.3	10.2	8.4	5.6	6.4	7.3
Dynasty IV	7.1	6.2	8.1	5.7				

CHAPTER 9

ANALYSIS OF SURVIVAL DATA

9.1 Special Features of Survival Data

The term *survival* in its literal sense concerns matters of life and death. Since the dawn of human civilization society has been aware that length and quality of life can be increased through, for instance, appropriate selection and preparation of food, isolation of those with incurable infectious diseases from the remainder of the community and the existence of peace rather than war. These principles were documented in the Old Testament for use by the Hebrew people. However it was not until 1662 that John Graunt constructed a basic life table using the weekly Bills of Mortality (lists of deaths) recorded in London during 1658. Starting from a specific number of births, Graunt estimated the number of individuals that would still be alive at specific ages. This document reflects the squalid conditions under which most Londoners lived at the time and is grim reading. According to Graunt's calculations only 25 out of 100 newborns survived to the age of 26.

This work triggered interest not only in life tables but also in possible factors that might influence human survival, such as type of occupation. It was realized that by linking the expectation of human life to carefully invested regular financial contributions from a large number of individuals provision could be made for a particular individual's dependants in the event of his or her untimely death. Life assurance became a cornerstone of society and the theory behind it, along with its application, has created the broad field of actuarial science.

In a medical context survival data usually consist of records for each patient of times following diagnosis of an illness or the application of a treatment to the occurrence of a nominated event, such as death, complete recovery, or duration of remission. Interest usually centres on the comparison of data sets corresponding to different treatments. For convenience we use the word treatment in a very broad sense. For example, the data in Table 9.1 below refer to groups diagnosed with two different forms of a disease. We shall refer to these as *treatments*.

In an engineering context we are often interested in time to failure of some product or machine or machine component, or time to completion of some process, or time to attain some goal such as a specified speed or pressure.

We mentioned after Example 1.2 in Section 1.1.2 that the data given there were only part of the data in the original paper. The complete set is given

Table 9.1 *Survival times in weeks for 10 symptomatic (S) and 28 asymptomatic (A) cases of lymphocytic non-Hodgkin's lymphoma. An * represents a censored observation.*

S	49	58	75	110	112	132	151	276	281	362*
A	50	58	96	139	152	159	189	225	239	242
	257	262	292	294	300*	301	306*	329*	342*	346*
	349*	354*	359	360*	365*	378*	381*	388*		

in Table 9.1. These exhibit the two key features of survival data that have stimulated the development of appropriate new tools. The features are

- Distributions of survival times of patients in clinical trials, or of times to failure for components in an engineering or industrial context are typically long-tailed or skewed to the right.

- Some units are often lost to a study before the response of interest (e.g., death of a patient or complete recovery, or failure of a machine) occurs.

In a medical context the asymmetric nature of the distribution reflects a common tendency for many treatments to be largely ineffective, or only moderately effective, with some patients, while markedly prolonging the time to the event of interest for others. Similarly, in an engineering context, many components may breakdown at or near the median time, while a few will have appreciably extended lifespans.

Censored observations may be due to a variety of causes. The main ones are

- The study is terminated before all subjects show the designated response.
- For one of the reasons given below some subjects withdraw from the study before showing the response.
- Data limitations may preclude observation of the designated response.

How censored data are best analysed depends crucially on which of these causes come into play, and how the censoring is related to a particular treatment when we are comparing survival times for several treatments.

In this chapter the tests we give will only be valid if censoring is not caused by a factor that is treatment dependent. Valid analysis is possible when censoring is treatment dependent, but the methods are more advanced than those given here.

One cannot tell if censoring is treatment dependent merely by looking at the data. At first sight the data in Table 9.1 appear to indicate treatment-dependent censoring since there is a greater proportion of censored observations in the *asymptomatic* group (12 among 28) than there is in the *symptomatic* group (1 among 10). However, this difference might simply be due to a decision to terminate the study on a fixed date. If the times recorded are those from the date of onset of first clinical evidence of the disease, this date will vary from patient to patient. Thus, some patients, irrespective of categories, will have been in the trial longer than others. This means the higher proportion of censored items in the asymptomatic cases may not be due to some factor which is inherent in this diagnosed form, but rather reflect the fact that because the asymptomatic cases have a tendency to survive longer, more will be in the study at a fixed calendar cut off date. Thus, the increased proportion of censored observations may be attributed to an administrative decision to terminate the study at a fixed date.

On the other hand, when two or more treatments are used in a trial to compare the efficacy of two drugs to treat the same illness or disorder, and one of these has serious side-effects that causes, or even forces, many patients to withdraw before the condition of interest (recovery or whatever) is observed, censoring would be treatment dependent.

Causes of censoring which often are not treatment dependent include the subject leaving the district, or dropping out of the study for other reasons (e.g., contracting another illness, death due to another illness or in an accident), failing to follow the treatment regime, forgetting to keep hospital appointments, etc. If it is reasonable to assume such factors are not treatment dependent, censoring might reasonably be regarded as random. As already indicated, the data itself cannot tell us if dropout is random or treatment dependent. Those conducting the experiment are best placed to answer that question.

It is evident that censored items do not contain as much information as complete data. However, they contain some information. For example, an observation censored at 362 weeks clearly has a longer survival time than one that survives for only 251 weeks, or indeed shows the response of interest at any time less than 362 weeks.

An example of censoring in an engineering or industrial context involving tyre failures is given by Davis and Lawrance (1989). Tyres that had not failed at the end of an experimental period were regarded as censored observations.

The characteristic long right tail of survival time distributions reduces the role of the normal distribution. Parametric methods for survival analysis often assume data are samples from exponential, Weibull or other long-tail distributions. We explore some nonparametric analogues and indicate how these take censorship into account.

Sometimes the exact time of occurrence of an event of interest is not known. For instance, in the United Kingdom many people choose to undergo a routine dental examination every six months. If a patient has a filling placed on a particular tooth and subsequently loses the filling but does not report the loss until the next routine inspection there is some uncertainly regarding the exact time that the filling became detached from the tooth.

Innes et al. (2006) report a dental crown survival study and discuss the estimation of the length of survival where the time of loss is not known exactly. Dental researchers customarily estimate this from the middle of the interval between the date of the last inspection with the filling present and the date of the following inspection, although some prefer to use the date on which the presence of the filling was last recorded. In practice the method of estimation may be a minor issue as the loss of a filling is rare (80 percent of amalgam fillings placed on adult teeth survive at least 10 years) and most adults would contact a dentist soon after the filling was lost and so provide an exact or almost exact failure time.

There are situations in which left, rather than right, censorship can occur. For example, in a study that considers survival times from the first symptoms of a particular type of cancer, the symptoms may not be detected until a medical examination at a later date, by which time they may be well developed.

In the remainder of this chapter we restrict any censoring that occurs to right censorship.

9.2 Modified Wilcoxon Tests

We discuss two modifications to the Wilcoxon test that may be used with censored data.

9.2.1 The Gehan–Wilcoxon Test

Gehan (1965a, 1965b) proposed generalized Wilcoxon-type tests for both one- and two-sample problems with various types of censoring. We consider the two-sample case. A censored observation clearly has a longer survival time than any unit that has failed at or before the time of censoring. However, Gehan regarded any two censored observations as giving no definite information on survival times relative to each other. Nor did he regard a censored observation as providing information on whether that subject does or does not have a greater survival time than that of a later failure. He scored these *no definite information* cases as ties. Example 9.1 illustrates the method.

Example 9.1

The problem. For the data in Table 9.1 test the hypothesis H_0: *the survival times distributions are identical* against H_1: $G(x) \leq F(x)$ *with strict inequality for some x*, where $F(u), G(v)$ are survival time cumulative distribution functions for the symptomatic and asymptomatic cases respectively.

Formulation and assumptions. A one-tail test is appropriate since H_1 is one-sided. Because of censoring we use Gehan–Wilcoxon scores.

Procedure. The test is easily explained and carried out using analogues of the WMW statistics U_m or U_n. To obtain the modified form of U_n for each observation in the first sample (symptomatic) that is a definite failure time (i.e., not a censored observation) we count the number that survive longer in the second sample (i.e., that fail or are censored at a later time) scoring 1 for each such case as in the WMW test; we also score 1/2 for each observation in the second sample that is censored earlier. For any censored observation in the first sample, we count the number of second sample obsevations that fail after the time that item was censored and add to this count the number of second sample items that have been censored at either an earlier, the same, or a later, time, regarding each as a tie and scoring it as $1/2$. The statistic U_n is the sum of all such scores over all first sample observations.

For the first observation, 49, the score is 28 because all second sample observations represent later failures. For the second observation, 58, the score is 26.5 since there are 26 later failures and one tied value. For the final entry 362* the score is 6 since there are no second sample values that definitely have a longer survival time, but there are 12 censored items in the second sample that are each regarded as a tie and scored as $1/2$. Carrying out the process for all observations in the first sample it is easily confirmed that

$$U_n = 28 + 26.5 + 26 + 25 + 25 + 25 + 24 + 16 + 16 + 6 = 217.5$$

Computer programs for an exact test give a one-tail $P \approx 0.004$.

Conclusion. There is strong evidence against H_0.

Comments. 1. We express H_1 in terms of one distribution dominating rather than as a median shift because an efficient treatment is likely to increase life expectancy not by a constant amount for all subjects, but by amounts which vary from subject to subject depending upon whether there is slow or rapid recovery or perhaps just an arrest of the development of a disease.

2. All censored values in this example are at least 300. For the asymptomatic cases survival times are sufficiently long for many patients to survive this time, whereas only one symptomatic case does.

3. The treatment of censored data in the Gehan–Wilcoxon test is conservative. Peto and Peto (1972) and others have suggested stronger, but realistic, assumptions about the life expectancy of units that provide only censored observations. These often lead to more powerful tests, especially if there is substantial censoring. One such test is described below.

Computational aspects. StatXact includes a program for a generalized Gehan–Wilcoxon test which differs slightly from that given here, but it usually gives similar results. For this example it gives $P = 0.0041$ for a one-tail test.

9.2.2 The Peto–Wilcoxon test

The Gehan–Wilcoxon test does not require strong assumptions about the form of the survival function. It also makes only mild assumptions about the life expectancy of censored units. As intuition suggests, when censoring is severe there will be an appreciable loss of power relative to tests that give more weight to censored data. This is because heavily censored data contribute many ties to U_n. Since large values of U_n correspond to small P-values, the effect of censoring may be to dilute the U_n we would have got if censoring had not occurred.

To explore how we might modify the Gehan–Wilcoxon test to overcome this loss of power we introduce the concept of a *survival function*. If the distribution of survival times, T, is $F(u)$ this implies that for any individual the probability of the response of interest (which for simplicity we call a *failure*) at or before time t is $\Pr(T \le t) = F(t)$. The opposite event is survival beyond t. The relevant probability is given by the survival function

$$H(t) = \Pr(T > t) = 1 - F(t).$$

The physical interpretation of $H(t)$ is that it is the proportion of the population that survives beyond time t. For survival time distributions $F(t) = 0$, if $t < 0$. This is often indicated by the shorthand $F(0-) = 0$. Since $F(t)$ is a monotonic increasing function of t and $F(\infty) = 1$ it is easy to see that $H(t)$ decreases monotonically from 1 to 0 as t increases from 0 to ∞.

In Section 2.3 we defined the sample distribution for a sample of n distinct values as

$$S_n(t) = \frac{\text{number of sample values} \le t}{n}.$$

The corresponding sample survival function is

$$R_n(t) = \frac{\text{number of sample values} > t}{n} = 1 - S_n(t).$$

Like $S_n(t)$, if there are no ties $R_n(t)$ is a step function with steps of height $1/n$ at each sample value, but now at each step the value decreases by $1/n$, remaining constant between each step. It represents the proportion of the sample surviving beyond time t.

It is sometimes a matter of practical convenience to modify the definition of $R(t)$ at the rth step t_r, $r = 1, 2, \ldots, n$ by assigning to it the mean value immediately before and immediately after that step, i.e., setting

$$R(t_r) = \frac{1}{2}\left\{ \left[1 - \frac{(r-1)}{n} \right] + \left[1 - \frac{r}{n} \right] \right\} = \frac{(n - r + 1/2)}{n}. \tag{9.1}$$

It is reasonable to look upon $R(t_r)$ defined this way as a score allocated to that observed t_r. This scoring system assigns to each observation the probability that it survives at least to time t_r. If we replace this score by

$$W(r) = n + \frac{1}{2} - nR(t_r)$$

it follows from (9.1) that $W(r) = r$, the usual rank score used in computing rank-sum statistics.

Thus, for samples of size m, n with no ties, and no censored observations, to test the hypothesis that the population survival distributions are the same against alternatives of a location shift or of dominance of survival times in one distribution, a Wilcoxon test is appropriate. The argument is easily extended to cover tied survival times when the survival function has a step of height s/n if there are s tied values at a given t.

The above approach suggests a way to formulate a test when there are censored data. Steps in $R(t)$ are still confined to failure times, but if there is censoring between two *consecutive* failures these steps are no longer equal. The modification to the WMW test proposed by Peto allocates scores to censored

data in a different way to those allocated to known failure times. We call this
the *Peto–Wilcoxon test*.

We illustrate the basic procedure with simple numerical illustrations at key
points. When there are censored data we estimate the proportion surviving at
each failure time by considering the proportion of the sample that we know for
sure survives beyond that time. The best known estimator of this proportion
is the *Kaplan–Meier estimator*, proposed by Kaplan and Meier (1958).

We denote by $R(t-)$, $R(t+)$ the values of $R(t)$ immediately before and
immediately after time t. These are identical unless t corresponds to one or
more failure times, t_r. If there are exactly s, $s \geq 1$ failures at time t_r the
Kaplan–Meier rule for calculating $R(t_r+)$ is

$$R(t_r+) = R(t_r-)\left[1 - \frac{s}{r(t_r)}\right] \tag{9.2}$$

where $r(t_r)$ is the number of failures *plus* the number of observations censored at or after time t_r. It may be described as the *number of units at risk*
immediately prior to any failures at time t_r.

Example 9.2

The problem. For the data 2   5   9*   11   11   22*   29*   47   51* determine the
value of $R(t)$ before and after each failure. An asterisk denotes a censored observation.

Formulation and assumptions. The Kaplan–Meier procedure is used to compute
relevant $R(t_r-), R(t_r+)$.

Procedure. There are steps at 2, 5, 11, and 47. At 2, $R(2-) = 1$. Thus by (9.2)
$R(2+) = 1(1 - 1/9) = 8/9$ since $s = 1$ and $r(t_r) = 9$. This implies a single step of
$1/9$ at $r = 2$. Noting that $R(2+) = R(5-)$ we get, by similar reasoning, $R(5+) =
(8/9)(1 - 1/8) = 7/9$, again indicating a step of $1/9$. Since steps only occur at
observed failure times we have $R(5+) = R(11-)$, whence $R(11+) = (7/9)(1-2/6) =
14/27$ since there are 2 failures at $t = 11$ and 6 at risk. Similarly at the next failure
at $t = 47$ we have $R(47+) = (14/27)(1 - 1/2) = 7/27$. All we can say about $R(t)$
for values of $t > 47$ is that at some unknown $t > 51$ when the observation censored
at 51 has failed, then $R(t)$ will fall to zero.

Conclusion. The relevant values are

$$R(2-) = 1, R(2+) = R(5-) = 8/9, R(5+) = R(11-) = 7/9,$$
$$R(11+) = R(47-) = 14/27, R(47+) = 7/27.$$

Remembering how we formed scores in the no-ties, no-censoring situation
an obvious thing to do at the steps is again to take the mean of each relevant
$R(t+)$ and $R(t-)$. All we know about a censored observation is that it will fail
between the time of censoring and infinity, when $R(t)$ is zero. Peto and Peto
suggested that a reasonable score to assign to a censored observation was the
mean of $R(t)$ at censoring and the final value zero, i.e., half the value of $R(t)$
at censoring. The reader should check that for the data in Example 9.2 the

value at the time of censoring of the observations censored at 22^* and 29^* is $R(11+) = 14/27$ giving each a score of $(1/2)(14/27) = 7/27$.

Example 9.3

The problem. Using the scoring system described above obtain appropriate scores for the data set in Example 9.2.

Formulation and assumptions. At each failure time t_r the score is the mean of the Kaplan–Meier scores $R(t_r-), R(t_r+)$. For each censored observation the score is one half of the then current $R(t)$.

Procedure. For the first observation, 2, the score is

$$s_1 = [(R(2-) + R(2+)]/2 = (1 + 8/9)/2 = 17/18.$$

By similar reasoning that corresponding to 5 is $15/18$. For the censored observation 9^* since the current value of $R(t)$ is then $R(5+) = 7/9$ the Peto score is $(1/2)(7/9) = 7/18$. Proceeding this way the reader should check that the remaining scores are $35/54, 35/54, 7/27, 7/27, 7/18, 7/54$. Denoting these scores by s_i, the transformation $W(r_i) = n + 1/2 - nR(t_r)$ becomes $W(r_i) = (19 - 18s_i)/2$ giving

$$W(r_i) = 1, 2, 6, 3.67, 3.67, 7.16, 7.16, 6, 8.33.$$

Conclusion. The $W(r_i)$ are appropriate scores for the given data if the data were relevant to a Peto–Wilcoxon test.

Comments. The sum of these scores is

$$W = 1 + 2 + 6 + 3.67 + 3.67 + 7.16 + 7.16 + 6 + 8.33 = 45.$$

This is the same as the sum of the ranks in a combined sample of 9 in a WMW test had there been no censoring.

We may apply the Peto–Wilcoxon test to the data in Table 9.1. Without appropriate software computing scores is tedious. Testimate calculates relevant scores that are linear transformation of those obtained by the method described above. They do, however, lead to the same exact or asymptotic tests. The one tail permutation test gives an exact $P = 0.0042$ in close agreement with that for the Gehan–Wilcoxon test. StatXact gives a test not quite identical to the Peto–Wilcoxon test based on a similar, but not exactly equivalent, scoring system proposed by Kalbfleish and Prentice (1980, p. 147). The modification takes account of ties by averaging the scores that would be given if the ties had been arbitrarily split, a procedure often used in other contexts. Details are given in the StatXact manual. The test is described in StatXact as a generalized Gehan–Wilcoxon test. For the Table 9.1 data it gives similar results to the Peto–Wilcoxon test.

The two tests usually give similar results unless censoring is severe as in an example on survival times for dogs given by Prentice and Marek (1979). There were 281 dogs in the combined samples, 83 in one and 198 in the other. All but 29 observations were censored. Such situations do occur in real investigations.

In formulating the Gehan–Wilcoxon and Peto–Wilcoxon tests there is an implicit assumption that under H_0 the survival time distribution for censored items is the same as that for uncensored ones. This may not be the case if some or all of the censored items have a characteristic unrelated to treatments that affect their life expectation. For example, if some have a history of drug addiction their life expectancy may be shorter than that of those not so addicted. However, if the allocation of such patients to treatments is random, and under H_0 the failure time distributions for such patients are the same for each treatment, but not necessarily the same as that for items not in this sub-category, the analysis is still valid.

The following illustration appears to contradict this statement. Consider the survival time data:

Sample I	9	25	35	38	41			
Sample II	10	12*	15*	22*	37	48	57	58

The one-tail Peto–Wilcoxon test for dominance in the second population gives $P \approx 0.0497$. However, if the censored items all have a lower life expectancy and it is later discovered that their failure times are respectively 26, 29 and 36 the WMW test gives a one tail $P = 0.218$.

Even though all censored items appear in one sample this need not imply censoring is treatment dependent, since with only three censored items there is a probability of 0.25 that these would all be in the same sample if these were both of equal size, and may even be greater for unequal sample sizes. In these circumstances the Gehan–Wilcoxon or Peto–Wilcoxon tests in fact protect us against making incorrect inferences when all censored observations having a different life expectancy fall in one sample. When some subgroup have a potentially different distribution for a measured variate precision can sometimes best be increased by introducing appropriate covariates. If censored observations form a subgroup that might, with hindsight, have been measured by an appropriate covariate (e.g., duration of drug addiction) the censoring may act as a surrogate for such a covariate.

9.3 Savage Scores and the Logrank Transformation

The prevalence of long-tail distributions, often not unlike an exponential distribution, has given that distribution a role in survival and failure time studies almost as central as that of the normal distribution in many other types of problem. While methods of analysis based on the normal distribution are often moderately robust against many minor (and even some major) departures from assumptions, tests based on the exponential distribution are often sensitive to even small departures from that distribution.

This has led to the development of distribution-free methods to overcome such problems. Savage (1956) introduced a transformation — now often referred to in a slightly extended form as the *logrank* transformation — in which ranks are replaced by what are often known as *Savage* or *exponential* scores. The transformation is essentially one to expected values of the order statistics of an exponential distribution with parameter $\lambda = 1$. This is somtimes referred to as the *unit* exponential distribution. If there are no ties and no censoring in two samples of m, n observations then the Savage score corresponding to the combined sample rank r, $(r = 1, 2, \ldots, N)$ is

$$s'_r = \frac{1}{N} + \frac{1}{N-1} + \ldots + \frac{1}{N-r+1}$$

where $N = m + n$.

It is easy to show that the sum of all s'_r is N and an equivalent score commonly used is

$$s_r = 1 - s'_r \qquad (9.3)$$

and these latter scores sum to zero. The sums of these scores for either sample are appropriate statistics in an exact permutation test which may be carried out using either specific tests for these scores, or they may be substituted for the original data in an arbitrary score permutation test program. StatXact includes a program that calculates scores automatically for a given data set making appropriate modifications for ties or censored data. Several different approaches are given in the literature for calculating appropriate scores in these circumstances.

We show here that two commonly used approaches are equivalent. To do so we need to look more closely at the structure of Savage scores in the no-tie and no-censoring case. We use the form (9.3) to look at the statistic S_m formed by the sum of these scores allocated to Sample I of size m. If Sample 2 has n observations the combined sample has $N = m + n$ observations. From (9.3)

$$s_t = 1 - \sum_{j=1}^{t} [1/(m + n - j + 1)],$$

so if S_m is the sum of all s_t in Sample I it follows that S_m is given by

$$S_m = m - \sum_{j=1}^{m+n} \frac{m_j}{m + n - j + 1}, \qquad (9.4)$$

where m_j is the number of all Sample I failures having rank j or higher. This is the number at risk immediately prior to failure at time $t_{(j)}$. In particular $m_1 = m$; also $m_j = 0$ when j exceeds the highest ranked observation in Sample I.

The denominator in the summation terms in (9.4) is the number of survivors in *both* samples immediately prior to failure at time t_j. This means that if we define n_j for sample 2 in a way analogous to the definition of m_j for Sample 1, it follows that the denominator term may be written $m_j + n_j$, whence (9.4) becomes

$$S_m = m - \sum_{j=1}^{m+n} \frac{m_j}{m_j + n_j}, \qquad (9.5)$$

Careful examination of (9.5) shows that the terms in S_m may be arranged in two groups. The first consists of m terms each corresponding to a failure in sample I and each taking the form

$$s_{1j} = 1 - \frac{m_j}{m_j + n_j}. \tag{9.6}$$

The second consists of n terms each corresponding to a failure in Sample II and each taking the form

$$s_{2k} = -\frac{m_k}{m_k + n_k}. \tag{9.7}$$

It is easily seen that (9.5) is the sum of all s_{1j}, s_{2k}.

Example 9.4

The problem. Calculate S_m for the following data using (i) the Savage scores given by (9.3) and (ii) by using (9.4) directly.

Sample I	19	24	31		
Sample II	26	29	33	34	37

Formulation and assumptions. Tests are based on the ordering of observations and it is convenient for (i) to work with overall rankings, though these are not needed for (ii).

Procedure. Ranking all data in ascending order gives

Sample I	1	2	5		
Sample II	3	4	6	7	8

For Sample I the Savage scores using (9.3) are $s_1 = 1 - 1/8$. $s_2 = 1 - 1/8 - 1/7$, $s_5 = 1 - 1/8 - 1/7 - 1/6 - 1/5 - 1/4$ whence

$$S_m = 3 - 3/8 - 2/7 - 1/6 - 1/5 - 1/4) = 3 - 1073/840 \approx 1.723.$$

To use (9.4) directly we do not need to use ranks. All we need to know at each failure time is the number of survivors in each sample immediately before that failure. For example, just before $t = 29$ there is 1 survivor in Sample 1 and 4 in Sample II giving a contribution in (9.4) with $j = 4$ and $m_j = 1$, i.e., a contribution $1/(8 - 4 + 1) = 1/5$ to the sum. The reader should check that the complete set of contributions to the sum starting with $j = 1$ and concluding with $j = m + n$ are $3/8$, $2/7$, $1/6$, $1/5$, $1/4$, $0/3$, $0/2$, $0/1$ and it immediately follows that this leads to the same value of S_m as that using (9.3).

The test using these scores is a locally most powerful test for certain hypotheses about what are called *proportional hazards models* relevant to survival functions that are called *Lehmann alternatives*. These include exponential and Weibull distributions.

Given a continuous distribution of survival times $F(u)$ the *hazard rate* $h(t)$ at time t is the limiting probability of failure in the small interval $(t, t + \delta t)$ conditional upon survival to time t. If $f(u)$ is the probability density function corresponding to $F(u)$, this conditional probability is

$$h(t) = \frac{f(t)}{1 - F(t)}.$$

The denominator is the survival function $H(t)$.

Two families with survival time cumulative distribution functions $F(u)$, $G(v)$ are *Lehmann alternatives* if their survial functions satisfy the relationship

$$1 - F(t) = [1 - G(t]^\alpha \qquad (9.8)$$

for some $\alpha > 0$. Differentiating (9.8) with respect to t we get

$$f(t) = \alpha[1 - G(t)]^{\alpha-1}g(t)$$

and writing $j(t)$ for the hazard function corresponding to $G(t)$ this implies

$$h(t) = \alpha j(t) \qquad (9.9)$$

It follows immediately that the ratio of the hazard rates is a constant α; hence the name *proportional hazards*.

One-parameter exponential survival time distributions satisfy (9.8) and (9.9). This follows since, if the distributions have means λ_1, λ_2, then

$$F(t) = 1 - \exp(-t/\lambda_1) \text{ and } G(t) = 1 - \exp(-t/\lambda_2)$$

whence

$$1 - F(t) = \exp(-t/\lambda_1) \text{ and } 1 - G(t) = \exp(-t/\lambda_2).$$

It follows that (9.8) holds with $\alpha = \lambda_2/\lambda_1$.

For a proportional hazards model we may want to test H_0: $\alpha = 1$, which implies identical survival time distributions or equal hazard rates, against a one or two tail alternative indicating a different value of α. Such tests lead naturally to the logrank test for no-tie and no-censoring situations, and we may extend this to deal with ties and censoring. Peto and Peto (1972) showed this to be a locally most powerful test for proportional hazard alternatives.

In the absence of ties or censoring if the sample sizes are m, n then under H_0: $\alpha = 1$ the hazard rate for the identical populations at any time t are the same, so immediately before the jth failure time $t_{(j)}$, if the numbers at risk in each sample are respectively m_j, n_j, then the probability that the jth failure is in Sample I is $m_j/(m_j + n_j)$. We now introduce a counter, d_j, such that $d_j = 1$ if the jth failure is in Sample 1 and $d_j = 0$ if the jth failure is in Sample II. Then

$$\Pr(d_j = 1 | m_j, n_j) = \frac{m_j}{m_j + n_j}$$

and

$$\Pr(d_j = 0 | m_j, n_j) = \frac{n_j}{m_j + n_j}.$$

This in turn implies that

$$e_j = \mathrm{E}(d_j | m_j, n_j) = \frac{m_j}{m_j + n_j} = \Pr(d_j = 1 | m_j, n_j)$$

Here the d_j (either 0 or 1) are the observed number of failures in Sample I at each of the observed failure times for the combined samples and e_j are the corresponding expected numbers. An intuitively reaonable test statistic is the sum of the differences between these observed and expected numbers, i.e.,

$$S_m = \sum_{j=1}^{m+n} d_j - \sum_{j=1}^{m+n} e_j = m - \sum_{j=1}^{m+n} \frac{m_j}{m_j + n_j}.$$

This is the same as (9.5). The terms for which $d_j = 1$ contribute components of the form (9.6) and those for which $d_j = 0$ contribute components of the form (9.7).

Modification for censoring or ties is relatively straightforward. We calculate conditional expectations relevant to each observed failure, replacing m_j, n_j by m'_j, n'_j where these are now the numbers still certainly at risk immediately before the jth failure. That is, we exclude all earlier failures and any observations that have previously been censored. If there are c_j tied failures at the time of the jth failure the corresponding d_j is the number of those failures in Sample I and the corresponding $e_j = c_j m'_j / (m'_j + n'_j)$. Because we only compute the components of S_m at actual failure times there are now fewer than $m + n$ components.

Example 9.5

The problem. Determine the logrank statistic S_m for testing H_0: $\alpha = 1$ against the alternative H_1: $\alpha \neq 1$ in a proportional hazards model given

Sample I	7	9	13*	16		
Sample II	8	10	14*	16	25*	27

Formulation and assumptions. We compute all relevant m'_j, n'_j, c_j and e_j in the way described above and use these values to compute the modified S_m.

Procedure. It is convenient to exhibit the computations in tabular form as in Table 9.2

The computations in Table 9.2 should be checked carefully. The censored observations do not appear directly in the table. Their presence is reflected by changes in m'_j and n'_j. The two failures when $t = 16$ only require one table entry because c_j takes account of the tie. Here

$$S_m = \sum d_j - \sum e_j = 3 - 1.8940 = 1.106.$$

Testimate has an exact permutation test using these scores and gives for these data $P = 0.1762$

Conclusion. There is no evidence against the hypothesis of identical survival distributions.

Comments. 1. The method used above does not provide scores for each observation irrespective of whether or not that observation is censored. Peto and Peto derived logrank scores for individuals using an estimator of the survival function proposed by Altshuler (1970). These scores are used in programs such as that in StatXact as

Table 9.2 *Computation of logrank statistic with censoring and ties.*

Failure time	m'_j	n'_j	c_j	d_j	e_j
7	4	6	1	1	0.4000
8	3	6	1	0	0.3333
9	3	5	1	1	0.3750
10	2	5	1	0	0.2857
16	1	3	2	1	0.5000
27	0	1	1	0	0.0000
Total				3	1.8940

appropriate logrank scores. We omit details, but in the notation we use their method lead to the following scores

- If the jth ordered observation is a failure, or if there are c_j tied failures at that time each is assigned the score

$$s_j = 1 - \sum_i \frac{c_i}{m_i + n_i}$$

where the summation is over all failures at or before $t_{(j)}$. When $c_i = 1$ there is a single failure. Earlier censored items are not included in this count.

- For all censored items between a failure at time $t_{(j)}$ and the next failure (if any) each is given the score

$$s_k = - \sum_i \frac{c_i}{m_i + n_i}$$

where the summation is over all failure times up to and including that at time $t_{(j)}$. Censored items prior to the first failure are recorded as zero.

2. Applying the process outlined in *Comment* 1 to the data in this example the reader should confirm that the scores are those given in Table 9.3.

From Table 9.3 we obtain the sum of the scores for the sample I units:

$$\begin{aligned} S_m &= 1 - 1/10 + 1 - 1/10 - 1/9 - 1/8 - 1/10 - 1/9 - 1/8 - 1/7 \\ &+ 1 - 1/10 - 1/9 - 1/8 - 1/7 - 1/2 = 3 - 4773/2520 = 1.106. \end{aligned}$$

This is the same total as we obtained using Table 9.2.

The scores in Table 9.3 may be used in an asymptotic test valid for large $m + n$ providing neither m nor n are particulary small. StatXact includes such a test but slightly different results are obtained because the program deals differently with ties, assuming initially that ties are separated by a small amount and then allocating to each tied value the mean of these separate scores. This has a carryover effect to other scores but the test results are generally very close to those obtained using the Altshuler scores, or using the

Table 9.3 *Logrank scores with censoring and ties.*

Sample	Response	Score
1	7	$1 - 1/10$
2	8	$1 - 1/10 - 1/9$
1	9	$1 - 1/10 - 1/9 - 1/8$
2	10	$1 - 1/10 - 1/9 - 1/8 - 1/7$
1	13*	$-1/10 - 1/9 - 1/8 - 1/7$
2	14*	$-1/10 - 1/9 - 1/8 - 1/7$
1	16	$1 - 1/10 - 1/9 - 1/8 - 1/7 - 1/2$
2	16	$1 - 1/10 - 1/9 - 1/8 - 1/7 - 1/2$
2	25*	$-1/10 - 1/9 - 1/8 - 1/7 - 1/2$
2	27	$1 - 1/10 - 1/9 - 1/8 - 1/7 - 1/2 - 1$

differences between the sum of observed and expected scores at each failure time.

The basic scoring systems behind the Gehan–Wilcoxon test, the Peto–Wilcoxon test and the logrank test all generalize to more than two samples. A concise formulation of these extensions is given by Desu and Raghavarao (2004, Section 4.7). StatXacT and some other software packages include programs for carrying out tests using the relevant scoring systems discussed in this chapter. At their simplest these are analogous to the Kruskal–Wallis test with the relevant scoring system replacing ranks and proceed along the general lines described in Section 7.2.4.

Modification to the scoring systems when there are more than two samples are usually relatively straightforward. For example, suppose we use an analogous scoring system to that given in the comments on Example 9.5. Then in the formulae for the scores s_j, s_k given there, the only modification is that the denominator sum $m_i + n_i$ is extended to cover the relevant observations in all k samples. We illustrate this for a simple example.

Example 9.6

The problem. Using the extension outlined above for the scoring system given in *Comments* in Example 9.5, test whether there is any evidence that the following samples indicate that the corresponding population survival distributions differ.

Sample I	4	16		
Sample II	9	19	23	51*
Sample III	11	49*	63	

Table 9.4 *Logrank scores with censoring and ties for 3 samples.*

Sample	Response	Score
I	4	$1 - 1/9 = 0.889$
II	9	$1 - 1/9 - 1/8 = 0.764$
I	11	$1 - 1/9 - 1/8 - 1/7 = 0.621$
III	16	$1 - 1/9 - 1/8 - 1/7 - 1/6 = 0.454$
II	19	$1 - 1/9 - 1/8 - 1/7 - 1/6 - 1/5 = 0.254$
II	23	$1 - 1/9 - 1/8 - 1/7 - 1/6 - 1/5 - 1/4 = 0.004$
III	49*	$-1/9 - 1/8 - 1/7 - 1/6 - 1/5 - 1/4 = -0.996$
II	51*	$-1/9 - 1/8 - 1/7 - 1/6 - 1/5 - 1/4 = -0.996$
III	63*	$1 - 1/9 - 1/8 - 1/7 - 1/6 - 1/5 - 1/4 - 1 = -0.996$

Formulation and assumptions. We obtain scores for a logrank test in the manner indicated above and use an analogue of the Kruskal–Wallis test using these scores.

Procedure. The scores are given in Table 9.4. Using appropriate software such as that in StatXact for what is there called the logrank (Peto & Peto) test gives $P = 0.2452$.

Conclusion. There is insufficient evidence to reject the hypothesis of identical populations. This is largely an indication that the populations are too small to detect any possible difference.

Comments. Use of alternative scoring systems with censored survival distributions often lead to somewhat different P-values. This is largely a consequence of differing assumptions made about the information given by a censored observation. For the data in this example Gehan–Wilcoxon scores give $P = 0.1746$.

Computational aspects. Several software packages provide programs using scoring systems suggested in this text — and also a range of alternative scoring systems not covered here. Many scoring systems used for procedures we have discussed may not be identical with those in Table 9.4, but are usually equivalent. For example the scores used in StatXact have opposite signs to those given in Table 9.4 but they lead to identical conclusions. In the case of data ties, different ways of handling these may result in slight differences to allocated scores but these usually have only a small effect on computed P-values.

Tarone and Ware (1977) generalized many of the procedures covered in this and the previous section. A full account of their test, highlighting its relationship to those described here, is given by Desu and Ragghavarao (2004, Section 2.6.3) and StatXact has a program to carry out the test.

A modern nonparametric approach to analysis of factorial designs with censored data is given by Akritas and Brunner (1997). A wide-ranging review of recent developments in nonparametric analysis of survival data is given by Akritas (2004).

9.4 Median Tests for Sequential data

In Section 6.2 we cautioned against use of the median test for small samples even in situations where its Pitman efficiency was relatively high. However this test may result in an appreciable saving of resources in survival studies when observations are made sequentially over time.

In many survival experiments observations are collected sequentially and are automatically ordered as times to failure not only within each of the samples of size m, n but also in the combined samples. If we only want to know whether the median times to failure appear to differ for the populations providing the samples Gastwirth (1968) showed that it may be possible to terminate the experiment before all $m + n$ failure times have occurred and still get the same P-value as one would get when the experiment was completed where we use either the median or the control median test. This may lead to an appreciable saving in experimental resources, especially if a control median test is used.

For the median test it follows from the arguments used in Section 6.2.2 that the critical value of the test statistic is the number in one of the samples either above or below the combined sample median, M, under H_0. Without loss of generality we might make this the number below M in Sample I. Further, for either the median or the control median test when m, n are fixed we may work out before we actually start the experiment the critical value for the number of Sample I values that must be below M for any assigned significance level P. This means that if the median test is used the experiment may be terminated when this number is exceeded in Sample I, because in that case we then accept H_0. If, on the other hand the critical number has *not* been exceeded by the time we have obtained $(m + n)/2 + 1$ observations we may then terminate the test with rejection of H_0. This follows because our observations are ordered, and therefore we know that any further observations in Sample 1 will be greater than M.

A similar argument holds for the control median. If we take Sample I as the control, the stopping rule is now that we stop if the critical value in Sample II is exceeded and accept H_0. Otherwise we stop when we have obtained $m/2 + 1$ values for then, because the observations are ordered, no more Sample II values will be below the control median.

It is clear that our decision would not in these circumstances be changed if the experiment were continued until all items had failed, since our test only requires numbers below the combined median M or the control median M_1 to be known.

Gastwirth showed that asymptotically the average number of failures that had to be observed for any given m, n before a decision is made is less for the control median test than it is for the median test. He also proposed a slight modification to the median test that reduces even further the expected number of observations needed before a decision can be made.

9.5 Choice of Tests

Sometimes, especially with small data sets, it is difficult to decide which of a number of possible test procedures is optimal. In the two independent sample case there might be a choice between a WMW test or one based on e.g., van der Waerden scores or logranks. Using one test when another is appropriate often results in loss of Pitman efficiency and of power. Gastwirth (1985) proposed tests applicable in a wide range of circumstances where we may be reasonably confident that one of a finite set of models is appropriate without clear evidence of which is optimal. The tests he proposed are robust over the family of possible models in the sense that no other test has asymptotically a higher minimum efficiency relative to the optimum test for each model in the family. Gastwirth called such tests *maximin efficiency robust tests* — now often abbreviated to *MERT*.

The theory involves some advanced ideas and the reader should consult Gastwirth's paper for a wide ranging theoretical discusssion. In essence, the tests are based on linear functions of test statistics used in the competing optimal tests for the different models. The appropriate linear function takes into account the correlations between the different statistics or the scores used in them. The method provides protection against a poor choice of test and is especially useful with censored data. MERT procedures generally have a relatively high efficiency when compared to the optimal test that would be used if the true model were known. StatXact provides facilities for the calculation and use of MERT scores. A related robust procedure, which they call MX, is described by Freidlin, Podgor and Gastwirth (1999).

Software like StatXact or Testimate let us carry out tests quickly with a wide range of scores so the safety net feature of MERT is perhaps less useful than it would be without these facilities. But in trying many tests one must bear in mind the caution given in Section 1.5 about unethical statistical behaviour. MERT and MX procedures are, however, more than just a safety net, and their robustness can prove invaluable when there is appreciable uncertainty about an appropriate model. Tests applicable to alternative model situations are also discussed by Andersen et al. (1982).

9.6 Fields of Application

Medicine

Studies of patient survival, remission or recovery times are a major source of censored data and have stimulated research into and development of increasingly refined statistical methods of analysis. The literature includes many examples both from clinical observations, and from experiments on animals.

Component Reliability.

A manufacturer of light bulbs may wish to compare the lifetimes of bulbs made with a new type filament with those having a standard filament. If most bulbs with the standard filament fail within 4000 hours an appropriate test may be to burn to destruction 100 bulbs with the new filament and 100 with the standard filament. The experiment might be terminated after 4000 hours and any bulbs that have not failed by that time would provide censored observations.

Causes of Failure.

Items such as car tyres or machine parts such as pumps may fail for one of several reasons. Samples may be tested to destruction for a finite time and both cause and time of failure noted for any that fail. Any that have not failed when the experiment is terminated are censored observations. In some cases one might expect a few items to fail due to irrelevant causes of an external nature, e.g., a pump might fail due to accidental contamination of the fluid it was designed to pump. Extensions of methods described in this chapter are available to study survival time characteristics for each cause of breakdown.

Material Quality.

Several different hard-wearing fabrics may be subjected to an abrasive treatment until a hole appears and the lifetimes recorded. Those that have not been holed by the time the experiment is completed are censored, as are any for which the test is not completed because of machine breakdowns, etc.

9.7 Summary

Survival data are often best described by long-tail distributions and may also be characterised by **censored observations**. Censored observations provide less information than complete observations, but various techniques enable us to use what information they contain. Censored observations may or may not be treatment dependent.

If censoring is not treatment dependent the **Gehan–Wilcoxon test** is a modification of the WMW test that makes mild assumptions about the information in censored observations. The **Peto–Wilcoxon** test makes stronger assumptions about this information, using the **Kaplan–Meier** estimator for the expected survival time of censored observations.

Savage or **logrank** scores are particularly appropriate for survival data and reflect the key role of the exponential distribution in simple survival models.

The **proportional hazards model**, in which the ratio of the hazard rates are constant, has been widely used in studies of survival data. In particular, one parameter exponential survival time distributions follow the proportional hazards model.

The **median** or the **control median test** allow early termination of an experiment where results are obtained sequentially in time, and the only interest is in possible median differences between two populations.

MERT tests provide a useful safety net when it is not clear which of several tests is most appropriate.

9.8 Exercises

9.1 The observations below indicate the ages at first tooth filling for a group of young people.

$$
\begin{array}{lcccccccc}
Boys & 3 & 5 & 6 & 8 & 11 & 12 & 15 & 16 \\
Girls & 4 & 7 & 9 & 10 & 14 & 17 & &
\end{array}
$$

Is there a difference in 'survival' times to this event between boys and girls?

9.2 Peto et al. (1977) give the following data for survival times in days of two groups of cancer patients under two different treatment regimes, A and B. An * denotes a censored observation. Investigate the evidence for a difference in survival times.

A	8	852*	52	229	63	8	1976*	1296*
	1460*	63	1328*	365*				
B	180	632	2240*	195	76	70	13	1990*
	18	700	210	1296	23			

*9.3 Burns (1984) gave the following data in a study of motion sickness incidence. One group was subjected to a motion *Type A* and the other to a motion *Type B* in a simulator. The data are times in minutes until each person first vomited. The experiment was stopped after 120 minutes and subjects who had not vomited at that time are censored. Some subjects requested that the experiment be stopped before they vomited and are also censored. Is there evidence the survival distributions differ?

Type A	30	50	50*	51	66*	82	92			+ 14 censored at 120
Type B	5	6*	11	11	13	24	63	65		
	65	69	69	79	82	82	102	115	+ 13 censored at 120	

*9.4 Pocock (1983) and Hand et al. (1994, Section 423) give data for survival times in months for patients suffering from a form of hepatitis for a control group and for patients treated with *prednisolone*. Patients marked * were still alive at the stated time and no further information is available about them. Is there evidence of a difference in the survival patterns?

Cont.	2	3	4	7	10	22	28	29	32	37	40
	41	54	61	63	71	127*	140*	146*	158*	167*	182*
Tr.	2	6	12	54	56*	68	89	96	96	125*	128*
	131*	140*	141*	143	145*	146	148*	162*	168	173*	181*

*9.5 Use (i) Gehan–Wilcoxon scores and (ii) logrank scores to test whether there is evidence that the following samples indicate a difference between the corresponding survival distributions.

Sample A	12	19	23	27*		
Sample B	18	29	45	61	97	102*
Sample C	29	45	81*	201		
Sample D	72	151*	189	207		

9.6 The data for a control sample (A) and a further sample (B) are obtained sequentially. The complete data set is given below. Use a control median test approach to determine the minimum number of observations needed before a decision could be reached to either accept or reject the hypothesis of identical medians in a one-tail test where the alternative hypothesis is that the B sample is from a population with median greater than the control, A, and the test is performed using a nominal significance level $\alpha = 0.05$. What decision is then reached?

A	2	5	9	14	21	27	33	37	42	49	53	59		
B	22	29	47	56	73	82	87	94	95	95	98	102	105	109
	112	115	116	117										

CORRELATION AND CONCORDANCE

10.1 Correlation in Bivariate Data

We often want measures that summarize the strength of relationships between variables. Correlation is a measure of the strength of association or dependence between two variables.

In the parametric context, the Pearson product moment correlation coefficient estimates the degree of linear association between two variables. The coefficient takes values between -1 and $+1$. The extremes are attained only when points lie exactly on a straight line. The extreme -1 implies a negative slope, and $+1$ a positive slope. If one of the variables tends to be large when the other is large and small when the other is small, the correlation is positive. If large values of one variable are associated with small values of the other the correlation is negative. If the two variables are independent of each other, the value of the correlation is zero. A nonlinear relationship can also produce a correlation value close to zero. It is prudent to plot a scatter diagram before calculating any correlation coefficient.

10.1.1 Pearson Product Moment Correlation Coefficient

If we have measurements on two variables (such as height and weight) for a sample of n individuals, these paired observations can be written

$$(x_1, y_1), (x_2, y_2), \dots, (x_n, y_n)$$

where the population distributions are X and Y. The sample *Pearson product moment correlation coefficient*, r, is defined as

$$r = \frac{\sum_{i=1}^{n}[(x_i - \overline{x})(y_i - \overline{y})]}{\sqrt{\sum_{i=1}^{n}(x_i - \overline{x})^2 \sum_{i=1}^{n}(y_i - \overline{y})^2}}, \tag{10.1}$$

which, for simplicity and computational convenience, is usually rearranged and written

$$r = \frac{c_{xy}}{\sqrt{c_{xx}c_{yy}}}$$

where

$$c_{xy} = \sum_i (x_i y_i) - \left(\sum_i x_i\right)\left(\sum_i y_i\right)/n.$$

and

$$c_{xx} = \sum_i x_i^2 - \left(\sum_i x_i\right)^2 \bigg/ n, \quad c_{yy} = \sum_i y_i^2 - \left(\sum_i y_i\right)^2 \bigg/ n.$$

It is easy to verify from (10.1) that r is unaltered by what are called location and scale changes (see also location and scale parameters, Section 6.4). That is, r is unaltered if all x_i are replaced by $(x_i - k)/s$ and all y_i are replaced by $(y_i - m)/t$ where k, m are any constants that represent location or centrality changes and s and t are any positive constants that represent scale changes. Another way of putting this is to say that r is invariant under linear transformations of the x and y. This property extends to the alternative correlation coefficients described in Sections 10.1.3 to 10.1.5. It also has important practical implications in some regression methods considered in Chapter 11.

In parametric inference the Pearson coefficient is particularly relevant to a bivariate normal distribution where the sample coefficient r is an appropriate estimate of the population correlation coefficient ρ. If $\rho = 0$ for a bivariate normal distribution this implies X and Y are independent and values of r close to zero support that hypothesis. Even when the sample $(x_1, y_1), (x_2, y_2), \ldots, (x_n, y_n)$ is assumed to come from a bivariate normal distribution, making inference about an unknown population correlation coefficient ρ based upon the sample coefficient r given by (10.1) is more difficult than that for inferences about the means or variances of the distributions of X or Y. If the joint distribution of X and Y is not bivariate normal, parametric inference about ρ is even harder.

Basic nonparametric correlation inference does not require the assumption of bivariate normality and can be applied to both paired observations of continuous data and to data consisting of ranks. These ranks may be the original data or they may be derived from continuous measurements.

Pitman (1937b) introduced a permutation test based on the Pearson coefficient for continuous data that may be used for inference. Like other permutation tests with continuous data such as those for the one- and two-sample problems the test is a conditional test and fresh computation of the permutation distribution is required for each data set. Modern computer software makes the computation feasible for small samples. For larger samples Monte Carlo or asymptotic approximations are available in, for example, StatXact. The appropriate test is based on the fact that if we fix the order of the x values then all $n!$ permutations of the y values are equally likely under an hypothesis of independence or lack of association between X and Y. Permutations giving values of r near to -1 or $+1$ indicate strong evidence against the hypothesis $H_0: \rho = 0$. Lack of robustness means the test, although of historical interest, is little used in practice. When normality assumptions hold, the Pitman efficiency of the test is 1. A simple numerical example is given in Sprent (1998, Section 9.1).

Table 10.1 *A data set for scores x, y for two questions Q1 and Q2 on a paper.*

Q1	x	1	3	4	5	6	8	10	11	13	14	16	17
Q2	y	13	15	18	16	23	31	39	56	45	43	37	0

10.1.2 Other Measures of Bivariate Association

Desirable properties of any correlation coefficient are that its values should be confined to the interval $(-1, 1)$ and that lack of association implies a value near zero. Values near $+1$ should imply a strong positive association, and values near -1 a strong negative association. For the Pearson coefficient, $r = \pm 1$ implies linearity, but for the rank coefficients we introduce below values of ± 1 need not, and usually do not, imply linearity in continuous data from which the ranks may have been derived. Rather, we are interested in what, for continuous data, is known as *monotonicity*. If x and y increase together, this is a monotonic increasing relationship. If y decreases as x increases the relationship is monotonic decreasing. For rank correlation, the value $+1$ implies strictly increasing monotonicity, the value -1 strictly decreasing monotonicity.

Rank correlation coefficients are also relevant to, and indeed were originally developed for use in, situations where there is no underlying continuous measurement scale, but where the ranks simply indicated order of preference expressed by two assessors for a group of objects, e.g., contestants in a diving contest, or different brands of tomato soup in a tasting trial.

We illustrate some basic concepts in nonparametric correlation using the artificial, but realistic, data in Table 10.1. This gives, for 12 candidates, the scores achieved in an examination paper consisting of a short question marked out of 20 and a long question marked out of 60. To a certain extent the two marks are positively associated. However, some students concentrated their revision on the material for Question 1 (Q1), which they answered well, and performed less well on Question 2 (Q2). One student only revised the material for Q1 and did not even attempt Q2. Figure 10.1 is a scatter diagram for these data. It shows the tendency for the Q2 marks to increase with the Q1 marks up to a certain point, beyond which the Q2 marks tend to decrease.

Other examples of this type of relationship can be found in a biological context. A measured response often increases up to a maximum as the level of a stimulus increases. After reaching this maximum, the response may level off or decrease. This might be due to a toxic effect associated with an overdose of a potentially beneficial treatment. For the data in Table 10.1, the Pearson coefficient is $r = 0.373$. This is not significantly different from zero at the 5 percent level even in a one-tail test (exact one-tail $P \approx 0.116$) if we assume the sample is from a bivariate normal distribution. That assumption is not realistic for these data. This lack of significance of the Pearson coefficient is counter to an intuitive feeling that there is evidence of a reasonably high degree

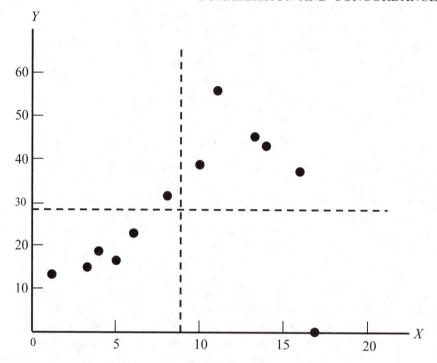

Figure 10.1 *Scatter diagram for data in Table 10.1. The broken lines at right-angles represent axes when the origin is transferred to the x, y sample medians.*

of association in the data. Not unexpectedly, because it lacks robustness, the Pitman permutation test for these data gives an exact one-tail $P = 0.112$, close to the value assuming normality.

In Sections 10.1.3 to 10.1.5 we consider three nonparametric measures of correlation and apply each to the data in Table 10.1. We present the coefficients in a way that shows relationships between these measures of correlation and certain concepts introduced in earlier chapters.

Exact tests are based on appropriate permutations of data, ranks or counts. For tests based on ranks the simplest permutation procedure is to fix the rank order associated with one variable, conventionally x, and to calculate the value of the chosen coefficient under all equally likely permutations of ranks of the other variable, y. The probabilities of the observed or more extreme values of the chosen coefficient, or some linear function of it, determine the size of the relevant critical region for testing for zero correlation.

10.1.3 The Spearman Rank Correlation Coefficient

Spearman (1904) proposed a rank correlation coefficient that bears his name. It is computationally equivalent to the Pearson coefficient calculated for ranks.

Ranks replace the original continuous data if these are given. The coefficient is often denoted by the Greek letter ρ (rho), and referred to as *Spearman's rho*. Because ρ is also used for the correlation coefficient in bivariate normal distributions (estimated by the Pearson coefficient r) to avoid confusion we denote the Spearman coefficient by ρ_s and the corresponding sample value by r_s. Formula (10.1) may be used to calculate r_s if we replace (x_i, y_i) by their ranks (r_i, s_i). When there are no ties, a simpler formula can be obtained by straightforward algebraic manipulation using properties of sums of ranks and sums of squares of ranks. The formula is

$$r_s = 1 - \frac{6T}{n(n^2 - 1)} \qquad (10.2)$$

where $T = \sum_{i=1}^{n}(r_i - s_i)^2$; i.e., the sum of the squares of the difference between the ranks for each sample pair. If the x, y ranks are equal for each pair, i.e., if $r_i = s_i$ for all i, then $T = 0$ and $r_s = 1$. For a complete reversal of ranks, tedious elementary algebra establishes that $r_s = -1$. If there is no correlation between ranks it can be shown that $E(T) = n(n^2 - 1)/6$, so that r_s has expected value zero. If the observations are a random sample from a bivariate distribution with X, Y independent we expect near-zero values for r_s.

Tables giving critical values at nominal 5 and 1 percent significance levels for testing H_0: $\rho_s = 0$ against one- and two-sided alternatives are widely available [see, e.g., Hollander and Wolfe, (1999, Table A31) or Neave (1981, p. 40)]. For not too large samples StatXact provides exact P-values for this test with the choice of an asymptotic or a Monte Carlo estimate of P for larger samples. Since r_s is a monotonic function of T when n is fixed we may use T itself as a test statistic and critical values of T have been tabulated for small n. It is preferable to use r_s because it gives a more easily appreciated indication of the level of association. It complies with the convention that possible values of a correlation coefficient should lie in the interval $[-1, 1]$.

Both this coefficient and one we develop in Section 10.1.4 are often used to test whether there is broad equivalence between ranks, or orderings, assigned by different assessors. Do two examiners concur in their ranking of candidates? Do two judges agree in the placings of n competitors in a diving contest? Do job applicants' ranks for manual skill based on a psychological test show a relationship to their rankings in a further test for mathematical skills?

These coefficients are also appropriate for tests of trend. Does Y increase (or decrease) as X increases? Such tests are often more powerful than the Cox–Stuart test described in Section 4.4.3.

Example 10.1

The problem. Compute the Spearman coefficient for the data in Table 10.1. Use the result to test H_0: $\rho_s = 0$ against the one-sided alternative H_1: $\rho_s > 0$.

Formulation and assumptions. The coefficient is computed using (10.2) and tables or appropriate software are used to assess the evidence.

Table 10.2 *Paired ranks and their differences d for the data in Table 10.1.*

x-rank	1	2	3	4	5	6	7	8	9	10	11	12
y-rank	2	3	5	4	6	7	9	12	11	10	8	1
d	−1	−1	−2	0	−1	−1	−2	−4	−2	0	3	11

Procedure. To compute T in (10.2) we replace the data in Table 10.1 by the ranks and compute the rank differences d as given in Table 10.2.

T is calculated by squaring the difference, d, and adding the squares. Thus

$$T = (-1)^2 + (-1)^2 + (-2)^2 + \ldots + (11)^2 = 1 + 1 + 4 + \ldots + 121 = 162,$$

whence $r_s = 1 - (6 \times 162)/(12 \times 143) = 0.4336$. One-tail critical values for significance at nominal 5 and 1 percent levels given by tables are 0.503 and 0.678. For this example StatXact gives an exact one-sided P = 0.081. Using an asymptotic approximation (see Section 10.1.6), $P = 0.080$.

Conclusion. Since $r_s = 0.4336$ is less than the value 0.503 required for significance at a nominal 5 percent level and corresponds to an exact $P \approx 0.08$ there is not strong evidence against a hypothesis of lack of monotonic association.

Comment. We indicated in Section 10.1.2 that the exact test based on the Pearson coefficient gave an associated $P = 0.112$, slightly greater than that obtained here. This again reflects the influence of the curved nature of the relationship between x and y for high Q1 marks. The conceptual relationship between the Spearman and the Pearson coefficients often results in both coefficients lacking robustness when a few observations are markedly out of line with the general pattern.

Computational aspects. Many general statistical packages have programs that compute r_s or T or an equivalent statistic, but often either leave one to look up tables, or else give only an asymptotic result to establish whether there is strong evidence against H$_0$. If no specific program is included for the Spearman coefficient one might use the paired ranks as data for a Pearson test. This will give the correct numerical value of the coefficient, but if such a program gives only an asymptotic P-value this will usually be calculated on the assumption that the data are a sample from a bivariate normal distribution and so it is not relevant.

The exact permutation distribution of r_s for small samples is not hard to compute. StatXact will do this. The basic idea is similar to that outlined in Section 10.1.2 for the permutation distribution of the Pearson coefficient. In practice one would be likely to use a facility like that in StatXact to compute the distribution, but if it were needed, we illustrate the procedure for four paired ranks in Example 10.2.

Example 10.2

The problem. Given four paired observations calculate the exact permutation distribution of the Spearman correlation coefficient r_s. Assume there are no ties.

Formulation and assumptions. It is easy to see that there are $4! = 24$ possible data pairings. For each such pairing we may compute the value of r_s. Since all pairings are equally likely under a hypothesis of no association, each has an associated probability of $1/24$.

Procedure. The number of possible pairings is obtained by noting that the y-rank associated with the x-data value that is ranked 1 may take any of the four values 1, 2, 3, or 4. Once we note which y-rank is associated with the x-rank 1 there are three remaining y-ranks that may be associated with the x-rank 2. This leaves 2 possible y-ranks for association with the x-rank 3, and only 1 y-rank for association with the x-rank 4. Thus, as already stated, there are $4 \times 3 \times 2 \times 1 = 4! = 24$ possible arrangements. These are given in Table 10.3 together with the value of r_s computed for each.

From the last column of Table 10.3 we easily confirm that the possible values of r_s and the numbers of times each occurs, n_s, are those given below. Since each of the 24 cases is equally likely to occur division of these numbers by 24 gives P_r, the probability of occurrence of each possible value of r_s under the hypothesis H_0: $\rho_s = 0$.

r_s	-1.0	-0.8	-0.6	-0.4	-0.2	0.0	0.2	0.4	0.6	0.8	1.0
n_s	1	3	1	4	2	2	2	4	1	3	1
P_r	1/24	1/8	1/24	1/6	1/12	1/12	1/12	1/6	1/24	1/8	1/24

Conclusion. The distribution of r_s under the null hypothesis is discrete and takes only 11 possible values in the interval $[-1, 1]$ with the given probabilities.

Comment. The distribution is clearly symmetric about zero, and this property holds for any n.

Computational aspects. The program for the Spearman coefficient in StatXact will generate the complete distribution for any specified value of n that is not too large. For $n = 10$ the discontinuities are appreciably smaller than those in this example. The statistic takes 166 distinct values in the interval $[-1, 1]$.

Ties in data are replaced by mid-ranks. To obtain exact P-values with ties requires a program like that in StatXact, although the value of r_s may be computed by applying the Pearson coefficient formula (10.1) to these mid-ranks. Thomas (1989) showed that if (10.2) is used to compute r_s when there are ties represented by mid-ranks it gives a value that is always greater than or equal to the correct value given by the Pearson formula. We consider an asymptotic approximation in Section 10.1.6.

10.1.4 The Kendall Correlation Coefficient

We use methods reminiscent of the Mann–Whitney scoring procedure for the WMW test to obtain another measure for monotonic association. Inferences involve a coefficient known as the *Kendall correlation coefficient*, or *Kendall's tau*, which is closely related to the Jonckheere–Terpstra test statistic.

Table 10.3 *Possible arrangements of y-ranks for four data pairs corresponding to the x-ranks 1, 2, 3, 4 for a set of four paired observations.*

r_x	1	2	3	4	
	r_y	r_y	r_y	r_y	r_s
Case number					
1	1	2	3	4	1.0
2	1	2	4	3	0.8
3	1	3	4	2	0.4
4	1	3	2	4	0.8
5	1	4	2	3	0.4
6	1	4	3	2	0.2
7	2	1	3	4	0.8
8	2	1	4	3	0.6
9	2	3	1	4	0.4
10	2	3	4	1	−0.2
11	2	4	1	3	0.0
12	2	4	3	1	−0.4
13	3	1	2	4	0.4
14	3	1	4	2	0.0
15	3	2	1	4	0.2
16	3	2	4	1	−0.4
17	3	4	1	2	−0.6
18	3	4	2	1	−0.8
19	4	1	2	3	−0.2
20	4	1	3	2	−0.4
21	4	2	1	3	−0.4
22	4	2	3	1	−0.8
23	4	3	1	2	−0.8
24	4	3	2	1	−1.0

In practice, as in the Jonckheere–Terpstra test, we need not replace continuous data by ranks, although the test may also be carried out using ranks. Computation is simpler if we assume the x data are arranged in ascending order. We may use either data in a form like that in Table 10.1 where the x are already in ascending order, or the corresponding ranks. It is convenient to use the labels "x-ranks" and "y-ranks" to refer respectively to the ranks associated with the first and second members of a bivariate pair (x, y), even when our basic data are themselves ranks only, and there is no specific underlying continuous variable. The latter is the case for taste preferences, or for two judges ranking a number of competitors in a sporting event like ice skating or high diving.

Denoting the ranks of x_i, y_i by r_i, s_i respectively, the Kendall coefficient is based on the principle that if there is association between the ranks of x and y, then, if we arrange the x-ranks in ascending order (i.e., so that $r_i = i$), the s_i should show an increasing trend if there is positive association and a decreasing trend if there is negative association. Kendall (1938) therefore proposed that after arranging observations in increasing order of x-ranks, we score each paired difference $s_j - s_i$ for $i = 1, 2, \dots, n - 1$ and $j > i$ as $+1$ if this difference is positive, and as -1 if negative. Kendall called positive and negative differences *concordances* and *discordances* respectively. Denoting the numbers of concordances and discordances by n_c, n_d respectively, Kendall's tau, which we denote by t_k, although historically it has usually been denoted by the Greek letter τ, is

$$t_k = \frac{n_c - n_d}{n(n-1)/2}. \tag{10.3}$$

We distinguish between this sample estimate t_k and the corresponding population value that we denote by τ_k.

If we are comparing ranks assigned by two judges to a finite set of objects such as ten different brands of tomato soup this is essentially the entire population of interest, so our calculated coefficient is a measure of agreement between the judges in relation to that population. On the other hand, if we had a sample of n bivariate observations from some continuous distribution (of measurements, say) and calculate t_k using (10.3) this is an estimate of some underlying τ_k that is a measure of the degree, if any, of monotonic association, or dependence between the variables X, Y in the population. Since there are $n(n-1)/2$ pairs $s_j - s_i$, if all are concordances $n_c = n(n-1)/2$ and $n_d = 0$, whence $t_k = 1$. Similarly, if all are discordances, then $t_k = -1$. If the rankings of x and y are independent we expect a fair mix of concordances and discordances and t_k should be close to zero.

When there are no ties in the rankings, $n_c + n_d = n(n-1)/2$. Although the coefficient is often referred to as a *rank* correlation coefficient it is clear that to compute it we only need to know the order of the relevant x and y and not the actual ranks as we do for the Spearman coefficient.

Example 10.3

The problem. Compute Kendall's tau for the data in Table 10.1.

Formulation and assumptions. As the x are ranked in ascending order, we may compute n_c from the y-ranks in Table 10.2. We may also obtain n_d this way, or deduce it from $n_d = n(n-1)/2 - n_c$.

Procedure. To count the number of concordances we inspect the y-ranks in Table 10.2, noting for each successive rank the number of succeeding ranks that are greater. The first y-rank is 2 and this is succeeded by 10 greater ranks, namely 3, 5, 4, 6, 7,

9, 12, 11, 10, 8. Similarly, the next y-rank, 3, gives 9 concordances. Proceeding this way it is easily verified that the total number of concordances is

$$n_c = 10 + 9 + 7 + 7 + 6 + 5 + 3 + 0 + 0 + 0 + 0 = 47.$$

Since $n = 12$ we easily deduce that $n_d = (12 \times 11)/2 - 47 = 19$. This may be verified by counting discordances directly from Table 10.2.

Conclusion. For the given data $t_k = (47 - 19)/66 = 0.4242$.

Comments. 1. Since the x are ordered in Table 10.1, the order of the corresponding y values is a one-to-one ordered transformation of the ranks in Table 10.2. Thus, we could equally well have computed the number of concordances and discordances by noting the signs of all differences $y_j - y_i$ in Table 10.1 as these signs correspond to those of the $s_j - s_i$ above.

2. As noted above, the counting procedure for concordances is reminiscent of the Mann–Whitney counts for the WMW test, and we show below equivalence to the Jonckheere–Terpstra test.

The x_i are in ascending order in Table 10.1. This simplifies computation. More generally, the number of concordances is the number of positive c_{ij} among all $n(n - 1)/2$ of the $c_{ij} = (y_j - y_i)/(x_j - x_i), i = 1, 2, \ldots, n - 1$ and $j > i$. The number of negative c_{ij} is the number of discordances.

Equivalence to a Jonckheere–Terpstra test is evident if we arrange the x_i in ascending order, as in Table 10.1 and regard the x_i (or the associated ranks) as indexing n ordered samples, each of 1 observation, where that observation is the corresponding y value. The number of concordances is the sum of the Mann–Whitney sample pairwise statistics used in the Jonckheere–Terpstra test in Section 7.2.5. Thus, in theory, any program for the exact Jonckheere–Terpstra test may be used for Kendall's coefficient to test H_0: *no association between x and y ranks* against either a one- or two-sided alternative. In practice, however, most programs for the Jonckheere–Terpstra test will, in this situation, only give exact tail probabilities for very small n. Asymptotic results, or Monte Carlo estimates, of the exact tail probability must be used with this approach if the samples are moderate to large. We consider more specific asymptotic results in Section 10.1.6.

Programs in general statistical packages often calculate t_k directly, some indicating significance levels (usually asymptotic), others leaving the user to determine these at a nominal level from tables. Extensive tables are given, for example, by Neave (1981, p. 40) and by Hollander and Wolfe (1999, Table A30). Some tables give equivalent critical values of $n_c - n_d$ for various n. However, calculating t_k requires little extra effort and gives a better feel for the degree of association implied by any correlation coefficient. When there are no ties tests may be based on n_c only, for it is clear from (10.3) that if n is fixed, t_k is a linear function of n_c. StatXact includes a program that computes exact P-values for the test of no association based on this coefficient for small to moderate sample sizes.

In passing it is worth noting one difference between the usual scoring systems for the Kendall coefficient and that used for Jonckheere–Terpstra, namely that whereas the former uses $1, 0, -1$ for concordances, ties and discordances, the latter uses $1, 1/2, 0$. This is not a fundamental difference, because providing certain obvious adjustments are made to relevant formulae, the scoring systems may be interchanged without altering conclusions.

Example 10.4

The problem. For the data given in Table 10.1 use Kendall's tau to test H_0: *no association* against H_1: *positive association between x and y.*

Formulation and assumptions. The appropriate test is a one-tail test of H_0: $\tau_k = 0$ against H_1: $\tau_k > 0$.

Procedure. In Example 10.3 we found $t_k = 0.4242$, We also showed that $n_c - n_d = 47 - 19 = 28$. Tables give nominal 5 and 1 percent critical values for significance when $n = 12$ in a one-tail test as 0.3929 and 0.5455. Corresponding values for $n_c - n_d$ are 26 and 36. More precisely, StatXact for these data, gave an approximate Monte Carlo estimate of the exact one-tail $P = 0.032$.

Conclusion. There is moderately strong evidence against the hypothesis of no association. In a formal significance test one would reject H_0 at a nominal 5 percent significance level, and accept a positive association between the ranks of x and y.

Comments. 1. Had the alternative hypothesis been H_1: $\tau_k \neq 0$, requiring a two-tail test, we would not have rejected H_0 at the 5 percent significance level.

2. The evidence of association provided by Kendall's tau is stronger than that provided by Spearman's rho. We give a likely explanation in Section 10.1.7.

Computational aspects. As indicated above, most general statistical programs do not give exact P-values or significance levels for relevant hypothesis tests. That in StatXact gives exact results for small samples, but a Monte-Carlo approximation is needed for moderate to large samples.

For small samples we may calculate the exact distribution of t_k in a way like that for the Spearman coefficient in Example 10.2.

Example 10.5

The problem. Calculate the exact permutation distribution of Kendall's t_k when $n = 4$, under the null hypothesis of no association.

Formulation and assumptions. If the x-ranks are in ascending order, then for each of the $4! = 24$ equally likely permutations of the y-ranks we count the number of concordances, n_c. When $n = 4$, $n_c + n_d = 6$ whence, given n_c, we easily find n_d. We then compute t_k using (10.3), and record the number of times each value of t_k occurs. Division of each number of occurrences by 24 gives the probability of observing each value.

Procedure. In Exercise 10.1 we ask for computation of t_k for each of the 24 possible permutations of 1, 2, 3, 4. For example, for the permutation 4, 2, 1, 3 we easily see that $n_c = 2$, whence (10.3) gives $t_k = (2 - 4)/6 = -0.333$. It is easy to verify that 4 other permutations give the same n_c, whence $\Pr(t_k = -0.333) = 5/24$.

Conclusion. Completing Exercise 10.1 gives the following probabilities associated with each of the 7 possible values of t_k:

t_k	1	0.67	0.33	0	−0.33	−0.67	−1
Probability	1/24	1/8	5/24	1/4	5/24	1/8	1/24

Comments. 1. The distribution of t_k is symmetric.

2. When $n = 4$, a *P*-value less than 0.05 can only be obtained in a one-tail test corresponding to $t_k = 1$ for which $P = 1/24 \approx 0.042$. Samples of this size are of little use in practice for inference.

Computational aspects. StatXact will calculate complete distributions for small samples. The Monte Carlo option in that program may be used to estimate *P*-values rapidly and with good accuracy.

Modifications are needed to calculate and use t_k when there are rank ties. With ties (10.3) no longer takes values $+1$ or -1 even when all mid-rank pairs lie on a straight line. Ties among the x or y are both scored zero in Kendall's coefficient because they are neither concordances nor discordances. Now (10.3) can no longer take the value ± 1 because $|n_c + n_d|$ is necessarily less than $n(n-1)/2$. It can be shown (see e.g., Kendall and Gibbons, 1990, Chapter 3) that we obtain a coefficient that takes values in the interval $[-1, 1]$ if we replace the denominator in (10.3) by $[(D - U)(D - V)]$ where .

$$D = n(n-1)/2, \ U = \sum u(u-1)/2, \ V = \sum v(v-1)/2$$

and u, v are the number of consecutive ranks in a tie within the x- and the y-ranks respectively, and the summations in U, V are over all sets of tied ranks. If there are no tied ranks obviously $U = V = 0$ and the modified denominator reduces to that in (10.3). We denote this modified statistic by t_b. It is often called *Kendall's tau-b*. If there is an exact linear relationship between mid-ranks t_b takes either the value $+1$ or -1. A simple example with heavy tying illustrates some of the above points.

Example 10.6

Two judges are each asked to arrange 12 Cabernet Sauvignon wines in order of preference. Both agree on which wine they rank 1 (or best). There are four wines that they both agree are almost as good, but neither can separate within that group, so each gives these a tied rank 3.5, while there are a further five wines that they both classify as fairly good but neither judge can separate the wines within that group. Each of these wines is thus given a mid-rank 8. Both also agree the remaining two wines are equally bad and thus are ranked 11.5. Thus the mid-ranks allocated are:

Judge A	1	3.5	3.5	3.5	3.5	8	8	8	8	8	11.5	11.5
Judge B	1	3.5	3.5	3.5	3.5	8	8	8	8	8	11.5	11.5

There is a linear relationship between these mid-ranks pairs since they all lie on a straight line through the origin with unit slope. Since the ranks awarded by Judge A (i.e., x-ranks) are ordered, n_c, the number of concordances, is obtained by counting the number of positive $s_j - s_i$, $j > i$. This is found by a direct, but careful, counting process, (see Exercise 10.12) to be 49, whence (10.3) gives $t_k = 0.7424$, since there are no discordances. Also in Exercise 10.12 we ask you to verify that $U = V = (4 \times 3 + 5 \times 4 + 2 \times 1)/2 = 17$, whence

$$t_b = 49/\sqrt{(66 - 17)^2} = 1.$$

Here t_b exhibits the desirable property of a correlation coefficient, i.e. $t_b = 1$, when there is complete ranking agreement, whereas t_k does not.

The next example covers a less extreme situation involving ties.

Example 10.7

The problem. Life expectancy showed a general tendency to increase during the nineteenth and twentieth centuries as standards of health care and hygiene improved. The extra life expectancy varies between countries, communities and even families. Table 10.4 gives the year of death and age at death for 13 males in clan McDelta buried in the Badenscallie burial ground in Wester Ross, Scotland (see Appendix 1). Is there an indication that life expectancy is increasing for this clan in more recent years?

Formulation and assumptions. If there is an increase in life expectancy those dying later will tend to be older than those dying earlier. Years of death, x, are already arranged in ascending order. So we may count the number of concordances and discordances by examining all pairs of y and scoring each as appropriate (1 for concordance, -1 for discordance, 0 for a tie).

Procedure. There are no ties in the x. The first y entry is 13. In the following entries this is exceeded in nine cases (concordances); there are two discordances and one tie. Similarly for the second y entry (83) there is one concordance, nine discordances and one tie. Proceeding this way we find that

$$n_c = 9 + 1 + 6 + 9 + 8 + 6 + 4 + 5 + 2 + 2 + 0 + 1 = 53$$

and

$$n_d = 2 + 9 + 4 + 0 + 0 + 1 + 2 + 0 + 2 + 1 + 2 + 0 = 23.$$

Table 10.4 *Year of death and ages of 13 McDeltas.*

Year	1827	1884	1895	1908	1914	1918	1924
Age	13	83	34	1	11	16	68
Year	1928	1936	1941	1964	1965	1977	
Age	13	77	74	87	65	83	

Since $n = 13$, (10.3) gives

$$t_k = \frac{53 - 23}{(13 \times 12)/2} = 0.3846.$$

There are two tied values at each of $y = 13$ and at $y = 83$. Using the formula for the denominator adjustment we have $D = 78$, $U = 0$ and $V = (2 \times 1)/2 + (2 \times 1)/2 = 2$. Hence

$$t_b = \frac{53 - 23}{\sqrt{78 \times (78 - 2)}} = 0.3896.$$

This small difference between t_k and t_b does not affect our conclusion. StatXact gives Monte Carlo estimates of the exact P and in three simulations of 10,000 samples gave estimated $P = 0.0359$, 0.0399 and 0.0381, suggesting $P < 0.04$.

Conclusion. There is reasonable evidence against the hypothesis H_0: *no increasing trend of life expectancy* and favouring the one-tail alternative that expectancy has increased during the period covered.

Comments. 1. If there are only a few ties and we ignore them and use (10.3) the test is conservative, for if we allow for ties by using t_b the denominator is less than that in (10.3).

2. A one-tail test is justified because we are interested in whether data for the McDelta clan follow established trends in most developed countries.

3. Care is needed in counting concordances if there are ties in both x and y. For example, for the data set:

x	2	5	5	7	9
y	5	1	3	8	1

the first pair $(2, 5)$ gives one concordance $(7, 8)$ and three discordances; the second pair $(5, 1)$ gives one concordance $(7, 8)$. There are no discordances because of the tied values in x at $(5, 3)$ and in y at $(9, 1)$. The next pair $(5, 3)$ gives one concordance $(7, 8)$ and one discordance $(9, 1)$.

Computational aspects. With more than a few ties, a program that gives at least a Monte Carlo estimate of the relevant tail probabilities is useful.

For moderately large samples with only a few ties, use of (10.3) and tables of critical values for the no-tie situation, or the use of asymptotic results given in Section 10.1.6 should not be seriously misleading.

We have so far established only that it is intuitively reasonable to expect Kendall's t_k to take values near zero when the (x_i, y_i) are sample values from a bivariate population where the random variables X and Y associated with that population are independent, implying a population $\tau_k = 0$ (though the converse may not be true). This leaves open the question of what precisely the sample Kendall coefficient is estimating when $\tau_k \neq 0$.

We now give a probabilistic interpretation of τ_k which leads naturally to t_k as a reasonable sample estimator of τ_k whether or not X and Y are independent and establishes the parameter τ_k as a reasonable measure of a certain type of dependence between X and Y. Further, the measure is distribution-free as it makes only the assumption that we are sampling from some unspecified bivariate distribution.

The argument is based on the fact that if (X_s, Y_s) and (X_t, Y_t), are independent samples from a population with some joint cumulative distribution function $F_{XY}(x, y)$, then each has that same distribution.

We consider some properties of a variable $B_{ts} = (Y_s - Y_t)/(X_s - X_t)$. If $B_{ts} > 0$ this implies that either the joint event $(Y_s > Y_t$ and $X_s > X_t)$ or the joint event $(Y_s < Y_t$ and $X_s < X_t)$ has occurred. These are mutually exclusive, so this in turn implies that

$$\Pr(B_{ts} > 0) = \Pr[(Y_s - Y_t)/(X_s - X_t) > 0] = \\ \Pr(Y_s > Y_t \text{ and } X_s > X_t) + \Pr(Y_s < Y_t \text{ and } X_s < X_t).$$

If also X and Y are independent we have

$$\Pr(Y_s > Y_t \text{ and } X_s > X_t) = \Pr(Y_s > Y_t) \times \Pr(X_s > X_t). \tag{10.4}$$

Because Y_s and Y_t both have the marginal distribution of Y it follows that $\Pr(Y_s > Y_t) = 0.5$. Similarly $\Pr(X_s > X_t) = 0.5$. Thus from (10.4) independence implies $\Pr(Y_s > Y_t$ and $X_s > X_t) = 0.5 \times 0.5 = 0.25$. Similarly, under independence $\Pr(Y_s < Y_t$ and $X_s < X_t) = 0.25$ which in turn implies that under independence

$$\Pr(B_{ts} > 0) = \Pr\left[\frac{Y_s - Y_t}{X_s - X_t} > 0\right] = \frac{1}{4} \times \frac{1}{4} = \frac{1}{2}.$$

If we now define the population Kendall τ_k to be

$$\tau_k = 2\Pr(B_{ts} > 0) - 1 \tag{10.5}$$

it follows that under independence $\tau_k = 2 \times 0.5 - 1 = 0$. A moment's reflection indicates that if the departures from independence are such that high values of Y tend to be associated with high values of X and low values of Y with low values of X then

$$\Pr(B_{ts} > 0) = \Pr\left[\frac{Y_s - Y_t}{X_s - X_t} > 0\right]$$

will be greater than 0.5, and will equal 1 if $Y_s > Y_t$ always implies $X_s > X_t$ in which case, from (10.5), $\tau_k = +1$. A similar argument shows that if high values of Y tend to be associated with low values of X and low values of Y with high values of X then τ_k will be negative. For a complete reversal of rank orders $\tau_k = -1$.

In the discussion following Example 10.3 we showed that if for all pairs of n sample values (x_i, y_i), (x_j, y_j) we formed quotients $b_{ij} = (y_j - y_i)/(x_j - x_i)$, $i = 1, 2, \ldots, n-1$, for all $j > i$, then a positive value of this quotient corresponds to a concordance and a negative value to a discordance. In the notation used in (10.3) $n_c/(n_c + n_d)$ provides a sensible estimate of

$$\Pr[(Y_s - Y_t)/(X_s - X_t) > 0]$$

in (10.5). Further, in the no-tie case since $n_c + n_d = n(n-1)/2$ it is easily verified that substitution of $n_c/(n_c + n_d)$ in (10.5) leads to (10.3). Kerridge (1975) gives an interesting example of an estimation problem with this probabilistic interpretation.

It should be realized that the type of dependence described above is not the only possible kind. For example, it is not uncommon for high values of Y to go with both high values of X and low values of X while intermediate values of Y go with intermediate values of X. At its simplest this might represent a near quadratic relationship between X and Y. The correlation measures described in this chapter are generally unsuitable for detecting this, or many other, possible kinds of nonlinear dependence.

10.1.5 A Median Test for Correlation

Blomqvist (1950, 1951) proposed a test that embodies similar concepts to the median test considered in Section 6.2. Suppose the medians of the marginal distribution of X, Y are respectively θ_x, θ_y. In a sample about half the observed x should be below θ_x and about half the y below θ_y. If X and Y are independent we expect a good mix of high, medium and low values of Y to be associated with the various values of X. If we knew θ_x and θ_y we could shift the origin in a scatter diagram of the observed (x_i, y_i) to (θ_x, θ_y). Then, under H$_0$: *X, Y are independent* we expect about one quarter of all points to be in each of the four quadrants determined by the new axes.

Usually we do not know θ_x, θ_y. Blomqvist proposed replacing them by the sample medians M_x, M_y as reasonable estimates. In Figure 10.1 axes with a new origin at (M_x, M_y) are shown by broken lines. Working anticlockwise from the first (or top right) quadrant the numbers of points in the respective quadrants determined by the new axes are 5, 1, 5, 1. The concentration of points in the first and third quadrants suggests dependence between X and Y. We may formalize a test of H$_0$ against a one-sided alternative of either positive or negative association, or a two-sided alternative of some association, following closely the procedure in Section 6.2.2. In general, if n is even, denoting the sample medians for X, Y by M_x, M_y and assuming no values coincide with these medians, we count the numbers of points a, b, c, d in each of the four quadrants determined after this shift of the scatter plot origin to (M_x, M_y). Because of sample median properties we get a 2×2 contingency table of the form shown in Table 10.5.

The marginal totals follow from the properties of the sample median and restriction to even n with no observation at either M_x or M_y. If n is odd at least one sample value for x and one for y will coincide with the medians, resulting in points at or on axes through the sample medians. Such points are omitted from the count. This necessitates modifications that we give at the end of this section.

In the case envisaged in Table 10.5, if X, Y are independent, then a, b, c and d each has expected value $n/4$. Marked departures from this value indicate association between X, Y. The appropriate test procedure then uses the Fisher exact test. We illustrate this in Example 10.8.

Table 10.5 *Contingency table for the median test for correlation.*

	Above M_x	Below M_x	Row total
Above M_y	a	b	$n/2$
Below M_y	c	d	$n/2$
Column total	$n/2$	$n/2$	n

First we consider what is an appropriate correlation measure having the desirable properties that it takes values in the interval $(-1, 1)$ with values near ± 1 indicating near-monotonic dependence. Independence results in values near zero (although the converse may not be true). Two possible coefficients are

$$r_d = (a + d - b - c)/n \qquad (10.6)$$

and

$$r_m = 4(ad - bc)/n^2 \qquad (10.7)$$

It is easily verified that in the special circumstances of Table 10.5, where the marginal totals imply that $a = d, b = c$ and $a+b = n/2$, both (10.6) and (10.7) have the desirable properties for a correlation coefficient and are equivalent, each taking the same set of $n/2 + 1$ possible values with the same associated probabilities under the null hypothesis of independence. While (10.6) has arithmetic simplicity, (10.7) generalizes more readily to other contexts, including the case n odd, so it is usually preferred.

Example 10.8

The problem. Calculate r_m (or r_d) for the data in Table 10.1 and test the hypothesis of independence against that of positive association between X and Y.

Formulation and assumptions. We find the sample medians M_x, M_y and deduce the entries in Table 10.5. The relevant coefficient values are computed using (10.6) or (10.7) and the test of significance is based on the Fisher exact test as in Example 6.7. A one-tail test is appropriate since the alternative is in a specified (positive) direction.

Procedure. For the data in Table 10.1 the sample medians are $M_x = 9, M_y = 27$, from which we deduce the values of a, b, c, d in Table 10.5. These are easily found by inspecting Figure 10.1 to be $a = d = 5$ and $b = c = 1$. Since $n = 12$ we easily calculate $r_m = 2/3 = r_d$. We use the Fisher exact test, the relevant tail probabilities being associated with the observed value $a = 5$ and the more extreme $a = 6$. Calculating these probabilities using (6.7) is easy even if an appropriate computer program is not available. The relevant probability is $P = 0.0400$.

Conclusion. There is fairly strong evidence against the hypothesis of independence and favouring that of positive association. The numerical value of the appropriate median correlation coefficient is $r_m = 0.667$.

Comments. 1. The Pearson correlation coefficient was not significant for these data. It is not robust against departures from normality of the type implicit in the observation (17, 0). However, r_m is robust and indeed we would reach the same conclusion if the point (17, 0) were replaced by any point in the fourth quadrant (positive x, negative y) relative to axes with origin at the sample medians if there were no other data changes. In general, the median correlation test is robust against a few major departures from a monotonic or near-monotonic trend. Unfortunately, it has low Pitman efficiency compared to the Pearson coefficient test when the relevant normality assumption holds.

2. The Blomqvist test is applicable no matter what the marginal distributions of X, Y may be.

Computational aspects. For all but very small samples a computer program such as that in StatXact or Testimate or many other packages is desirable for the Fisher exact test. The asymptotic result based on the chi-squared test is generally reliable for large samples.

If n is odd, or generally if there are values equal to one or both sample medians, the marginal totals in a table like Table 10.5 will no longer be $n/2$. The Fisher exact test is then used with the marginal totals observed. If a correlation coefficient is required it is best to use a modification of r_m, namely

$$r_m = \frac{ad - bc}{\sqrt{(a+b)(c+d)(a+c)(b+d)}}$$

where a, b, c, d are the observed cell values. This reduces to (10.7) when all marginal totals are $n/2$.

The coefficient r_m is easy to calculate, but in practice it is used less often than the Spearman or Kendall correlation coefficients.

The concept of partitioning bivariate data by the median that is inherent in Blomqvist's approach has been extended to partitioning based on empirical or sample quantiles by Borkowf et al (1997).

Other nonparametric tests for correlation have been suggested. Gideon and Hollister (1987) proposed a measure that is also relatively easy to calculate, although this advantage is partly negated for small n by the fact that the coefficient takes only relatively few possible values and the permutation distribution under the null hypothesis is less easily established than that for the Kendall or Spearman coefficients. The authors give fairly extensive tables of nominal critical values.

Dietz and Killen (1981) extended the concept of Kendall's tau to a multivariate test for trend applicable to a pharmaceutical problem.

10.1.6 *Asymptotic Results*

For large n, tests using the Spearman and Kendall coefficients may be based on the distribution of functions of r_s and t_k that have an asymptotic standard normal distribution.

Detailed discussions of the distribution of r_s under H_0: $\rho_s = 0$ are given by Gibbons and Chakraborti (2004, Section 11.3) and by Kendall and Gibbons

(1990, Section 4.14). Convergence to normality tends to be slow, so the asymptotic result should not be used for values of $n < 30$, and some writers suggest an even higher value. For large n it is usually assumed that $z = r_s\sqrt{n-1}$ has approximately a standard normal distribution under H_0. This approximation is identical in form to one often used for the Pearson coefficient. A better approximation that is often reasonable for values of n as low as 10 for both the Pearson and Spearman coefficients suggested in Kendall and Gibbons and elsewhere is

$$t_{n-2} = \frac{r\sqrt{n-2}}{\sqrt{1-r^2}}.$$

This has approximately a t-distribution with $n - 2$ degrees of freedom under H_0. This approximation is used in StatXact for asymptotic estimation of a P-value for both the Pearson and Spearman coefficients.

For Kendall's tau the exact distribution of t_k under H_0: $\tau_k = 0$ is difficult to obtain. A full discussion is given in Gibbons and Chakraborti (2004, Section 11.2) and by Kendall and Gibbons (1990, Section 4.8) who show under H_0 that $E(t_k) = 0$ and $Var(t_k) = 2(2n + 5)/[9n(n - 1)]$ and that for reasonably large n

$$z = \frac{3t_k\sqrt{n(n-1)}}{\sqrt{2(2n+5)}} \tag{10.8}$$

has approximately a standard normal distribution. This is the basis of the asymptotic P-value determination in StatXact and in many general statistical programs. Modifications are needed for ties and Gibbons and Chaktraborti discuss these, although a few ties may make little difference. Some writers suggest that the asymptotic result for the Kendall coefficient should not be used if $n < 15$ so it would not be recommended for the data in Example 10.7. However, in that example where $t_k = 0.3846$ we find $z = 1.83$ implying that for a one-tail test $P = 0.0336$, broadly in line with the Monte Carlo approximations to the exact P in Example 10.7 despite the small sample size.

A difficulty in obtaining confidence intervals for the population coefficients, ρ_s and τ_k using asymptotic results is that the limits may lie outside the closed interval $[-1, 1]$. Another is that the distributions of the sample statistics are not simple for non-zero ρ_s or τ_k. In particular, the variances are not the same as those under a zero value hypothesis. The problem is discussed further by Kendall and Gibbons (1990, Chapters 4 and 5) who give useful approximations.

10.1.7 A Comparison of the Spearman and Kendall Coefficients

Statisticians are often asked which of the coefficients — Kendall or Spearman — is to be preferred. There is no clear-cut answer. They seldom lead to markedly different conclusions, though Examples 10.1 and 10.4 show this is possible when one ranked pair is markedly out of line with the general trend. For instance, in Example 10.1 one pair of ranks, i.e., (12, 1), contributes

$(12 - 1)^2 = 121$ to a total $T = 162$. This suggests r_s may be sensitive to such an outlier.

An alternative way of calculating the coefficients clarifies the relationship between them. We assume there are no ties and that the pairs of ranks are arranged in ascending order of x-ranks as in Tables 10.1 and 10.2. Writing s_k for the rank of y_k we define $s_{ij} = s_j - s_i$, $i = 1, 2, 3, \dots, n - 1$ for all $j > i$. It is easily seen that the number of concordances in Kendall's t_k equals the number of positive s_{ij}, and the number of discordances equals the number of negative s_{ij}.

We do not prove it, but tedious algebra shows that if we denote by n_{cs} the sum of the values of the positive s_{ij} (the values of differences that are concordant) and by n_{ds} the sum of the values of the negative s_{ij} (the values of differences that are discordant), then in the no-tie case $n_{cs} + n_{ds} = n(n^2 - 1)/6$ and r_s is given by

$$r_s = \frac{n_{cs} - n_{ds}}{n(n^2 + 1)/6}. \tag{10.9}$$

This has a formal similarity to (10.3). We may write the s_{ij} as an upper triangular matrix

$s_2 - s_1$	$s_3 - s_1$	$s_4 - s_1$	.	.		.	$s_n - s_1$
	$s_3 - s_2$	$s_4 - s_2$	.	.		.	$s_n - s_2$
		.	.	.		.	.
			.	.	.		.
				$s_{n-1} - s_{n-2}$		$s_n - s_{n-2}$	
							$s_n - s_{n-1}$

For the data in Table 10.2 the matrix (see Exercise 10.3) is

1	3	2	4	5	7	10	9	8	6	−1
	2	1	3	4	6	9	8	7	5	−2
		−1	1	2	4	7	6	5	3	−4
			2	3	5	8	7	6	4	−3
				1	3	6	5	4	2	−5
					2	5	4	3	1	−6
						3	2	1	−1	−8
							−1	−2	−4	−11
								−1	−3	−10
									−2	−9
										−7

To compute t_k using this array we count the number of positive entries, $n_c = 47$, and the number of negative entries, $n_d = 19$. These were the values obtained in Example 10.3. To obtain n_{cs} we add all positive entries in the matrix. In Exercise 10.3 we ask you to verify that $n_{cs} = 1 + 3 + 2 + 4 + \dots + 3 + 2 + 1 = 205$. Similarly, adding all negative entries gives

$n_{ds} = 1 + 2 + 1 + 4 + \ldots + 10 + 2 + 9 + 7 = 81$. Substitution in (10.9) gives

$$r_s = 6 \times \frac{205 - 81}{12 \times 143} = 0.4336,$$

agreeing with the value in Example 10.1.

Examining the triangular array helps explain why, in this example, r_s plays down any correlation relative to that indicated by t_k. A few negative s_{ij} in the last column make relatively large contributions to the sum n_{ds}. This is attributable to the fact that the low rank value of the 12th and last observation, $s_{12} = 1$, is out of line with the high ranks assigned to its near neighbours.

Many other aspects of parametric correlation have nonparametric equivalents, including that of partial correlation. A brief introduction to that topic is given in Section 15.5.2.

10.1.8 Efficiency Power and Sample Size

Various results concerning efficiency, power and sample sizes needed to ensure a required power when specific alternative hypotheses hold for both Kendall and Spearman coefficients are scattered throughout the literature. One important result is that when sampling from the bivariate normal distribution both the Kendall and Spearman coefficients have the same Pitman efficiency of 0.912 relative to the Pearson coefficient. The Kendall coefficient tends to do rather better than the Pearson coefficient when sampling from long-tailed symmetric distributions and has Pitman efficiency of 1.266 relative to the Pearson coefficient for a bivariate double exponential distribution. Noether (1987a) gives the following asymptotic formula for the sample size needed to ensure power $1 - \beta$ in a one-tail test of H_0: $\tau_k = 0$ against H_1: $\tau_k = \tau_1$ where τ_1 has some fixed nonzero value

$$n \approx \frac{(z_\alpha + z_\beta)^2}{9\tau_1^2}.$$

Here z_α, z_β have the fairly obvious standard normal tail probability meanings assigned to them in (5.3). Some further results on power for the Kendall coefficient are given by Hollander and Wolfe (1999, pp. 375–376).

10.2 Ranked Data for Several Variables

If more than two observations are ranked for each of a set of n experimental units we often want to test for evidence of concordance between rankings of the units. We indicated in Section 7.3.1 that the Friedman test may be applied to rankings of objects (e.g., preferences for different varieties of raspberry, placings in a gymnastics contest by different judges, ranking of candidates by different examiners) to test whether there is evidence of consistency between those making the rankings.

Table 10.6 *Ranking of five objects by four judges.*

Judge	A	B	C	D
Object				
I	1	5	1	5
II	2	4	2	4
III	3	3	3	3
IV	4	2	4	2
V	5	1	5	1

Kendall, independently of Friedman, proposed the use of a function of the Friedman statistic which is often referred to as the *Kendall coefficient of concordance* and tabulated some small-sample critical values relevant to testing the hypothesis that the rankings were essentially random against the alternative of evidence of consistency. Kendall regarded his coefficient of concordance as an extension of the concept of correlation to more than two sets of rankings. In the bivariate case we may use the Kendall correlation coefficient to measure both agreement (positive association) and disagreement (negative association). Concordance, whether measured by Kendall's original statistic, or the Friedman modification, is one-sided in the sense that rejection of the null hypothesis indicates positive association. For example, if four judges A, B, C, D rank five objects in the order given in Table 10.6 we would not with this test reject the hypothesis of no association because the Friedman statistic (7.5) would take the value zero.

There is here complete agreement between judges A and C and between judges B and D: but the latter pair are completely at odds with judges A and C. The Kendall and the Friedman statistics do not detect such patterns.

An example indicates Kendall's approach in developing his concordance test.

Example 10.9

The problem. There are six contestants in a diving competition. Three judges each independently rank their performance in order of merit (1 for best, 6 for worst). The rankings allocated by each judge are given in Table 10.7. Is there evidence of consistency between judges?

Formulation and assumptions. Clearly the judges are not in complete agreement, but there is a reasonable consensus that competitor III is a good performer. There is fairly strong support for competitor I, while competitors IV and VI are thought to perform poorly. A low rank total (last column) indicates a performance considered by the judges overall to be better than that of a competitor who attains a high rank

Table 10.7 *Ranks awarded to six competitors by three judges in a diving contest.*

Judge	A	B	C	Rank total
Competitor				
I	2	2	4	8
II	4	3	3	10
III	1	1	2	4
IV	6	5	5	16
V	3	6	1	10
VI	5	4	6	15

total. Had the judges been inconsistent, effectively equivalent to allocating ranks at random, one would expect all rank sums to be nearer to the average value of $63/6 = 10.5$. Had there been complete agreement between judges it is easily verified that the six competitors rank sums would be some permutation of 3, 6, 9, 12, 15, 18.

Kendall's coefficient of concordance is based on the sum of squares of deviations of the competitors' rank sums from the mean or expectation under random allocation.

If we denote this sum of squares by S then it is intuitively reasonable, and can indeed be shown, that S has a maximum when the judges are in complete agreement, and a minimum of zero when all rank sums equal the mean of 10.5 (which would imply some tied ranks in the situation considered in Table 10.7). Arguing along these lines Kendall proposed the statistic

$$W = \frac{S}{\text{Max}(S)}.$$

He called W a *coefficient of concordance*. This takes the maximum value 1 when there is complete agreement, and the minimum value zero when there is no agreement. It will also take the value zero in a situation like that in Table 10.6 where there are contrary opinions held by pairs of judges. Tables of critical values are available, but any program that gives exact P-values for the Friedman test may be used also to obtain P for the Kendall statistic.

Procedure. From Table 10.7 we see that

$$S = (8 - 10.5)^2 + (10 - 10.5)^2 + \ldots + (15 - 10.5)^2 = 99.5.$$

If there are no ties it is easy to show that $\text{Max}(S) = m^2 n(n^2 - 1)/12$ if there are m judges and n competitors, which in this case gives $\text{Max}(S) = 157.5$, whence $W = 99.5/157.5 = 0.632$. StatXact has a program for this test. Any Friedman test program will give the same exact P-value, namely $P = 0.062$.

Conclusion. There is no strong evidence of concordance between the judges.

Comment. The weakness of the evidence indicates the fairly low power of the test against alternatives other than a very high level of agreement. The judges are particularly erratic in their assessment of competitor V.

Computational aspects. As indicated above P-values are obtainable from a program for the Friedman test, but this will not calculate the value of W directly.

We do not prove it, but it can be shown that the statistic T for the Friedman test given by (7.5) and W satisfy the relationship

$$W = \frac{T}{m(n-1)}. \tag{10.10}$$

Some modifications to the procedure in the above example are needed for ties. These are similar to those for ties in the Friedman test. If we compute T for the tied Friedman case, as described in Section 7.3.2 we may derive the correct W using (10.10).

An intuitively reasonable measure of concordance between rankings of competitors by several judges is one based on the mean of pairwise rank correlation coefficients. It can be shown that for m judges the mean, R_s, of all $m(m+1)/2$ such Spearman coefficients is such that $R_s = (mW - 1)/(m - 1)$. It follows that when $R_s = 1$ (i.e., when all pairwise rankings are in complete agreement) then $W = 1$. When $W = 0$, then $R_s = -1/(m - 1)$ which can be shown to be the least possible value for R_s. Examination of the overall pattern of pairwise rank correlation coefficients is useful for detecting patterns of the type illustrated for an extreme case in Table 10.6 where the pairwise Spearman coefficients are easily shown to take the value $+1$ twice and the value -1 four times, thus exhibiting a definite pattern where $W = 0$. This indicates that, like rank correlation in the bivariate situation, $W = 0$ when there is no association pattern, but the converse may not hold.

10.3 Agreement

All measures so far described in this chapter relate to the concept of association between variables. With positive association between two variables if one variable takes a large value then so will the other. This does not necessarily mean, however, that the scores on some scale for both variables are identical. If, in addition to positive association, scores on both variables tend to be the same then we say that there is a high level of agreement. Agreement is not the same as association. It is possible for a correlation coefficient to be very high, whilst at the same time, the agreement is very low. For instance, scores for one variable may be consistently higher than the score for the other. Agreement in this sense can be assessed for categorical variables.

This concept of agreement is particularly important in assessing observer variation. Suppose that two dentists examine the X-ray of a particular tooth for the presence or absence of dental caries on two separate occasions. Although the information available from the X-ray remains the same, there are two possible types of variation with such examinations. A dentist re-examining the X-ray of a tooth some time later may make a different judgment (e.g., caries absent rather than caries present). This is known as *intra-observer* variation. Also, a colleague may make a different decision about the same tooth, leading

Table 10.8 *A general contingency table for assessing agreement between two observers.*

Obs B		1	2	. . .	k	Total
	1	n_{11}	n_{12}	. . .	n_{1k}	n_{1+}
	2	n_{21}	n_{22}	. . .	n_{2k}	n_{2+}
	.		.	. . . .	.	.
Obs A	.		.	. . . .	.	.
	.	.	.	. . . .	.	.
	k	n_{k1}	n_{k2}	. . .	n_{kk}	n_{k+}
		n_{+1}	n_{+2}	. . .	n_{+k}	n

to *inter-observer* variation. For both types of variation, the same question can be asked, i.e., how well do the two sets of data agree with each other?

If there are relatively few possible categories (ordered or nominal), the most obvious measure of agreement between two sets of data is the proportion of cases in which agreement occurs. This has the clear drawback that a substantial amount of agreement can occur by chance alone. We need a *chance-corrected* measure of agreement. A simple and widely used measure is the *kappa statistic*, developed by Cohen (1960).

10.3.1 The Kappa Statistic for Agreement between Two Assessors

Two observers allocate each of n patients to one of k possible categories. We denote the number of patients allocated to category i by Observer A and to category j by Observer B by n_{ij}, the total number of patients allocated to category i by Observer A by n_{i+}, and the total number of patients allocated to category j by Observer B by n_{+j}. Table 10.8 shows the data in a $k \times k$ contingency table.

Dividing the number of observations, n_{ij}, for the cell in the ith row and jth column by the total number of patients, n, gives the proportion p_{ij} of patients for that cell. The observed proportion of agreement is the proportion of patients for which i and j are equal (i.e., on the diagonal of the table) which is given by

$$p_o = (n_{11} + n_{22} + \ldots + n_{kk})/n = \sum_{i=1}^{k} p_{ii}.$$

The expected proportion of agreement, p_e is determined using the appropriate row and column totals in a similar manner to that used for expected numbers in the familiar chi-squared test (Sections 12.1.1 and 12.2.2) so that

$$p_e = \sum_{i=1}^{k} p_{i+}p_{+i}$$

where $p_{i+} = n_{i+}/n$ and $p_{+i} = n_{+i}/n$.

The kappa statistic is defined as:

$$\kappa = \frac{\text{Observed proportion} - \text{Expected proportion}}{1 - \text{Expected proportion}},$$

or symbolically as

$$\kappa = \frac{p_o - p_e}{1 - p_e}.$$

The kappa coefficient takes values between -1 and 1, with the value 1 for perfect agreement, zero for the level of agreement that would be expected by chance, and negative values for less than chance agreement, i.e., apparent disagreement. Landis and Koch (1977) suggested benchmarks for kappa, e.g., a score over 0.8 indicates good agreement, 0.6 to 0.8 indicates substantial agreement and 0.4 to 0.6 moderate agreement. Assuming that the patients are assessed independently of each other, and that the assessors operate independently, we can test the null hypothesis that for the population $\kappa = 0$. An exact test for small samples is available in StatXact together with a Monte Carlo approximation for moderate sized samples and an asymptotic result for large samples. Fleiss, Lee and Landis (1979) give an asymptotic formula for the standard error of kappa assuming H_0 holds, namely

$$\text{se}(\kappa) = \frac{\sqrt{p_e + p_e^2 - \sum_{i=1}^{k} p_{i+}p_{+i}(p_{i+} + p_{+i})}}{(1 - p_e)\sqrt{n}}$$

The null hypothesis is tested by assuming that if it holds, then $Z = \kappa/[\text{se}(\kappa)]$ has a standard normal distribution.

Fleiss (1981) gives an asymptotic formula for a confidence interval for κ based on the estimated standard error of its maximum likelihood estimate.

Table 10.9 *Need of treatment for a series of patients as assessed by two dentists.*

Dentist B	*Treatment needed*	*Treatment not needed*	*Total*
Dentist A			
Treatment needed	40	5	45
Treatment not needed	25	30	55
Total	65	35	100

Example 10.10

The problem. Two dentists inspected 100 patients and classified them as either *requiring treatment* or *not requiring treatment*. Table 10.9 shows the decisions by the two dentists as to whether or not they thought treatment was required. Use the kappa statistic to calculate the chance-corrected agreement for this sample and test the null hypothesis that $\kappa = 0$.

Formulation and assumptions. It is assumed that patients are classified independently of each other and that the dentists do not confer. Cohen's kappa is calculated using the formulae for observed and expected proportions of agreement given above. The null hypothesis may be tested using an exact test if relevant software is available or the asymptotic formula for the standard error.

Procedure. We first display the data in terms of observed proportions:

Dentist B	*Treatment needed*	*Treatment not needed*	*Total*
Dentist A			
Treatment needed	0.40	0.05	0.45
Treatment not needed	0.25	0.30	0.55
Total	0.65	0.35	1.00

With only two assessors and two categories the formulae for the observed and expected proportions of agreement are straightforward. The observed proportion of agreement is $0.4 + 0.3 = 0.7$. However, some of this agreement could have been expected by chance, and this is calculated from the row and column totals. Under chance agreement, the expected proportion of cases in the *yes/yes* cell of the table is given by $0.65 \times 0.45 = 0.2925$. The expected proportion of cases in the *no/no* cell is given by $0.35 \times 0.55 = 0.1925$.

The total proportion of *expected* agreeing cases is therefore $0.2925 + 0.1925 = 0.485$. Hence

$$\kappa = (0.7 - 0.485)/(1 - 0.485) = 0.417.$$

The asymptotic formula for the standard error of kappa under the null hypothesis gives:

$$se(\kappa) = \frac{\sqrt{0.485 + 0.485^2 - 0.45 \times 0.65(0.45 + 0.65) - 0.55 \times 0.35(0.55 + 0.35)}}{(1 - 0.485)\sqrt{100}}$$

$$= 0.09215,$$

whence

$$Z = \kappa/se(\kappa) = 0.417/0.09215 \approx 4.53.$$

Reference to a standard normal distribution indicates $P < 0.0001$. StatXact confirms this, and indicates that the same is true for an exact P.

Conclusion. There is moderate chance-corrected agreement ($\kappa = 0.417$) between the two dentists. The very small P-value shows overwhelming evidence against the null hypothesis.

Comments. 1. Superficially, the observed agreement appears satisfactory, but once chance agreement is taken into account the level of agreement is less impressive.

2. The strong evidence against the null hypothesis should come as no surprise. One would expect dentists, who receive several years of training, to have considerable agreement with each other. However, the fact that kappa is only 0.417 should be a cause for concern.

3. An asymptotic 95 percent confidence interval for kappa is $(0.256, 0.578)$. This is rather wide, indicating that even a sample of 100 patients does not provide a great deal of information about the population value for kappa. Studies of observer agreement frequently involve complex and time-consuming methods for assessing patients or specimens. As a consequence, samples are generally small (typically no more than 50) and the benchmarks given by Landis and Koch (1977) are in practice therefore not always helpful.

Computational aspects. One can calculate kappa along with the asymptotic P-value for the null hypothesis $\kappa = 0$ using SPSS or Stata. SPSS additionally gives the standard error for the calculated kappa value, from which the 95 percent confidence interval for kappa can be obtained. Similar information, together with an exact P-value or a Monte Carlo estimate for large samples is also given by StatXact. The asymptotic standard error given by StatXact is slightly different from that given above, being based on the maximum likelihood estimate of κ rather than on the null hypothesis value.

10.3.2 Extensions of Cohen's Kappa Statistic

If there are more than two categories, some types of disagreement between the two assessors might be particularly serious. For instance, in assessing a patient's state of health, disagreeing between responses of "well" and "poor" are more serious than disagreeing between "well" and "moderately well". In the calculation of weighted kappa (Cohen, 1968), different types of disagreement are given different weights.

The kappa coefficient has been given in a conditional form by Light (1971). It has been extended to allow for more than two assessors by Conger (1980) and by Posner et al (1990). A straightforward method of calculating multiple kappa is to take the arithmetic mean of the kappa values obtained by taking pairs of observers in turn. However, the formulae for hypothesis testing and the calculation of confidence intervals are involved (Fleiss, 1981). Stata can be used to calculate kappa for several observers and to test the null hypothesis $\kappa = 0$.

Kappa has also been developed for ordinal data by Fleiss (1978) and continuous data by Rae (1988). In certain situations kappa is equivalent to the intraclass correlation coefficient (Fleiss and Cohen, 1973; Rae, 1988). Sample size calculation for the case of two assessors has been investigated by Cantor (1996).

Table 10.10 *Need of treatment for a series of patients as assessed by two doctors at Centre h.*

Doctor B	Treatment needed (1)	Treatment not needed (0)	Total
Doctor A			
Treatment needed (1)	a_{11h}	a_{12h}	$a_{1.h}$
Treatment not needed (0)	a_{21h}	a_{22h}	$a_{2.h}$
Total	$a_{.1h}$	$a_{.2h}$	$a_{..h}$

Table 10.11 *Frequency of rating combinations in the k centres.*

Category	Ratings	Centre 1	Centre 2	...	Centre k	Total
1	(1, 1)	n_{11}	n_{12}	...	n_{1k}	$n_{1.}$
2	(1, 0) or (0, 1)	n_{21}	n_{22}	...	n_{2k}	$n_{2.}$
3	(0, 0)	n_{31}	n_{32}	...	n_{3k}	$n_{3.}$
	Total	N_1	N_2	...	N_k	N

10.3.3 Testing the Equality of Cohen's Kappa Statistic in Several Populations

A multicentre study of health care usually involves the selection and assessment of a group of patients at several different locations. The measurement of inter-observer agreement may indicate considerable variation between centres (Crane et al., 2003). If the level of agreement is summarised by the kappa statistic it is appropriate to test the null hypothesis that the population kappa values are equal.

Donner et al. (1996) describe methods of dealing with this problem for the case of two categories. These include the large sample approach sketched out here, which is based on the chi-squared distribution.

Suppose that there are k centres and at each a pair of doctors assess the need of a particular treatment for a series of patients. The contingency table for the hth centre ($h = 1, 2, \ldots, k$) will take the form shown in Table 10.10.

Writing $n_{1h} = a_{11h}$, $n_{2h} = a_{12h} + a_{21h}$, $n_{3h} = a_{22h}$ and $a_{..h} = N_h$ the information from the k centres can be displayed as in Table 10.11.

We now define for the hth centre

$$\hat{\kappa}_h = 1 - \frac{n_{2h}}{2N_h\hat{\pi}_h(1 - \hat{\pi}_h)}$$

where

$$\hat{\pi}_h = \frac{2n_{1h} + n_{2h}}{2N_h}.$$

For the test of the null hypothesis $H_0: \kappa_1 = \kappa_2 = \ldots = \kappa_k$ one calculates for each centre $(h = 1, 2, \ldots, k)$

$$\hat{V}_h = \frac{1 - \hat{\kappa}_h}{N_h} \left[(1 - \hat{\kappa}_h)(1 - 2\hat{\kappa}_h) + \frac{\hat{\kappa}_h(2 - \hat{\kappa}_h)}{2\hat{\pi}_h(1 - \hat{\pi}_h)} \right]$$

to obtain

$$\hat{W}_h = \frac{1}{\hat{V}_h}.$$

The test statistic required is

$$\hat{\chi}_V^2 = \sum_{h=1}^{k} \hat{W}_h (\hat{\kappa}_h - \tilde{\kappa})^2$$

where

$$\tilde{\kappa} = \frac{\sum_{h=1}^{k} \hat{W}_h \hat{\kappa}_h}{\sum_{h=1}^{k} \hat{W}_h}.$$

This is referred to a chi-squared distribution with $k - 1$ degrees of freedom.

Example 10.11

The problem. A group of 601 young adults living in Limache, central Chile, were investigated regarding their experience of symptoms of asthma (Smeeton et al., 2006). Initially the participants completed an asthma questionnaire that included an item about whether they had experienced symptoms of wheeze in the previous 12 months (*yes* or *no*). Around 18 months later the participants answered the same series of questions, but on this occasion a trained interviewer gave a physical demonstration of how symptoms of wheeze (and other characteristics of asthma) would manifest themselves. For each individual, the response regarding wheeze recorded from the initial questionnaire was compared to that given following the demonstration. The level of agreement between the two sets of responses was assessed using the kappa statistic.

One issue of interest was whether the level of agreement was influenced by the individual's standard of living. Each participant indicated whether or not they owned the following household items: a gas fuelled water heating device, personal computer, refrigerator, washing machine and microwave oven. Individuals were categorised by how many of these items they possessed into *low* (0–1), *intermediate* (2–3) or *high* (4–5). The contingency tables for the question regarding wheeze experienced in the previous 12 months stratified by the number of items owned are shown in Table 10.12. The information for household items owned was incomplete for one adult, reducing the total number to 600.

Formulation and assumptions. It is assumed that individuals did not confer with each other and with an 18-month period between the two assessments any memory effect will be minimal. There are $k = 3$ levels for the standard of living and Cohen's

Table 10.12 *Questionnaire and demonstration responses regarding personal symptoms of wheeze in the last 12 months, stratified by number of specified items owned.*

0–1 Items

Demonstration	Symptoms (1)	No symptoms (0)	Total
Questionnaire			
Symptoms (1)	14	23	37
No symptoms (0)	5	54	59
Total	19	77	96

2–3 Items

Demonstration	Symptoms (1)	No symptoms (0)	Total
Questionnaire			
Symptoms (1)	45	104	149
No symptoms (0)	23	213	236
Total	68	317	385

4–5 Items

Demonstration	Symptoms (1)	No symptoms (0)	Total
Questionnaire			
Symptoms (1)	5	34	39
No symptoms (0)	9	71	80
Total	14	105	119

---

kappa is calculated by the method shown in Example 10.10. The asymptotic chi-squared test of H_0: $\kappa_1 = \kappa_2 = \kappa_3$ is then performed.

Procedure. From Table 10.12 we find $\kappa_1 = 0.323$, $\kappa_2 = 0.227$, $\kappa_3 = 0.019$, which indicates that there might be lower agreement for the group owning the most items. Applying the test of H_0: $\kappa_1 = \kappa_2 = \kappa_3$ Table 10.12 gives:

$$n_{11} = 14, n_{21} = 28, n_{31} = 54, N_1 = 96; n_{12} = 45, n_{22} = 127, n_{32} = 213, N_2 = 385;$$

$$n_{13} = 5, n_{23} = 43, n_{33} = 71, N_3 = 119.$$

Hence $\hat{\pi}_1 = 0.292, \hat{\pi}_2 = 0.282$, $\hat{\pi}_3 = 0.223$, from which, after lengthy calculation, it follows that $\hat{W}_1 = 90.370$, $\hat{W}_2 = 351.755$, $\hat{W}_3 = 130.032$, finally giving $\chi^2 = 7.191$ with 2 degrees of freedom. This indicates a *P*-value of 0.027.

Conclusion. There is moderate evidence against the null hypothesis. For the less prosperous groups some agreement may exist between responses from the two formats, but for the group having a high standard of living there is no agreement whatsoever.

Comments. 1. Donner et al. (1996) showed using Monte Carlo simulation that the large sample test applied here is appropriate if the size of each group exceeds 100. The smallest group in this example is very close to this threshold, so the approximate test should give reasonable results.

2. There is some evidence of an association between the prevalence of asthma and deprivation in the medical literature. It could be reasoned that individuals from a less prosperous background might know more people who suffer from asthma and so have a clearer understanding of the symptoms involved.

3. It would be difficult to generalise these findings to more economically developed countries where most of the population would have access to all these items.

4. It might be argued that the level of agreement between the two formats was influenced by treatment received during the intervening eighteen months. However, at the time of the study appropriate treatment for asthma was virtually unavailable in this area of Chile so once asthma was diagnosed (usually during childhood) symptoms would generally have continued.

Computational aspects. Methodological developments are relatively recent and there are few examples of the application of these techniques to practical problems. We are not aware of any programs in general statistical packages designed to test the equality of kappa values.

10.4 Fields of Application

Political Science

Leaders of political parties may be asked to rank issues such as the economy, health, education, transport and current affairs in order of importance. In comparing orderings given by leaders of two parties, rank correlations may be of interest. If leaders of more than two parties are involved the coefficient of concordance may be appropriate, or we may prefer to look at pairwise rank correlations, because leaders of parties at different ends of the political spectrum may tend to produce reversed, or partly reversed, rankings.

Psychology

A psychologist might show prints of 12 different paintings separately to a twin brother and sister and ask each to rank them in order of preference. We might use Blomqvist's, Spearman's or Kendall's coefficient to test for consistency in the rankings made by brother and sister.

Business Studies

Market research consultants list factors that may stimulate sales, e.g., consistent quality, reasonable guarantees, keen pricing, efficient after-sales service, clear operating instructions, etc. They ask a manufacturers' association and a consumers' association each to rank these characteristics in order of importance. A rank correlation coefficient will indicate the level of agreement between manufacturers' and consumers' views on relative importance.

Personnel Management

A personnel officer ranks 15 sales representatives on the basis of total sales by each during one year. His boss suggests he should also rank them by numbers of customer complaints received about each. A rank correlation could be used to see how well the rankings relate. A Pearson correlation coefficient might be appropriate if we had for each sales representative figures both for sales and precise numbers of customer complaints. Note that a positive correlation between numbers of sales and numbers of complaints need not imply that a high rate of complaints stimulates sales!

Horticulture

Leaf samples may be taken from each of 20 trees and magnesium and calcium content determined by chemical analysis. A Pearson coefficient might be used to see if levels of the two substances are related, but this coefficient can be distorted if one or two trees have levels of these chemicals very different from the others. Such influential observations are not uncommon in practice. A rank correlation coefficient may give a better picture of the correlation. If a third chemical, say cobalt, is also of interest, a coefficient of concordance might be appropriate.

Medicine

For many forms of cancer alternative forms of primary treatment are surgery, chemotherapy or radiotherapy. Which is preferred for a particular patient depends on factors like age, general health status, location and stage of development of the tumour, etc. Two consultants may or may not agree on which option is to be preferred for individual patients. If each of two consultants make independent assessments of the appropriate treatment for each of a group of N patients, then Cohen's kappa may be used as a measure of agreement.

Histopathology

A number of biopsy specimens are graded for cancer severity as mild, moderate or severe by a consultant histopathologist. A medical student working independently grades a different group of specimens. After six weeks they repeat the process on the specimens that they graded previously. The consistency of each individual's grading (degree of intra-observer variation) could be assessed using Cohen's kappa. To compare the level of consistency for the consultant and the medical student, it would be useful to test the equality of the true kappa values.

10.5 Summary

The **Pearson product moment correlation coefficient** is traditionally used as a measure of association for samples from continuous distributions. When normality is not assumed inferences may be based on a permutation test but the procedure is not robust.

The most widely used measures of rank correlation are the **Spearman rank correlation coefficient** (rho) (Section 10.1.3) and the **Kendall rank correlation coefficient** (tau) (Section 10.1.4). For the former (10.2) and for the latter (10.3) apply for the no-tie case. Modifications to both are needed with ties and asymptotic results for large samples are given in Section 10.1.6.

The **Blomqvist median coefficient** (Section 10.1.5) is usually estimated by (10.6) or preferably by (10.7) with appropriate modification for ties.

For multivariate ranked data **Kendall's coefficient of concordance** (Section 10.2) is equivalent to the Friedman test statistic for ranked data in randomized blocks given in Section 7.3.1.

The **Cohen kappa statistic** (Section 10.3.1) is a widely applicable useful measure of agreement between independent assessors on possible courses of action. Some caution is needed in interpreting the statistic. Rejection of the null hypothesis that any indication of agreement is due to chance does not necessarily imply that the level of agreement is strong. A test for equality of Cohen's kappa in several populations is discussed in Section 10.3.3.

10.6 Exercises

10.1 Compute the probabilities associated with each possible value of Kendall's t_k under the hypothesis of no association when $n = 4$ to verify the results quoted in Example 10.5.

10.2 Compute the probabilities analogous to those sought in Exercise 10.1 for Spearman's r_s statistic.

10.3 Verify the numerical values in the triangular array of s_{ij} given in Section 10.1.7 for the data in Table 10.2 and also the values of n_{cs} and n_{ds}.

*****10.4** Reanalyse the data in Table 10.4 for McDelta clan deaths using Spearman's rho. How do your conclusions compare with those based on Kendall's tau?

*****10.5** In Table 10.4 ages at death are ordered by year of death. Use the Cox–Stuart trend test (Section 4.4.3) to test for a time trend in life spans. Do the results agree with those based on the Kendall and on the Spearman coefficients? If not, why not?

10.6 A china manufacturer is investigating market response to seven designs of dinner set. The main markets are the British and American. To get some idea of preferences in the two markets a survey of 100 British and 100 American women is carried out and each woman is asked to rank the designs in order of preference from 1 for favourite to 7 for least acceptable. For each country the 100 rank scores for each design are totalled. The design with the lowest total is assigned rank 1, that with the next lowest total rank 2, and so on. Overall rankings for each country are:

Design	A	B	C	D	E	F	G
British rank	1	2	3	4	5	6	7
American rank	3	4	1	5	2	7	6

Calculate the Spearman and Kendall correlation coefficients. Is there evidence of a positive association between orders of preference?

*10.7 The manufacturer in Exercise 10.6 later decides to assess preferences in the Canadian and Australian markets by a similar method and the rankings obtained are:

Design	A	B	C	D	E	F	G
Canadian rank	5	3	2	4	1	6	7
Australian rank	3	1	4	2	7	6	5

Calculate the Spearman and Kendall correlation coefficients. Is there evidence of a positive association between orders of preference?

10.8 Perform an appropriate analysis of the ranked data for all four countries in Exercises 10.6 and 10.7 to assess the evidence for any overall concordance. Comment on the practical implications of your result.

*10.9 In a pharmacological experiment involving β-blocking agents, Sweeting (1982) recorded for a control group of dogs, cardiac oxygen consumption (MVO) and left ventricular pressure (LVP). Calculate the Kendall and Spearman correlation coefficients. Is there evidence of correlation?

Dog	A	B	C	D	E	F	G
MVO	78	92	116	90	106	78	89
LVP	33	33	45	30	38	24	44

10.10 Bardsley and Chambers (1984) gave numbers of beef cattle and sheep on 19 large farms in a region. Is there evidence of correlation?

Cattle	41	0	42	15	47	0	0	0	56	67
Sheep	4716	4605	4951	2745	6592	8934	9165	5917	2618	1105
Cattle	707	368	231	104	132	200	172	146	0	
Sheep	150	2005	3222	7150	8658	6304	1800	5270	1537	

*10.11 Paul (1979) discusses marks awarded by 85 different examiners to each of 10 scripts. The marks awarded by 6 of these examiners were:

Script Examiner	1	2	3	4	5	6	7	8	9	10
1	22	30	27	30	28	28	28	28	36	29
2	20	28	25	29	28	25	29	34	40	30
3	22	28	29	28	25	29	33	29	33	27
4	24	29	30	28	29	27	30	30	34	30
5	30	41	37	41	34	32	35	29	42	34
6	27	27	32	33	33	23	36	22	42	29

Use rank tests to determine (i) whether the examiners show reasonable agreement on ordering the scripts by merit and (ii) whether some examiners tend to give consistently higher or lower marks than others.

10.12 For the tied data in Example 10.6 confirm that correct values have been computed for all terms required to calculate t_b.

*10.13 For the data in Example 10.7 explore the use of the asymptotic approximations for the Spearman coefficient given in Section 10.1.6.

*10.14 Timber veneer panels are classified by experts as either de luxe (DL), standard (S) or reject (R) depending upon the number and type of faults detected. To check for consistency of classification between two observers each is asked to allocate the same set of 150 panels to the categories they consider relevant. Use Cohen's kappa to assess whether the following findings for a pair of observers indicate reasonable agreement between them. Comment critically on the interpretation of your results.

	Grade	DL	Obs A S	R	Total
	DL	7	2	1	10
Obs B	S	4	96	18	118
	R	1	10	11	22
	Total	12	108	30	150

10.15 For the data in Example 10.11 the information for agreement between the two methods of questioning with regard to wheeze in the last 12 months was tabulated by gender and is shown below. Calculate the two kappa values. Is there evidence that the population kappa values differ between males and females?

Males

Demonstration	*Symptoms* (1)	*No symptoms* (0)	*Total*
Questionnaire			
Symptoms (1)	27	58	85
No symptoms (0)	15	136	151
Total	42	194	236

Females

Demonstration	*Symptoms* (1)	*No symptoms* (0)	*Total*
Questionnaire			
Symptoms (1)	37	104	141
No symptoms (0)	22	202	224
Total	59	306	365

CHAPTER 11

BIVARIATE LINEAR REGRESSION

11.1 Fitting Straight Lines

Correlation is mainly concerned with qualitative aspects of possible relationships. For example, in looking at measurements of height and weight we might be interested in the possibility of positive association between the two variables, and if there is association whether in broad terms the relationship is linear, monotonic, linear in ranks, etc.

Regression is concerned with the quantitative aspects of relationships such as determining the slope and intercept of a straight line that in some sense provides a best fit to given data. Correlation and regression are closely related. For example, in straight-line regression a test of zero slope is equivalent to a test of zero correlation. Values of $+1$ or -1 for the Pearson product moment correlation coefficient tell us that all the observed points lie on a straight line. Other relationships between correlation and regression emerge in this chapter. In particular, in straight-line regression we shall be interested in correlations between what are usually referred to as residuals (departures of the observed y_i from their values predicted by the fitted regression equation) and the x_i. This concept, which we explain more fully below, is basic to both classic least squares regression, where it is based on the Pearson correlation coefficient, and to nonparametric regression, where it may be based on that coefficient, or on the Spearman or Kendall coefficient.

Least squares is the classic method of fitting a straight line to bivariate data. The method has optimal properties subject to well-known independence and homogeneity assumptions. If we add certain normality assumptions there are standard procedures for hypothesis testing and estimation. Regression provides useful tools for forecasting and prediction.

Given a set of bivariate observations (x_i, y_i) there are many cases where the assumptions needed to validate least squares procedures do not hold. This may result in misleading or invalid inferences. Least squares is relatively insensitive to some types of departures from basic assumptions, but it is strongly influenced by others. This has led to the development of a group of techniques known as regression diagnostics. We do not pursue that approach but both Atkinson (1985) and Cook and Weisberg (1982) provide comprehensive treatments. McKean, Sheather and Hettmansperger (1990) describe regression diagnostics for rank-based methods. An alternative to using classic least squares with regression diagnostics is to use distribution-free methods.

Our treatment follows closely the approach by Maritz (1995, Chapter 5), but we omit much of the theory given there.

11.1.1 Least Squares

The classic bivariate least squares regression model will be familiar to many readers. We briefly discuss aspects that help one understand the rationale behind many distribution-free approaches. In the classic parametric approach it is assumed that for each of a set of n given observed x_i — which may be either random variables or a set of fixed values that may or may not be chosen in advance — we observe some value y_i of a random variable, Y_i, which has the properties that its mean depends upon (i.e., is conditional upon) the value of x_i in such a way that

$$E(Y_i|x_i) = \alpha + \beta x_i \tag{11.1}$$

while the variance of Y_i is independent of x and for all x_i has the value

$$\text{Var}(Y_i) = \sigma^2,$$

where α, β, σ^2 are unknown. The straight-line relationship between $E(Y|x)$ and x of the form $E(Y|x) = \alpha + \beta x$ between the conditional mean of Y and any x defines the *regression* of Y on x. The line has slope β and intercept α on the y-axis. The notation $Y_i|x_i$ is the conventional notation for an event or variable Y_i having some property conditional upon a specified x_i.

For many inference purposes it is assumed also that the conditional distribution of $Y_i|x_i$ is $N(\alpha + \beta x_i, \sigma^2)$. A classic regression problem is to estimate α, β, and sometimes also σ^2, given a set of n paired observations $(x_1, y_1), (x_2, y_2), \ldots, (x_n, y_n)$ where, for a given x_i, the y_i is an observed value of the random variable Y_i featured in (11.1). These conditions hold if each (x_i, y_i) satisfies a relationship

$$y_i = \alpha + \beta x_i + \epsilon_i,$$

where each ϵ_i is an unobserved value of a $N(0, \sigma^2)$ random variable and the ϵ_i are independent of each other, and also of x_i. The portion $\alpha + \beta x_i$ is the systematic, or *deterministic*, part of the model for y_i and the ϵ_i is the *random* element.

Least squares obtains as estimates of α, β, the values a, b that minimize

$$S = \sum_{i=1}^{n} (y_i - a - bx_i)^2 \tag{11.2}$$

Denoting these estimates by $\hat{\alpha}, \hat{\beta}$ the straight line $y = \hat{\alpha} + \hat{\beta}x$ is called the *least squares regression* of y on x.

If, for any line $y = a + bx$ we denote the y-coordinate corresponding to $x = x_i$ by $\hat{y}_i$ (i.e., $\hat{y}_i = a + bx_i$), then the differences between the observed and predicted values, i.e., $e_i = y_i - \hat{y}_i$, $i = 1, 2, \ldots, n$, are called the *residuals* with respect to that line. Equation (11.2) may be written $S = \sum_{i=1}^{n} e_i^2$, The least squares estimators of α, β are so-called because they minimize the sum of squares of residuals. To obtain $\hat{\alpha}$, $\hat{\beta}$ we differentiate S separately with respect to a and b. These derivatives are set equal to zero to obtain the so-called *normal equations* that are solved for a, b.

The solutions are $\hat{\alpha} = \bar{y} - \hat{\beta}\bar{x}$ where $\bar{y} = (\sum_i y_i)/n$, $\bar{x} = (\sum_i x_i)/n$ and

$$\hat{\beta} = \frac{\sum_i(x_i - \bar{x})(y_i - \bar{y})}{\sum_i(x_i - \bar{x})^2} \tag{11.3}$$

With the assumptions above, classic least squares regression is a generalization of classic tests for comparison of treatment means. In particular, a test of H_0: $\beta = 0$ is equivalent to a test of equality of a set of sample means. This follows from (11.1) if we regard the x_i as indicators or labels attached to samples. If there are n paired observations (x_i, y_i) each y_i corresponding to a particular x_i may be looked upon as an observed value from the sample labelled by that x_i. The total number of samples, m, may be any number between 2 and n and the numbers of observations, n_j, in sample $j, j = 1, 2, \ldots, m$ are subject to the constraint $n_1 + n_2 + \ldots + n_m = n$. In particular, if no two x_i are equal there are n samples each of one observation and at the other extreme if there are only two distinct x_i there are two samples with n_1 and $n_2 = n - n_1$ observations respectively. In this last case we may label the first sample by $x = 0$ and the second sample by $x = 1$ and the first sample values are

$$y_{11}, y_{12}, \ldots y_{1n_1}$$

with mean m_0 and the second sample values are

$$y_{21}, y_{22}, \ldots y_{2n_2}$$

with mean m_1. It can then be shown using (11.3) that in this case $\hat{\beta} = m_1 - m_0$, the sample mean difference used in the familiar t-test for equality of two treatment means (Exercise 11.1). Thus the test of equality of means is in this case identical to testing H_0: $\beta = 0$.

In the more general case of m samples a test for $\beta = 0$ is a test for identity of all m population means.

It is well-known that for the classic least squares model the estimator of β does not depend upon α, but that of α depends upon β through $\hat{\beta}$ since $\hat{\alpha} = \bar{y} - \hat{\beta}\bar{x}$. From this expression for $\hat{\alpha}$ we see that the fitted equation may be written, without reference to the intercept, as

$$y = \bar{y} - \hat{\beta}(x - \bar{x}) \tag{11.4}$$

implying that the line passes through the point $(\bar{x}, \bar{y})$.

A graphical interpretation is that the slope of the fitted line is unaltered by a change of origin. In particular, if we shift the origin to the bivariate mean $(\bar{x}, \bar{y})$ and write $x' = x - \bar{x}$, $y' = y - \bar{y}$ equation (11.4) becomes

$$y' = \hat{\beta}x'$$

and (11.3) reduces to

$$\hat{\beta} = \frac{\sum_i x_i' y_i'}{\sum_i x_i'^2}.$$ (11.5)

The form (11.5) still holds if we only shift the origin to the mean of the x_i and use the original y_i.

In Section 10.1 we pointed out that the Pearson correlation coefficient was unaltered by linear transformations of (x, y) of the form $x' = (x - k)/s$ and $y' = (y - m)/t$ where s, t are both positive. We have just seen that in regression although the estimate of α is affected by a change in origin, that of β is not, i.e., transformations of the form $x' = (x - k)$, $y' = (y - m)$ do not affect the estimate of β. However, it is easy to establish using (11.3) that the transformation $x' = (x - k)/s$ and $y' = (y - m)/t$ alters both the true value of β and its least squares estimate by the same scale factor s/t, changing β to $\beta' = s\beta/t$. However, we do not make scale changes in this chapter.

We shall make use of another property of the correlation coefficient in permutation tests that we drew attention to in Section 10.1.1. This is, since all other quantities in (10.1) remain constant under permutation, we may base permutation tests on the statistic $T = \sum_i x_i y_i$ in the case of the Pearson coefficient as an alternative to using the sample correlation coefficient r. A corresponding property carries over to the Spearman coefficient with ranks replacing the x_i, y_i.

Dropping the normality assumption for the Y_i makes joint inferences about α and β difficult, so we consider first the estimation of β only. The problem then reduces to one of making inferences about differences in medians or means of samples labelled by the different x_i. For any independent Y_i, Y_j associated with distinct x_i, x_j we drop the normality assumptions on the conditional distributions of $Y|x$ and assume now only that for any x_i, x_j they have distributions $F_i(Y_i|x_i)$, $F_j(Y_j|x_j)$ that differ only in their centrality measure. We shall here take this to be the *median*. This will coincide with the mean for symmetric conditional distributions for which the mean exists.

For the straight line regression model the median of $F_i(Y_i|x_i)$ is $\mathrm{Med}(Y_i|x_i) = \alpha + \beta x_i$ and that of $F_j(Y_j|x_j)$ is $\mathrm{Med}(Y_j|x_j) = \alpha + \beta x_j$. These are the analogues of (11.1) and the difference between the medians is easily seen to be

$$\mathrm{Med}[(Y_j - Y_i)|x_j, x_i] = \beta(x_j - x_i).$$

Assuming that differences between distributions are confined to a median difference implies that for all i

$$D_i(\beta) = Y_i - \beta x_i$$

are identically and independently distributed with median α. Since the D_i are therefore independent of the x_i it follows that they are uncorrelated with the x_i. As we have pointed out, the x_i can be altered by addition or subtraction of a constant without affecting either the slope or its estimate. In particular, it will often simplify the algebra if we adjust the x_i by adding an appropriate constant to make $\sum_i x_i = 0$. In graphical terms this implies shifting the origin to a point on the x-axis corresponding to the mean $\bar{x}$.

Writing $d_i = y_i - \beta_0 x_i$, an intuitively reasonable test of the hypothesis $H_0\colon \beta = \beta_0$ against a one- or two-sided alternative is a test for zero correlation between the x_i and the d_i since we know that if H_0 holds then $D_i(\beta_0)$ is uncorrelated with the x_i and the d_i are then observed values of the variable $D_i(\beta_0)$. Whether the most appropriate test is based on the Pearson, Spearman, Kendall or some other coefficient will depend upon what assumptions are made about characteristics of the distribution $F(Y|x)$, (e.g., whether long-tailed, symmetric or asymmetric, etc.). We call the d_i *residuals* for convenience, but this is an unorthodox use of the term which more conventionally refers to the $e_i = y_i - \alpha_0 - \beta_0 x_i$ where α_0 is some hypothesized value of α. An appropriate statistic based on the Pearson coefficient for the test of zero correlation is

$$T(\beta_0) = \sum_i x_i d_i = \sum_i x_i(y_i - \beta_0 x_i). \tag{11.6}$$

To estimate β an intuitively reasonable estimation procedure takes the form of solving for b the equation

$$T(b) - \sum_i x_i \sum_i d_i = 0. \tag{11.7}$$

This follows from the form of c_{xy} given below (10.1), since equating c_{xy} to zero ensures a zero value for the sample correlation coefficient because it makes the numerator in (10.1) zero.

Recalling that we may adjust the x_i to have mean zero, i.e. so that $\sum_i x_i = 0$ without affecting our estimate of β, we assume this has been done so that (11.7) simplifies to

$$T(b) = \sum_i x_i(y_i - bx_i) = 0 \tag{11.8}$$

with solution $\hat{\beta} = (\sum_i x_i y_i)/(\sum_i x_i^2)$. It is easily verified that this is identical to the solution given by (11.3) with the constraint $\bar{x} = 0$.

We have made no assumptions about the distribution of the D_i except that they are identical for all i, so that for hypothesis testing a permutation test for a zero Pearson correlation coefficient based on the sample values (x_i, d_i) is appropriate. As we have indicated in other cases a difficulty with inferences based on raw data using a Pitman permutation test is that the results tend to be similar to those based on equivalent theory assuming normality even when the normality assumption is clearly violated. In other words the method lacks robustness.

This is illustrated for a small data set in Example 11.1. There are many situations where data like these arise, although usually in larger data sets where the complexity of computation may mask a clear understanding of what is happening.

Example 11.1

The problem. The water flow in cubic metres per second (y) at a fixed point in a mountain stream is recorded at hourly intervals (x) following a snow-thaw starting at time $x = 0$.

Hours from start (x)	0	1	2	3	4	5	6
Flow rate (y)		2.5	3.1	3.4	4.0	4.6	5.1	11.1

During previous thaws there has often been a near straight-line relationship between time and flow. Use the method of least squares to fit a straight line to the data and a permutation test based on the Pearson correlation coefficient to test the null hypothesis H_0: $\beta = 1$ (implying that on average flow increases by 1 cubic metre per second over a period of 1 hour) against the alternative H_1: $\beta \neq 1$.

Formulation and assumptions. We have seen that the estimate of β for classic least squares and permutation based least squares are identical and may be obtained using (11.3), or from (11.5) after replacing x by $x' = x - 3$, or from any least squares regression program. The requested hypothesis test is performed by computing the P-value associated with the Pearson coefficient between the x_i and $d_i = y_i - \beta x_i$ after setting $\beta = 1$.

Procedure. Direct substitution of the values of x and y in the appropriate formula gives $\hat{\beta} = 1.1070$. It is also easily verified that the classic least squares estimate of α is $\hat{\alpha} = 1.508$. To test the hypothesis H_0: $\beta = 1$ against a two-tail alternative we calculate for each data pair $d_i = y_i - x_i$ and compute the correlation coefficient between the x and d values. Ideally, for such a small sample an exact permutation test program should be used to compute the P-value appropriate to the hypothesis H_0: $\rho = 0$ implying that the x_i and the d_i are uncorrelated. Relevant values of the x_i and d_i are:

x_i	0	1	2	3	4	5	6
$d_i = y_i - x_i$	2.5	2.1	1.4	1.0	0.6	0.1	5.1

For these data StatXact gives $r = 0.1392$ for the sample Pearson coefficient, and the exact two-tail $P = 0.7925$.

Conclusion. The least squares line of best fit is $y = 1.508 + 1.107x$. Since H_0: $\beta = 1$ has an associated permutation test $P = 0.7925$ the data certainly do not provide evidence against H_0.

Comments. 1. Figure 11.1 shows the data points and fitted line. The fit is unsatisfactory largely because the point $(6, 11.1)$ seems to be an outlier relative to the other points. If we omit that point it is easily verified that the other points are well fitted by an amended least squares regression line which is now $y = 2.491 + 0.517x$.

If the point $(6, 11.1)$ is omitted the P-value associated with the test of H_0: $\beta = 1$ against H_1: $\beta \neq 1$ reduces to $P = 0.0028$, providing strong evidence (as one would expect from an inspection of Figure 11.1) against H_0. A larger sample, perhaps including readings between 5 and 6 hours, and after six hours, would give a clearer indication of the degree of association and the shape of the relationship after the first five hours.

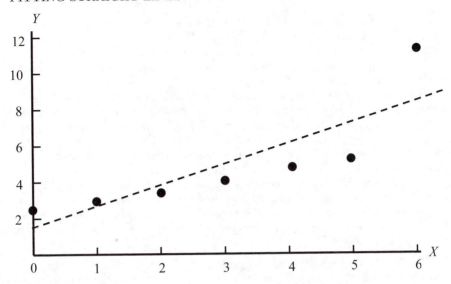

Figure 11.1 *Scatter diagram and classic least squares line of best fit to data in Example 11.1.*

2. Unfortunately, both the theory and practice of obtaining exact permutation-based confidence intervals for β present difficulties. Possible procedures are discussed by Maritz (1995, Section 5.2) and by Sprent (1998, Section 10.2). Using the asymptotic results given there a trial and error hypothesis testing approach may then be used to refine the asymptotic limits in programs such as that for the Pearson coefficient in StatXact. For these data (including the outlying pair (6, 11.1)) the 95 percent confidence interval for β is (0.495, 1.984) which is not very different from the interval (0.231, 1.984) given by classic parametric least squares. Both clearly demonstrate the tendency for the upper limit (as is the point estimate of β itself) to be pulled upward by the final data point.

3. The sudden departure at the 6-hour reading from the earlier near-linear relationship might result from a rapidly increasing rate of thaw after 5 hours, perhaps one caused by a temperature or wind change or other meteorological factors. If so, this may have important practical implications because readings like these are often used as a basis for warnings of likely flooding at points downstream from the recording station.

4. If there are outlying values, and there is no reasonable explanation for them, such values may have been incorrectly recorded. In the context considered here it is unlikely that the last observation is an error unless there had been some mishap with the meter used to measure flow. Another possibility might be that the recorded y values are the mean of two readings taken, say, 1 minute apart at each hour to average out any sudden surges, and that for the last reading somebody added the two but forgot to divide by 2 to get the mean. In practice the observers should know if such an error were possible. In other contexts, such as biological experiments, it is not uncommon for a recorded y to be the mean of two or more observations. A further possible explanation for the last reading is that a lead digit has been omitted

and that the final point should be (16, 11.1); this lies much closer to the straight-line relationship suggested by the remaining points. In the context of this example, this might occur if, for instance, readings were suspended overnight and the final reading was the first on the following morning. Again, in practice it should be possible to check if this were the source of an error.

5. Data like these are common in environmental studies, especially following incidents, natural or accidental, that have an impact over an extended period. For example, each y reading could be parts per million of a toxic gas in the atmosphere x hours after an explosion at a chemical plant, or the y could be some measure of radioactivity recorded at daily (or weekly) intervals after an accident at a nuclear power station.

6. The assumption that consecutive ϵ_i are independent might be queried in this example, but unless some physical condition intervenes, such as a block in the waterway restricting flow over an appreciable period, it is probably reasonable to assume that over intervals of 1 hour the random elements in flow are for practical purposes effectively independent.

Computational aspects. Even simple tests based on the Pearson coefficient require suitable computing facilities. Asymptotic tests may be unreliable if normality assumptions break down seriously. Bootstrapping methods (a topic discussed in Section 14.3) are often used in this context.

Better procedures for dealing with data like that in Example 11.1 are often based on ranks.

11.1.2 Regression Associated with Spearman's Rho

In practice the procedures to be described in Section 11.1.3 are easier to apply and are more widely used than those in this section. However, the method introduced here illustrates some key features of distribution-free regression and helps to clarify the relationship between procedures.

The obvious analogue to the Wilcoxon rank sum test procedure is to replace the d_i in (11.6) by $\mathrm{rank}(d_i)$. This leads to simplifications of the exact test and estimation procedures when the x_i are equally spaced, for inferences may then be based on the distribution of Spearman's rho, since, if the x_i are equally spaced they are linear functions of the ranks and hence the statistic

$$T_s(\beta_0) = \sum_i x_i \mathrm{rank}(d_i) = \sum_i x_i \mathrm{rank}(y_i - \beta_0 x_i) \qquad (11.9)$$

is appropriate for a test of zero Spearman correlation. It is convenient for estimation problems, though not essential, to assume also that any origin change needed to make the mean of the x_i zero has been made. The test is not conditional on either the observed x or y because for any hypothesized β_0 and fixed n only the paired set of ranks (or linear transformations of these ranks) from 1 to n are needed for the permutation reference set.

Example 11.2 shows that a program providing an exact test for Spearman's $\rho_s = 0$ may be used to test H_0: $\beta = \beta_0$ against a one- or two-sided alternative.

Point estimation of β is now less straightforward than it is for least squares, but it is still possible and confidence intervals based on the permutation test are relatively easy to obtain if appropriate software is available.

Estimation is based on an analogue of (11.8) derived from (11.9) and we seek b satisfying the equation

$$T_s(b) = \sum_i x_i \mathrm{rank}(y_i - bx_i) = 0. \tag{11.10}$$

A difficulty is that there is in general no value of b for which (11.10) holds exactly because $T_s(b)$ is a step function in b. This is because the ranking of the d_i only changes as b passes through a value where two or more d_i, d_j are equal, i.e., for b such that for some i, j, we have $y_i - bx_i = y_j - bx_j$. Denoting this value of b by b_{ij}, it follows that $b_{ij} = (y_j - y_i)/(x_j - x_i)$. Thus $T_s(b)$ does not change in value for any b lying between successive values of b_{ij} for $i = 1, 2, \ldots, n - 1, j > i$. It is easy to show [see, e.g., Sprent (1998, Section 10.3)] that $T_s(b)$ is a nonincreasing step function in b and that if for different (i, j) the corresponding b_{ij} are equal, this implies three or more observations are either collinear, or the lines joining the relevant pairs of points have identical slopes. The steps will then not all be of equal height.

Some implications of the properties of $T_s(b)$ for hypothesis testing and estimation, including confidence intervals, are best demonstrated by an example.

Example 11.2

The data in Example 11.1 with the x_i adjusted to have mean zero become

x	-3	-2	-1	0	1	2	3
y	2.5	3.1	3.4	4.0	4.6	5.1	11.1

It is convenient to record the pairwise slopes b_{ij} as a triangular array or upper triangular matrix similar to that used for the s_{ij} in Section 10.1.7. A typical entry is

$$b_{26} = \frac{5.1 - 3.1}{2 - (-2)} = 0.500$$

shown in bold in the array below. Statistical software to calculate the b_{ij} is becoming increasingly common. Minitab will produce pairwise slopes in the order of the columns in an upper triangular matrix. For our data these are:

0.600	0.450	0.500	0.525	0.520	1.433
	0.300	0.450	0.500	**0.500**	1.600
		0.600	0.600	0.567	1.925
			0.600	0.550	2.367
				0.500	3.250
					6.000

Repeated values among the b_{ij} imply collinearity or parallelism of some pairwise joins. Table 11.1 is easily, if tediously, formed using (11.9) although the complete table is not needed for most inferences. Since $T_s(b)$ remains constant between successive values of b_{ij} we only need to evaluate (11.9) for one value β_0 of b in each interval between successive values. When b coincides with some b_{ij} there are ties in the ranks of the d_i and using mid-ranks for these gives a value of the corresponding

Table 11.1 *Intervals for b-values corresponding to each possible $T_s(b)$ arising from the data in Example 11.2. The values of $T_s(b)$ when $b = b_{ij}$ are the means of the values in adjacent b intervals. Values of r_s corresponding to each $T_s(b)$ are also given.*

b-interval	mid-b value	$T_s(b)$	r_s
$-\infty$ to 0.300		28	1.000
0.300 + to 0.450−	0.375	27	0.964
0.450 + to 0.500−	0.475	23	0.821
0.500 + to 0.520−	0.510	12	0.429
0.520 + to 0.525−	0.5225	7	0.250
0.525 + to 0.550−	0.5375	3	0.107
0.550 + to 0.567−	0.5583	1	0.036
0.567 + to 0.600−	0.5835	−2	−0.071
0.600 + to 1.433−	1.0165	−7	−0.250
1.433 + to 1.600−	1.5165	−13	−0.464
1.600 + to 1.925−	1.7625	−18	−0.643
1.925 + to 2.367−	2.146	−22	−0.786
2.367 + to 3.250−	2.8085	−25	−0.893
3.250 + to 6.000−	4.625	−27	−0.964
6.000 + to ∞		−28	−1.000

$T_s(b)$ which is the mean of its values for b immediately greater than and immediately less than that b_{ij}. As an illustration of the computation of $T_s(b)$ for a given b consider that for the interval between $b_{16} = 0.520$ and $b_{15} = 0.525$. It suffices to set $b = 0.523$, say, and to evaluate

$$T_s(0.523) = \sum_i x_i \text{rank}(y_i - 0.523x_i).$$

For the given data we obtain

x_i	−3	−2	−1	0	1	2	3
y_i	2.5	3.1	3.4	4.0	4.6	5.1	11.1
$d_i = y_i - 0.523x_i$	4.07	4.15	3.92	4.0	4.08	4.05	9.53
rank(d_i)	4	6	1	2	5	3	7

whence

$$T_s(0.523) = (-3) \times 4 + (-2) \times 6 + (-1) \times 1 + 0 \times 2 + 1 \times 5 + 2 \times 3 + 3 \times 7 = 7.$$

The remaining values of $T_s(b)$ in column 3 of Table 11.1 are obtained in a similar manner. Since there is no unique b corresponding to each possible value of $T_s(b)$ it is sometimes convenient to associate with each such value the mid-b value. This is the midpoint of the interval of b values associated with that $T_s(b)$. For example, we have just shown that $T_s(b) = 7$ is associated with b in the interval (0.520, 0.525), so the associated mid-b value is 0.5225. These mid-b values are given in column 2 of Table 11.1.

The maximum value, 28, of $T_s(b)$ corresponds to a Spearman sample coefficient $r_s = 1$. Since T_s is a linear function of r_s if we divide any $T_s(b)$ by 28 we get a corresponding value of the equivalent r_s. The relevant values are given in column 4 of Table 11.1. We show below that this is useful for determining confidence intervals for β, but first we consider point estimation of β. Maritz proposed an estimate obtained by regarding the step function $T_s(b)$ as an approximation to a continuous function. He then used linear interpolation between the mid-b values for the closest $T_s(b)$ on either side of zero. In this example we interpolate between $T_s = 1$, $b = 0.5583$ and $T_s = -2$, $b = 0.5835$ to obtain a point estimate corresponding to $T_s = 0$. This is

$$b = 0.5583 + (0.5835 - 0.5583)/3 = 0.567.$$

It is fortuitous that this is identical with the median of the b_{ij} although an estimate obtained in this way will usually be close to that median. In practice, if only a point estimate of β is required, one need calculate only a few entries in Table 11.1 for values of b in intervals close to $\mathrm{med}(b_{ij})$, then use linear interpolation.

If an exact test program for the Spearman coefficient is available hypothesis tests about β are straightforward. For example, to test H_0: $\beta = 1$ against H_1: $\beta \neq 1$ we may use the values of x_i and d_i computed in Example 11.1 and insert these in a Spearman test program. Not surprisingly, (as can be verified from Table 11.1), this gives $r_s = -0.250$ and the corresponding two-tail $P = 0.5948$ so there is clearly no substantial evidence against H_0. If no program is available for an exact test, tables like Table A31 in Hollander and Wolfe (1999) clearly indicate that only values of r_s greater than about 0.7 in magnitude provide any acceptable evidence against H_0.

Given a facility like that in StatXact for generating the complete distribution of r_s under H_0: $\rho_s = 0$ it is relatively easy to obtain confidence intervals for β. Using this distribution, one finds that for $n = 7$ the one-tail P associated with $r_s \geq 0.786$ is $P = 0.024$. From Table 11.1 we note that because tied ranks occur for certain values of b there is no upper tail b corresponding exactly to $r_s = 0.786$, but in the lower tail when $r_s = -0.786$ the corresponding mid-b value is 2.146. The smallest value of b that will just produce an $r_s = -0.786$ is a value slightly above 1.925, say, 1.926. In the upper tail of r_s values we see that the largest value of b that will produce a value of r_s of at least 0.786 is b slightly less than 0.500, say, 0.499. Thus the b-interval (0.500, 1.925) is an appropriate $(1 - 2 \times 0.024)100 = 95.2$ percent confidence interval.

If no program giving exact P-values is available, the values of r_s corresponding to significance at many conventional levels are available from tables like Table A31 in Hollander and Wolfe which, for a sample of 7, gives $\Pr(r_s \geq 0.786) = 0.025$. Using this value leads to the same conclusions for an approximate 95 percent confidence interval.

The interval obtained here is slightly shorter than that given in *Comment* 2 on Example 11.1. More interestingly, the point estimate $\hat{\beta} = 0.567$ is much smaller than the full-data least squares estimate and reasonably close to the least squares estimate 0.517 obtained if the rogue point (6, 11.1) is omitted. This indicates that the suspect point is less influential on the point estimate of β although it still exerts an upward pressure on the confidence limit. Here the confidence limits, unlike those in classic parametric least squares estimation, are no longer symmetric about the point estimator, but they reflect more realistically the upward pressure of the point $(6, 11.1)$ on estimates of β.

Clearly if confidence intervals are required at or near conventional levels only a few values of $T_s(b)$ corresponding to large values of $|T_s(b)|$ need be calculated.

An asymptotic procedure for approximate confidence intervals usually works well for larger n. This is discussed briefly in Sprent (1998, Section 10.3). Since inference procedures discussed in this section are generalizations of the WMW test it is not surprising that they have the same Pitman efficiencies.

If the x_i are not equally spaced $T_s(b)$ given by (11.9) is no longer a linear function of Spearman's rho. Then the method given in the next section is generally preferable. Some writers have suggested that inferences about β might be based on a statistic that replaces x_i by rank(x_i) in (11.9), i.e.,

$$S(b) = \sum_i \text{rank}(x_i)\text{rank}(y_i - bx_i). \tag{11.11}$$

This is equivalent to using $T_s(b)$ for equally spaced x for hypothesis testing, but care is needed for estimation since the mean of the ranks of the x_i are no longer zero, but adjustments can be made for this. However, for unequally spaced points, transformation to ranks is not a linear transformation of the x_i. This precludes use of $S(b)$ for making reliable inferences about a true β for the original data.

11.1.3 Regression Associated with Kendall's Tau

In the previous section we saw that regression inference procedures based on Spearman's rho are complicated if the x_i are not equally spaced. Equal spacing is not needed for inferences based on Kendall's tau, since τ_k depends only upon the order of the observations, not on magnitudes of differences between data values. To simplify presentation when the x_i are all different, we assume that $x_1 < x_2 < \ldots, < x_n$. We consider briefly the case when some x_i are equal in Section 11.1.4.

The statistic used for inference about β is

$$T_t(b) = \sum_i \text{sgn}[d_{ij}(b)] \tag{11.12}$$

where

$$d_{ij}(b) = (y_j - a - bx_j) - (y_i - a - bx_i) = y_j - y_i - b(x_j - x_i),$$

and summation of the signs of the d_{ij} is over all $i = 1, 2, \ldots, n-1$ and $j > i$. Since the x_i are all different and in ascending order, it follows that $T_t(b)$ is the numerator in the expression (10.3) for Kendall's tau composed of the numbers of concordances minus the number of discordances in the data pairs (x_i, d_i) where $d_i = y_i - bx_i$.

Use of a statistic equivalent to $T_t(b)$ was first proposed by Theil (1950). The procedure is widely known as Theil's method. Sen (1968) highlighted the relationship to Kendall's tau, so we refer to it as the *Theil–Kendall method*.

$T_t(b)$ is a linear function of the sample estimator of Kendall's tau, and using arguments similar to those in Sections 11.1.1 and 11.1.2 an appropriate point estimator of β is obtained by setting $T_t(b) = 0$. Since the x_i are in ascending order it is easy to see that $T_t(b)$ is unaltered if we replace $d_{ij}(b)$ by $b_{ij} - b$ where, as in Section 11.1.2, $b_{ij} = (y_j - y_i)/(x_j - x_i)$. Then $T_t(b) = 0$ if we choose $b = \text{med}(b_{ij})$, for in that case the number of positive and the number of negative d_{ij} will be equal. As was the case for $T_s(b)$, it is easy to see that $T_t(b)$ only changes in value when b passes through a value of b_{ij}, and since the statistic is the numerator term in Kendall's tau it follows that if all b_{ij} are distinct as b increases from $-\infty$ to ∞, $T_t(b)$ is a step function decreasing by steps of 2 from $n(n-1)/2$ to $-n(n-1)/2$. Division of $T_t(b)$ by $n(n-1)/2$ yields Kendall's t_k which may be used as an equivalent statistic. If the b_{ij} are not all distinct some of the steps will be multiples of 2. In particular, if a b_{ij} occurs r times it induces a step of height $2r$. The value of $T_t(b)$ at any b_{ij} may be regarded as the mean of the values immediately above and below that b_{ij}.

Tests of hypotheses about β are straightforward using either Kendall's t_k, or the equivalent $T_t(b)$ as the test statistic. If we know the exact distribution of the sample Kendall coefficient when $\tau_k = 0$, a confidence interval for β may be obtained. The procedures are illustrated in Example 11.3.

Example 11.3

The problem. Given the data in Example 11.1, i.e.,

x	0	1	2	3	4	5	6
y		2.5	3.1	3.4	4.0	4.6	5.1	11.1

use the Theil–Kendall method (i) to test the hypothesis H_0: $\beta = 1$ against the alternative H_1: $\beta \neq 1$; (ii) to obtain a point estimate of β and a confidence interval giving at least 95 percent coverage.

Formulation and assumptions. To test the hypothesis in (i) we compute $T_t(1)$ and use an exact test program to assess evidence for or against H_0. To obtain a point estimate of β we determine the median of the b_{ij} defined above. For a confidence interval we then compute $T_t(b)$ and the corresponding t_k for all b. Using the exact distribution of t_k when $\tau_k = 0$, or appropriate tables if the exact distribution when $n = 7$ is not available, we compute the relevant confidence interval for β.

Procedure. The b_{ij} may conveniently be arranged in a triangular matrix with the first row elements $b_{12}\ b_{13}\ \ldots\ b_{1n}$, second row elements $b_{22}\ b_{23}\ \ldots\ b_{2n}$ and so on, to give

0.600	0.450	0.500	0.525	0.520	1.433
	0.300	0.450	0.500	0.500	1.600
		0.600	0.600	0.567	1.925
			0.600	0.550	2.367
				0.500	3.250
					6.000

which is identical to the array introduced in Example 11.2. Inspection shows that the point estimate of β, i.e., $\text{med}(b_{ij})$ is 0.567 because there are 10 greater, and 10 lesser,

Table 11.2 *Intervals for b-values corresponding to each possible $T_t(b)$ or the corresponding t_k for the data in Example 11.3. The values of $T_t(b)$ when $b = b_{ij}$ are the means of the values in adjacent b intervals.*

b-interval	$T_t(b)$	t_k
$-\infty$ to 0.300	21	1.000
0.300 + to 0.450−	19	0.905
0.450 + to 0.500−	15	0.714
0.500 + to 0.520−	7	0.333
0.520 + to 0.525−	5	0.238
0.525 + to 0.550−	3	0.143
0.550 + to 0.567−	1	0.048
0.567 + to 0.600−	−1	−0.048
0.600 + to 1.433−	−9	−0.428
1.433 + to 1.600−	−11	−0.524
1.600 + to 1.925−	−13	−0.619
1.925 + to 2.367−	−15	−0.714
2.367 + to 3.250−	−17	−0.810
3.250 + to 6.000−	−19	−0.905
6.000 + to ∞	−21	−1.000

b_{ij}. To test the hypothesis $\beta = 1$ we subtract $b = 1$ from each b_{ij} and find this implies 15 discordances (negative values) and 6 concordances whence $T_t(1) = 6 - 15 = -9$ and when $n = 7$ this implies $t_k = -9/21 = 0.4286$. Corresponding to this value of t_k StatXact gives an exact one-sided $P = 0.1194$. This is doubled for a two-tail test, so there is no convincing evidence against H_0.

Using the arguments outlined before this example and noting that there are respectively 2, 4 and 4 tied values of b_{ij} at 0.450, 0.500, 0.600 we can establish the values of T_t for all values of b. These are given in Table 11.2 together with corresponding values of t_k obtained by dividing each $T_t(b)$ by 21.

To obtain an approximate 95 percent confidence interval for β we must first determine a value of $|T_t|$ with a one tail P as close as possible to, but not exceeding, $P = 0.025$. The exact distribution when $n = 7$ given by StatXact indicates that $P = 0.015$ when $T_t = 15$ while $P = 0.035$ when $T_t = 13$. If the StatXact program is not available these values may be obtained from tables such as those in Kendall and Gibbons (1990, Appendix Table 1) or Hollander and Wolfe (1999, Table A30). The latter give the relevant probabilities for the statistic T_t. From Table 11.2 we deduce that the shortest $(1 - 2 \times 0.15)100 = 97$ percent interval based on $|T_t| = 15$ will be $(0.500-, 1.925+)$, or the open interval $(0.500, 1.925)$. See also *Comment 3* below.

Conclusion. The point estimate of β is 0.567. A nominal 95 (actual 97) percent confidence interval for β is $(0.500, 1.925)$.

Comments. 1. The point estimate of β is identical to that obtained in Example 11.2 using Spearman's r_s and the nominal 95 percent confidence interval is virtually the same despite minor differences in the exact coverage.

2. Comparing the methods used in this and the previous section we see that the Theil–Kendall procedure is easier to compute. More importantly, in practice (though not relevant to this specific example) the restriction to equally spaced x_i is no longer needed.

3. To find a confidence interval we do not need to form Table 11.2, since the observation $T_t = 15$ implies $n_c - n_d = 15$, and since $n_c + n_d = 21$ this implies $n_c = 18$ and $n_d = 3$. Thus we should reject values of β greater than or equal to the three largest b_{ij} or less than or equal to the three smallest. Inspection of the matrix of b_{ij} given under *Procedure* shows that the third largest and third smallest b_{ij} are respectively 2.367 and 0.450. Rejecting these three, implies the limits 0.500 and 1.925 given above.

Computational aspects. More extensive tables are available for Kendall's tau than is the case for most statistics relevant to nonparametric inference. This to some extent relieves the need for readily available software either for the Theil–Kendall method, or indeed for inferences about Kendall's tau.

The close relationship between Kendall's t_k and T_t makes it easy to carry over asymptotic approximations between them. Using results for τ_k it is easily shown (see, e.g., Maritz, 1995, Section 5.2.5) that $E(T_t) = 0$ and that $Var(T_t) = n(n-1)(2n+5)/18$. For large n the distribution of $Z = T_t/[Var(T_t)]$ is approximately standard normal. Asymptotic inferences can then be made in the usual way. Even for $n = 7$ the approximation is sometimes not seriously misleading (Exercise 11.4) but caution is advisable with so small a sample.

We showed in Section 10.1.4 that the Kendall procedure is a special case of the Jonckheere–Terpstra method. This in turn is a generalization of the Mann–Whitney formulation of the WMW procedure, whereas the method using Spearman's rho is a generalization of the Wilcoxon formulation of the WMW procedure. Although the generalizations are not exactly equivalent, it is not unreasonable when applying them to regression problems to expect them to lead to broadly similar conclusions as we saw in Examples 11.2 and 11.3.

11.1.4 Some Alternative Approaches

Having noted that the Theil–Kendall procedure is a generalization of the WMW method, it is interesting to explore the possibility of procedures that generalize those of a sign test. Theil proposed one which is usually called the *abbreviated Theil method*. It uses a small independent subset of the b_{ij} and is not recommended for sample sizes smaller than about 14. Even for larger n the full Theil–Kendall method is preferable if adequate computing facilities or tables are available.

For n even the only b_{ij} we use are $b_{i,i+n/2}$ where $i = 1, 2, \ldots, n/2$ while for n is odd we use only $b_{i,i+(n+1)/2}$ where $i = 1, 2, \ldots, (n-1)/2$. These estimators all involve different data pairs, and hence are independent, so test and estimation procedures based on the sign test may be used. The point estimator of β is the median of the reduced set of b_{ij} indicated above. Tests and confidence intervals are easily derived from the corresponding sign-test procedures.

Example 11.4

The problem. We give below the modal length (y cm) of samples of Greenland turbot of various ages (x years) based on data given by Kimura and Chikuni (1987). Fit a straight line using the abbreviated Theil method and obtain a 95 percent confidence interval for β.

Age (x)	4	5	6	7	8	9	10	11	12
Length y	40	45	51	55	60	67	68	65	71
Age (x)	13	14	15	16	17	18	19	20	
Length y	74	76	76	78	83	82	85	89	

Formulation and assumptions. For the 17 points we calculate the subset of pairwise slopes $b_{10,1}, b_{11,2}, \ldots, b_{17,8}$ and obtain their median. A confidence interval for β is obtained in a manner analogous to that for a median based on the sign test.

Procedure. Since the x_i are equally spaced in this example the denominators in each of the required b_{ij} are all equal to $x_{10} - x_1 = 13 - 4 = 9$ so we need only compute the relevant $y_j - y_i$, obtain their median, and divide that by 9, to obtain an estimate of β. Thus, $y_{10} - y_1 = 76 - 40 = 34$, $y_{11} - y_2 = 76 - 45 = 31$, etc. The complete set of differences is 34, 31, 25, 23, 23, 15, 17, 24. The median of these is 23.5, so $\hat{\beta} = 23.5/9 = 2.61$. Using the argument in Section 3.4.2 for a nominal 95 percent confidence interval for the above set of 8 differences between the y-values we use appropriate B(8, 0.5) distribution probabilities (Exercise 11.5) to find that an exact 93 percent confidence interval for the median of these differences is the interval (17, 31). Dividing the limits by 9 gives the corresponding interval for β, i.e., (1.89, 3.44). Similarly the interval (1.67, 3.78) is a 99.2 percent interval. With so small a sample the discontinuities in coverage of possible intervals are quite marked.

Conclusion. An appropriate estimate for β is $\hat{\beta} = 2.61$ and a 93 percent confidence interval for β is (1.89, 3.44).

Comment. Using the program in Minitab for generating all pairwise slopes for these data one can show that the full Theil–Kendall procedure gives an estimator $\hat{\beta} = \text{med}(b_{ij}) = 2.65$. We may also establish that a 95.8 percent confidence interval is (2.2, 3.1) following the procedure outlined in *Comment* 3 in Example 11.3. This follows because the one-tail $P = 0.021$ when $n = 17$ corresponds to $T_t = 50$, implying $n_c - n_d = 50$. We deduce that to establish the relevant confidence interval we reject the 43 smallest and 43 largest b_{ij}. Inspection of the paired values establishes the above interval. If one has a facility to generate all b_{ij} we recommend the full rather than the abbreviated Theil-Kendall method for samples of this size. A comparison of the lengths of the confidence intervals given by the two methods indicates an appreciable loss of efficiency when using the abbreviated method. This is not surprising, because the abbreviated method only uses 8 of the 136 items of information (the b_{ij}) used in the full method. It does, however, use an independent set, whereas there are correlations in the complete set.

Many variants of procedures considered in this chapter for slope estimation appear in the literature. Another sign-test analogue related to the Spearman coefficient rather than the Kendall coefficient is briefly discussed in Sprent (1998, Section 10.5). As alternatives to procedures involving correlation between x_i and rank (d_i) such as that discussed in Section 11.1.2, Adichie (1967) considered possible transformations of the rank(d_i) such as that to van der Waerden or to normal scores. So far as we know these have not been widely used in practice and one might expect their performance to show little improvement on the direct use of ranks.

It will be clear that rank-based methods, especially those such as the Theil–Kendall method depend heavily on the b_{ij}. The use of median estimators based on these pairwise slope estimators introduces robustness to point estimators in the presence of rogue observations, or observations that might indicate inadequacy of the model. In Example 11.3, for instance, the observation $(6, 11.1)$ which is so obviously out of line with the other observations, influences only the b_{ij} that involve that point, i.e., those in the last column of the triangular matrix of b_{ij}, and these have little influence on med(b_{ij}) as a point estimator of β. They do, however, exert an influence on the confidence limits and raise the upper limit above what it would be with no such out-of-line observation.

If there are no suspect observations one might feel that more weight should be given to those b_{ij} associated with larger values of $x_j - x_i$. This is what classic least squares does, for it can be shown that that estimator is a weighted mean of the b_{ij} with weights proportional to $(x_j - x_i)^2$. Jaeckel (1972) recommended taking as a point estimator of β the median of the weighted b_{ij} with weights $w_{ij} = (x_j - x_i)/[\sum_{i<j}(x_j - x_i)]$. The procedure has some optimum properties when there are no outliers, but these and other weighting schemes that have been proposed may be less satisfactory than, for instance, the Theil–Kendall method if there are rogue observations. In simulation studies Hussain and Sprent (1983) found that the Theil–Kendall method performed almost as well as least squares when relevant assumptions held for the latter and that it showed a marked improvement in efficiency for long-tail error distributions, whereas in the latter situation weighted medians performed no better than, or sometimes less well than, Theil–Kendall.

We indicated in Section 11.1.3 that a problem arises with the Theil–Kendall method when there are tied values of x_i. The b_{ij} corresponding to such a pair becomes infinite. For only a small proportion of ties there will be little loss of efficiency if such points are replaced by one data entry with the y value set equal to the mean of the y values for all the points with that tied x value. This is usually preferable to artificial tie-breaking devices such as splitting the ties by making arbitrary small changes to tied x-values to separate them. That process may lead to large or even bizarre b_{ij} associated with such splits because of the small denominators. An alternative with greater appeal is to exclude all comparisons between points with a common x_i value, but to consider all joins of each of these points to points with other values x_j, i.e., where $x_j \neq x_i$. An extreme example is when there are only two distinct x values x_1 and x_2, and repeated y at each. The problem is then equivalent to the two-sample problem for means or medians of the y. If we consider the slopes of all pairwise joins between y values associated with the two x it is easy to see that these are similar to the differences computed for the Hodges–Lehmann estimator in Section 6.1.4. Indeed, they are simply these differences divided by the constant $x_2 - x_1$, so the procedure is exactly equivalent to the WMW test.

11.1.5 Joint Estimation of Slope and Intercept

Estimating slope is often the main aim of a regression analysis but when we want to use the fitted line for forecasting or prediction we must also estimate α. We are then able to predict using the estimate of $E(Y|x) = \alpha + \beta x$ for some specified x. The intercept corresponds to the choice $x = 0$. Whereas estimation of β is unchanged if we replace x by $x' = x - h$ where h is any constant, this is not the case for α.

Even in classic parametric least squares estimates of α and β are in general correlated. Only in the special case when the mean of the x_i is zero and the estimator of α reduces to $\hat{\alpha} = \overline{y}$ is this estimate uncorrelated with that for β and then the customary normality assumption implies independence, so in that case we can make inferences about α without having to worry about what is now a *nuisance* parameter β. Unfortunately, when we move to distribution-free methods this simplification does not hold because even when the mean of the x_i is zero, the statistics used to estimate α and β are correlated.

Many of the implications are discussed in detail by Maritz (1995, Section 5.3) and more informally by Sprent (1988, Section 10.6). A broad class of statistics relevant to joint estimation of α and β were discussed by Adichie (1967). These lead to inferences which in practice call for iterative methods of estimation starting with an estimate of β that is used to obtain an estimate of α. This estimate of α is used to get a revised estimate of β, and so on. The process usually converges after a few cycles. Some of the statistics used have links with, and similarities to, those used for estimation of slopes in Sections 11.1.2 and 11.1.3.

We give two variants of a method used in practice that often gives satisfactory estimates of α providing we start with what we accept as a good estimate of β. For example, for the data in Example 11.1 the estimate $\hat{\beta} = 0.567$ obtained by the methods used in Sections 11.1.2 or 11.1.3 is a reasonable slope estimate for a line through all points other than $(6, 11.1)$ if that is what is of interest.

The assumptions in Section 11.1.2 imply that all $D_i = y_i - \beta x_i$ (as defined in Section 811.1.1) have identical distributions with median α. We do not know β, but if $\hat{\beta}$ is a reasonably good estimate of β it is not unreasonable to assume that $d_i = y_i - \hat{\beta}x_i$ will have a distribution with a median close to α. An obvious procedure is to compute all n values of d_i and to take the median of these d_i as an estimate of α. This is equivalent to the optimum estimator based on the sign test. If one makes the further assumption that the D_i are symmetrically distributed about α the Hodges–Lehmann estimator based on the Walsh averages of the d_i is arguably more appropriate. We stress however that these results are only approximate and will obviously be influenced by the choice of $\hat{\beta}$.

Example 11.5

The problem. Given the data in Example 11.1, i.e.,

Hours from start (x)	0	1	2	3	4	5	6
Flow rate (y)		2.5	3.1	3.4	4.0	4.6	5.1	11.1

and assuming a reasonable estimate of β is $\hat{\beta} = 0.567$ obtain an estimate of the intercept α.

Formulation and assumptions. We compute all $d_i = y_i - 0.567x_i$ and take as our estimate either (i) the median of the d_i or (ii) the Hodges–Lehmann estimator, which is the median of the Walsh averages.

Procedure. The d_i are respectively 2.500, 2.533, 2.266, 2.299, 2.332, 2.265, 7.698. The median of these 7 values is 2.332, there being three smaller and three larger values. The Hodges–Lehmann estimator is easily obtained from any program that computes the Walsh averages, or a program such as the dedicated Hodges–Lehmann estimation program in StatXact. For these data the estimate (the median of the Walsh averages) is 2.399.

Conclusion. Two possible estimates of α are 2.332 or 2.399.

Comments. 1. One may argue that the median of the d_i should be the preferred estimator because the extreme value $d_7 = 7.698$ suggests that the symmetry assumption needed to justify Hodges–Lehmann estimation may not hold. The line of best fit using the Theil–Kendall estimator of slope and the median of the d_i as intercept estimator is shown in Figure 11.2.

2. It is tempting to apply either the sign-test method or Hodges–Lehmann procedure to obtain a confidence interval for α. However, this may not be appropriate because we use only an estimate of β since we do not know its true value. It is easily shown that alternative choice of the estimate of β may profoundly influence both the pattern of the d_i as well as the estimate based upon them. For instance, in Example 11.1 we established that if the last point were ignored the least squares estimate of β would be $\hat{\beta} = 0.517$. For this the values of the d_i are 2.500, 2.583, 2.366, 2.449, 2.532, 2.515, 7.998 with median 2.515 reflecting a general upward shift in the d_i. The pattern changes even more markedly if we assume $\hat{\beta} = 0.5$ when the d_i, apart from the last, are even more concentrated, being 2.5, 2.6, 2.4, 2.5, 2.6, 2.6 and 8.1. Indeed, to obtain reasonable approximate confidence intervals for α one must use a more sophisticated approach. As indicated above, some that are used require an iterative approach for the joint estimation of α and β. One possible approach is given by Sprent (1998, Examples 10.7 and 10.8) where it is shown that in this example an

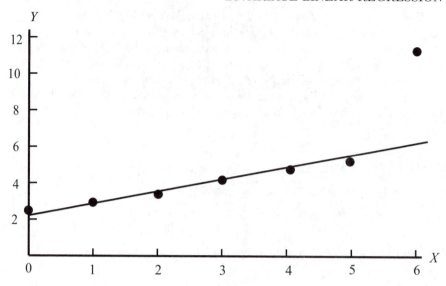

Figure 11.2 *The Theil–Kendall regression line fitted to the data in Example 11.5.*

approximate interval based on the estimate $\beta = 0.567$ may not be unreasonable as iteration does not improve the joint estimates of slope and intercept.

Some alternative proposals for estimation of α and the more general problem of estimating med$(Y|x)$ are discussed and by Maritz (1995, Section 5.3) and by Sprent (1998, Section 10.6).

11.1.6 Comparison of Slopes for Several Lines

Suppose we have k mutually independent sets of paired observations of variables (x, y) where the ith set contains n_i observations $(i = 1, 2, \ldots, k)$ and $k \geq 2$. Assume also that the jth observation $(j = 1, 2, \ldots, n_i)$ in the ith set [i.e., (x_{ij}, y_{ij})] satisfies the relationship

$$Y_{ij} = \alpha_i + \beta_i x_{ij} + \epsilon_{ij}$$

where the ϵ_{ij} are realized values of a $N(0, \sigma^2)$ random variable, and the α_i, β_i and σ^2 are all unknown. This model specifies k regression lines all with the same error distribution. The model has been widely studied in parametric analysis, in particular in the context of testing the hypothesis that all the β_i are equal against an alternative of some inequality. Equality of the β_i implies all k lines are parallel. An extended hypothesis of equality of all α_i in addition to equality of all β_i implies the k regression lines are identical.

If we drop the normality assumptions on the distribution of the ϵ_{ij} and assume only that they all have median zero, and are otherwise identically distributed appropriate methods of estimation of β within each of the k data sets include those given in Sections 11.1.2 to 11.1.4. The importance of the pairwise slopes denoted in the methods in those sections by b_{ij} (not to be confused with the i, j subscripts as in x_{ij} in this section) here suggests that a comparison of these for the different data sets might produce a useful base for tests for parallelism. Unfortunately this intuitively reasonable approach is complicated by the lack of independence between pairwise slopes within any one set. This complication does not arise, however, in the case of the abbreviated Theil method. If this is applied to each data set the pairwise slopes involved are independent within that set, and the independence of observations in different sets implies they are independent of the pairwise slopes obtained by the same method in the other sets. If, in addition, the x are equally spaced, all pairwise slopes have the same distribution within a particular set. If the slopes of all sets are the same, they will have identical distributions in all sets. If the β_i are not all equal (i.e., if the lines are not all parallel) the distributions of the pairwise slopes for each set will differ only in their medians. Thus, in this very restricted case, an appropriate test for parallelism against the alternative of at least one different slope is the Kruskal–Wallis test (or the WMW test when $k = 2$).

We do not know the Pitman efficiency of the above procedure, but it may not be high because the abbreviated Theil procedure does not make full use of all relevant information in the data.

More sophisticated tests for parallelism are discussed by Sen (1969). A very general test that is asymptotically distribution-free under broad assumptions developing some of Sen's ideas in combination with proposals due to Adichie (1984) is described with an example by Hollander and Wolfe (1999, Section 9.5).

11.1.7 Efficiency Power and Sample Size

It is well-known in classic least squares straight line regression that the standard error of the estimator $\hat{\beta}$ of β is highly dependent upon the positioning of the x_i. If one has a choice of the x_i before data are collected one aim should be to select these to make this standard error as small as possible. If this is done the confidence interval for β at any chosen level will be the shortest available. It is also well known that for least squares the shortest confidence interval is obtained in the case when n, the sample size, is even by setting half the x_i at the least possible value, and the other half at the greatest possible value.

In practice, this ideal design for an experiment to estimate β may be impossible because of experimental constraints. Also, other factors may have to be taken into consideration. One of these could be the need to test whether a straight line is adequate to describe the systematic component of the model. To test this one or more x values intermediate between the two extreme values must be included.

In many practical situations there are major restrictions on the available x values. For example, if the study is designed to determine whether a straight

line provides a reasonable relationship between weight and age of chickens
from shortly after hatching to maturity one may have to select a sample from
available chickens. There may be doubt about whether a straight-line relationship is adequate, so one would be unwise to choose only day old chickens
and fully mature birds. However, it would be helpful if one could reduce the
standard error of the estimate of β by choosing, if possible, relatively large
numbers of young, and of mature, birds and just a few intermediate in size.
There may well be practical limitations on the extent to which this can be
done. For example, there may be only a few mature birds available. For valid
inferences if more than one bird is available at a given x and only some of
these are included, those included should be chosen at random.

The often quite profound influence of design (in the sense of choosing x)
on the precision of estimates for β carries over to nonparametric situations.
Suppose, for example, that the data in Example 11.1 had been generated by
adding errors to some Y generated by the systematic relationship $Y = x + 2$.
Thus, for instance, for the first point (0, 2.5) an 'error' 0.5 has been added
to the systematic $Y = 0 + 2 = 2$. It is easily shown for the remaining points
(see Exercise 11.10) that the errors are $0.1, -0.6, -1, -1.4, -1.9, 3.1$. Suppose
that we retain $Y = x + 2$ as the systematic part of our model, but replace
the values $x = 4, 5, 6$ in Example 11.1 by values $x = 99, 100, 101$ and impose
upon the true Y given by $Y = x + 2$ the same errors as those just given.
For example we replace the point (5, 5.1) by the point (100, 100.1) since,
when $x = 100$, $Y = 100 + 2 = 102$ and the associated error is -1.9, whence
$y = 102 - 1.9 = 100.1$. Proceeding this way our new data set is

x	0	1	2	3	99	100	101
y	2.5	3.1	3.4	4.0	99.6	100.1	101.1

In Exercise 11.10 we ask you to verify that the point estimate of β given
by the Theil–Kendall method is now 0.985 and that a nominal 95 percent
confidence interval for β given by that method is (0.50, 1.037). Although the
lower limit is the same as that obtained in Example 11.3, the upper limit is
appreciably lower, and the point estimator is now near the top end of the
interval. The reduction in width of the interval implies an increase in power
attributable to the design change of altering dramatically three of the x-values.
We strongly urge readers to spend a few minutes considering the implications
of these findings. In particular, one should consider whether the model with
systematic part $Y = x + 2$ and the given errors seems a reasonable model in
relation to the amended data values, and why the point estimate of β changes
so dramatically between that in Example 11.3 and that obtained here.

We drew attention in earlier sections to a relationship between some linear
regression models and the procedures used to analyse them and methods such
as WMW and the sign test. Where relevant, results for Pitman efficiency
under various error structures carry over to regression. However, power studies
become very complicated not only because, as in the simpler situations, they
are often highly dependent upon what distributional assumptions are made,

but as we have just indicated, the power is strongly influenced by design factors in choosing x, as well upon the size of the sample. Further, changes in sample size may well introduce new x-values that may or may not in themselves enhance the power. In view of the complexity we do not pursue the matter further here. Other useful comments on efficiency are given by Hollander and Wolfe (1999, Section 9.8).

11.2 Fields of Application

Before fitting any bivariate regression curve it is wise to plot the points. This simple application of EDA will indicate whether a straight line is adequate for fitting or prediction purposes and whether we are likely to need a robust method of fitting due to presence of outliers, or whether a more sophisticated curve (perhaps a polynomial or a hyperbola) might be more appropriate.

Medicine

One of the variables used in assessing possible childhood asthma is the forced expiratory volume in one second (FEV1). The child concerned is asked to breathe in deeply and then breathe out as quickly as possible. If the volume of air released in the first second is relatively small the child may have respiratory problems. FEV1 is related to height in that taller children would be expected to breathe out air more quickly; the FEV1 value recorded is therefore corrected for the child's height. A regression of FEV1 against height for a group of children with normal breathing could be used to give reference values against which a child with suspected asthma could be checked. As FEV1 increases with age as well this should also be taken into account.

Agriculture

A farmer may be interested in the relationship between the yield of milking cows (litres per day) and their sale value at a cattle market auction. If records of milk production are available for each cow, information from previous sales could be used in a regression of sale value against milk production to predict likely sale prices in the future.

Store Planning

A supermarket chain has opened several dozen convenience stores within petrol stations situated by main roads. Sites for future stores are being considered. If hourly traffic counts and annual turnover are available for each current store, and traffic counts for potential future sites, a regression of annual turnover against traffic count might assist in choosing sites that would return a good yield on future investment.

Car Engine Efficiency

Cars from a range of makes could be tested for fuel consumption at a specific speed against engine size. Although fuel consumption would be expected to rise with engine size, cars with an actual consumption rate well below that predicted by a regression model would be identified as more economical or *environmentally friendly* vehicles.

11.3 Summary

Least squares regression may be appropriate providing errors are independent and are symmetrically distributed all with the same variance, even without a normality assumption. However, although inferences may be based on the relevant permutation distribution there is a lack of robustness against certain departures from homogeneity of error distributions.

For equally spaced x_i, rank based methods computing **Spearman's rho** (Section 11.1.2) between the y_i and $d_i = y_i - bx_i$ may be more robust than classic least squares.

The Theil–Kendall method (Section 11.1.3) often gives similar results to methods based on Spearman's rho. It is easier to apply, and is not restricted to equally spaced x_i. The point estimator is the median of the slopes of all pairwise joins of data points with different x coordinates.

An abbreviated Theil procedure (Section 11.1.4) uses only a small subset of slopes that join independent pairs of points. Confidence intervals based on the Theil–Kendall method depend on the distribution of Kendall's tau for the complete procedure, and upon the sign test for the abbreviated procedure.

Estimates of the **intercept** may be strongly influenced by the choice of a reliable estimate of the slope and iterative procedures are often appropriate.

Power of any estimation procedures for slope is strongly influenced by design factors such as the spacing of the x-values.

11.4 Exercises

11.1 Show that if a sample of n_1 observations has mean m_1 and another sample of $n_2 = n - n_1$ has mean m_2 and the first sample values are indexed by $x = 0$, and the second sample values by $x = 1$, then $\hat{\beta}$ given by (11.3) equals $m_2 - m_1$.

11.2 For the data in Example 11.1 verify that the sum of residuals e_i and the sum of the products of residuals with the corresponding x_i is zero. Explain why this is an inevitable consequence of the least squares estimation procedure.

*11.3 The numbers of rotten oranges (y) in 10 randomly selected boxes from a large consignment are counted after they have been kept in storage for a stated number of days (x). Use the Theil–Kendall method to compute the slope of a straight line fitted to these data and obtain an appropriate estimate of the intercept.

x	3	5	8	11	15	18	20	25	27	30
y	2	4	7	10	17	23	29	45	59	73

Plot these data and the fitted line. Does the fit seem reasonable?

11.4 Using the asymptotic result relevant to the Theil–Kendall method given in Section 11.1.3 obtain an approximate 95 percent confidence interval for the data used in Example 11.3. Compare this with the exact permutation-based interval obtained in that example.

*11.5 Verify the confidence limits quoted in Example 11.4 using the abbreviated Theil method.

11.6 Comment critically on and explain the meaning of the following statement: Given n observations (x_i, y_i) where the x_i are equally spaced, the Cox–Stuart test for trend (Section 4.4.3) applied to the y_i is essentially equivalent to testing whether the abbreviated Theil estimator is or is not consistent with zero slope.

*11.7 Mattingley (1987) gives the following data based on the US census of agriculture which show at approximately 10-year intervals from 1920 to 1980 the percentages of U.S. farms with tractors and farms with horses. Explain why it would be pointless, or wrong, to fit a linear regression for tractor percentage on horse percentage to these data.

Percent tractors	9.2	30.9	51.8	72.7	89.9	88.7	90.2
Percent horses	91.8	88.0	80.6	43.6	16.7	14.4	10.5

11.8 Gat and Nissenbaum (1976) give ammonia concentration (y mg l^{-1}) at various depths (x m) in the Dead Sea. Fit a linear regression for concentration on depth using the Theil–Kendall method and obtain an approximate 95 percent confidence interval for β.

x	25	50	100	150	155	187	200	237	287	290	300
y	6.13	5.51	6.18	6.70	7.22	7.28	7.22	7.48	7.38	7.38	7.64

Do you consider a straight-line fit satisfactory for these data?

*11.9 Katti (1965) gave data for weight of food eaten (x) and weight gain (y) for 10 pigs fed on one type of food (A) and for 10 fed on a second type (B). Use the abbreviated Theil method to fit linear regressions to each, and test whether the hypothesis that the true slopes β_1, β_2 are equal is supported.

A	x	575	585	628	632	637	638	661	674	694	713
	y	130	146	156	164	158	151	159	165	167	170
B	x	625	646	651	678	710	722	728	754	763	831
	y	147	164	149	160	184	173	193	189	200	201

11.10 In Section 11.1.7 we gave the data set

x	0	1	2	3	99	100	101
y	2.5	3.1	3.4	4.0	99.6	100.1	101.1

Verify that the y values are consistent with the explanation given in that section of the model used to obtain them. Also use the Theil–Kendall method to obtain a point estimate of β together with a nominal 95 percent confidence interval.

11.11 The Phenological Reports (1891–1948) of the Royal Meteorological Society give the average dates in days from first January of first flowering over that period

for various plants in several locations in the United Kingdom. We give below these
dates for 10 plants in Southwest England and in Northern Scotland. Draw a scatter
diagram for these data and in the light of any evidence as to a possible relationship
and the presence of any outliers fit what you consider an appropriate linear regression
and comment on the adequacy of fit or otherwise. If a further species has an average
flowering date of 122 in Southwest England, what estimate would you give for the
average flowering date in Northern Scotland?

	SW England	N Scotland
Hazel	38	56
Coltsfoot	57	76
Wood Anemone	85	103
Hedge Garlic	108	131
Hawthorn	129	150
White Ox-Eye	145	160
Dog Rose	156	168
Greater Bindweed	185	201
Harebell	193	195
Ivy	265	292

CHAPTER 12

CATEGORICAL DATA

12.1 Categories and Counts

Data may consist of counts of the number of units (people, institutions, towns, countries, items, etc.) with given attributes. It is often convenient to present these in one-, two-, three- or higher-dimensional tables usually referred to as one-way, two-way, three-way, etc., *contingency tables*. Each dimension, or way, corresponds to a classification into categories representing attributes.

Attributes may be explanatory (e.g., dose levels of a drug, the names of several different drugs, gender, psychiatric diagnoses, ethnic groups, income levels). Alternatively, they may be responses (e.g., side-effects of drugs classified as none, slight, moderate, severe, blood pressure levels after administration of a drug, examination grades). Attributes are often qualitative. If there is no natural ordering they are described as *nominal* (e.g., psychiatric diagnoses, ethnic groups). Attributes that may be arranged in a natural order are described as *ordinal* (e.g., reactions to a drug classified as slight, moderate, severe; grouping by age under 50 and age 50 and over).

This and the following chapter deal with problems that at first sight appear different from any previously considered, yet many are solved using procedures we have already developed. The link is possible because we can re-express many problems met earlier in an equivalent contingency table format.

This chapter is mainly about two-way tables consisting of two or more rows and columns. Typically, each row represents either a level of an explanatory attribute or a level of response, and each column represents a level of response.

Table 12.1 has two rows and five columns. We may want to know whether the data indicate that the incidence rate of side-effects differs between drugs. The null hypothesis of independence is often expressed as one of *no association between row and column categories*. Generally, this means that the population distribution of the possible column outcomes is the same for each row category.

In Table 12.1 the row categories are nominal and explanatory. Had each row referred to different dose levels of the same drug they would have been ordinal explanatory categories. The column categories are ordinal responses that reflect increasing seriousness.

We have already met situations where we formed contingency tables deduced from measurement data by counting the numbers of measurements falling into defined categories. For instance, Table 7.4 involves counts from six samples of numbers in each of two categories (above and below an overall sample median

Table 12.1 *Numbers of patients showing side-effects for two drugs.*

| | Side-effect level | | | | |
	None	Slight	Moderate	Severe	Fatal
Drug A	23	8	9	3	2
Drug B	42	8	4	0	0

M). The column categories there were ordinal responses ($<$ M or $>$ M). The row categories (samples) were explanatory and nominal.

We could also present the data in Example 6.1 as a 2×21 contingency table. The rows correspond to Group A and Group B. Each column corresponds to one of the times taken to carry out the calculations, times being arranged in ascending order. The numbers in the cells of this table are either 0 or 1. An entry 1 is placed in the cell in the first row of a time column if someone from Group A took that time to complete the calculations and similarly a 1 is entered in the appropriate cell in the second row for those in Group B. All other cell entries are zero. This leads to Table 12.2. We show why this is useful in Sections 12.3.1 and 13.2.6.

In Example 6.5 a contingency table format was used for what were for practical purposes tied data. There a contingency table was a natural way to present the data.

The correspondence between tables of measurements and counts of units or

Table 12.2 *The data in Example 6.1 presented in a contingency table. For display purposes the table is divided, but it should be read as a 2×21 table.*

Time	17	18	19	21	22	23	24	25	26	27	28
Group A	1	1	1	0	1	1	0	1	1	0	0
Group B	0	0	0	1	0	0	1	0	0	1	1

Time	29	30	31	32	33	34	35	36	39	41
Group A	1	0	1	0	1	0	0	0	0	0
Group B	0	1	0	1	0	1	1	1	1	1

Table 12.3 *Outcome for patients who receive a prescription for a particular antibiotic.*

	Otitis media			
	Worse	The same	Better	Total
Bronchitis				
Worse	13	5	6	24
The same	1	19	4	24
Better	4	0	8	12
Total	18	24	18	60

subjects exhibiting each measurement is a useful tool for showing relationships between methods.

Many tests for lack of association applicable to nominal categories in both rows and columns may still be applied when either or both of row and column categories are ordered. Generally, however, more powerful procedures exist to take account of ordering.

Sometimes both rows and columns represent response categories. For instance, a particular antibiotic can be used to treat both bronchitis and otitis media (infection of the middle ear) in the same patient (Hornbrook, Hurtado and Johnson, 1985). Table 12.3 shows a data set where the antibiotic is prescribed to 60 patients who have both health problems. Its effect is noted on changes (worse, the same, better) in *bronchitis* (rows) and *otitis media* (columns). Such a variation in outcome might occur in practice, as a substantial minority of the patients may not take the prescribed antibiotic leading to a worsening in their condition. Since, for each illness, the three outcomes are mutually exclusive, all subjects can be classified into one and only one combination of row and column categories.

This may not be the case in a different study; in asthma the outcomes of interest might be no symptoms, cough and wheeze. It is possible for individuals to have both cough and wheeze so a fourth category of "both" would be needed to take this into account.

In Table 12.3 both row and column categories are ordinal; in the study of cough and wheeze in asthma outlined above there is no natural ordering and these categories are nominal.

We have shown row and column totals and the total number of patients, 60, in Table 12.3. This is a common practice because totals play a prominent part in analyses.

Table 12.4 *Expected numbers of patients in Table 12.3 under null hypothesis of independent outcomes for bronchitis and otitis media.*

| | Otitis media | | | |
	Worse	The same	Better	Total
Bronchitis				
Worse	7.2	9.6	7.2	24
The same	7.2	9.6	7.2	24
Better	3.6	4.8	3.6	12
Total	18	24	18	60

12.1.1 Some Basic Concepts

Before giving a formal treatment of the analysis of contingency tables we illustrate some basic concepts using the antibiotic treatment data in Table 12.3. Here the appropriate null hypothesis asserts that there is no association between the changes in the severity of bronchitis and changes in the severity of otitis media. So, in the population of such patients who receive this antibiotic, the proportions of patients in the categories for outcome in otitis media are the same whatever the change (if any) in the severity of the bronchitis. Under this null hypothesis, the expected number of patients in a given cell can be calculated using the row and column totals. For instance, if outcomes in bronchitis and otitis media are indeed unrelated we would expect $(18 \times 24)/60 = 7.2$ patients to deteriorate in both conditions. A table of expected values can be constructed for the cells, as shown in Table 12.4.

If the sample reflected the null hypothesis exactly, the observed and expected values in each cell would be the same apart from the rounding needed to produce integer values. Is it likely that the absolute differences observed — ranging in value from 0.4 to 9.4 — could have come about purely by sampling fluctuation if the null hypothesis is true for the population?

The classic way to address this problem is to calculate the chi-squared statistic we introduce in Section 12.2.2. This summary measure assesses the differences between observed and expected values. High values indicate strong evidence against H_0, but a number close to zero supports H_0. This statistic takes into account the sample size, and the numbers of rows and columns (large tables tend to produce larger values of the statistic). Most statistical packages calculate this statistic along with a P-value to indicate the strength of the evidence against the null hypothesis.

For the data in Table 12.3 association appears likely. Indeed, the calculated chi-squared value is large with corresponding $P < 0.0001$. Changes in condition in *otitis media* and *bronchitis* seem to be positively associated. Since row and column categories are ordered it is possible to get a more powerful test than the above if this ordering is taken into account.

12.1.2 Some Models for Counts in $r \times c$ Tables

Books on categorical data analysis in both parametric and nonparametric contexts include Bishop, Fienberg and Holland (1975), Breslow and Day (1980), Plackett (1981), Agresti (1984, 1996, 2002), Christensen (1990), Everitt (1992), Lloyd (1999) and Simonoff (2003).

In general r and c are used to denote the numbers of rows and columns in a table, referred to as an $r \times c$ (in speech "r by c") table. For Table 12.1, $r = 2$ and $c = 5$. It is convenient to represent a general $r \times c$ table in the form of Table 12.5. The entry n_{ij} at the intersection of the ith row and jth column, referred to as cell (i, j), is the number of items having that combination of row and column attributes. The total count for row i is denoted by n_{i+}, i.e., $n_{i+} = \sum_j n_{ij}$. Similarly, n_{+j} is the total count for column j, and

$$N = \sum_i n_{i+} = \sum_j n_{+j} = \sum_{i,\,j} n_{ij} = n_{++}$$

is the total count over all cells. Some writers use $n_{i.}$ and $n_{..}$ instead of n_{i+} and n_{++}.

The methods we develop are nonparametric in the sense that, although several parametric models may give rise to an observed distribution of counts in a contingency table, the inference procedures often do no depend upon which model is relevant, or upon the precise values of certain parameters, since these are not specified in our hypotheses. We avoid the problem of specifying parameter values by a device called *conditional inference*. This makes use of properties of sufficient and ancillary statistics, a topic beyond the scope of this introductory text. Readers familiar with the method of maximum likelihood

Table 12.5 *A general $r \times c$ contingency table.*

		Column categories						
		1	2	3	.	.	c	Total
	1	n_{11}	n_{12}	n_{13}	.	.	n_{1c}	n_{1+}
	2	n_{21}	n_{22}	n_{23}	.	.	n_{2c}	n_{2+}
Row	.	.	.	.	.	.	.	.
categories	.	.	.	.	.	.	.	.
	r	n_{r1}	n_{r2}	n_{r3}	.	.	n_{rc}	n_{r+}
	Total	n_{+1}	n_{+2}	n_{+3}	.	.	n_{+c}	N

will find an excellent account of inference theory for categorical data in Agresti (2002, Chapter 3). Agresti (1996) deals at a less technical level with both classic and modern methods of categorical data analysis.

When making inferences we nearly always regard row and column totals as fixed. This is the conditional feature, implied by conditional inference. It is not, in general, a consequence of the experimental setup. Sometimes neither row nor column totals are fixed *a priori* by the nature of the experiment. In other cases one may be fixed, the other not. Sometimes both are fixed.

For example, if we compare two drugs, A and B, that will result in either an improvement (a success, denoted by S) or no improvement (a failure, denoted by F) we may have 19 patients and allocate 9 of these at random to receive drug A. The remaining 10 receive drug B. We observe these results:

	S	F	Total
Drug A	7	2	9
Drug B	5	5	10
Total	12	7	19

$$(12.1)$$

Row totals are fixed by the numbers allocated to each drug. If the 19 patients are the entire population of interest, then under the hypothesis, H_0, that the drugs are equally effective we are justified in regarding the total number of successes, 12, as fixed irrespective of how we allocate patients to the groups of 9 and 10 corresponding to the two drugs. Just as in Example 2.1, it is extreme outcomes under randomization like none (or 9) of the 12 successes being among the 9 patients allocated to drug A that make us suspect the hypothesis of equal effectiveness is implausible. We may compute the probability of each possible outcome under H_0 when the marginal totals are fixed.

The situation is different when we sample our patients as two groups of 9 and 10 from a larger population. Even under H_0, if we took another pair of samples of 9 and 10 (the fixed row totals representing our designated sample sizes) it is unlikely that the combined number of successes for these different patients would again be 12. We might get an outcome like that in (12.2).

	S	F	Total
Drug A	6	3	9
Drug B	3	7	10
Total	9	10	19

$$(12.2)$$

However, the hypothesis of independence does not involve the actual probability of success with each drug, but stipulates only that the probability of success, π, say, is the same for each drug. In practice, we usually do not know the value of π.

However, if we assume there is such a common value for both drugs, we can obtain an estimate based on the total number of successes, 12 from 19, in our sample of 19 in our observed situation, i.e., that in (12.1). We can use this information to compute the expected number of successes in sub-samples of 9 and 10 from the total of 19 under the assumption that H_0 is true. Again, big discrepancies between observed and expected numbers of successes suggests association, i.e., that the probability of success for drug B is not the same as that for drug A. In effect, we are looking at a permutation distribution once more, but only in possible samples which have fixed column totals (in this case 12, 7 respectively) as well as fixed row totals. Such conditional inference is perfectly valid. It is the entries in the body of any such table with fixed marginal totals that provide the evidence for or against independence. As indicated above, the column marginal totals in this example tell us something about the likely value of π, the probability of success if H_0 is true. They contain little information as to whether there is really a common probability, π, of success for each drug. The individual cell values contain virtually all the information on this latter point. Similar arguments apply for regarding all marginal totals in an $r \times c$ table as fixed for nearly all the models we consider in this and the next chapter.

Methods for inference that do not require all marginal totals to be fixed have been proposed and are used. However, there are many statisticians who feel strongly that conditional inference with fixed marginal totals is the most appropriate way to test for independence (although independence is not the only thing we may want to test in contingency tables).

Conditional inference is convenient, valid, and overcomes technical problems with hypothesis tests in this context. For enlightened discussions about this approach and those that do not assume marginal totals are fixed see Yates (1984), Upton (1992) and a wide-ranging paper by Agresti (1992), also that by Freidlin and Gastwirth (1999).

Experimentally, we might arrive at the outcome in (12.1) without pre-fixing even the row totals. For example, the data might have come from what is called a retrospective study, by simply going through hospital records of all patients treated with the two drugs during the last 12 months and noting the responses. Our method of testing for no association using conditional inference would be the same, i.e., conditional upon the observed row and column totals.

For elucidation, we describe briefly several models used to describe counts, indicating where each might be appropriate, and show that each leads to the same conditional inference procedures. We confine this introductory discussion largely to 2×2 tables. Readers who are not familiar with the multinomial and Poisson distributions and the concept of conditional and marginal distributions may wish to omit the remainder of this section. We describe applications in Section 12.2.

A typical model for the situation exemplified by (12.1), where the row classifications are explanatory and the column classifications are responses, is to assume that each row has associated with it independent binomial distributions and that the cell entries are the numbers of occurrences of a response A and a response B for a $B(n_{1+}, p_1)$ distribution in the first row, and for a $B(n_{2+}, p_2)$ distribution in the second row. In the case of (12.1) we have $n_{1+} = 9$ and $n_{2+} = 10$. Under this model for each cell we obtain probabilities p_{ij} for the observed numbers n_{ij} as follows:

$$p_{11} = \Pr(n_{11} = 7) = \binom{9}{7} p_1^7 (1 - p_1)^2, \qquad p_{12} = 1 - p_{11},$$

$$p_{21} = \Pr(n_{21} = 5) = \binom{10}{5} p_2^5 (1 - p_2)^5, \qquad p_{22} = 1 - p_{12}.$$

The hypothesis of no association implies H_0: $p_1 = p_2$ but the associated common value π, say, is not specified. Independence of response from patient to patient, and the additive property of binomial variables, imply that under H_0 the probability of observing a total of 12 favourable outcomes in 19 observations (the column total line) is

$$p_{+1} = \binom{19}{12} \pi^{12} (1 - \pi)^7.$$

Since the classifications of individuals are independent, it follows from standard definitions of conditional and independent probabilities that the probability of observing 7 successes $(= n_{11})$ in 9 $(= n_{1+})$ trials, conditional upon having observed a total of 12 successes in 19 trials, is

$$p^* = p_{11} p_{21} / p_{+1}.$$

Thus, when $\pi = p_1 = p_2$

$$\begin{aligned} P^* &= [\binom{9}{7} \pi^7 (1 - \pi)^2 \times \binom{10}{5} \pi^5 (1 - \pi)^5] / [\binom{19}{12} \pi^{12} (1 - \pi)^7] \\ &= [\binom{9}{7} \times \binom{10}{5}] / \binom{19}{12} \\ &= [(9!) \times (10!) \times (12!) \times (7!)] / [(19!) \times (7!) \times (2!) \times (5!) \times (5!)] = 0.180. \end{aligned}$$

As indicated in Section 6.2.2, for a general 2×2 table with cell entries

$$\begin{matrix} a & b \\ c & d \end{matrix}$$

this generalizes to

$$p^* = \frac{(a+b)!(c+d)!(a+c)!(b+d)!}{(a+b+c+d)!\,a!b!c!d!} \qquad (12.3)$$

and this does not involve π.

Equation (12.3) represents the frequency, or probability mass, function for the *hypergeometric distribution*. We saw in Section 6.2.2 that when we fix marginal totals, P^* depends essentially only on a, since, given a, then b, c and d are automatically fixed to give the correct marginal totals. The fact that fixing a determines all cell entries is summarized by saying that a 2×2 contingency table has *one degree of freedom*. The value of P^* given by (12.3) is relevant to testing for independence (no association). As explained in Section 6.2.2 the test based on the hypergeometric distribution is the *Fisher exact test*. We described its application to a 2×2 table in Example 6.7. By calculating P^* for all possible values of a we may set up critical regions in the tail of the distribution relevant to assessing the strength of evidence against H_0 in either one- or two-tail tests, as we did in Example 6.7.

We sometimes want estimates of the expected values of each n_{ij} under H_0. If we knew π we could obtain exact expected values, but when π is unknown we base our estimates on the marginal totals, using these to estimate π. Since n_{+1} has expected value $N\pi$, we estimate π by $\hat{\pi} = n_{+1}/N$ and the expected value of n_{11} is $m_{11} = n_{1+}\hat{\pi} = (n_{1+}n_{+1})/N$. Similarly, we establish the general result $m_{ij} = (n_{i+}n_{+j})/N$. These expected values sum to the marginal totals.

More generally, if there is association in a 2×2 table, then in (12.1) for example, if the binomial probabilities were known to be p_1, p_2 the expected values in that case would be

$$\begin{matrix} 9p_1 & 9(1-p_1) \\ 10p_2 & 10(1-p_2) \end{matrix} \qquad (12.4)$$

The quotients, $p_1/(1-p_1)$ and $p_2/(1-p_2)$, represent the odds on the outcome *success* for each row. A quantity often used as an indicator of association is the *odds ratio*, defined as

$$\theta = \frac{p_1/(1-p_1)}{p_2/(1-p_2)} = \frac{p_1(1-p_2)}{p_2(1-p_1)}$$

From (12.4) the odds ratio is also the ratio of cross products of the cell expected frequencies m_{ij}, i.e., $\theta = (m_{11}m_{22})/(m_{12}m_{21})$. If $p_1 = p_2$ then $\theta = 1$. This is also true for the estimated expected values calculated under H_0, i.e., $m_{ij} = (n_{i+}n_{+j})/N$. If, as usual, we do not know p_1, p_2 we may estimate θ by the sample odds ratio, which is defined as $\theta^* = (n_{11}n_{22})/(n_{12}n_{21})$. In Section 13.2 we give tests for association, including a test for whether θ^* is consistent with H_0: $\theta = 1$.

If $p_1 = 3/4$ and $p_2 = 1/2$ it is intuitively obvious that we expect more successes with drug A than with drug B and from (12.4) we find $m_{11} = 6.75$ and $m_{21} = 5$. These expected values no longer add to the column total as do the expected values under H_0, This should cause no surprise as we have a situation where the probabilities of success are known and different for each drug. For these p_1, p_2 values $\theta = [(3/4) \times (1/2)]/[(1/2) \times (1/4)] = 3$. If we interchange probabilities, so that $p_1 = 1/2$ and $p_2 = 3/4$ we find $\theta = 1/3$. Both these cases reflect the same degree of association, but in opposite directions. This is demonstrated more clearly if we use $\phi = \log\theta$ as a measure of association, for now

$$\log 3 = -\log(1/3)$$

and generally

$$\log \theta = -\log(1/\theta).$$

Possible values of ϕ range from $-\infty$ to ∞. When there is no association $\phi = 0$. The above relationships are true for logarithms to any base, but most formulae for testing and estimation use natural logarithms with the exponential base e. These are sometimes denoted by *ln* rather than *log*, but we use the latter.

The binomial model given above is not appropriate if data are for two response classifications, like those in the table (12.5) giving numbers of patients responding in specified ways to a drug.

		Blood cholesterol level			
		Increased	*Unchanged*	*Total*	
Blood	*Higher*	14	7	21	(12.5)
Pressure	*Lower*	11	12	23	
	Total	25	19	4	

Here all four cells are on a par in the sense that all we can say is that, with an unknown probability p_{ij}, any of the $N = 44$ subjects may be allocated to cell (i, j). The multinomial distribution is an extension of the binomial distribution with independent allocation of N items to $k > 2$ categories with specified probabilities for each category. Here we have 4 categories, one corresponding to each cell. The probabilities are subject to the constraint $p_{11} + p_{12} + p_{21} + p_{22} = 1$, since each person is allocated to one and only one cell. We write p_{1+}, p_{2+} for the marginal probabilities that items will fall in row 1 or row 2. Likewise, we specify marginal probabilities p_{+1}, p_{+2} for columns. For independence between rows and columns, standard theory gives the relationship $p_{ij} = p_{i+}p_{+j}$ for $i, j = 1, 2$.

Even if there is independence we still do not know the value of the p_{ij}, but this presents no problem if we use conditional inference. Assuming no association (i.e., independence), we base our estimates on marginal totals, estimating p_{i+} by n_{i+}/N and p_{+j} by n_{+j}/N. Since $p_{ij} = p_{i+}p_{+j}$ if there is independence, we estimate p_{ij} by $\hat{p}_{ij} = n_{i+}n_{+j}/N^2$. It follows from properties of the multinomial distribution that we may then obtain expected estimated counts as $m_{ij} = N\hat{p}_{ij} = n_{i+}n_{+j}/N$, exactly as in the two-sample binomial model. We proceed to tests as in that model. It is easy to show that under independence the odds ratio, here defined as $\theta = (p_{11}p_{22})/(p_{12}p_{21})$, is unity (hint: put $p_{11} = p_{1+}p_{+1}$, etc.). If the p_{ij} are estimated under H_0 in the way indicated above, using these estimates, or the estimated expected frequencies m_{ij} gives $\theta = 1$. The sample odds ratio is $\theta^* = (n_{11}n_{22})/(n_{12}n_{21})$.

Another model often relevant is one where counts in each cell have independent Poisson distributions with a mean (Poisson parameter) m_{ij} for cell (i, j). If $\theta = (m_{11}m_{22})/(m_{12}m_{21}) = 1$, there is no association between rows and columns. In general, we do not know the m_{ij}, but it can be shown that, again under H_0, these may be estimated from marginal totals in the way we do for the row-wise binomial or the overall multinomial models.

Extension of the three models above to $r \times c$ tables is reasonably straightforward. For $c > 2$, binomial sampling within rows generalizes to multinomial sampling within rows. For this, or the overall multinomial sample, or the Poisson count situation, under the null hypothesis of no association the expected cell frequencies are estimated in the notation of Table 12.5 by $m_{ij} = n_{i+}n_{+j}/N$, $i = 1, 2, \dots, r$ and $j = 1, 2, \dots, c$. An odds ratio may be defined for any two rows i, k and any two columns j, l as

$$\theta_{ijkl} = (m_{ij}m_{kl})/(m_{il}m_{kj}),$$

and $\theta_{ijkl} = 1$ for all i, j, k, l if there is no association. The sample odds ratios are $\theta^*_{ijkl} = (n_{ij}n_{kl})/(n_{il}n_{kj})$.

The Fisher exact test extends to $r \times c$ tables in a way described in Section 12.2.1.

12.2 Nominal Attribute Categories

Tests developed in this section are appropriate when row and column categories are both nominal. They are often used for tests of no association when attributes are ordered, but tests described in Section 12.3, or some in Chapter 13 are then often more appropriate.

We describe three commonly used procedures. Confusion exists in the literature because each often (but not invariably) leads to essentially equivalent permutation distributions for 2×2 tables, but to different asymptotic test statistics even in this case. For general $r \times c$ tables both the exact permutation distribution of the test statistics and the asymptotic test statistics differ. One should not exaggerate the importance of these differences, for unless the choice of a precise significance level is critical, it is seldom that the three possible exact tests lead to markedly different conclusions. However, there may be considerable differences between exact and asymptotic tests in some circumstances. Numerical values of the three test statistics often differ considerably, even for moderately large samples, but each statistic has asymptotically a chi-squared distribution with $(r-1)(c-1)$ degrees of freedom under the hypothesis of independence (no association) between row and column classifications.

The three tests are

- The *Fisher exact test* (also known as the Fisher–Irwin or as the Fisher–Freeman–Halton test).

- The *Pearson chi-squared test*.

- The *Likelihood ratio test*.

Practice has often been to use the Fisher exact test when asymptotic theory is clearly inappropriate and either of the others when asymptotic results are thought to be acceptable. Despite differences in numerical value of the relevant test statistics, and the fact that exact permutation distribution theory differs in each case, there seems no compelling reason to depart from this practice.

Tests based on exact permutation theory, except in the case of 2×2 tables (where the Fisher exact test is commonly used) are not practical without relevant software. StatXact allows exact permutation tests based on all three methods and some facilities are also included in Testimate and in more general software. This means there is now little excuse for using asymptotic theory when it is of doubtful validity. This applies particularly to sparse contingency tables in which there are many low cell counts, even though N may be large.

This last situation is common in medical applications where responses may be recorded on a limited number of diseased patients because resources are in short supply, or responses in certain categories are few in number (see, e.g., Exercise 12.26).

12.2.1 The Fisher Exact Test

The Fisher exact test for 2×2 tables based on the hypergeometric distribution specified in (12.3) was described by Fisher in various editions of his famous book *Statistical Methods for Research Workers* dating from the 1930s. It was extended to $r \times c$ tables by Freeman and Halton (1951). In the notation of Table 12.5, the generalized hypergeometric distribution relevant to an $r \times c$ table gives the probability of an observed configuration under H_0, conditional upon fixed marginal totals as

$$P^* = \frac{\prod_i (n_{i+}!) \prod_j (n_{+j}!)}{n! \prod_{i,j} (n_{ij}!)}. \tag{12.6}$$

The P-value for testing no association is the sum of all probabilities for configurations with the given marginal totals having the same, or lower, probability.

Example 12.1

The problem. Scarlet fever is a childhood infection that among other symptoms gives rise to severe irritation of the nose, throat and ears. In a developing country where the infection is a problem, six districts A to F were chosen. In each district, patients were located, and parents were asked to state the site at which they thought their child's irritation was worst. Numbers in each response category were:

				District			
	A	B	C	D	E	F	Total
Nose	1	1	0	1	8	0	11
Throat	0	1	1	1	0	1	4
Ears	1	0	0	0	7	1	9
Total	2	2	1	2	15	2	24

Is there evidence of association between districts and sites of greatest perceived irritation?

Formulation and assumptions. We use the Fisher exact test for independence and the program in StatXact or some other statistical package.

Procedure. StatXact gives the probability of observing this configuration under H_0: *no association* to be 0.0001, and the probability of this or a less probable configuration is $P = 0.0261$. It also gives a probability of 0.0039 of obtaining a configuration with the same probability as that observed, indicating that more than one configuration has this probability.

Conclusion. Since $P = 0.0261$ there is fairly strong evidence against H_0.

Comments. 1. It is evident that the main cause of association is the relatively high number of responses of greatest irritation at the nose or ears (but not throat) in district E. This district has responses out of line with the general trend, whereas the other districts produce reasonably consistent results for all three sites of irritation. It may be that in district E more importance is placed on visual appearance compared with the other five districts.

2. The finding of dependence is not surprising, so it is relevant to note that asymptotic theory here does not indicate significance at a conventional 5 percent level. We discuss the more commonly used asymptotic statistics in Sections 12.2.2 and 12.2.3, but see also the remarks below under *Computational aspects.*

3. While it is impractical to compute manually all relevant permutation distribution probabilities, it is not difficult to verify that for these data (12.6) gives $P^* = 0.000058$, as indicated (after rounding) by StatXact (see Exercise 12.1).

4. We are not testing whether incidence levels differ between districts, but whether the proportions at irritation sites differ between districts (or equivalently, whether the proportions for districts differ between irritation sites). Had there, in addition, been eight cases of greatest irritation in the throat in district E we would have found little evidence against the hypothesis of no association, although clearly the numbers of cases at all three irritation sites would be much higher for district E.

5. The test of association is essentially two-tail, the situation being broadly analogous in this aspect to that in the Kruskal–Wallis test where we test for differences in centrality which may be in any direction, or in different directions in populations corresponding to different samples.

Computational aspects. 1. A program like that in StatXact is invaluable in situations where asymptotic results may be misleading. Although seldom used in practice, a statistic is given in StatXact which is a monotonic function of P^* given by (12.6). Asymptotically, this statistic has a chi-squared distribution with $(r-1)(c-1)$ degrees of freedom. This statistic, denoted by F^2, takes the value 14.30 for this example, and $\Pr(\chi^2 > 14.30) = 0.1597$. This is in sharp contrast to the result from the exact test, for we do not establish significance at any conventional level with the asymptotic test. In practice, the asymptotic tests given in Sections 12.2.2 and 12.2.3 are more widely used.

2. The Fisher exact test for $r \times c$ tables is available in Stata. If the table has many cells and the sample size is large the computation of the P-value may take several minutes. Many general packages include the Fisher exact test for 2×2 tables only.

Example 12.1 is not atypical. The StatXact manual gives very similar data for location of oral lesions in a survey of three regions in India. The asymptotic test has $(r-1)(c-1)$ degrees of freedom in an $r \times c$ table because, if we allocate arbitrarily cell values in, say, the first $r-1$ rows and $c-1$ columns the last row and column cell entries are fixed to ensure consistency with the row and column totals.

12.2.2 The Pearson Chi-Squared Test

An alternative statistic for testing independence of row and column categories is the Pearson chi-squared statistic. Its appeal is

- Ease of computation.

- Fairly rapid convergence to the chi-squared distribution as count size increases.

- Intuitive reasonableness.

An exact test is based on the permutation distribution of this statistic over all sample configurations having the fixed marginal totals. Computation of P-values is no easier than it is for the Fisher exact test and only for 2×2 tables are the exact tests often equivalent.

Denoting, as before, the expected number in cell (i, j) by $m_{ij} = n_{i+}n_{+j}/N$, then $(m_{ij} - n_{ij})^2$ is a measure of departure from expectation under H_0. It is reasonable to weight this measure inversely by the expected count. This is because a departure of 5 from an expected count of 100 is less surprising than a departure of 5 from an expected count of 6 for any of the models given in Section 12.1.2. Pearson (1900) proposed the statistic

$$X^2 = \sum_{i,j} \left[\frac{(m_{ij} - n_{ij})^2}{m_{ij}} \right] \tag{12.7}$$

as a test for evidence of association.

For computational purposes it is easier to use the equivalent form

$$X^2 = \sum_{i,j} \left[\frac{(n_{ij})^2}{m_{ij}} \right] - N. \tag{12.8}$$

Example 12.2

The problem. For the data in Example 12.1 calculate X^2. Test for significance using the asymptotic result that X^2 has a chi-squared distribution.

Formulation and assumptions. We calculate each $m_{ij} = n_{i+}n_{+j}/N$, then use formula (12.8) for X^2.

Procedure. Using the row and column totals for the data in Example 12.1 we obtain the following table of expected frequencies. An explanation of how to get these was given in Section 12.1.1.

			District				
	A	*B*	*C*	*D*	*E*	*F*	*Total*
Nose	11/12	11/12	11/24	11/12	55/8	11/12	11
Throat	1/3	1/3	1/6	1/3	5/2	1/3	4
Ears	3/4	3/4	3/8	3/4	45/8	3/4	9
Total	2	2	1	2	15	2	24

Using (12.8) we find

$$X^2 = [1^2/(11/12) + 1^2/(11/12) + \ldots + 7^2/(45/8) + 1^2/(3/4)] - 24 = 14.96.$$

With 10 degrees of freedom, chi-squared tables show that significance at the 5 percent level requires $X^2 \geq 18.31$. In fact, $\Pr(\chi^2 \geq 14.96) = 0.1335$.

Conclusion. There is little evidence against H_0.

Comment. An asymptotic result may be unsatisfactory with many sparse cells, but here that using X^2 behaves marginally better than that based on F^2 (Example 12.1). See also Exercise 12.26.

Computational aspects. Nearly all standard statistical packages include a program for tests based on X^2 using the asymptotic result. StatXact also computes $P = 0.1094$ using the exact permutation distribution for X^2. This not only shows that the Fisher exact test and the exact version of the Pearson chi-squared tests are not equivalent, but suggests that the tests may differ in power.

For 2×2 tables the Fisher exact test, and the exact tests based on X^2 are often equivalent, though the asymptotic test statistics generally differ. It is customary to use the latter in asymptotic tests, and when doing so a better approximation is obtained by using a continuity correction commonly known in this context as *Yates' correction*. With this correction we compute

$$X^2 = \sum_{i,j} \left[\frac{(|n_{ij} - m_{ij}| - 1/2)^2}{m_{ij}} \right],$$

i.e., we subtract $1/2$ from the magnitude of each difference in the numerator before squaring.

An alternative form useful for computation is

$$X^2 = \frac{N[(|n_{11}n_{22} - n_{12}n_{21}|) - N/2]^2}{n_{1+}n_{+1}n_{2+}n_{+2}}. \tag{12.9}$$

Omitting the term $N/2$ gives the X^2 value without continuity correction. The correction is used only for 2×2 tables.

Example 12.3

The problem. A new drug (A) is only available in a pilot experiment for 6 patients who are chosen at random from a total of 50 patients; the remaining 44 receive the standard drug (B). Of those receiving drug A only one shows a side-effect, while 38 of those receiving drug B do. Test the hypothesis of no association between side-effects and drug administered.

Formulation. Using the information above we form the following 2×2 table:

	Side-effect	None	Total
Drug A	1	5	6
Drug B	38	6	44
Total	39	11	50

We use (12.9) to compute X^2 and test using chi-squared with 1 degree of freedom.

Procedure. Substituting in (12.9) gives

$$X^2 = \frac{50[|(1 \times 6 - 5 \times 38)| - 25]^2}{6 \times 39 \times 44 \times 11} = 11.16.$$

Tables, or software for the chi-squared distribution, indicate $P = \Pr(\chi^2 \geq 11.16) = 0.0008$.

Conclusion. There is strong evidence against H_0: *no association.*

Comment. The permutation X^2 test indicates $P = 0.0012$. Without Yates' correction $X^2 = 14.95$ which overestimates the evidence against H_0.

Computational aspects. Most general statistical packages include a program for the Pearson chi-squared test for $r \times c$ tables, often using Yates' correction, or providing it as an option, for 2×2 tables.

12.2.3 The Likelihood Ratio Test

An alternative statistic for a test of association is the likelihood ratio

$$G^2 = 2 \sum_{i,j} n_{ij} \log(n_{ij}/m_{ij}) \tag{12.10}$$

where *log* is the natural logarithm, i.e., to the base e. Asymptotically G^2 has a chi-squared distribution with $(r-1)(c-1)$ degrees of freedom if there is no association. For the data in Example 12.1, $G^2 = 17.33$. Although this does not indicate significance at a conventional 5 percent level, it is a better approximation to the exact P than that given by X^2. The test has some optimal properties for the overall multinomial sampling model and convergence to the chi-squared distribution under H_0 is often faster than that of X^2.

An exact permutation theory test may be based on G^2 and, except for 2×2 tables, the results will generally differ from those for exact tests based on F^2 or X^2. For the data in Example 12.1 the exact test based on G^2 gives $P = 0.048$.

In each of the three tests for the data in Example 12.1 the asymptotic test is an unsatisfactory approximation. Although the three exact permutation distribution tests are not equivalent, two give strong indications of evidence that there is association and all give lower P-values than those for asymptotic theory. The sample has a preponderance of sparse cell entries. Although such situations do arise in practice, in many contingency tables very few cells have low expected values under H_0.

There is a variety of conflicting advice in the literature about when it is safe to use asymptotic results with G^2 or X^2. Generally speaking, the larger r and c are, the less we need worry about a small proportion of sparse cell counts. Programs such as StatXact or Testimate that provide both exact and asymptotic P-values, so that these may be compared, are invaluable for those who frequently meet contingency tables with sparse data.

Since the tests in Section 12.2 apply to nominal categories, reordering rows or columns does not affect the value of the test statistic. As already indicated, for ordered categories there are more appropriate tests that take account of any natural ordering in rows and columns.

12.3 Ordered Categorical Data

We consider first nominal row explanatory categories and ordered column response categories; then ordered row explanatory and ordered column response classifications, or both ordered row and ordered column response categories.

12.3.1 Nominal Row and Ordered Column Categories

Table 12.6 repeats Table 12.1 with the addition of row and column totals. The explanatory row categories, type of drug, are nominal but column responses are ordinal. Here the Wilcoxon–Mann–Whitney (WMW) test is appropriate. However, the data are heavily tied, so, e.g., the 65 responses tied as *none* are allocated the mid-rank 33, the 16 responses tied as *slight* are given the mid-rank of ranks 66 to 81, i.e., 73.5, and so on. Using the Mann–Whitney formulation obviates the need to specify mid-ranks. We simply count the numbers of recipients of drug B showing the same or more severe side effects than that for each recipient of drug A, scoring an equality as 0.5 and an excess as 1. Thus, the 23 responses *none* for drug A each give rise to 42 ties, scoring 21

Table 12.6 *Numbers of patients showing side-effects for two drugs.*

	\multicolumn{5}{c}{*Side-effect level*}	
	None	*Slight*	*Moderate*	*Severe*	*Fatal*	*Total*
Drug A	23	8	9	3	2	45
Drug B	42	8	4	0	0	54
Total	65	16	13	3	2	99

and $8 + 4$ higher ranks, contributing a total score $23(0.5 \times 42 + 8 + 4) = 759$ to the Mann–Whitney U_m.

Example 12.4

The problem. For the data in Table 12.6 compute the Mann–Whitney statistic, and test for evidence of a difference in levels of side-effects.

Formulation and assumptions. We compute either U_m or U_n in the way described above, and which was also illustrated in Example 6.5. Because the sample is large we might use an asymptotic result, but if a suitable computer program is available, with so many ties an exact test is preferable.

Procedure. We calculate

$$U_m = 23(0.5 \times 42 + 8 + 4) + 8(0.5 \times 8 + 4) + 9(0.5 \times 4) = 841.$$

Using (6.2) (i.e., ignoring ties) gives $Z = 2.63$, whereas the appropriate modification of (6.6) gives $Z = 3.118$. Using StatXact, the exact permutation test gives a two-tail $P = 0.0019$, equal to that using $Z = 3.118$.

Conclusion. With sample sizes 45 and 54, having allowed for ties, an asymptotic result should be reasonable and this indicates strong evidence against H_0 since $P < 0.01$. This is confirmed by the exact $P = 0.0019$.

Comment. If we ignore ordering and use the Fisher exact test this gives an exact $P = 0.0127$ while for the asymptotic Pearson chi-squared test $P = 0.0193$. These results are in line with our general remark that we gain power by taking account of ordering. Nevertheless, for some configurations we may observe marginally smaller P-values when we ignore ordering.

Computational aspects. Since asymptotic results appear reasonable here, any statistical package for the WMW test that uses mid-ranks and gives the significance level based on asymptotic theory that takes account of ties should be satisfactory. A program such as the one in StatXact does, however, give additional confidence in that it compares not only exact and asymptotic results for the WMW test, but also allows us to perform the Fisher exact test or a Pearson chi-squared test (exact or asymptotic) and compare results.

In Section 6.2.1 we recommended the Fisher exact test when using a median test to compare treatments. Since we had ordered categories, why did we not recommend the WMW test if we have a program for exact permutation distribution tests? It turns out that it does not matter which we use for 2×2 tables, for the exact tests are then equivalent, although the asymptotic forms of the two are not quite the same.

The position is more complicated for the median test in Section 7.2.6, where we have more than two samples. We discuss this in Section 12.3.2.

Nearly all tests for centrality in earlier chapters can be formulated as tests involving contingency tables with fixed marginal totals. This is a key to efficient computation of exact permutation distributions and is how StatXact handles data for these tests. We gave an example in Table 12.2 for a data set encountered earlier. In Example 2.4 we developed a permutation test for what we now know to be a specific example of the WMW test. In that example we had sets of 4 and 5 patients ranked 1 to 9. In contingency table format we regard the drugs as the row classifications and the ordered ranks 1 to 9 as column (response) classifications. We count the number of items allocated to each row and column classification; e.g., this count will be 0 in row 1, column 4 if no patient given drug A was ranked 4. It will be 1 in row 2 column 5 if a person given drug B is ranked 5. There were no ties in Example 2.1, so if we observed the ranks 1, 2, 3, 6 for the four patients given drug A (implying those receiving drug B were ranked 4, 5, 7, 8, 9) we may construct the contingency table:

				Rank					
	1	2	3	4	5	6	7	8	9
Drug A	1	1	1	0	0	1	0	0	0
Drug B	0	0	0	1	1	0	1	1	1

In this table the row totals are 4, 5 (the sample sizes). In this no-tie situation each column total is 1. In the Wilcoxon formulation of the WMW text the test statistic is $S_i = \sum_j j n_{ij}$, where either $i = 1$ or $i = 2$. In this context j is a *column score* which here is the rank, and the n_{ij} take the value 0 or 1.

The permutation distribution in Example 2.1 is obtained by permuting all possible cell entries in the above table consistent with the fixed row and column totals, and calculating S_i for each permutation. Rank is only one possible column score. We may replace the score j in column j by other appropriate ordered scores s_j. We then obtain in the form $S_i = \sum_j s_j n_{ij}$, where $i = 1$ or $i = 2$, various other statistics discussed in Chapter 6 for testing centrality. For example, the s_j may be the van der Waerden scores or the ordered raw data, the latter giving rise to the raw scores Pitman permutation test. Choices of column scores in $2 \times c$ tables for many well-known tests are discussed by Graubard and Korn (1987).

In the case of ties in the WMW test, cell entries correspond to the number of ties in the relevant sample (row) at any given mid-rank and the column scores are the appropriate mid-rank values. Each column total is the total number of ties assigned that mid-rank score. Permutation is over all configurations giving the fixed row and column totals.

In a no-tie situation, under H_0 we expect a good mix of ones and zeros over the cells of the $2 \times c$ table. A concentration of ones at opposite corners of the table indicates association and possible rejection of H_0. For example, a configuration with the marginal totals above that indicates rejection is

					Rank				
	1	*2*	*3*	*4*	*5*	*6*	*7*	*8*	*9*
Drug A	0	0	0	0	0	1	1	1	1
Drug B	1	1	1	1	1	0	0	0	0

We look more fully at the approach used here in Section 13.2.6.

When there are more than two nominal explanatory categories, each corresponding to a row and the column responses are ordinal the obvious extension from the WMW test is to the Kruskal–Wallis test (with ties if appropriate). Unless some samples are very small, or there is a concentration of many entries in one or two columns, (heavy tying) asymptotic results are usually satisfactory. Except for fairly small samples, even programs like StatXact cannot cope with the exact permutation distribution for the Kruskal–Wallis test, but the Monte Carlo approximation available in StatXact works well in practice.

Example 12.5

The problem. The following information on times to failure of car tyres due to various faults is abstracted from a more detailed and larger data set kindly made available to us by Timothy P. Davis. The classified causes of failure are:

- *A* Open joint on inner lining
- *B* Rubber chunking on shoulder
- *C* Loose casing low on sidewall
- *D* Cracking on the sidewall
- *E* Other causes

One cause of failure, cracking of the tread wall, is omitted from this illustrative example, for it gave a markedly different pattern of times to failure from that for other faults. Depending on one's viewpoint, the row categories A, B, C, D, E might be regarded as nominal explanatory or as nominal response categories. In the sense that they are different types of failure they represent different responses. On the other hand, each tyre failing for a particular reason might be regarded as one of a sample of tyres doomed to fail in that way. Then if we are interested in times to failure at which different faults occur we might regard the different faults as explanatory variables.

For illustrative purposes in this example we take the latter stance, although the analysis would still be appropriate if we regarded these as distinct response categories. The column classification is time to failure recorded to the nearest hour, grouped into intervals < 100, 100 − 199, 200 − 299, 300+ hours.

| | Time to failure (hrs) | | | |
	< 100	100–199	200–299	300+
Cause				
A	2	6	9	0
B	0	1	7	0
C	4	10	6	0
D	1	6	5	0
E	0	8	10	2

Formulation. We perform a Kruskal–Wallis test with five samples. Mid-ranks are calculated regarding all units in the same column as ties. Thus, the seven ties in the column labelled < 100 are all given the mid-rank 4, and so on. The Kruskal–Wallis statistic is calculated as in Section 7.2.2 and the statistic has asymptotically a chi-squared distribution with $r - 1 = 4$ degrees of freedom. Computation is tedious without a computer program.

Procedure. The computation outlined above gives the value 10.44 for the Kruskal–Wallis statistic, and tables or software indicate $P = \Pr(\chi^2 \geq 10.44) = 0.0336$.

Conclusion. There is reasonably strong evidence of association between rows and columns, implying that different types of fault do not all have the same median times to occurrence.

Comment. Despite heavy tying, in the light of our finding in earlier chapters, one hopes the asymptotic result is reasonable here. That this is so was confirmed by the StatXact Monte Carlo procedure for estimating the exact P. This gave an estimated $P = 0.0293$, with the 99 percent confidence interval for the true P being (0.0256, 0.0330). This suggests the asymptotic result is conservative, falling just above the upper confidence limit for the permutation distribution result.

Computational aspects. Many packages compute the Kruskal–Wallis statistic, so the asymptotic test is readily available. StatXact will compute exact P-values with small numbers of rows and columns but computation may fail on some PCs, or prove very time consuming for examples with a moderate number of ties. The Monte Carlo method is relatively fast. The estimate above of $P = 0.0293$, was based on 14 000 simulations. Increasing the number of simulations shortens the confidence interval for the true P.

In a no-tie situation with r samples and a total of N observations the Kruskal–Wallis test may be formulated in contingency table terms as an $r \times N$ table with rows corresponding to samples and all cell entries 1 or 0. The columns are ordered by ranks, and the observation of rank j is recorded as 1 in the cell in the jth column in the row corresponding to the sample in which it occurs. All other entries in that column are zero in the no-tie case. The row totals are the sample sizes n_i and each column total is 1. Ranks are the scores used to generate the exact permutation distribution. The statistic is a function of the rank sums for each row. Permutation is over all cell entries in the $r \times N$ table consistent with row and column totals.

12.3.2 Ordered Row and Column Categories

We consider first the case where row categories are ordinal explanatory categories; columns are responses. One appropriate test of no association is then the Jonckheere–Terpstra test with each row corresponding to one of the ordered samples. Asymptotic results may be reasonable if allowance is made for ties. We illustrate the need to allow for ties in a situation not uncommon in drug testing, where side-effects may not occur often, but may be serious if they do.

Example 12.6

The problem. Side-effects (if any) experienced by patients at increasing dose levels of a drug are classified as none, mild, moderate or severe. Do the following data indicate that incidence or severity of side-effects increase with dose level?

	None	*Side-effects* Slight	*Moderate*	*Severe*
Dose				
100 *mg*	50	0	1	0
200 *mg*	60	1	0	0
300 *mg*	40	1	0	0
400 *mg*	30	1	1	2

Formulation. The Jonckheere–Terpstra statistic is relatively easy to compute as an extension of the procedure used for the Mann–Whitney formulation of the WMW test in Example 12.4, scoring the relevant ties in any column as 1/2. Those in the first column make a substantial contribution to the total.

Procedure. Without a computer program it is tedious, but not difficult to compute the relevant statistic, denoted by U in Example 7.3. It is

$$U = 50[(60+40+30)/2+7]+60[(40+30)/2+6)]+40(30/2+4)+5+3.5+3+2.5 = 6834.$$

The asymptotic test in Section 7.2.5, with no correction for ties, gives $\Pr(U > 6834) = 0.2288$. The two-tail probability is $P = 0.4576$, so there is no evidence of association. The hypothesis H_0: *side-effect levels are independent of dose* is acceptable.

Conclusion. We would not reject H_0 even in a one-tail test, which might be justified on the general grounds that increasing dose levels of a drug are more likely to produce side-effects if indeed there are any.

Comments. 1. A clinician is unlikely to be satisfied with this finding. Side-effects are not common, but there is to a clinician (and even to the layman) an indication that these are more likely and more severe at the highest dose level. Either the evidence is too slender to reach a firm conclusion or our test is inappropriate. In particular, can we justify an asymptotic result that does not allow for ties when there is so high an incidence of these?

2. The maximum safe dose of a drug is often determined by giving the drug to independent groups of healthy volunteers in increasing doses, the first group receiving the smallest dose. Once a serious side-effect has been observed the study is stopped; to continue the investigation with healthy volunteers would be unethical. Higher doses of the drug may have to be used with genuine patients in order to achieve an improvement. The potential gains would then have to be balanced against the likely side-effects.

Computational aspects. Some general statistical packages include the asymptotic Jonckheere–Terpstra test without allowance for ties. We show below that correcting for ties may have a dramatic effect. StatXact includes an asymptotic test that allows for ties.

Example 12.7

The problem. Use the exact Jonckheere–Terpstra test on the data in Example 12.6 to determine if there is acceptable evidence that side-effects become more common as dose increases.

Formulation. The permutation test requires computation of probabilities for the Jonckheere–Terpstra statistic for all permutations of the contingency table in Example 12.6 subject to the same marginal totals and with the same or a lesser probability than that observed.

Procedure. StatXact provides a program for these computations which indicates $\Pr(U \geq 6834) = 0.0168$.

Conclusion. There is clear evidence against H_0 with a one-tail $P = 0.0168$.

Comment. This is an illustration of a situation where an exact permutation test leads to a markedly different conclusion to a commonly used asymptotic test despite a large sample size.

We mentioned in Section 7.2.5 an amended formula for $\text{Var}(U)$ which allows for ties. The formula is adapted from one given in Kendall and Gibbons (1990, p. 66), and also in Lehmann (1975, 2006). The notation here matches that in Table 12.5.

We calculate

$$\text{Var}(U) = U_1/d_1 + U_2/d_2 + U_3/d_3 \tag{12.11}$$

where

$$U_1 = N(N-1)(2N+5) - \sum_i n_{i+}(n_{i+}-1)(2n_{i+}+5) - \\ \sum_j n_{+j}(n_{+j}-1)(2n_{+j}+5)$$
$$U_2 = [\sum_i n_{i+}(n_{i+}-1)(n_{i+}-2)][\sum_j n_{+j}(n_{+j}-1)(n_{+j}-2)]$$
$$U_3 = [\sum_i n_{i+}(n_{i+}-1)][\sum_j n_{+j}(n_{+j}-1)]$$
$$d_1 = 72, \ d_2 = 36N(N-1)(N-2), \ d_3 = 8N(N-1).$$

When there are no ties, all $n_{+j} = 1$ and $U_2 = U_3 = 0$, and Var(U) reduces, after algebraic manipulation, to the form given in Section 7.2.5. In Example 12.6 the relevant values are $N = 188$, $n_{1+} = 51$, $n_{2+} = 61$, $n_{3+} = 42$, $n_{4+} = 34$, $n_{+1} = 180$, $n_{+2} = n_{+3} = 3$, $n_{+4} = 2$. Substituting in (12.11) reduces Var(U), from 415.36 (ignoring ties) to 145.18 (Exercise 12.2). The corresponding $Z = 2.125$, and standard normal distribution tables give $\Pr(Z \geq 2.125) = 0.0168$, agreeing with the exact permutation result obtained in Example 12.7.

In clinical trials situations like that in Examples 12.6 and 12.7 are common. They arise not only for relatively rare side-effects of drugs but also in other situations where some responses are rare, e.g., incidence of lung cancer in populations exposed to adverse environmental factors. We may observe several thousand people in each explanatory category, yet only a few develop lung cancer.

In Section 7.2.6 we recommended the Fisher exact test for the median test for $r > 2$ samples. This ignores the fact that the responses $< M$ and $\geq M$ are ordered. If the sample classifications are nominal an alternative is the Kruskal–Wallis test. If the samples are ordered so that the natural alternative hypothesis is some systematic increase or decrease in the median, we may do better using the Jonckheere–Terpstra procedure. With only two columns (ordered response categories) there will often be little difference between results using the Fisher test or Kruskal–Wallis or (if appropriate) the Jonckheere–Terpstra procedure. If only asymptotic results are used in this last test a correction for ties is important. In the special case of only two ordered response categories we show below that the Jonckheere–Terpstra test is equivalent to the WMW test.

Example 12.8

The problem. Exercise 7.15 gave data supplied by Chris Theobald for blood platelet counts and spleen sizes for 40 patients. The investigators were interested in whether the proportion of patients with an abnormally low platelet count increased with increasing spleen size. If a count below 120 is abnormal the data in Exercise 7.15 give the contingency table below. Does the Jonckheere–Terpstra test indicate an association between spleen size and platelet count?

	Blood platelet count	
	Abnormal	Normal
Spleen size		
0	0	13
1	2	10
2	5	6
3	3	1

Formulation and assumptions. Exercise 7.15 indicated that platelet counts tend to decrease as spleen size increases. Ties were minimal as the actual counts were given. Here we must take ties into account in an asymptotic test.

Procedure. Proceeding as in Examples 12.6 and 12.7, we find $U = 183$ and $E(U) = 287.5$. Using (12.11) we find the standard deviation of U is $\sqrt{\text{Var}(U)} = 30.67$, whence $Z = -3.407$. From tables of the standard normal distribution we find

$$\Pr(Z \leq -3.407) = Pr(Z \geq 3.407) = 0.0003.$$

Conclusion. There is strong evidence that the proportion of individuals with abnormal counts increases as spleen size increases.

Comments. 1. Computing Z by the formula in Section 7.2.5 ignoring ties gives $Z = -2.551$, indicating a one-tail $P = 0.0054$ (compared with $P = 0.0003$ obtained above). An exact test using StatXact with these data gave $P = 0.0002$ for a one-tail test, in close agreement with the asymptotic result taking ties into account.

2. In general, ties have a greater effect on $\text{Var}(U)$ when they occur *between* samples (in columns) than if they occur largely *within* one sample (in a row).

3. Regarding counts of less than 120 as abnormal is arbitrary. Using an extended data set, Bassendine et al. (1985) considered counts below 100 as abnormal. With such arbitrary decisions about categories, one choice may lead to a result significant at a formal level such as $\alpha = 0.05$ or $\alpha = 0.01$ while another does not. In Section 12.4.2 we discuss a similar problem arising with grouped continuous data.

4. Since we have only two columns, they remain ordered if we interchange them. We might look upon *abnormal* and *normal* as designating samples from two populations, and spleen sizes as rankings for those populations, and apply the WMW test as in Example 12.4. Doing so, we get exactly the same results as we do with Jonckheere–Terpstra.

The expression for $\text{Var}(U)$ in (12.11) is unaltered if we interchange rows and columns in an $r \times c$ table. However, the statistic U and the mean $E(U)$ are affected by this change. It is not difficult to show that if we write U_1 for the Jonckheere–Terpstra statistic for the original $r \times c$ table and U_2 for the corresponding statistic when rows and columns are interchanged then $U_2 = U_1 + (\sum_i n_{i+} - \sum_j n_{+j})/4$ and $E(U_2) = E(U_1) + (\sum_i n_{i+} - \sum_j n_{+j})/4$.

The asymptotic tests using U_1 and U_2 are identical. Although U_1 and U_2 differ by the constant value given above, their permutation distributions are otherwise identical, i.e., for all k

$$\Pr(U_1 = k) = Pr[U_2 = k + (\sum_i n_{i+} - \sum_j n_{+j})/4].$$

This equivalence may seem surprising as the Jonckheere–Terpstra test was introduced as a test for a monotone trend between samples. However, if there are no ties in p samples with $n_1, n_2, \dots , n_p$ observations respectively the problem may be stated in a contingency table format with p rows and $N = \sum_i n_i$ columns. In each column all cells except one have zero entries. The remaining cell has an entry unity. The entries in row i are constrained so that the sum of the cell entries that are unity (all others being zero) is n_i. If we interchange rows and columns the equivalent test problem is one in which we have N samples each of one observation and these take only p different values of which there are n_1 tied at the lowest value, n_2 tied at the next lowest, and so on. An extreme case is that in which we have N samples, each of one observation and none of these is tied. We may represent this situation by an $N \times N$ table with all column totals and all row totals equal to 1. This implies that in any row or column there is one cell value of unity and all others are zero. In this case $U_2 = U_1$. As pointed out in Section 10.1.4 the Jonckheere–Terpstra test procedure is here equivalent to that for Kendall's tau; U_1 or U_2 are the numbers of concordances in this no-tie situation, and this is sufficient to determine Kendall's t_k.

Any situation with ties, and where the concept of linear rank association implicit in Kendall's tau is relevant, may be looked upon as generating an $r \times c$ matrix which shrinks an $N \times N$ matrix of zeros and ones by super-imposing rows with identical x values and superimposing columns with identical y. In particular, if the ith sample has n_i observations we shrink n_i adjacent rows to a single row by adding all column entries in those n_i adjacent rows. The x variable values in this situation are effectively labels that distinguish the (ordered) samples. We have pointed out that in the no-tie case the Jonckheere–Terpstra statistic is equivalent to the number of concordances in Kendall's t_k. With ties, the two statistics use slightly different counting systems. The Kendall statistic counts concordances as unity, ties as zero and discordances as -1. for the Jonckheere–Terpstra statistic, concordances count as 1, ties as $1/2$ and discordances as zero. With appropriate adjustments to certain formulae the two counting systems are interchangeable and, used correctly, lead to the same conclusions.

If rows and columns are both response categories we may be interested in whether there is a patterned association (akin to a correlation) between row and column classifications in that high responses in one classification tend to be associated with high responses in the other, or high responses in one are associated with low responses in the other. Again the Jonckheere–Terpstra test is relevant.

Several statistics are used as measures of association but none is completely satisfactory in all circumstances. For some of those suggested, an exact Jonckheere–Terpstra test is appropriate to test for significant association in the form of a monotonic trend. However, for non-monotonic associations such tests may fail to detect patterns of association. For example, in an ordered-categories contingency table with cell entries

$$\begin{array}{ccc} 0 & 17 & 0 \\ 5 & 0 & 5 \\ 3 & 0 & 3 \end{array} \qquad (12.12)$$

there is a clear — but obviously not monotonic — association. It is easy to verify (Exercise 12.4) that the asymptotic Jonckheere–Terpstra statistic Z takes the value 0, giving $P = 0.50$ for a one-tail test. Apart from a small discontinuity effect the exact test confirms this probability.

A desirable property of a measure of monotonic association, akin to that of a correlation coefficient, is that it takes the value 1 when there is complete positive association, the value zero if there is no rank association, and the value -1 if there is inverse association, e.g., high category row classifications are associated with low category column classifications and vice-versa. An appropriate test for linear rank association using such statistics is often the Jonckheere–Terpstra test, but to measure degree of association there is a case for having a statistic that behaves like a correlation coefficient. We discuss one such coefficient here. This, and other measures of association are described in considerable detail by Kendall and Gibbons (1990, Chapter 3) and by Siegel and Castellan (1988, Chapter 9). These and other authors give guidance on relevant asymptotic tests which take account of the tied nature of the data.

The *Goodman–Kruskal gamma statistic* is usually denoted by G. We count the number of concordances and discordances between row and column classifications as for Kendall's t_k, but make no allowance for ties. When both row and column classifications are ordered, for each count in cell (i, j) there is concordance between that count and the count in any cell below and to the right. Thus, if N_{ij}^+ denotes the sum of all counts below and to the right of cell (i, j), which itself has count n_{ij}, the total number of concordances is

$$C = \sum_{i,j} n_{ij} N_{ij}^+, \ 1 \le i \le r - 1, \ 1 \le j \le c - 1.$$

This resembles the count for the Jonckheere–Terpstra test, except that for counts in cell (i, j) we ignore ties in column j. Each of these counted as $1/2$ in Jonckheere–Terpstra. Similarly, denoting the sum of all elements to the left of and below the cell (i, j) by N_{ij}^-, the total number of discordances is

$$D = \sum_{i,j} n_{ij} N_{ij}^-, \ 1 \le i \le r - 1, \ 2 \le j \le c.$$

The Goodman-Kruskal statistic is $G = (C - D)/(C + D)$. If $D = 0$ then $G = 1$. If $r \geq c$, then $G = 1$ if all entries in any column except the last are zero, unless $i = j$; in the last column they must be zero if $i < c$. If $c \geq r$, $G = 1$ if all entries in any row except the last are zero, unless $i = j$. In the last row they must be zero if $j < r$. If $r = c$, $G = 1$ if all entries for which $i \neq j$ are zero. In Exercise 12.5 we ask for confirmation that $G = 1$ for each of the tables

$$
\begin{array}{cccc}
 & & & \\
3 & 0 & 0 & 0 \\
0 & 4 & 2 & 7
\end{array}
\qquad
\begin{array}{ccc}
7 & 0 & 0 \\
0 & 1 & 0 \\
0 & 0 & 6 \\
0 & 0 & 7 \\
0 & 0 & 5
\end{array}
$$

It is easy (Exercise 12.6) to obtain analogous conditions for $G = -1$. When there is a good scatter of counts over all cells, C and D will not be very different and G will be close to zero. Near-zero values of G are also possible if association is not monotonic. It is easily verified (Exercise 12.7) that $C = D$, hence $G = 0$, in the highly patterned contingency table in (12.12).

Like Kendall's tau (as distinct from Kendall's tau-b), G takes no account of ties in both x and y, yet unlike Kendall's tau it may equal 1 even when there is heavy tying because the denominator is $C + D$. This falls well short of the denominator in Kendall's tau for the no-tie case. In Section 10.1.4 we used Kendall's t_b in preference to t_k when there are ties. This is a compromise between ignoring them (as in t_k) and deleting their effect if in the same row or column as we do in G. Calculation of t_b for contingency tables is straightforward and follows the method outlined in Section 10.1.4.

Other measures of association include Somers' D, proposed by Somers (1962). It is closely related to the Goodman–Kruskal statistic and to Kendall's tau. This and some other appropriate procedures for measuring association, are described by Siegel and Castellan (1988, Chapter 9). We concur with their comment that "all of them should be useful when appropriately applied." We describe other approaches to exploring association in contingency tables in Chapter 13.

12.4 Goodness-of-Fit Tests for Discrete Data

The Pearson chi-squared test and the likelihood ratio test (Sections 12.2.2 and 12.2.3) may be looked upon as tests of goodness-of-fit of data in a contingency table to a model of independence (i.e., no association) between row and column classifications. The asymptotic chi-squared test is also widely used as a goodness-of-fit test of data to any discrete distribution. This topic is covered in most standard courses in statistics. We give only a brief resume here, drawing attention to a few points that are sometimes not covered specifically in introductory treatments.

Tests of goodness-of-fit to any discrete distribution (often appropriate for counts) are usually based on the Pearson chi-squared statistic, X^2, in (12.7).

Table 12.7 *Recorded last digit of age at death, Badenscallie males.*

Digit	0	1	2	3	4	5	6	7	8	9
Freq	7	11	17	19	9	13	9	11	13	8

Expected values are calculated for a hypothesized distribution which may be binomial, negative binomial, Poisson, uniform or some other discrete distribution. Sometimes one or more parameters must be estimated from the data. If we have r counts, or cells, the test will have $r - 1$ degrees of freedom if no parameters are estimated. If a parameter (e.g., p for the binomial, or λ for the Poisson distribution) is estimated from data, one further degree of freedom is lost for each parameter estimated. While tests of the Kolmogorov–Smirnov type are sometimes also applied for discrete distributions, generally the test criteria are no longer exact and the tests are often inefficient.

Example 12.9

The problem. It is often suggested that recorded ages at death are influenced by two factors. The first is the psychological wish to achieve ages recorded by *decade*, e.g., 70, 80, 90, and that a person nearing the end of a previous age decade who knows his or her days may be numbered will by sheer willpower or perhaps even by a change in lifestyle (e.g., by stopping smoking or abstaining from alcohol) strive to attain such an age before dying. If so, ages at death with final digits 0, 1 should be more frequent than those with higher final digits. A second factor is misstatement of age, elderly people may be imprecise about giving their age, tending to round it; e.g., if they are in the mid-seventies they tend to say they are 75 if anywhere between about 73 and 77. Similarly, a stated age of 80 may correspond to a true age a year or two above or below. If these factors operate, final digits in recorded ages at death would not be uniformly distributed. Table 12.7 gives final digits at age of death for the 117 males recorded at the Badenscallie burial ground (see Appendix 1). Is the hypothesis H_0: *any digit is equally likely* acceptable?

Formulation and assumptions. For 117 deaths the expected number of occurrences of each of the 10 digits, if all are equally likely, is 11.7. Since the denominator in X^2 is always the expected number, $m = 11.7$, we may calculate X^2 by taking the differences between the observed numbers n_i and 11.7, squaring these, adding and finally dividing the total by 11.7.

Procedure. The differences $n_i - m$ are respectively

$$-4.7, \ -0.7, \ 5.3, \ 7.3, \ -2.7, \ 1.3, \ -2.7, \ -0.7, \ 1.3, \ \ -3.7.$$

These sum to zero – a well-known property of deviations from the mean. The sum of their squares is 136.1. Division by the expected number $m = 11.7$ gives $X^2 = 136.1/11.7 = 11.63$. This is well below the value 16.92 for χ^2 that corresponds to

$P = 0.05$ when there are $10 - 1 = 9$ degrees of freedom. Indeed, it corresponds to $P = 0.235$. StatXact gives the exact permutation $P = 0.239$.

Conclusion. There is no real evidence against H_0.

Comments. 1. In posing the problem we suggested a possible build-up of digits 0, 1 and perhaps also 5. The observed variation in digit frequency is not in line with such a pattern. Any build-up is around 2 and 3. As the overall result is not significant we must not read too much into this.

2. In this data set ages at death ranged from less than 1 to 96 years. In particular, there is some evidence of an increased risk of death for very young children. In all, 41 of the 117 recorded ages at death were less than 60 years. The influence of elderly people is therefore diluted.

3. Survival until a notable family event (e.g., the birth of a grandchild) may be a stronger incentive to an elderly person than the completion of an age decade.

4. Misstatement of age occurs across a wide range of ages for a variety of reasons, discussed in detail by Cox (1978).

12.4.1 Goodness-of-Fit with an Estimated Parameter

We often meet data that might belong to a particular distribution such as the Poisson distribution with unknown mean. We may use the statistic X^2 to test this, although for the Poisson and some other distributions alternative parametric tests may be more powerful.

For the asymptotic test using X^2 a difficulty arises if some expected numbers are small. Traditional advice is to group such cells to give a group expected number close to 5 (but this is only a guide). There is a corresponding reduction in degrees of freedom.

Example 12.10

The problem. A factory employs 220 people. The numbers experiencing $0, 1, 2, 3, \ldots$ accidents in a given year are recorded

Number of accidents	0	1	2	3	4	5	≥ 6
Number of people	181	9	4	10	7	4	5

Are these data consistent with a Poisson distribution?

Formulation and assumptions. The maximum likelihood estimate, $\hat{\lambda}$, of the Poisson parameter λ, is the mean number of accidents per person. The expected numbers having r accidents are $E(X = r) = 220\hat{\lambda}^r exp(-\hat{\lambda})/r!$. This follows because, for a Poisson distribution with parameter λ, $\Pr(X = r) = \lambda^r exp(-\lambda)/r!$.

Procedure. To obtain the mean number of accidents per person we multiply each number of accidents by the number of people having that number of accidents, add these products and divide the sum by 220, giving

$$\hat{\lambda} = [(0 \times 181) + (1 \times 9) + (2 \times 4) + (3 \times 10) + (4 \times 7) + (5 \times 4) + (6 \times 5)]/220 \approx 0.568.$$

This is approximate because we treat "six or more" as exactly six. In practice, this has little influence on the result, but it is a limitation we should recognize. To get expected numbers an iterative algorithm may be used, noting that

$$E(X = r + 1) = E(X = r) \times \hat{\lambda}/(r+1).$$

The expected numbers are

Number of accidents	0	1	2	3	4	5	6 or more
Number of people	124.7	70.8	20.1	3.8	0.5	0.1	0

Grouping results for 3 or more accidents gives an associated total expected number 4.4. We calculate X^2 using (12.8) as

$$X^2 = (181^2/124.7) + (9^2/70.8) + (4^2/20.1) + (26^2/4.4) - 220 = 198.3.$$

There are 2 degrees of freedom, since we have four cells in the final test and we have estimated one parameter, λ.

Conclusion. It is clear, even if using only conventional chi-squared tables, that there is extremely strong evidence that the data are not consistent with a Poisson distribution since, with 2 degrees of freedom, $\Pr(\chi^2 \geq 13.81) = 0.001$.

Comment. Accident data seldom follow a Poisson distribution. They would if accidents occurred entirely at random and each person had the same probability of experiencing an accident, and having one accident did not alter the probability of that person having another. In practice, a better model is one that allows people a differing degree of accident proneness. The negative binomial distribution assumes that individual proneness has a gamma distribution within the population. This is very flexible and has been shown to give a good fit to numbers of episodes of mental illness experienced by adults in a fixed period of several years (Smeeton, 1986). If an individual experiences an accident, the probability of a further accident may be reduced if he or she becomes more careful, or increased if concentration or confidence is lowered, making that person more accident-prone. These factors may act differently from person to person, or for individuals exposed to different risks. Multiple accidents, in which a number of people are involved in one incident, also affect the distribution of numbers of accidents per person in some accident data.

Computational aspects. Nearly all standard statistical packages include a program for chi-squared goodness-of-fit tests.

12.4.2 Goodness-of-Fit for Grouped Data

The chi-squared test is sometimes applied to test goodness-of-fit of grouped data from a specified continuous distribution, commonly a normal distribution with specified mean and variance, or with these estimated from the data. This is not recommended unless the grouped data are all that are available and grouping is on some natural basis. If the grouping is arbitrary, one arbitrary grouping may result in rejection of a hypothesis while some other grouping may not.

A situation where a test for normality based on Pearson's X^2 might be used is that for sales of clothing of various sizes. A large retailer might note numbers of sales of ready-made trousers with nominal leg lengths (cm) 76, 78, 80, etc. The implication is that customers requiring some ideal leg length between 75 and 77 cm will purchase trousers with leg length 76, those requiring leg lengths between 77 and 79 will purchase trousers of length 78, and so on. The sizes are the centre value for each group. To test whether sales at nominal lengths are consistent with a normal distribution with unspecified mean and variance, these parameters must be estimated from the grouped data and 2 degrees of freedom deducted to allow for this.

Given complete sample data from a continuous distribution, it is better to use goodness-of-fit tests of the Kolmogorov/Lilliefors type as appropriate, or other tests relevant to particular continuous distributions. If grouping is used, and tests are to be based on X^2, Kimber (1987) points out that when both grouped and ungrouped data are available anomalies may arise unless parameter estimates are based on the grouped data.

12.5 Extension of McNemar's Test

There is an asymptotic chi-squared approximation equivalent to the normal approximation to the binomial for McNemar's test (Section 5.2). For the data in Table 5.3, for example, we tested, in effect, whether 9 successes and 14 failures (or vice versa) are consistent with a binomial distribution with $p = 0.5$ and $n = 9 + 14 = 23$. We may perform a goodness-of-fit test using X^2. The expected numbers of successes or failures is 11.5, and the test statistic is $X^2 = (2.5)^2/11.5 + (2.5)^2/11.5 = 1.09$. This is compared with the relevant critical value for chi-squared with 1 degree of freedom. Recalling that the relevant cells in a 2×2 table for the McNemar test are the off-diagonal cells n_{12}, n_{21}, it is easily verified that with these numbers of successes or failures ($n_{12} = 9$ and $n_{21} = 14$ in Table 5.3) then

$$X^2 = (n_{12} - n_{21})^2/(n_{12} + n_{21}).$$

In Section 7.4 we introduced Cochran's test for comparing several treatments with binary (0,1) responses. When $t = 2$ it can be shown (Exercise 12.19) that Cochran's Q is equivalent to the form given above for X^2. In this sense the Cochran test is a generalization of McNemar's test.

Another generalization of McNemar's test is from a 2×2 contingency table to an $n \times n$ table in which we test for off-diagonal symmetry. Bowker (1948) proposed a relevant test. The test might be used if there is a new formulation of a drug which it is hoped will reduce side-effects. If we have a number of patients who have been treated with the old formulation, and records are available of any side-effects we might, if there are not ethical reasons to preclude this course, now treat each of these patients with the new formulation and note incidence of side-effects. Table 12.8 shows a possible outcome for such an experiment.

Table 12.8 *Side-effects with old and new formulation of a drug.*

| | | Side-effect levels — new formulation | | | |
		None	*Slight*	*Severe*	*total*
Side-effect	*None*	83	4	3	90
levels — old	*Slight*	17	22	5	44
formulation	*Severe*	4	9	11	24
	Total	104	35	19	158

In the table we see that each off-diagonal count below the diagonal exceeds that in the symmetrically placed cell above the diagonal, e.g., $n_{31} = 4$ is greater than $n_{13} = 3$, etc. This gives the impression that the trend is toward less severe side-effects with the new formulation, although a few who suffered no side-effects with the old formulation do show some with the new formulation.

Bowker proposed a test statistic to determine whether at least one pair of probabilities associated with the symmetrically placed-off diagonal cells differ in an $n \times n$ table. It is a generalization of the X^2 statistic used in McNemar's test and takes the form

$$X^2 = \sum \frac{(n_{ij} - n_{ji})^2}{n_{ij} + n_{ji}} \qquad (12.13)$$

where the summation is over all i from 1 to $n - 1$ and $j > i$. Under the null hypothesis of symmetry the asymptotic distribution of X^2 in (12.13) is chi-squared with $n(n-1)/2$ degrees of freedom.

Example 12.11

The problem. Do the data in Table 12.8 provide evidence that side-effects are less severe with the new formulation of the drug?

Formulation and assumptions. The Bowker statistic given by (12.13) is appropriate. Here $n_{12} = 4$, $n_{13} = 3$, $n_{21} = 17$, etc. The null hypothesis is that for all off-diagonal counts in the table the associated probabilities are such that all $p_{ij} = p_{ji}$. The alternative is that for at least one such pair $p_{ij} \neq p_{ji}$.

Procedure. Substitution of the relevant values in (12.13) gives

$$X^2 = (4 - 17)^2/21 + (3 - 4)^2/7 + (5 - 9)^2/14 = 9.33.$$

Under the null hypothesis, X^2 has a chi-squared distribution with 3 degrees of freedom, and appropriate tables or software establishes that with 3 degrees of freedom $P = \Pr(\chi^2 \geq 9.33) = 0.026$.

Conclusion. There is evidence of a differing incidence rate for side-effects under the two formulations. From Table 12.8 it is clear that this difference is towards less severe side-effects under the new formulation.

Comment. Marascuilo and McSweeney (1977, Section 7.6) discuss an alternative test based on marginal totals due to Stuart (1955; 1957). Agresti (2002, Chapter 10) discusses other generalizations of McNemar's test.

12.6 Fields of application

We indicate a few situations where tests for independence for categorical data may be appropriate.

Rail Transport

A train company may be interested in whether its image among standard class passengers is different from that among first class passengers. It may ask samples of each to grade service received as excellent, good, fair or poor, setting up a 2×4 table of ordered response numbers.

Television Viewing

A public service broadcasting channel competes with a commercial service. Samples of men and women are asked which they prefer. The results may be expressed in a 2×2 table, any differences between the genders in preference ratings being of interest. An extended test might involve four samples — adult males, adult females, teenage males, teenage females, leading to a 2×4 table.

Drug Addiction

A team of doctors compares 3 treatments allocated at random to subjects for curing drug addiction. For each subject, withdrawal symptoms are classed as severe, moderate, mild or negligible. The resulting 3×4 table of counts is tested to see if there is evidence of an association between treatments and severity of withdrawal symptoms. If the treatments can be ordered (e.g., by degree of physiological changes induced) a test for association of the Jonckheere–Terpstra type might be used. In the absence of such ordering, a Kruskal–Wallis test would be appropriate.

Sociology

A sociologist may be interested in variations in the perception of stigma related to asthma within white, Black Caribbean and Asian ethnic groups. A 2×3 table could be used in a test for lack of association between ethnic group and presence or absence of perceived stigma.

Public Health

After a contaminated food episode on a jumbo jet some passengers show mild cholera symptoms. The airline wants to know if those previously inoculated show a higher degree of immunity. They find out which passengers have and have not been inoculated; records also show how many in each category exhibit symptoms, so they can test for association.

Rain Making

In a low-rainfall area rain-bearing clouds are 'seeded' to induce precipitation. Randomly selected clouds are either seeded or not seeded on a sequence of occasions and we observe after each whether or not local rainfall occurs within the next hour, the results being expressed in a 2×2 table, a test being made for evidence of association.

Educational Research

Children are shown a video of a roller-coaster ride. This may reassure a child who is frightened of such rides but arouse fear in one who was initially unconcerned. McNemar's test may be used to indicate whether the video does more harm than good.

Medicine

To test whether a drug is effective it and a placebo are given to patients in random order. It is noted which, if either, they claim to be effective. Some claim both are effective, some neither, some one but not the other. McNemar's test could be used to see if results favour the placebo, the new drug, or neither.

The following are examples of goodness-of-fit tests.

Genetics

Genetic theory may specify the proportions of plants in a cross that are expected to produce blue or white flowers, round or crinkled seeds, etc. Given a sample in which we know the numbers in each colour/seed-shape combination, we may use a chi-squared goodness-of-fit test to see if these are consistent with theoretical proportions.

Sport

It is often claimed that the starting positions in horse racing, athletics or rowing events may influence the probability of winning. If we know the starting positions and winners for a series of rowing events in each of which there are six starters we might test the hypothesis that the numbers of wins from each starting position is consistent with a uniform distribution.

Horticulture

The positions at which leaf buds or flowers form on a plant stem are called nodes. Some theories suggest a negative binomial distribution for the node number (counted from the bottom of the stem) at which the first flower forms. Given data, a chi-squared goodness-of-fit test is appropriate.

Commerce

A car salesman may doubt the value of advertising. Each week he advertises in either 0, 1, 2, 3 or 4 newspapers. After a long period he might compare weekly sales with the number of advertisements. He would expect a uniform distribution for sales if advertising were worthless. However, in this situation it might be necessary to adjust for trends in sales over time. The effect of this might be minimized if weeks were chosen at random for each number of advertisements.

12.7 Summary

Tests for independence between row and column classifications in contingency tables are usually based on conditional inference, regarding the marginal totals as fixed (Section 12.1).

For $r \times c$ tables with nominal categories exact tests for association are commonly based on the **Fisher exact test** (Section 12.2.1), while asymptotic tests are usually based on the **Pearson chi-squared statistic**, X^2 (Section 12.2.2) or the **likelihood ratio statistic**, G^2 (Section 12.2.3). Exact permutation tests may also be based on X^2 and G^2. While the permutation distributions differ from that for the Fisher exact test (except in 2×2 tables) conclusions for all three approaches are often broadly in line.

For categorical tables with ordered rows and columns, tests equivalent to the **Jonckheere–Terpstra test** may be used (Section 12.3.2). Adjustments for ties are important if asymptotic theory is used. If only columns are ordered the **Kruskal–Wallis test** (Section 12.3.1) may be relevant. The **Goodman–Kruskal gamma statistic**, G, (Section 12.3.2) is closely related to Kendall's tau as a measure of association.

The **chi-squared goodness-of-fit test** (Section 12.4) is appropriate for testing goodness-of-fit to discrete distributions. It uses the Pearson X^2 statistic, the degrees of freedom depending on the number of parameters that are estimated from the data.

McNemar's test for off-diagonal symmetry in 2×2 tables may be carried out using a statistic (Section 12.5) which has an asymptotic chi-squared distribution. This may be extended to **Bowker's test** for off-diagonal symmetry in $n \times n$ tables.

12.8 Exercises

12.1 Use (12.6) to verify that the probability of the configuration in Example 12.1 given by the Fisher exact test is $P = 0.000058$.

12.2 Using the modified form of Var(U) for the Jonckheere–Terpstra test given in (12.11) show, for the data used in Examples 12.6 and 12.7, that the standard deviation of U is 145.18, whereas it is 415.36 if ties are ignored.

12.3 Confirm the results quoted in Example 12.8 for association between blood platelet count and spleen size.

12.4 Verify for the contingency table (12.12) that the asymptotic Jonckheere–Terpstra statistic Z takes the value zero. [It is not necessary to calculate $\text{Var}(U)$ to confirm this.]

12.5 Confirm that the Goodman–Kruskal statistic G takes the value $G = 1$ for each of the contingency tables near the end of Section 12.3.2 (p. 374).

12.6 Determine conditions similar to those given for the case $G = 1$ to ensure that the Goodman-Kruskal statistic takes the value $G = -1$.

12.7 Verify that $C = D$ and $G = 0$ for the Goodman–Kruskal statistic for the table (12.12).

12.8 In a psychological test for pilot applicants, each is classed as extrovert or introvert, and is subjected to a test for flying aptitude that he or she may pass or fail. Do the results suggest an association between aptitude and personality type?

	Introvert	Extrovert
Pass	14	34
Fail	31	41

*12.9 A manufacturer of washing machines issues instructions for their use in English for the UK and U.S. markets, French for the French market, German for the German market and Portuguese for the Portuguese and Brazilian markets. The manufacturer conducts a survey of randomly selected customers in each of these markets and asks them to classify the instructions (in the language appropriate to that country) as excellent, reasonable, or poor. Do the responses set out below indicate the instructions are more acceptable in some countries than in others?

	Excellent	Reasonable	Poor
UK	42	30	28
USA	20	41	19
France	19	29	12
Germany	26	22	12
Portugal	18	31	21
Brazil	31	42	7

Note: It is sensible to consider UK/USA separately — also Portugal/Brazil — because differences in idiom may affect understanding of the instructions: e.g., British houses have taps, American houses have faucets.

12.10 In Palpiteria all who visit a doctor must pay. The data below for a random sample of wage earners show incomes in palpiliras (P), the country's unit of currency. Do they substantiate a politician's claim that charges inhibit poorer people from seeking medical aid?

	Income		
Time since last visit to doctor	Over 10 000P	5000 – 10 000P	Under 5000P
Under 6 months	17	24	42
6–12 months	15	32	45
Over 12 months	27	142	271
Never been	1	12	127

Your test should take into account the fact that the row and column categories are both ordered.

*12.11 Would your conclusions in Exercise 12.10 have been different if the data had been in only two income groupings, under 5000P and 5000P or more?

12.12 Prior to an England v Scotland football match 80 English, 75 Scottish and 45 Welsh supporters are asked who they think will win. Do the numbers responding each way indicate that the proportions expecting each side to win are influenced by nationality?

	English	Scottish	Welsh
English win	55	38	26
Scottish win	25	37	19

*12.13 A machine part is regarded as satisfactory if it operates for 90 days without failure. If it fails in less than 90 days it is unsatisfactory and this results in a costly replacement operation. A supplier claims that for each part supplied the probability of a satisfactory life is 0.95. Each machine requires 4 of these parts and all must be functional for satisfactory operation. To test the supplier's claim, a buyer runs each of 100 machines initially fitted with 4 new parts for a 90-day test period. The numbers of original parts (0–4) still functioning after 90 days are recorded for each machine as follows:

No. surviving	0	1	2	3	4
No. of machines	2	2	3	24	69

Do these results substantiate the supplier's claim?

12.14 It is claimed that a typesetter makes random errors at an average rate of 3 per 1000 words set, giving rise to a Poisson process. One hundred randomly chosen sets of 1000 words from his output are examined and the number of mistakes in each are counted. Are the results below consistent with the above claim?

No. of errors	0	1	2	3	4	5	6	7
No. of samples	6	11	26	33	12	6	4	2

12.15 It is often claimed that the probability of a horse winning a race on a circular or elliptic track is influenced by its starting position. Do the following data for starting positions of the winners of 140 eight-horse races support this claim? Position number 1 is on the inside of the track and number 8 on the outside.

Position	1	2	3	4	5	6	7	8
Wins	24	27	21	16	18	12	8	14

12.16 Some researchers believe that hospital-based surveys fail to reflect the true feelings of patients because of a fear that negative responses may affect their future care adversely. To investigate this hypothesis, 200 outpatients visiting a hospital clinic complete a questionnaire that includes the item "Are you satisfied overall with the standard of care that you have received at this hospital?" Each patient sees one interviewer from a group of five. The interviewers are as follows, and each wears an identity card that states their job title or role:

A: A hospital doctor
B: A professional survey interviewer
C: A local politician
D: A hospital nurse
E: A member of a self-help patient group.

Each interviewer sees 40 patients and the results are:

Interviewer	A	B	C	D	E
Number of yes	38	30	23	34	15

Assess the evidence that the type of person conducting the interview may influence responses.

*__12.17__ The medical records of a group of patients who have experienced a haemorrhagic stroke (brain haemorrhage) are searched for evidence of diabetes prior to their stroke. Tabulated by ethnic group (*white* or *Black Caribbean*) and by diabetic status (*no record of diabetes* or *treated for diabetes by insulin or tablet* or *untreated diabetes* the numbers were

	White	Black Caribbean
No diabetes	279	52
Treated diabetes	6	4
Untreated diabetes	15	5

Investigate whether there is a relationship between ethnicity and history of diabetes (if any) for these patients treating the categories as (i) unordered (ii) ordered, with untreated diabetes being of most concern.

12.18 To measure abrasive resistance of cloth, 100 samples of a fabric are each subjected to a 10–minute test under a series of 5 scourers, each of which may or may not produce a hole. The number of holes (0 to 5) is recorded for each sample. Are the data consistent with a binomial distribution with $n = 5$ and p estimated from the data? (Hint: Determine the mean number of holes per sample. If this is x then an appropriate estimate of p is $x/5$.)

No. of holes	0	1	2	3	4	5
No. of samples	42	36	14	3	4	1

12.19 The McNemar test data in Table 5.3 on climbs can be reformulated in a way that makes the Cochran test given in Section 7.4 appropriate with $t = 2$. Denoting a success by 1, and a failure by 0, we may classify each of the 108 climbers' outcomes for the first and second climb as either 0 or 1 in a 2×108 table. Show that the Cochran Q statistic is in this case identical with the McNemar X^2 statistic given in Section 12.5.

12.20 Aitchison and Heal (1987) give numbers of OECD countries using significant amounts of only 1, 2, 3 or 4 fuels in each of the years 1960, 1973, 1983. Are the proportions in the different categories of use changing significantly with time?

Think carefully about an appropriate test and the interpretation of your findings. The data are in no sense a random sample from any population of countries. The categories are ordered, but do you consider that the monotonic trends in association implicit in the conditions to validate, say, a Jonckheere–Terpstra test are likely to be relevant for these data?

		Year	
No. of fuels	*1960*	*1973*	*1983*
1	7	10	1
2	13	11	13
3	5	4	9
4	0	0	2

*12.21 Marascuilo and Serlin (1979) report a survey in which a number of women were asked whether they considered the statement "The most important qualities of a husband are determination and ambition" to be true or false. The respondents were asked the same question at a later date. Numbers making the possible responses were as follows:

First response	*Second response*	*Numbers*
True	*True*	523
True	*False*	345
False	*True*	230
False	*False*	554

Is there evidence that experience significantly alters attitudes of women toward the truth of the statement?

12.22 Manfredini et al. (1994) provide the following data on the 71 males admitted to a hospital for attempted suicide by self-poisoning, in Ferraro, Italy between 1989 and 1991. Their admission times grouped into intervals of 4 hours were:

Time	0200 − 0559	0600 − 0959	1000 − 1359
Number	4	8	13
Time	1400 − 1759	1800 − 2159	2200 − 0159
Number	11	22	13

Does the pattern of admissions vary according to the time of day?

12.23 Jarrett (1979) gives the following data for numbers of coal mine disasters involving 10 or more deaths between 1851 and 1962.

Day of week	Sun	Mon	Tue	Wed	Thu	Fri	Sat
Number	5	19	34	33	36	35	29

Month	Jan	Feb	Mar	Apr	May	Jun
Number	14	20	20	13	14	10
Month	Jul	Aug	Sep	Oct	Nov	Dec
Number	18	15	11	16	16	24

Test whether accidents appear to be uniformly spread over days of the week and over months of the year. What are the implications? Do they surprise you?

*12.24 Dansie (1986) gives data for a survey in which 800 people were asked to rank 4 makes of car A, B, C, D in order of preference. The number of times each rank combination was specified is given below in brackets after each order. Do the data indicate that preference may be entirely random? Is there a significant preference for any car as first choice?

ABCD(41), ABDC(44), ACBD(37), ACDB(36), ADBC(49), ADCB(41),
BACD(38), BADC(38), BCAD(25), BCDA(22), BDAC(33), BDCA(25),
CABD(31), CADB(26), CBAD(40), CBDA(33), CDAB(33), CDBA(35),
DABC(23), DACB(39), DBAC(30), DBCA(21), DCAB(26), DCBA(34).

12.25 Noether (1987b) asked students to select by a mental process what they regarded as random pairs from the digits 1, 2, 3, repeating that process four times. Noether recorded frequency of occurrences of the last digit pair written down by each of 450 students. The results were:

| | | First digit | | |
		1	2	3
	1	31	72	60
Second digit	2	57	27	63
	3	53	58	29

What would be the expected numbers in each cell of the above table if pairs were truly random? Test whether one should reject the hypothesis that the students are choosing digits at random. How do you interpret your finding?

*12.26 For the following 2×12 table calculate the Pearson X^2 and the likelihood ratio G^2 statistics. Explain the basic cause of any difference between their values. If suitable computer software is available, determine the exact tail probabilities corresponding to each test statistic.

0	0	0	0	0	0	0	0	0	0	1	1
3	4	17	2	5	1	8	6	4	11	3	0

ASSOCIATION IN CATEGORICAL DATA

13.1 The Analysis of Association

When we reject the hypothesis of independence in contingency tables we often want to analyse the nature of association. In Section 12.3.2 we considered one aspect of association in tables with ordered row and column categories, recommending the Jonckheere–Terpstra test for monotonic association. We consider further aspects of association in this chapter. Our treatment is indicative rather than comprehensive. Approaches used to assess association range from parametric modelling to distribution-free methods. Parametric or semi-parametric approaches are usually formulated to reflect knowledge about the mechanism generating the counts. One such model is the logistic regression model for binary responses such as success and failure where the proportions responding depend upon levels of one or more explanatory variables. General descriptions of this model are given by Agresti (1996, Chapter 5 and 2002, Chapter 5) and a detailed treatment covering many applications is presented in Cox and Snell (1989). We return to this model in Section 15.4.

In Chapter 12 we considered two-way $r \times c$ tables. Three or more way tables arise in practice. These may require detailed, and sometimes subtle analysis, to elucidate patterns of association. The techniques used range from extensions of some of the nonparametric models outlined for a few special cases in this chapter, to analyses based on generalized linear models. A detailed account of the latter is given in McCullagh and Nelder (1989); Dobson (2001) gives a more elementary treatment illustrated by many practical applications.

A common way of presenting data from a three-way classification is by means of two-way cross-sectional tables. Table 13.1 is an illustration. The counts cover 197 patients in a study of coronary heart disease. The first categorization is into presence (CHD) or absence (no CHD) of the disorder. The second classification is at three blood cholesterol levels (A, B, C), and the third is into five blood pressure groups (I to V).

Table 13.1 shows two cross-sections of the second and third classifications determined or *stratified* by each of the two levels of the first classification (disease present or absent). Marginal totals relevant to each cross-section and a set of totals for each blood pressure group across the two strata are also given. The data may be presented in other ways (Exercise 13.1).

Tests for complete independence extend from two- to multiway tables, and may also be applied to two-way cross-sectional tables, but these analyses are

Table 13.1 *Categorization of 197 patients in a study of coronary heart disease.*

		Blood pressure group					
		I	II	III	IV	V	Total
	Cholesterol						
	A	2	3	1	0	4	10
CHD	B	2	1	5	3	0	11
	C	4	7	8	6	2	27
	Total (1)	8	11	14	9	6	48
	A	16	14	11	8	6	55
No CHD	B	22	18	5	3	2	50
	C	15	13	10	5	1	44
	Total (2)	53	45	26	16	9	149
	Total (1 + 2)	61	56	40	25	15	197

usually insufficient (and often inefficient) because we are more interested in exploring the nature of associations.

A particular type of three-way classification that has received considerable attention is the $k \times 2 \times 2$ table. The method of analysis depends on what questions are of interest and on the nature of the categories in each classification. In particular, it is often productive to look upon such tables as representing k strata, each stratum consisting of a 2×2 table, a situation we consider in Sections 13.2.2 to 13.2.5.

For 2×2 tables an appropriate measure of association is the *odds ratio* or, often better, the logarithm of the odds ratio which we introduced in Section 12.1.2.

13.2 Some Models for Contingency Tables

13.2.1 The Loglinear Model

In any $r \times c$ table if there is independence the expected frequency in cell (i, j) is $m_{ij} = n_{i+}n_{+j}/N$, where n_{i+}, n_{+j} are the ith row and jth column totals and N is the total of all counts in the table (Section 12.1.1). Taking logarithms

$$\log m_{ij} = \log n_{i+} + \log n_{+j} - \log N \qquad (13.1)$$

This means that under independence, for each cell the logarithm of the expected number is a linear function of the logarithms of the row, column and grand totals.

This is reminiscent of the additive linear model (i.e., one without interactions) described in Section 8.3, with the parameters there being replaced by logarithms of certain contingency table marginal totals. With the model in Section 8.3 the observed yield for any unit differed from the expected yield by an added amount usually regarded as a random error or departure from expectation.

By analogy, even when there is no association (i.e., independence) between row and column classifications in an $r \times c$ contingency table in general m_{ij} will not equal the observed count n_{ij} for cell (i, j). For the model based on (13.1) we may look upon the difference $\log n_{ij} - \log m_{ij} = \log(n_{ij}/m_{ij})$ as an additive error or departure term when the model for no association is accepted as adequate. This quantity appears in the likelihood ratio statistic G^2 given in (12.10).

Association in contingency tables has analogies with interaction in factorial treatment structures in the analysis of variance and these analogies are exhibited in extensions of (13.1) that allow for association. A detailed treatment is beyond the scope of this book, but we illustrate the basic ideas for 2×2 and $2 \times 2 \times 2$ contingency tables and briefly indicate some extensions to simple cases for $r \times c$ tables.

In Section 12.1.2 we established that for a 2×2 contingency table the condition for no association (independence) between row and column categories was that the odds ratio $\theta = (m_{11}m_{22})/(m_{12}m_{21}) = 1$, or equivalently, that $m_{11}m_{22} = m_{12}m_{21}$. Taking logarithms we get for the loglinear model

$$\log m_{11} + \log m_{22} - \log m_{12} - \log m_{21} = 0 \tag{13.2}$$

as the criterion for no association.

For observed counts n_{ij} the empirical odds ratio $(n_{11}n_{22})/(n_{12}n_{21})$ does not usually equal 1. A test of association (or interaction) in a contingency table becomes one of determining whether this ratio differs sufficiently from 1 to provide strong evidence against H_0: $\theta = 1$, or equivalently, whether $\log n_{11} + \log n_{22} - \log n_{12} - \log n_{21}$ differs significantly from zero; i.e., whether we prefer

$$H_1 : \log m_{11} + \log m_{22} - \log m_{12} - \log m_{21} = I, \qquad I \neq 0 \tag{13.3}$$

to H_0: $I = 0$. Here (13.3) is the analogue of (8.2).

This is called a first-order, or two-factor, interaction model, and represents the only possible kind of interaction with a 2×2 factorial treatment structure. The models in Section 8.1 and that in (13.3) extend respectively to $2 \times 2 \times 2$ factorial experiments and to $2 \times 2 \times 2$ contingency tables.

In the context of a loglinear model the m_{ijk} are the expected counts based on marginal totals in a way described below in Example 13.1. For $2 \times 2 \times 2$ tables the no association (no interaction) model corresponds to that of independence

Table 13.2 *Survival numbers for early and late planted long and short cuttings.*

	Planted early		Planted late	
	Alive	Dead	Alive	Dead
Cutting length				
Long	156	84	84	156
Short	107	133	31	209

Table 13.3 *Survival numbers for long and short cuttings planted early and late.*

	Long cuttings		Short cuttings	
	Alive	Dead	Alive	Dead
Planting time				
Early	156	84	107	133
Late	84	156	31	209

between classifications and in terms of odds ratios may be written

$$\frac{m_{111}m_{221}}{m_{121}m_{211}} = \frac{m_{112}m_{222}}{m_{122}m_{212}} = 1$$

Dependence, or association, may be first or second order (first- or second-order interaction in the loglinear sense). For a first-order interaction

$$\frac{m_{111}m_{221}}{m_{121}m_{211}} = \frac{m_{112}m_{222}}{m_{122}m_{212}} = k, \quad k \neq 1 \tag{13.4}$$

and for a second-order interaction

$$\frac{m_{111}m_{221}}{m_{121}m_{211}} \neq \frac{m_{112}m_{222}}{m_{122}m_{212}}$$

By taking logarithms analogies with the factorial treatment structure discussed in Section 8.1 becomes obvious. The reader should consult a specialist text such as Bishop, Fienberg and Holland (1975), Fienberg (1980), Plackett (1981), Agresti (1984, 1996, 2002) or Everitt (1992) for a detailed discussion of loglinear models and their use in measuring association.

We now consider association when we have a set of k cross-sectional tables each 2×2 with $k \geq 2$ which may be formed from or combined as a $k \times 2 \times 2$ table. How to answer some typical questions that may be asked will be illustrated using two specific data sets, the first given in Tables 13.2 and 13.3 and the second in Table 13.4. That in Tables 13.2 and 13.3 is a classic set first discussed by Bartlett (1935). The data are numbers of surviving (alive)

Table 13.4 *Side-effects of two drugs grouped by ages.*

| Age group (yr) | Drug | Side-effect status | |
		None	Some
20–29	A	8	1
	B	11	1
30–39	A	14	2
	B	18	0
40–49	A	25	3
	B	42	2
over 50	A	39	3
	B	22	6

and nonsurviving (dead) plum rootstock cuttings in each of four batches of 240 cuttings. The four batches were subjected to treatments having a 2×2 factorial structure. The first factor was *time of planting* — either *early* or *late* — and the second was *length of cutting* — either *long* or *short*. There are two possible responses at each level of each factor – *alive* or *dead*. Tables 13.2 and 13.3 represent two useful cross-sectional tables for comparing survival rates. Both contain the same information, but it is given in different order in each. In Table 13.2 the results are stratified by time of planting, since each of the component 2×2 tables (left and right) is for a different planting time. In Table 13.3 stratification is by length of cutting.

In Table 13.4 we effectively break down a $4 \times 2 \times 2$ table into four 2×2 tables where the $k = 4$ strata correspond to different age groups. In the 2×2 tables within each age group the explanatory variables are two drugs designated as A and B. The responses to each are the column categories *side-effect* or *no side-effect*. The k age groups forming the strata are also explanatory variables, often referred to by statisticians as *covariates*.

For the data in Tables 13.2 and 13.3 the Fisher exact test, the Pearson chi-squared test or the likelihood ratio test all indicate association in each cross-sectional table.

A question of interest about association in the basic $2 \times 2 \times 2$ table is whether there is evidence of only first-order association, or whether there might be second-order association. Remember that for the 2×2 cross-sectional tables independence implies that the expected odds ratios must each be unity, while first order association implies they should be equal, but the common value is not unity, while second-order association indicates they should not be equal.

For the data in Table 13.4 the possibilities are more varied. However, the ones likely to be of most interest are whether or not there is evidence of association, and if there is any association, whether or not it appears to be similar in nature, i.e., implies a common odds ratio, over all strata.

Further questions of interest arise if the odds ratio appears to change between strata. For instance, in Table 13.4 do the odds ratios of the cross-sectional 2×2 tables relating drugs and side-effects tend to increase with the stratification covariate *age*. Sometimes, with small numbers in individual strata, a test for independence applied separately in each strata may give no evidence to indicate association. However it is then wise to consider whether, collectively, there may be some evidence of association if we use appropriate tests. We consider the case of only two strata in Section 13.2.2 and the more general case of $k \geq 2$ strata in Sections 13.2.3 to 13.2.5.

13.2.2 Analysis of Two Cross-Sectional 2×2 Tables

Although some more general methods given in the next section apply to the case $k = 2$, once it has been established that there is evidence of association the main point of interest is often whether a first-order model adequately describes the association. When $k = 2$ this implies that (13.4) holds. We do not know the values of the expectations m_{ijk} for (13.4) but we may obtain from the data estimates $\hat{m}_{ijk}$ of the expected counts, where these are chosen to satisfy the condition

$$\frac{\hat{m}_{111}\hat{m}_{221}}{\hat{m}_{121}\hat{m}_{211}} = \frac{\hat{m}_{112}\hat{m}_{222}}{\hat{m}_{122}\hat{m}_{212}} \tag{13.5}$$

The $\hat{m}_{ijk}$ no longer have the simple values that they have for the separate cross-sectional tables under the hypothesis of independence. However, it is easily verified that if we retain the condition that all marginal totals in the complete $2 \times 2 \times 2$ table are fixed, there is only one degree of freedom and once we determine one of the $\hat{m}_{ijk}$ in (13.5) all the others are fixed. After these expected values are determined an associated P-value may be found using either a likelihood ratio test or a Pearson chi-squared test. Exact permutation tests are possible using these statistics, but the asymptotic chi-squared distribution results may be used for large samples. In practice, the likelihood ratio statistic G^2 has some advantages when we generalize from $k = 2$ to $k > 2$. Then there is more than one degree of freedom. We show in Section 13.3.2 that this latter statistic may be partitioned into additive single degree of freedom components in a sense that is not in general true for the Pearson X^2 statistic.

Example 13.1

The problem. Determine whether the data in Table 13.2 are consistent with a model that specifies only first order association.

Formulation and assumptions. We postulate a first-order association model (13.4) after first establishing that a no association or independence model is not appropriate. If (13.5) holds the condition that the marginal totals are fixed means that relationship can be used to calculate a value x for $\hat{m}_{111}$ under H_0: *association is first-order only.* The G^2 statistic may then be calculated using (12.10) and summing over all 8 cells of the two 2×2 subtables in Table 13.2.

Procedure. For convenience we reproduce below the data in Table 13.2. The Fisher exact test, a Pearson chi-squared test or a likelihood ratio test applied to the two 2×2 constituent tables shows clear evidence of association with all $P < 0.0001$.

	Planted early		Planted late	
	Alive	Dead	Alive	Dead
Cutting length				
Long	156	84	84	156
Short	107	133	31	209

If we denote $\hat{m}_{111}$ by x it is easily shown that the fixed marginal totals condition implies $\hat{m}_{121} = 240 - x$, $\hat{m}_{211} = 263 - x$, $\hat{m}_{221} = x - 23$, $\hat{m}_{112} = 240 - x$, $\hat{m}_{122} = x$, $\hat{m}_{212} = x - 125$, $\hat{m}_{222} = 365 - x$ and thus (13.5) is satisfied if x is chosen so that

$$\frac{x(x - 23)}{(240 - x)(263 - x)} = \frac{(240 - x)(365 - x)}{x(x - 125)}$$

This cubic equation in x may be solved numerically with an appropriate computer algorithm. The relevant solution is one that gives expected numbers that are all positive and turns out to be $x = \hat{m}_{111} = 161.1$. The remaining expectations may be calculated using the relationships given above and are summarized as

	Planted early		Planted late	
	Alive	Dead	Alive	Dead
Cutting length				
Long	161.1	78.9	78.9	161.1
Short	101.9	138.1	36.1	203.9

Calculating the likelihood ratio statistic G^2 using (12.10), where summation is over all 8 cells, gives $G^2 = 2.30$ (Exercise 13.2) which, with 1 degree of freedom implies $P = 0.1294$. The Pearson chi-squared statistic is $X^2 = 2.27$ giving $P = 0.1319$.

Conclusion. There is insufficient evidence to reject the hypothesis that there is only a first-order interaction.

Comments. 1. Accepting that there is only a first-order interaction is equivalent to saying that the odds ratios for the cross-sectional tables do not differ significantly.

2. We would reach exactly the same conclusion had we worked with the cross-sectional tables in Table 13.3 (Exercise 13.3). This is because each table contains essentially the same information.

3. We give an alternative analysis of these data in Example 13.2 and other methods developed in Section 13.2.3 to 13.2.5 for the case $k \geq 2$ may also be applied to this example if they are relevant.

Computational aspects. General statistical packages, e.g. SPSS, are tending to increase coverage of loglinear models giving at least asymptotic results. StatXact includes exact permutation tests for several of the methods described in the next section.

We pointed out in Section 12.1.2 that the odds ratio, or its logarithm, contains all the information on association in a 2×2 table. Example 13.1 showed one way of using that information. We also pointed out in Section 12.1.2 that for asymptotic tests normality is approached more rapidly if we use $\phi = \log \theta$ rather than θ itself.

If we use natural logarithms and denote the empirical log odds ratio by ϕ^*, i.e., $\phi^* = \log[(n_{11}n_{22})/(n_{12}n_{21})]$ then (see e.g., Agresti 2002, Chapter 3) ϕ^* is asymptotically normally distributed under the hypothesis of independence with mean zero and

$$\text{Var}(\phi^*) = 1/n_{11} + 1/n_{12} + 1/n_{21} + 1/n_{22}. \qquad (13.6)$$

If any $n_{ij} = 0$, both ϕ^* and $\text{Var}(\phi^*)$ become infinite. This difficulty is avoided if we replace all n_{ij} ($i, j = 1, 2$) by $n_{ij} + 0.5$ in (13.6) and differences in inference based on this modification are small when most n_{ij} are moderate or large. Association implies $\phi = \phi_0 \neq 0$ and that ϕ^* is asymptotically normally distributed with mean ϕ_0 and variance given by (13.6).

If we have two 2×2 tables with expected odds ratios θ_1 and θ_2 then H_0: $\theta_1 = \theta_2$ implies $\theta_2/\theta_1 = 1$ or $\phi_2 - \phi_1 = 0$. Thus, under H_0 it follows that asymptotically

$$Z = \frac{\phi_1^* - \phi_2^*}{\sqrt{\text{Var}(\phi_1^*) + \text{Var}(\phi_2^*)}} \qquad (13.7)$$

has a standard normal distribution.

Example 13.2

The problem. For the data in Table 13.2 show that the hypothesis H_0: $\theta_1 = \theta_2$ implying equal association (or independence if both equal 1) is plausible.

Formulation and assumptions. We evaluate $\phi_k^* = \log \theta_k^*$ for $k = 1, 2$ and base the test on (13.6) and (13.7)

Prodedure. From Table 13.2 we find

$$\phi_1^* = \log[156 \times 133/(107 \times 84)] = 0.8365,$$

$$\phi_2^* = \log[84 \times 209/(31 \times 156)] = 1.2893$$

$$\text{Var}(\phi_1^*) + \text{Var}(\phi_2^*) = (2/156 + 2/84 + 1/107 + 1/133 + 1/31 + 1/209) = 0.0905$$

whence

$$Z = (1.2893 - 0.8365)/\sqrt{(0.0905)} = 1.51.$$

Conclusion. Since $Z = 1.51$ corresponds to a two-tail $P = 0.1310$ we have no strong evidence against H_0.

Comments 1. This result is in line with that in Example 13.1 where we found $G^2 = 2.30$. Under H_0 the distribution of Z^2 is chi-squared with 1 degree of freedom. The value $Z^2 = 1.51^2 = 2.27$ is close to that of G^2.

2. In view of the large counts in each cell, having accepted the hypothesis that the odds ratios do not differ, it is reasonable to use the normal approximation to test for each table the hypothesis $\phi_i = 0$, $i = 1, 2$ implying independence. The standard test statistic $Z = \phi_i^* / \sqrt{\text{Var}(\phi_i^*)}$ shows strong evidence that there is association (Exercise 13.4).

3. The method used here avoids the need to estimate the expected values $\hat{m}_{ijk}$ needed in Example 13.1.

4. An asymptotic confidence interval for the true difference $\phi_2 - \phi_1$ may be obtained in the usual way. For example, the approximate 95 percent interval is of the form

$$\phi_2^* - \phi_1^* \pm 1.96 \sqrt{[\text{Var}(\phi_1^*) + \text{Var}(\phi_2^*)]}.$$

For the above data the interval is $(-0.1369, 1.0423)$. Because zero is in the interval a hypothesis of equal odds ratios is, as we found above, acceptable.

13.2.3 Analysis of 2×2 Tables for Several Strata

We saw in Section 12.1.2 that binomial sampling with two levels of an explanatory variable is a model often associated with a 2×2 table. A typical situation is one where presence or absence of side-effects (responses) is noted for each of two medicines (explanatory categories). Another example arises in a production process where two types of machine (explanatory categories) are being investigated in manufacturing runs from each of which the numbers of satisfactory and of flawed items (responses) are noted. We may introduce further explanatory variables, or covariates, leading to situations like that exemplified in Table 13.4 where each of the $k = 4$ covariates is an age group. These groupings are usually referred to as strata. In that table each stratum consists of a 2×2 table of counts classified by drugs and side-effects. An informal inspection of the table suggests that side-effect incidence tends to increase slightly with age for drug B, but there is no such tendency for drug A. Apart from this there looks to be little difference in side-effect incidence between the drugs.

Two obvious questions that may be of interest are:

- Is there an association between drugs and side-effects?
- If there is an association does it change with age, i.e., does the odds ratio change between strata?

A simplistic approach is to test for evidence of association within each stratum. If there is no evidence we accept the hypothesis of independence. A disadvantage is that if the counts in some cells are small the tests may have low power against alternatives that specify a slight association. Intuitively one feels that if the information could be combined in some way for all strata the larger sample size may detect such associations if these are similar for all strata. Care is needed in such analyses, for they may give misleading results, especially if the degree of association, or type of association, differs between strata, a matter we explore further in Section 13.3.1.

In Examples 13.1 and 13.2 we considered the special case of only two strata (i.e., $k = 2$). The large numbers in each cell enabled us to detect strong evidence of association within each stratum.

The situation is different for the data in Table 13.4, where exact tests applied to each of the four strata specified by age show no evidence of association. However, we shall see in Example 13.4 that the suspicion aroused by a visual inspection of the data that there are more side-effects for drug B among older patients is not without justification, although the evidence for this is not strong.

We proceed by first assessing whether there is evidence that the odds ratios are equal for all strata. If there is strong evidence that these ratios are not all equal, a test for independence across all strata (i.e., that the within-strata odds ratios $\theta_1, \theta_2, \ldots, \theta_k$ have a common value unity) is then not appropriate, since we have already decided they are probably not all equal.

Thus, a sensible procedure is to test first whether the hypothesis of equality of odds ratios holds. Only if there is little evidence against this should we test whether the common value may be unity.

The methods used in Examples 13.1 and 13.2 for the case $k = 2$ do not generalize easily to the more general $k > 2$. Further, we gave only asymptotic results and these are not appropriate if there are many cells with small counts. They are unlikely to be satisfactory for the data in Table 13.4, for example.

The approaches using asymptotic and exact results are somewhat different so we consider them separately.

13.2.4 Asymptotic Tests about Odds Ratios in k 2×2 Tables

For any $k \geq 2$ three common questions are

- Is there evidence of association (i.e., $\theta_s \neq 1, s = 1, 2, \ldots, k$)?

- If there is association are all θ_s equal?

- If all θ_s are equal how do we estimate the common value?

Several asymptotic tests help answer these questions. When these are not adequate, exact permutation tests are possible using packages like StatXact. We outline some basic ideas behind these procedures, quoting relevant formulae without derivation. Some asymptotic tests are provided in general statistical packages and may even be conducted manually in simple cases. Many inference procedures in this, and the next, section rely heavily on the fact that once we have the observed or expected count for cell $(1, 1)$ in any 2×2 table this fixes the observed, or expected, odds ratio if marginal totals are fixed.

We have pointed out that it is desirable to test whether a hypothesis of equality of odds ratios is tenable before testing whether that common ratio may be unity. If we reject the first possibility, the second is automatically rejected.

Data often suggest that if there is any association then it may be similar within all strata; e.g., all the empirical odds ratios $\theta_s = (n_{11s}n_{22s})/(n_{12s}n_{21s})$ might take values slightly greater than 1.

Care is needed in choosing an estimate of a common odds ratio appropriate for testing the hypothesis that there is indeed such a common ratio. With different stratum sizes as in Table 13.4, and perhaps different allocations of numbers to rows or columns within strata, a simple average of all stratum odds ratios is a naïve approach.

Mantel and Haenszel (1959) proposed the estimator

$$\theta^* = \frac{\sum_s (n_{11s}n_{22s}/N_s)}{\sum_s (n_{12s}n_{21s}/N_s)} \tag{13.8}$$

This is a weighted mean of the strata odds ratios θ_s with weights $n_{12s}n_{21s}/N_s$ where N_s is the total count for the cells in stratum s. The motivation for choosing these weights is that when all θ are close to unity they are nearly the reciprocals of the variances of the θ_s.

Breslow and Day (1980) proposed a test for whether data were consistent with an estimated common odds ratio given by (13.8). We first compute the expected count m_{11s} in cell $(1, 1, s)$ for each of the k stratum conditional upon the common odds ratio being θ^* given by (13.8). This amounts to choosing for each stratum s an $x = m_{11s}$ which is the positive root not exceeding n_{1+s} or n_{+1s} of the equation

$$\frac{x(N_s - n_{1+s} - n_{+1s} + x)}{(n_{1+s} - x)(n_{+1s} - x)} = \theta^*. \tag{13.9}$$

If odds ratios are equal across all strata the differences

$$|m_{11s} - n_{11s}|, \quad s = 1, 2, \ldots, k$$

should all be small and Breslow and Day (1980, Chapters 4 and 5) discuss appropriate tests for homogeneity of odds ratios in detail. They proposed a statistic

$$M_{BD}^2 = \sum_{s=1}^{k} \frac{(m_{11s} - n_{11s})^2}{\text{Var}(n_{11s})} \tag{13.10}$$

where

$$\text{Var}(n_{11s}) = \left[\frac{1}{m_{11s}} + \frac{1}{m_{12s}} + \frac{1}{m_{21s}} + \frac{1}{m_{22s}} \right]^{-1}. \tag{13.11}$$

Asymptotically M_{BD}^2 has a chi-squared distribution with $k - 1$ degrees of freedom under H_0: *all θ_s are equal.* The expected frequencies in the expression for $\text{Var}(n_{11s})$ are subject to the fixed marginal totals for stratum s given m_{11s}.

If we accept H_0 we may obtain a confidence interval for the common value θ which we estimated by θ^*. Asymptotically, the distribution of $\phi^* = \log \theta^*$ approaches normality more rapidly than that of θ^*. Also $E(\phi^*) = \phi = \log \theta$. Several estimates have been proposed for $\text{Var}(\phi^*)$. That recommended in StatXact was proposed by Robins, Breslow and Greenland (1986). It has been shown to work well for small k and large counts or for large k even if some of the 2×2 tables are fairly sparse. The expression for this estimate is complicated but we give it for completeness. For each stratum s we define

$$a_s = (n_{11s} + n_{22s})/N_s, \quad b_s = (n_{12s} + n_{21s})/N_s,$$
$$c_s = n_{11s}n_{22s}/N_s, \quad d_s = n_{12s}n_{21s}/N_s,$$

then write

$$c_+ = \sum_s c_s, \quad d_+ = \sum_s d_s$$

and

$$P = \left[\sum_s (a_s c_s)\right]/(2c_+^2), \quad Q = \left[\sum_s (a_s d_s + b_s c_s)\right]/(2c_+ d_+),$$
$$R = \left[\sum_s (b_s d_s)\right]/(2d_+^2)$$

then

$$\text{Var}(\phi^*) = P + Q + R.$$

Approximate 95 percent asymptotic confidence limits for ϕ are

$$\phi^* \pm 1.96\sqrt{\text{Var}(\phi^*)}. \tag{13.12}$$

For all but small data sets (where asymptotic procedures are not appropriate) a computer program is highly desirable for testing and estimation, but we illustrate salient features of the computation in Example 13.3. Approximate confidence limits for θ can be obtained by back-transformation.

Everitt (1992, Section 2.8.1) develops an alternative formula, basing the interval on θ^* directly.

Example 13.3

The problem. For the data in Table 13.5 explore whether the hypothesis of equality of odds ratios is acceptable. If it is, obtain an approximate 95 percent confidence interval for the common odds ratio θ.

Formulation and assumptions. The material sources form three strata ($k = 3$), and the two machine types, are explanatory categories. The output status categories represent two responses. The solution requires the following steps: (1) estimation of a common odds ratio using (13.8); (2) calculation of M_{BD}^2 from (13.10) for use in an asymptotic test of H_0: *all θ_i are equal* against H_1: $\theta_i \neq \theta_j$ for at least one $i \neq j$; (3) if H_0 is accepted, a confidence interval for the common value θ is obtained by back-transformation from the interval for ϕ given by (13.12).

Table 13.5 *Production count for two machines with three raw material sources.*

		Output status	
Material source	*Machine*	*Satisfactory*	*Faulty*
A	*Type I*	42	2
	Type II	33	9
B	*Type I*	23	2
	Type II	18	7
C	*Type I*	41	4
	Type II	29	12

Procedure. We recommend using widely available software for this test, but we outline the main steps in the computation. The reader should verify all quoted numerical values (Exercise 13.5). Using (13.8) gives

$$\theta^* = \frac{(42 \times 9)/86 + (23 \times 7)/50 + (41 \times 12)/86}{(33 \times 2)/86 + (18 \times 2)/50 + (29 \times 4)/86} = 4.702$$

The expected value m_{111} is the relevant root of (13.9), which here is

$$\frac{x(86 - 75 - 44 + x)}{(75 - x)(44 - x)} = 4.702.$$

This gives $x = m_{111} = 41.68$. The other root $x = 100.54$ exceeds both row and column marginal totals. Similarly, we find $m_{112} = 23.07$ and $m_{113} = 41.25$. Then (13.11) gives

$$\text{Var}(n_{111}) = (1/41.68 + 1/2.32 + 1/33.32 + 1/8.68)^{-1} = 1.6660,$$

and in like manner $\text{Var}(n_{112}) = 1.3181, \text{Var}(n_{113}) = 2.4550$, whence (13.10) gives

$$M_{BD}^2 = 0.32^2/1.6660 + 0.07^2/1.3181 + 0.25^2/2.455 = 0.09.$$

Assuming an asymptotic chi-squared distribution with $k - 1 = 2$ degrees of freedom there is clearly no evidence against H_0. Indeed, the fit is remarkably good ($P = 0.96$). To obtain the relevant confidence interval we note that $\phi^* = \log \theta^* = 1.5480$. Tedious but otherwise straightforward computation gives $\text{Var}(\phi^*) = 0.1846$ whence (13.12) gives the 95 percent confidence limits for ϕ as

$$1.5480 \pm 1.96\sqrt{0.1846}$$

giving an interval (0.7059, 2.3901). Taking natural antilogarithms, gives an interval (2.0256, 10.9145) for θ.

Conclusion. We accept the hypothesis of a common odds ratio, and estimate it (after sensible rounding) to be $\theta = 4.70$ with a 95 percent confidence interval (2.03, 10.92).

Comments. 1. The confidence interval is wide and is not symmetric about the point estimate. This reflects the markedly skew distribution of θ referred to in Section 12.1.2.

2. Since independence, or lack of association, implies $\theta = 1$ there is strong evidence of association, since $\theta = 1$ is well outside the above confidence interval.

Computational aspects. 1. The above results are confirmed by the program in StatXact, but the estimate and confidence limits given there are the reciprocals of those given here. This is because StatXact defines the odds ratios as reciprocals of those we give. This leads to similar inferences because, as we pointed out, ratios θ and $1/\theta$ represent the same degree of association, only in opposite directions.

2. Some software may give a different confidence interval to that obtained here, because, as indicated above, there are alternative estimates of variance to that given by the Robins, Breslow and Greenland formula.

The method in Example 13.3 tests whether a common odds ratio across all strata is an acceptable hypothesis. If it is we may then decide, with the aid of a confidence interval, whether that common value may be $\theta = 1$ implying independence. Another widely used test for independence is the Mantel–Haenszel test proposed by Mantel and Haenszel (1959), although Cochran (1954) proposed an effectively identical test. The latter is described by Everitt (1992, Section 2.7.2). We give the test in the Mantel–Haenszel formulation as we find this intuitively appealing because of similarities to the Breslow–Day test given above.

Basically the test requires the calculation in each stratum of the cell $(1, 1)$ expectations under the hypothesis of independence in that stratum, i.e., $m_{11s} = n_{1+s}n_{+1s}/N_s$. The statistic is

$$M^2 = \frac{(m_{11+} - n_{11+})^2}{\sum_s \text{Var}(n_{11s})} \tag{13.13}$$

where $m_{11+} = \sum_s m_{11s}$, $n_{11+} = \sum_s n_{11s}$ and

$$\text{Var}(n_{11s}) = \frac{n_{1+s}n_{+1s}n_{2+s}n_{+2s}}{N_s^2(N_s - 1)}$$

Asymptotically (13.13) has a chi-squared distribution with 1 degree of freedom under H_0: *all $\theta_s = 1$*.

If the expected and observed values in the cell of each table differ appreciably, and these differences are all in the same direction this suggests odds ratios are consistently above, or consistently below, 1 and large values of M^2 are likely. Circumstances may arise where differences between expected and observed values in cell $(1, 1)$ for each table are in an opposite direction in some tables to that in others. This may lead to small values of M^2 even when there is clear evidence of association — albeit in opposite directions — within some strata. For this reason the Mantel–Haenszel test is not recommended without first using a test such as the Breslow–Day test for equality of odds ratios. However, if the asymptotic theory is accepted for the Breslow–Day test we may, as shown above, compute approximate confidence intervals for a common odds ratio and accept or reject a hypothesis of independence on the basis of these, making the Mantel–Haenszel test somewhat superfluous. Since it is widely used we illustrate how it works.

Example 13.4

The problem. For the data in Table 13.4 compute M^2, and assuming all $k = 4$ strata have the same expected odds ratio θ is it reasonable to accept the hypothesis $H_0: \theta = 1$?

Formulation and assumptions. Assuming all θ_s are equal we compute M^2 using (13.13) and obtain a P-value assuming a chi-squared distribution with 1 degree of freedom.

Procedure. We find $m_{111} = 9 \times 19/21 = 8.14$, $m_{112} = 15.06$, $m_{113} = 26.06$, $m_{114} = 36.60$, whence $m_{11+} = 8.14 + 15.06 + 26.06 + 36.60 = 85.86$ and $n_{11+} = 8 + 14 + 25 + 39 = 86$. Also $\mathrm{Var}(n_{111}) = (19 \times 9 \times 12 \times 2)/(21^2 \times 20) = 0.4653$, $\mathrm{Var}(n_{112}) = 0.4832$, $\mathrm{Var}(n_{113}) = 1.1213$, $\mathrm{Var}(n_{114}) = 1.9096$, whence

$$\sum_s \mathrm{Var}(n_{11s}) = 0.4653 + 0.4832 + 1.1213 + 1.9096 = 3.9794,$$

so that

$$M^2 = (0.14)^2/3.9794 = 0.0049.$$

Conclusion. The extremely small value, 0.0049, means there is no evidence against the hypothesis of independence.

Comments. 1. Some writers recommend a continuity correction subtracting 0.5 from $|m_{11+} - n_{11+}|$ before squaring. This may avoid over-weighing of evidence against H_0 in borderline cases.

2. The small differences between corresponding m_{11s} and n_{11s} point toward independence. However, as indicated above situations arise where these differences may be large, and of opposite sign in different strata, but the numerator in (13.13) remains small. This is why we prefer the Breslow–Day test with confidence intervals for a common θ if that hypothesis is accepted. For the data in this example it can be shown (Exercise 13.6) that $M_{BD}^2 = 6.497$ with an associated $P = 0.0898$, so there is no substantial evidence against equality of odds ratios, although room for a little doubt. The point estimator of the odds ratio is $\theta^* = 1.0334$ with a 95 percent confidence interval (0.4036, 2.6462). In the light of these findings it is not surprising that the Mantel–Haenszel test strongly supports independence. The Breslow–Day test, however, does raise a small doubt about the hypothesis of equality of odds ratios.

13.2.5 Exact Tests for Equality of Odds Ratios in k 2×2 Tables

Zelen (1971) gives an exact permutation test for a common odds ratio. Gart (1970) gives a method for obtaining confidence intervals if a hypothesis of a common odds ratio is accepted. Unlike many other nonparametric tests, the asymptotic procedures described in the previous section are not based on the asymptotic properties of the statistics used in the exact test.

As computing facilities such as those provided by StatXact are needed to apply the exact methods, we only sketch their nature.

Zelen's test for homogeneity of the ratios is based on a property of the hypergeometric probabilities arising in the Fisher exact test for 2×2 tables. Subject to the condition that all n_{11s} sum over s to the observed marginal total n_{11+}, the product of the hypergeometric probabilities over all possible k 2×2 tables satisfying this condition is a maximum when the observations strongly support a common odds ratio θ whatever the value of θ, and this product decreases steadily as that support weakens. Under the hypothesis that all odds ratios are equal, the actual probability P^* of observing any particular set of within-stratum outcomes in which the cell $(1, 1)$ entries sum to the observed n_{11+} is given by the product of the k hypergeometric probabilities for those outcomes, divided by the sum of all such products over all possible cell $(1,1)$ values consistent with marginal totals for the relevant 2×2 tables, and which sum to n_{11+}. If we denote the probability for our observed tables by P_0^*, the corresponding P-value is the sum of the probabilities P^* associated with all outcomes for which $P^* \leq P_0^*$. If the test is applied to the data in Table 13.5 StatXact gives the probability associated with the observed outcome as $P_0^* = 0.008213$ and the relevant $P = 0.0689$ compared with the asymptotic Breslow–Day value $P = 0.0898$ given in Comment 2 on Example 13.4.

The procedure proposed by Gart for estimating the common odds ratio if the Zelen test supports that hypothesis uses a different approach. We noted in the asymptotic Breslow–Day procedure that if we estimate expected values under an assumption that the common odds ratio takes a particular value θ^*, say, that in general if m_{11s} is the expected value in cell $(1, 1)$ of stratum s then $\sum_s m_{11s} \neq n_{11+}$. Large discrepancies indicate that such a value of θ^* is an unsatisfactory value for the common odds ratio. There may be no value of θ^* for which this discrepancy is not large if the hypothesis that there is a common odds ratio is not supported.

Gart's procedure for obtaining a confidence interval for a common odds ratio is in essence to compute a statistic T which is the sum of the observed n_{11s}. One then considers its position in the distribution of all possible values that this statistic might take if each observed n_{11s} were replaced by any possible value consistent with the marginal totals for the sth 2×2 table; i.e., the values are no longer constrained to sum to n_{11+}. We may then find values θ^-, θ^+ that give values of T which would just be accepted as hypothetical estimates of the common odds ratio in a test at the $100(\alpha/2)$ percent significance level. These provide $100(1 - \alpha)$ percent confidence limits for θ. Modifications are needed if the data indicate acceptance of either a zero or infinite common odds ratio.

Details of a number of alternative procedures for analysis of stratified 2×2 tables are given in the StatXact 7 manual, Chapter 15.

For the data in Table 13.5 the exact 95 percent confidence interval given by Gart's method is $(1.9149, 12.306)$, slightly longer than the asymptotic interval $(2.0256, 10.9145)$ obtained in Example 13.3. The latter depends on the validity of certain variance estimates, involves a back-transformation from natural logarithms and assumes a reasonable rate of convergence to normality.

From the practical viewpoint the difference between the asymptotic result and that for exact permutation theory should not cause alarm, bearing in mind that there are also questions of discontinuity in the distribution of Gart's T statistic so the interval may not be an exact 95 percent interval.

13.2.6 The Linear-by-Linear Association Model

The linear-by-linear association model is a relatively simple, yet versatile, log-linear model with only one additive interaction term to represent association. For an $r \times c$ table a score u_i is allocated to row $i, i = 1, 2, \dots, r$ subject to constraints $u_1 \leq u_2 \leq \dots \leq u_r$ and a score v_j to column $j, j = 1, 2, \dots, c$ subject to constraints $v_1 \leq v_2 \leq \dots \leq v_c$. If m_{ij} is the expected count in cell (i, j) the model may be written

$$\log m_{ij} = \text{independence terms} + \beta u_i v_j, \tag{13.14}$$

where β is a constant.

The model reduces to that for independence (no association) between row and column categories when $\beta = 0$. It is sometimes convenient to use the modified form

$$\log m_{ij} = \text{independence terms} + \beta(u_i - m_u)(v_j - m_v), \tag{13.15}$$

where m_u, m_v are the means of the u_i and of the v_j respectively. It is easy to see that (13.15) is equivalent to (13.14) if, in the latter, we replace u_i, v_j by $u_i^* = u_i - m_u$ and $v_j^* = v_j - m_v$.

A number of standard nonparametric test statistics have exact permutation distributions based on this model. It is relevant to tables of counts where both row and column categories are ordered and when, if the classifications are not independent, they show a monotonic trend across row and column categories. When the scores are ranks they are often referred to as *unit interval scores* because the difference between the scores associated with any pair of neighbouring rows, or of neighbouring columns, is in each case unity.

The parameter β may be estimated to reflect the extent of any relevant departure from independence. We do not discuss estimation of β, but indicate what we mean by relevant departures in Example 13.5. Cells for which $|\beta(u_i - m_u)(v_j - m_v)|$ is large show the greatest departure from independence.

Example 13.5

The problem. For the case $r = 5$, $c = 7$ with rank scores, i.e., $u_i = i$, $v_j = j$, determine for each cell in the $r \times c$ table the contribution to $\log m_{ij}$ of the interaction term $\beta(u_i - m_u)(v_j - m_v)$ in (13.15).

Formulation and assumptions. In this example $m_u = 3$ and $m_v = 4$, so the interaction term in cell (i, j) in (13.15) is $\beta(i - 3)(j - 4)$.

Procedure. For any β the interaction term in each cell is given in Table 13.6.

Table 13.6 *Interaction terms in* $\log m_{ij}$ *for rank scores in a* 5×7 *contingency table.*

Column score	-3	-2	-1	0	1	2	3
Row score							
-2	6β	4β	2β	0	-2β	-4β	-6β
-1	3β	2β	β	0	$-\beta$	-2β	-3β
0	0	0	0	0	0	0	0
1	-3β	-2β	$-\beta$	0	β	2β	3β
2	-6β	-4β	-2β	0	2β	4β	6β

Conclusion. If $\beta > 0$, $\log m_{ij}$ is greater than its value under independence for cells at the top left and bottom right of Table 13.6, and less for cells at the top right or bottom left. If $\beta < 0$ the situation is reversed. If $\beta = 0$ there is no association.

The name linear-by-linear association reflects the characteristic that within any row, say row i, for any chosen u_i, v_j the interaction contribution across columns is a linear function of the v_j with slope $\beta(u_i - m_u)$. Similarly, within any column j the interaction contribution across rows is a linear function of the u_i with slope $\beta(v_j - m_v)$.

The relevance of this model to permutation theory is that many permutation tests depend upon a statistic, S, that may be written

$$S = \sum_{i,j} u_i v_j n_{ij} \tag{13.16}$$

with appropriate choices of u_i, v_j. Independence ($\beta = 0$) corresponds to the null hypothesis; we reject H_0 if our observed outcome, expressed in the appropriate contingency table format, indicates a general pattern of the form exemplified for a particular case in Table 13.6 when $\beta \neq 0$. This corresponds to extreme values of S given by (13.16), while intermediate values indicate lack of association.

Example 13.6

The problem. In Example 6.1 we applied the WMW test to a data set we presented in a contingency table format in Table 12.2. Show that if, in that format, we use row scores 0, 1 corresponding to Group A and Group B respectively, and column scores corresponding to the ranks of the observations in the combined groups, then S is a statistic for the Wilcoxon rank sum test.

Formulation. In Table 12.2 the columns were ordered by increasing times taken to complete the task. We now allocate to the columns in that table the rank scores 1, 2, 3, . . . , 21 and compute the Statistic S in (13.16) using these scores.

Table 13.7 *The information in Table 12.2 with times replaced by corresponding ranks for column scores and 0, 1 assigned as row scores. For display purposes the table is divided, but it should be read as a 2 × 21 table.*

Time ranks	1	2	3	4	5	6	7	8	9	10	11
Group A	1	1	1	0	1	1	0	1	1	0	0
Group B	0	0	0	1	0	0	1	0	0	1	1

Time ranks	12	13	14	15	16	17	18	19	20	21
Group A	1	0	1	0	1	0	0	0	0	0
Group B	0	1	0	1	0	1	1	1	1	1

Procedure. With the above formulation, by referring to Table 12.2 it is easy to obtain the relevant Table 13.7 with the scores shown. It follows immediately that S given by (13.16) is the sum of the ranks for Group B and here $S = 155$.

Conclusion. The Wilcoxon rank sum test is one for independence in a linear-by-linear association model if the above row and column scores are used.

Comments. 1. If we interchange Group A and Group B, i.e., swap rows and retain the same row scores, the statistic S becomes the sum of the Group A ranks, the alternative Wilcoxon rank sum statistic used in Example 6.1. Hence, in this two-sample situation, the ordering of the groups does not matter. There is an opposite association with the two different orderings.

2. In the context of the WMW test independence, or lack of association, corresponds to a situation when there is no difference between group medians. It is associated with a broad scatter of zeros and ones in the cells of a table like Table 13.7. There is strong linear-by-linear association if the ones (corresponding to an observation) tend to concentrate towards the left or right in the first row and towards the opposite extreme in the second row, a situation we saw in a simpler case in Section 12.3.1.

The Spearman rank correlation coefficient provides a more informative illustration. If there are n paired observations without ties the rank outcome pattern can be presented as an $n \times n$ contingency table with rows and columns representing x, y ranks respectively. If an x-rank i is paired with a y-rank j we enter 1 in cell (i, j). All other entries in row i or in column j are 0. The row and column totals are all 1. We illustrate the pattern for three cases when $n = 5$.

$$
\begin{array}{llccccc}
\text{Case A} & x \ ranks & 1 & 2 & 3 & 4 & 5 \\
& y \ ranks & 4 & 2 & 5 & 3 & 1 \\
\text{Case B} & x \ ranks & 1 & 2 & 3 & 4 & 5 \\
& y \ ranks & 2 & 1 & 3 & 5 & 4 \\
\text{Case C} & x \ ranks & 1 & 2 & 3 & 4 & 5 \\
& y \ ranks & 1 & 2 & 3 & 4 & 5
\end{array}
$$

These are represented as 5×5 tables in (13.17) (13.18), (13.19)

Case A	y ranks	1	2	3	4	5	
	x ranks						
	1	0	0	0	1	0	
	2	0	1	0	0	0	(13.17)
	3	0	0	0	0	1	
	4	0	0	1	0	0	
	5	1	0	0	0	0	

Case B	y ranks	1	2	3	4	5	
	x ranks						
	1	0	1	0	0	0	
	2	1	0	0	0	0	(13.18)
	3	0	0	1	0	0	
	4	0	0	0	0	1	
	5	0	0	0	1	0	

Case C	y ranks	1	2	3	4	5	
	x ranks						
	1	1	0	0	0	0	
	2	0	1	0	0	0	(13.19)
	3	0	0	1	0	0	
	4	0	0	0	1	0	
	5	0	0	0	0	1	

In case A, the 1s are well scattered over rows and columns. In case B there is a discernable tendency for the 1s to drift toward the top left and bottom right, hinting at some association between x- and y-ranks. Case C clearly represents the closest possible association and corresponds to $r_s = 1$. The resemblance of these tables to scatter diagrams (with conventional axes rotated clockwise through $90°$) is apparent if we imagine the 0s to be deleted and the 1s to be replaced by dots.

It is implicit from our remarks in Section 10.1.1 about permutation tests for the Pearson coefficient that the sum $\sum_i r_i s_i$ may be used here as a test statistic for the Spearman coefficient in an exact permutation test. It immediately follows that if we take the ordered x-ranks and y-ranks as scores for a linear-by-linear association test in contingency tables of the form (13.17) to (13.19) that S given by (13.16) equals $\sum_i r_i s_i$. The relevant permutation distribution is over all such possible 5×5 tables with row and column sums of unity. It is easy to see that there are $5!$ such tables in each case.

Similar arguments for n paired observations (x_i, y_i), with no ties in either variable, show that the statistic $S = \sum_i x_i y_i$ appropriate for an exact permutation test for the Pearson product moment correlation coefficient (Section 10.1), is equivalent to (13.16) with scores $u_i = x_{(i)}, v_j = y_{(j)}$. The relevant permutation is over all $n \times n$ tables with row and column sums all unity.

The models extend easily to tied situations. For example, using mid-ranks for the Spearman coefficient with ties, consider the case:

x ranks	1.5	1.5	3	4	5
y ranks	1	4	4	2	4

The relevant contingency table with mid-rank scores is now:

y ranks	1	2	4	Total
x ranks				
1.5	1	0	1	2
3	0	0	1	1
4	0	1	0	1
5	0	0	1	1
Total	1	1	3	5

To generate the relevant permutation distribution we calculate S using mid-rank scores for all permutations of this table with the given marginal totals. Alternatively, we may use the centralized scores $u_i - m_u, v_j - m_v$.

StatXact has a program to compute relevant P-values for any linear-by-linear association model based on permutation of such tables subject to fixed marginal totals for arbitrarily assigned scores. This is a powerful tool but it is important to choose relevant and sensible scores. We consider such scores in a reanalysis of the data in Example 12.6. The tests are essentially applications of the permutation test for a Pearson product moment correlation coefficient with many ties and scores (x, y values) chosen to reflect what may be regarded as appropriate measures of *distance* between categories.

Example 13.7

The problem. Given the data on side-effect of dose levels of a drug in Example 12.6, i.e.,

	Side-effects			
Dose	None	Slight	Moderate	Severe
100 mg	50	0	1	0
200 mg	60	1	0	0
300 mg	40	1	0	0
400 mg	30	1	1	2

allocate appropriate scores and perform a linear-by-linear association test for evidence of association between dose level and side-effects.

Formulation and assumptions. One reasonable choice of scores would be row and column ranks. For rows this corresponds to dose levels in 100 mg units. For columns it represents an ordering of side-effects consistent with limited information. Another possibility would be to allocate column scores in a way a clinician might interpret the data. He or she might accept 1 and 2 as reasonable scores for no and slight side-effects, but regard moderate side-effects as 10 times as serious as slight ones, and severe side-effects as 100 times as serious as slight ones, giving logical column scores $v_1 = 1$, $v_2 = 2$, $v_3 = 20$, $v_4 = 200$.

The StatXact linear-by-linear association program will perform the relevant tests using either of these (or any other chosen) scoring system. If no suitable program is available we give below an asymptotic result. This should be used with some reservations when, as here, there are many cells with small counts.

Procedure. Using the StatXact linear-by-linear association program with row and column rank scores gives a one-tail $P = 0.0112$ and doubling this gives a two-tail $P = 0.0224$. With the alternative column scores 1, 2, 20, 200 the corresponding tail probabilities are $P = 0.0071$ and $P = 0.0142$.

Conclusion. A one-tail P-value is relevant as it is logical to expect side-effects to increase rather than decrease as dose increases. Whichever scoring system we use there is strong evidence that side-effects increase with dose.

Comments. 1. In Example 12.6 the Jonckheere–Terpstra test indicated strong evidence of association. As pointed out in Section 12.3.2 that test is equivalent to one using Kendall's tau. Using ranks for row and column scores is more like using Spearman's rho without taking ties into account, while other row and column scores give a test equivalent to that for a Pearson coefficient with those scores playing the roles of x, y values.

2. One might have performed a test exactly equivalent to that for Spearman's rho by allocating mid-rank scores for ties. Because there are 51 patients tied at

the 100 mg dose, the score for row 1 would then be 26, that for row 2 would be $(52 + 112)/2 = 82$, and so on. A valid criticism of the use of these mid-ranks in this highly tied situation is that the mid-ranks depend heavily on the number of subjects allotted to each treatment (here dose levels). There must be some unease about such a scoring system, especially if the numbers allocated to each treatment differ substantially. There is a better case for basing scores on the nature of treatments or responses rather than upon the number of patients allocated to each treatment. Possible unsatisfactory influences of mid-rank scores with heavy tying and unbalanced row and column totals have been pointed out by a number of writers including Graubard and Korn (1987).

3. In a group of tests on the same data each using different but plausible scores, the P-values often differ more substantially than they did in this example. An important caveat about any scoring system is that it should be chosen for its relevance to the problem at hand and not on a hint inspired by an inspection of the data that suggests that a particular scoring system may enhance the prospects of getting a small P-value. The speed of modern computers tempts one to try many analyses for the same data. The temptation is best avoided for linear-by-linear association tests by choosing a scoring system before data are obtained. Once again we refer the reader to the penultimate paragraph in Section 1.5.

We do not prove it, but it can be shown that for S given by (13.16)

$$E(S) = [(\sum_i u_i n_{i+})(\sum_i v_j n_{+j})]/N,$$

and

$$\mathrm{Var}(S) = \frac{[\sum_i u_i^2 n_{i+} - (\sum_i u_i n_{i+})^2][\sum_j v_j^2 n_{+j} - (\sum_j v_j n_{+j})^2]}{N^2(N-1)}$$

and asymptotically $Z = [S - E(S)]/\sqrt{\mathrm{Var}(S)}$ has a standard normal distribution.

Example 13.8

The problem. Apply an asymptotic test to the data in Example 13.7 using rank scores for rows and columns.

Formulation and assumptions. $E(S)$ and $\mathrm{Var}(S)$ are computed by the formulae given above. The value of Z is calculated.

Procedure. The ranks scores are $u_i = i$, $v_j = j$. We easily verify that $n_{1+} = 51$, $n_{2+} = 61$, $n_{3+} = 42$, $n_{4+} = 34$, $n_{+1} = 180$, $n_{+2} = 3$, $n_{+3} = 3$, $n_{+4} = 2$, $N = 188$.

For the given data simple calculations give $S = 484$, $E(S) = 469.7$ and $\mathrm{Var}(S) = 35.797$ whence $Z = 2.39$, and $P = \Pr(Z > 2.39) = 0.0084$.

Conclusion. There is strong evidence against H_0.

Comment. This result is in line with that for the exact test. In Exercise 13.7 we ask you to confirm that with the second choice of scores the asymptotic $P = 0.0101$. Despite many ties, and many small cell counts, in this example the asymptotic results are not misleading.

Many other tests already encountered may be formulated as linear-by-linear association tests. If the rank column scores in Example 13.6 are replaced by raw data scores this leads to a Pitman-type test. Other alternatives might use the van der Waerden scores or, if appropriate, logrank scores.

A test for trends in $r \times 2$ contingency tables known as the *Cochran–Armitage test* may also be formulated as a linear-by-linear association test. The test applies basically to a situation where the rows represent explanatory variables and the columns two mutually exclusive and exhaustive outcomes. Typically, the rows are ordered explanatory variables such as age groups or increasing doses of a drug. It is assumed that columns represent binomial responses, the first for an event A, the second for an event B. For the ith row, $i = 1, 2, \ldots, r$ for each of the n_{i+} units in the counts in that row the probability of the event A is p_i and of the opposite event B is $q_i = 1 - p_i$.

The Cochran–Armitage test was proposed independently by Cochran (1954) and Armitage (1955). It is used to test the hypothesis H_0: $p_1 = p_2 = \ldots = p_r$ against H_1: $p_1 \leq p_2 \leq \ldots \leq p_r$ (or a similar inequality with all signs reversed) where at least one inequality is strict. Effectively, Cochran and Armitage proposed a test which is a linear-by-linear association test with column scores 0, 1 and row scores $1, 2, 3, \ldots, r$. In Example 13.9 we apply it to a set of data given by Graubard and Korn (1987).

Example 13.9

The problem. Table 13.8 indicates whether or not congenital sex organ malformation was found among children born to mothers whose stated alcohol consumption (number of drinks per day) during pregnancy fell within various ranges. There is a slight indication that the probability that malformation is absent may decrease with increasing maternal alcohol consumption, since the maximum likelihood estimates of p_i, for each row are $p_1 = 17066/(17066 + 48) = 0.9972$ and similarly $p_2 = 0.9974$, $p_3 = 0.9937$, $p_4 = 0.9921$, $p_5 = 0.9737$.

Formulation and assumptions. Our alternative to equality of the p_i is one of a decreasing trend. We perform an exact Cochran–Armitage test as a linear-by-linear association test with row scores 1, 2, 3, 4, 5 and column scores 0, 1.

Procedure. Using StatXact the exact one-tail $P = 0.1046$.

Conclusion. There is not strong evidence against H_0: *all p_i are equal.*

Comments. 1. A one-tail test is appropriate on the grounds that there is no medical or other evidence to suggest a beneficial effect of increased alcohol uptake in this context. The situation may be different for other health issues. There is some evidence, for example, that moderate consumption of red wine may protect against the onset of heart disease.

2. The choice of row scores is arbitrary. There may be a temptation to use mid-rank scores for rows as an intuitively reasonable scoring system. The disadvantage noted in Comment 2 on Example 13.7 is more marked in this example. It is easily verified that the mid-rank scores for rows are 8557.5, 24365.5, 32013, 32473 and 32555.5. The final three rows — where most of the changes in malformation probability, if

Table 13.8 *Numbers of children with or without malformations in relation to mother's stated alcohol consumption (Graubard and Korn, 1987).*

| | Malformation | |
	Absent	Present
Mother's alcohol intake (drinks per day)		
0	17066	48
< 1	14466	38
1–2	788	5
3–5	126	1
≥ 6	37	1

any, would seem to take place — get almost identical scores. For this scoring system $P = 0.2861$.

3. Temptation to experiment with many scoring systems should be avoided. When no other system is obviously appropriate, the default scoring system $1, 2, 3, 4\ldots$ has much to commend it. However, if there are clinical grounds for adopting some other scoring system it is sensible to do this. For this example Graubard and Korn suggest using the mean alcohol intake for row scores, i.e., 0, 0.5, 1.5, 4 and 7 where the choice of 7 is arbitrary but not unreasonable. This results in an exact one-tail $P = 0.0168$, suggesting that there is indeed some evidence against H_0. Sprent (1998, Section 13.5) suggests yet another scoring system that might well get clinical support. It gives a P-value not very different from that using the mean intake scores. This is comforting because it suggests that the analysis is not too sensitive to differences between scoring systems that may appear realistic under what most people would regard as rational assumptions.

Computational aspects. Availability of programs like that in StatXact for exact linear-by-linear association tests with arbitrary scores adds flexibility to the Cochran–Armitage test as originally proposed with restricted row scores $1, 2, 3, 4, \ldots$.

13.2.7 Capture-Recapture Techniques

Capture-recapture analysis was originally proposed to estimate the number of animals in a population when it was impracticable or even impossible to do this using a complete count or census.

For example, to estimate the total number of squirrels in a wood one might use a benign form of trapping to capture 25 animals which are then marked for identification and released to the wild. At a later date a further 40 animals might be trapped and these might include 5 marked, i.e., recaptured, animals, the remaining 35 being unmarked. If the unknown number of animals in the wood is N the first sample marking procedure may be looked upon as dividing the population into two sub-populations — the first consisting of the 25 marked animals; the second of the $N - 25$ unmarked animals. If all animals have the same probability of being included in the first and second trapping then the ratio of the number of marked animals to unmarked animals in the second sample is an intuitively reasonable estimate of the proportions in the population.

Equating these ratios, i.e., setting $25/(N - 25) = 5/35$, and solving gives $N = 200$. It is easily verified that this is equivalent to using the ratio of marked to total populations, i.e., $25/N = 5/40$. More generally, if the first, or capture, sample size is n_1 and the second, or recapture, sample size is n_2 and in this second sample there are b marked animals then we estimate N by equating the ratios $n_1/N = b/n_2$ giving the estimate $N = n_1 n_2/b$.

The assumptions that all animals have an equal probability of being included in each sample, and that the population size N remains constant between the capture and the recapture sample are crucial, but may be unrealistic in practice. Some animals may be more inquisitive than others, and thus more likely to be trapped. Sometimes the probability of a marked animal being retrapped may be decreased by some aspect of the marking process such as being frightened by the initial trapping or marking procedure. Or it may be increased because the bait used was a desirable food item. These factors negate the equal probability assumption. Further, the population size may vary between samples due to births, deaths or migration.

We show below that when the model leading to the simple estimate given above is valid, approximate confidence limits for the true population size may be obtained by formulating the problem using an incomplete 2×2 contingency table. When the basic assumptions needed for validity of the simple estimate break down it is well-known to users of capture-recapture methods that some of these can be dealt with by taking more than two samples. This leads to a multiple capture-recapture process, where in each succeeding sample, one identifies which animals have been captured previously, and at which sample or samples this happened. Analysis of such data is often based on loglinear models to describe interactions or associations.

Capture-recapture methods have been developed to take into account the use of covariates. The most straightforward situation is with a single categorical covariate. One suggested approach is to stratify the data according to the covariate (e.g., if gender is the covariate stratify into males and females), estimate the total number of cases for each stratum and combine these to obtain an overall estimate (Hook and Regal, 1995). More sophisticated models that take into account continuous covariates are described by Tilling and Sterne (1998). The simple estimate of population size is biased if genuine covariate effects are ignored

A detailed treatment is beyond the scope of this elementary text, but we outline the basic ideas behind a contingency table approach. We see in Example 13.10, and in the subsequent discussion, that the basic ideas of capture-recapture analysis have been taken beyond the estimation of animal

Table 13.9 *A contingency table for a simple capture-recapture model. Interrogation marks indicate an entry to be estimated.*

| | Present in second sample | | |
	Yes	No	Total
Present in first sample			
Yes	b	$n_1 - b$	n_1
No	$n_2 - b$	?	?
Total	n_2	?	?

populations. They are now employed in wider contexts. The use of loglinear models for capture-recapture analysis is described by Bishop, Fienberg and Holland (1975) while Seber (1982) provides a comprehensive review of basic capture-recapture procedures. Cormack (1989) gives a detailed description of practical uses of loglinear models in this context.

For the two-sample procedure outlined at the start of this section, we may express results in a 2×2 table like Table 13.9 where rows represent numbers from the total population of unknown size N that are included, or are not included, in the first sample; columns represent similar information for the second sample.

It is easily seen that in the standard notation n_{ij} for the count in cell (i, j), the relevant numerical entries in Table 13.9 are $n_{11} = b$, $n_{12} = n_1 - b$, $n_{21} = n_2 - b$, and the first row and column totals are respectively n_1, n_2. In any particular application these all have known numerical values. However, we do not know n_{22}, which, in our example, is the number of squirrels that are not observed in either sample. This means we do not know the second row and column marginal totals, nor do we know the grand total, N. The unobservable number n_{22} is called a *structural zero*.

If we denote the unknown population number by N, an intuitively reasonable estimate of N is obtained by choosing it so that the observed number in each cell n_{ij} equals the expected number $m_{ij} = n_{i+}n_{+j}/N$ under the assumption of independence.

Remember that in a 2×2 table, knowing m_{11} fixes all other m_{ij} to ensure correct marginal totals. We know $n_{11} = b$ and from Table 13.9 it follows for any unknown total N that $m_{11} = n_1 n_2/N$. Setting $m_{11} = n_{11} = b$ implies an estimate $\hat{N} = n_1 n_2/b$. This is what we got at the start of this section.

This approach may look like using a sledgehammer to crack a nut, but it opens up useful ways to obtain a confidence interval for N. It also has implications for developing loglinear models allowing for interactions when applied to more than two samples. Sekar and Deming (1949) derived an expression for the asymptotic variance of $\hat{N}$, namely

$$\mathrm{Var}(\hat{N}) = \frac{n_{1+}n_{+1}n_{12}n_{21}}{n_{11}^3}, \tag{13.20}$$

and asymptotic 95 percent confidence limits for N are

$$\hat{N} \pm 1.96\mathrm{Var}(\hat{N})$$

For populations not much bigger than the combined sample sizes, these asymptotic limits are sometimes unsatisfactory, giving a lower limit less than the total count of distinct units in the two samples.

Regal and Hook (1984) suggest an alternative approach that uses the likelihood ratio statistic G^2. If the capture–recapture estimate of the total number of cases is used in the 2×2 table, G^2 is zero, since the method effectively equates observed to expected numbers under independence. They suggested forming a 95 percent confidence interval that included all N for which the fit of the independence model is adequate in a significance test at the 5 percent level. As G^2 is associated with a chi-squared distribution with one degree of freedom, values of N giving a $G^2 < 3.84$ form the 95 percent confidence interval. Unlike the interval calculated from the standard error, this gives a confidence interval that is asymmetric around $\hat{N}$, with the lower limit being closer to $\hat{N}$. It is also necessarily no less than the total observed distinct number of population members.

Capture-recapture methods play a valuable role in the needs assessment of the most marginalized members of society such as the homeless (Fisher et al., 1994) and drug abusers (Hickman et al, 2006). In addition they can, as the following example shows, be used to correct for underestimation of the prevalence of a medical problem in human populations when several incomplete registers (each corresponding to a capture or a recapture sample of patients) exist (Hook and Regal, 1982; Smeeton et al., 1999).

Example 13.10

The problem. Guillain–Barr syndrome is a paralyzing neurological condition due to inflammation of the peripheral nerves. In the UK patients are listed on various registers, none of which is complete. Rees et al. (1998) estimated the number of such cases in the south east of England. Two lists were obtained, one from the British Neurological Surveillance Unit (BNSU), another from hospital activity analysis (HAA). There were 23 patients on the BNSU list, 68 were on the HAA list and 17 were on both. Find an estimate of the total number of cases in the south east of England, along with a 95 percent confidence interval.

Formulation and assumptions. The relevant data can be presented in a 2×2 table in the format of Table 13.9:

| | Present in HAA list | | |
	Yes	No	Total
Present in BNSU list			
Yes	17	6	23
No	51	?	?
Total	68	?	?

There are only two registers, so for the simple capture-recapture model inter-actions cannot be investigated. The two lists are assumed to be independent. We have $n_{+1} = 68$, $n_{1+} = 23$ and $n_{11} = 17$. The appropriate estimate of N is $\hat{N} = 68 \times 23/17 = 92$.

The asymptotic variance of this estimate given by (13.20) is

$$\text{Var}(\hat{N}) = (23 \times 68 \times 51 \times 6)/17^3 = 97.41.$$

This leads to 95 percent confidence limits for N of $92 \pm 1.96\sqrt{97.41}$ which lead, after integer rounding toward the estimate $\hat{N}$, to the interval (73, 111). This is clearly unacceptable, since we know from the lists there are at least $68 + 23 - 17 = 74$ cases.

Using the likelihood ratio statistic G^2 gives a lower limit for N of 79, and an upper limit of 121. These limits do not include the total number of known cases, and the corresponding interval is in that respect at least more appropriate than the earlier one.

Conclusion. The estimated total number of cases is 92. On the basis of the likelihood ratio 95 percent confidence interval the likely number of hidden cases of Guillain–Barr syndrome in this area of England is between $79 - 74 = 5$ and $121 - 74 = 47$.

Comments 1. With small samples, caution is needed when using the asymptotic standard error method to obtain a 95 percent confidence interval for N. The lower limit of the interval should always be checked against the number of known cases to see if it makes sense.

2. Confidence intervals for small samples will almost certainly be wide, and in this example may not shed much light on the number of cases missed by both lists. Sometimes the upper limit of the confidence interval may exceed any reasonable estimate expected by individuals working in the field of study.

3. If it was thought that the degree of overlap for the two registers might differ between males and females, gender could be used as a categorical covariate. The number of cases would be estimated for males and females separately, the overall total being the sum of the two estimates. In this example little would be gained from this approach as the sample size is so small that the confidence intervals for the subgroups would be too wide to give any real information.

4. Additional registers are invaluable for improving the estimate of N. Using these extra samples, loglinear models allowing interaction terms can be investigated. Rees et al. had access to two further registers — a research database containing 22 cases, and 5 death certificates giving this syndrome as a cause of death. This brought the total number of known cases to 79, with an estimate for N of 98. The new

likelihood ratio 95 percent confidence interval for N was (86, 120). As one would expect, further registers increased the number of known cases. Note that the 95 percent confidence interval is less wide. The upper limit for N is slightly smaller. An interaction term was required for an acceptable model. This casts doubt on the unavoidable assumption of independence in our example.

5. A problem arises if the count b of numbers in both samples is zero, for the estimate of N using the above approach is infinite. Adjustments are available in these circumstances, but it is usually better, when possible, to overcome the problem by taking larger samples.

Computational aspects. Nearly all programs that provide the chi-squared test for 2×2 tables quickly give the likelihood ratio statistic G^2, and these at least provide a basis for a trial and error process for forming a confidence interval based on this statistic. An improved interval might be based on that for the Fisher exact test.

The contingency table approach generalizes to three or more samples in a multiple capture-recapture program. For three samples the outcomes may be presented in an incomplete $2 \times 2 \times 2$ table, where the three dimensions refer to the respective samples and the categories in each are presence or absence. We write the observed cell count in cell (i, j, k) as n_{ijk} where a subscript value 1 implies a population unit is observed in the corresponding sample and a subscript 2 that it is not observed. Thus, for example, n_{121} is the number of animals, patients, or whatever observed in both samples 1 and 3, but not in sample 2. Similarly, n_{111} is the number observed in all three samples. Because we do not know N there is one unobservable or structural zero cell count corresponding to the unknown n_{222}, the number in the population that are never seen.

The expected numbers m_{ijk} in any cell, other than the one with a structural zero, can be modelled by a loglinear model that allows for possible interactions due to factors such as varying probabilities of capture for individual animals within or between samples, population changes due to births, deaths, immigration or emigration, etc. The detail is beyond the scope of this book, and the interested reader should refer to Cormack (1989).

In the rest of this section we make some general remarks about the method that indicate the versatility of the approach.

The modeling process differs from that considered in Section 13.2.1 due to the lack of a specific count in cell (2, 2, 2). This means that in the three-sample case we can no longer make inferences about, or test for, a second-order interaction. We can, however, test hypotheses about first-order interactions. If samples are taken on k occasions the results can be presented in a 2^k contingency table with one structural zero cell. Inferences may be made about interactions up to order $k - 2$.

The loglinear models appropriate to such analyses are a special case of the important class of models known as *generalized linear models*. How well a model fits the data can be examined using the likelihood ratio test statistic G^2, which is often referred to as the *deviance*.

For the independence model with k samples, G^2 is associated with a chi-squared distribution having $2^k - k - 2$ degrees of freedom. If the P-value from this goodness-of-fit test is low, the model gives an inadequate fit to the data, so a more complex model is sought. If an extra interaction term is added to the model, the number of degrees of freedom is reduced by one. To create models that are as simple as possible, it is usual to accept the extra term only if the decrease in the deviance is statistically significant ($P < 0.05$). In other words, the deviance has to decrease by at least 3.84. If a model provides a very poor fit this is invariably the case. As extra terms are added, the P-value rises, i.e., the fit is better.

The estimate for N should be given with an appropriate confidence interval. Asymptotic formulae for calculating confidence intervals in the case of multiple samples have been developed (Bishop, Fienberg and Holland, 1975; Selvin, 1995). However, as with the basic two-samples-only situation, these may lead to confidence intervals that contain values less than the number of known cases. To avoid this problem, Regal and Hook (1984) suggest the calculation of G^2_{min}, the minimum value of the likelihood ratio statistic for the loglinear model under consideration. This is obtained when using the capture-recapture estimate for N. Using similar reasoning to the two-sample case, the appropriate confidence interval for that particular model contains all those values of N for which G^2 is no more than $G^2_{min} + 3.84$.

In applying capture-recapture methods to real data, it is sometimes necessary to introduce a subjective element in the model selection process. A more complex model is sometimes preferred if this allows the P-value of the overall fit to rise above 0.05, even if the decrease in the deviance is modest. One should bear in mind, however, that additional terms in a model may lead to wider confidence intervals for N.

A more complex model may need to be rejected even if there is a statistically significant decrease in the deviance if the confidence interval contains implausible values from an ecological or clinical point of view (Hay, 1997).

Smeeton et al. (1999) used capture-recapture methods in order to estimate the prevalence of congenital malformation of the heart in the first year of life. All such infants are severely ill and doctors would in the light of experience question a predicted rate of more than 20 cases per 1000. In a practical study, a model with three interaction terms gave an estimate of 14.5 cases per 1000 and a 95 percent confidence interval for the rate from 5.7 to 79.9 cases per 1000. Such findings would not be taken seriously by the medical profession. The removal of one of the interaction terms reduced the estimate to 6.8 cases per 1000 and the upper limit of the confidence interval to a much more realistic rate of 13.3 cases per 1000.

A further discussion of interval estimation of population size using overlapping lists or records is given by Regal and Hook (1999).

Capture-recapture analysis of constant populations can be performed with SPSS which allows definition of the missing cell value referred to as a structural zero. The estimate for N is calculated from the coefficients of the loglinear model. If N varies and birth, death and migration need to be taken into account, then the package GLIM can be used in the way described in Cormack (1989).

13.3 Combining and Partitioning of Tables

Intuitively one may feel that if several tables refer to the same row and column categories then combining these into one table to get larger counts in cells may be helpful. It might, for example, increase the power of tests for detecting association. In other circumstances, one may feel that some restructuring or partitioning of tables, perhaps by combining certain rows and columns, or breaking down into strata, may reveal information not easily discernable in the original table. Combining or partitioning should be done with caution. Detailed discussions are given by Agresti (1996, 2002). The treatment here covers only certain aspects of these topics to give some indication of when they can be useful and points out some of the pitfalls that may go with incautious use of the methods.

13.3.1 Simpson's Paradox and Combining Tables

Suppose we are given the information in Table 13.10 for numbers of individuals in urban and rural areas responding to a standard cough medicine and a new medicine where, for each person, it is recorded whether there is no effect or if it cures the cough.

Visual inspection of Table 13.10 shows that the proportion of cures in both urban and rural areas is higher with the new medicine than it is with the standard. This is confirmed if we carry out the Fisher exact test on each table separately giving $P < 0.0001$ for urban areas and $P = 0.0017$ for rural areas. Both provide strong evidence against a hypothesis that the medicines are equally effective. If we now ignore the split into urban and rural areas and combine the data for each giving the 2×2 table

	No effect	Cure
Standard medicine	837	444
New medicine	1167	557

Table 13.10 *Responses to two cough medicines in urban and rural areas.*

	Urban		Rural	
	No effect	Cure	No effect	Cure
Standard medicine	498	103	339	341
New medicine	1042	367	125	190

and apply the Fisher exact test to this table we find $P = 0.0946$. So much for a hope that combining the information would enhance the evidence of association. Indeed, if we look at this table we see that the cure rates do not look very different for the two medicines, roughly a third of all those treated being cured, irrespective of which medicine they are given. If anything, the standard medicine appears to do slightly better.

This is a situation where the individual tables indicate association, and here association in the same direction. Yet the combined table does not indicate association, but suggests that if there were any association it might be in the opposite direction. This illustrates *Simpson's Paradox*, first described formally by Simpson (1951). The paradox may arise when we combine data for two strata where different models hold, especially when, as here, the sample sizes in each stratum are different. In the above example, the models for urban and rural areas differ because the proportion of cures for each medicine in these areas are markedly different. Indeed, the percentages in each calculated from Table 13.10 are

	Urban	*Rural*
Standard medicine	17.14	50.15
New medicine	26.05	60.32

The moral is that one should only combine tables if one is confident the relevant identical distributional model is valid for each table. It is probably only worth attempting combinations if counts are small in corresponding cells in all tables that are to be combined.

The paradox fascinates both statisticians and the inquisitive layman. Entering *Simpson's paradox* in a web search engine will guide one to many examples.

13.3.2 Partitioning of Tables

We saw in Sections 12.2.2 and 12.2.3 that both the Pearson X^2 statistic, and the likelihood ratio G^2 statistic have asymptotically a chi-squared distribution with $(r-1)(c-1)$ degrees of freedom. It is well-known that a variate having a chi-squared distribution with ν degrees of freedom may be expressed as a sum of ν independent variates each with 1 degree of freedom. The G^2 statistic may be partitioned similarly into components, each of which has asymptotically a chi-squared distribution with 1 degree of freedom. Rules for partitioning are complicated in the general $r \times c$ case. There is usually more than one possible partitioning. Necessary conditions for the single degree of freedom components to be additive are given by Agresti (2002, Sections 3.3) and are:

- Subtable degrees of freedom must sum to the degrees of freedom of the original table.
- Each cell count in the original table must appear in one and only one subtable.
- Each marginal total of the original table must be a marginal total for one and only one subtable.

Lancaster (1949) suggested partitioning an $r \times c$ table into $(r-1)(c-1)$ separate 2×2 tables each with 1 degree of freedom having the form

$$\sum_{s<i}\sum_{t<j} n_{st} \quad \sum_{s<i} n_{sj}$$
$$\sum_{t<j} n_{it} \quad\quad n_{ij}$$

For example, for a 3×3 table

$$\begin{array}{ccc} n_{11} & n_{12} & n_{13} \\ n_{21} & n_{22} & n_{23} \\ n_{31} & n_{32} & n_{33} \end{array}$$

this partitioning gives the four tables

(i)

$$\begin{array}{cc} n_{11} & n_{12} \\ n_{21} & n_{22} \end{array}$$

(ii)

$$\begin{array}{cc} n_{11}+n_{21} & n_{12}+n_{22} \\ n_{31} & n_{32} \end{array}$$

(iii)

$$\begin{array}{cc} n_{11}+n_{12} & n_{13} \\ n_{21}+n_{22} & n_{23} \end{array}$$

(iv)

$$\begin{array}{cc} n_{11}+n_{12}+n_{21}+n_{22} & n_{13}+n_{23} \\ n_{31}+n_{32} & n_{33} \end{array}$$

It is easy to verify (Exercise 13.14) that this partitioning satisfies the above necessary conditions. The partitioning is not unique. For example, if rows and columns of the original table are permuted the chi-squared statistic for the whole table is unchanged, but clearly the partitioning above will be altered numerically. Although, under the null hypothesis of independence, asymptotically the chi-squared statistics for subtables are independent and additive, the Pearson X^2 statistics computed for each of the 4 tables (i)–(iv) above do not, in general, add to the corresponding statistic for the original table.

However, if the likelihood ratio statistic G^2 is used, the components are additive.

Partial partitioning, where not all components correspond to single degrees of freedom, is also possible. In practice one should choose a partitioning that helps to answer relevant questions about association.

Example 13.11

The problem. Two drugs used in chemotherapy are tested on 100 patients, 60 receiving drug A and 40 receiving drug B. Numbers treated with each drug who exhibit the presence or absence of two specific side-effects, hair loss (HL) and visual impairment (VI) are as follows:

	Side-effect status			
	HL	VI	HL + VI	None
Drug A	9	4	16	31
Drug B	3	16	2	19

Partition G^2 to examine association between single side-effects, between single and double side-effects, and between overall side-effect status for the two drugs.

Formulation and assumptions. The necessary conditions for additivity given above are satisfied if we compare column 1 with column 2, then columns $1 + 2$ with column 3, then columns $1 + 2 + 3$ with column 4. These components represent appropriate tests for association (i) between single side-effects for each drug, (ii) between a single side-effect (either HL or VI) and a double side-effect (both HL and VI) and (iii) at least one side-effect and no side-effect.

Tests for each may be carried out using the G^2 statistic in either an exact or asymptotic test if the latter is felt appropriate.

Procedure. The relevant partitioned 2×2 tables are

(i)	HL	VI
Drug A	9	4
Drug B	3	16

(ii)	HL or VI	HL + VI
Drug A	13	16
Drug B	19	2

(iii)	Side-effect	None
Drug A	29	31
Drug B	21	19

It is easily verified (Exercise 13.15) that for the complete table $G^2 = 22.13$ and for the partitioned 2×2 tables that for (i) $G^2 = 9.72$, for (ii) $G^2 = 12.24$ and for (iii) $G^2 = 0.17$. The sum of these components is $9.72 + 12.24 + 0.17 = 22.13$, the G^2 value for the complete 2×4 table. StatXact may be used to compute both exact and asymptotic P-values for the original table and each of the 2×2 components and gives the following values:

	2×4 table	(i)	(ii)	(iii)
Exact P	0.0001	0.0035	0.0010	0.8384
Asymptotic P	0.0001	0.0018	0.0005	0.6830

Conclusion. The extremely low P-value for the full table indicates strong evidence of association between drugs and side-effect status. The low P for subtable (i) reflects the fact that if there is only one side-effect it is more likely to be hair loss for drug A, and visual impairment for drug B. The low P for subtable (ii) reflects the fact that occurrence of both side-effects is more common with Drug A, while the high P for subtable (iii) shows there is no evidence of association between drugs and the incidence of some side-effect, i.e., that both drugs seem equally likely to give rise to some side-effects.

Comments. 1. The action a clinician might take in the light of these findings depends upon the importance of each side-effect. If visual impairment were slight and temporary it might be rated less serious than hair loss and this would indicate a preference for drug B. If visual impairment were severe and permanent it would be more serious than hair loss. If both side effects are moderate, two side-effects may be regarded as more serious than just one.

2. The above partitioning is not unique; we might reorder the columns; e.g., place column 4 first and then combine columns in the order used above. The order we have used leads to more logical interpretation of the components of G^2.

3. We have used a particular case of a rule for partitioning G^2 into $c - 1$ additive components each with 1 degree of freedom when we have a $2 \times c$ table. The rule is

to form first the subtable with columns 1 and 2, next the subtable with column 1 + 2 as the first column and column 3 as the second column, next the subtable with the sum of columns 1, 2, 3 and column 4, and so on until the final table uses the sum of columns $1, 2, 3, \ldots, c - 1$ and column c.

4. Although the Pearson X^2 statistic and G^2 have the same asymptotic distributions under H_0, if X^2 is partitioned as above the component values of the Pearson statistic will not in general sum to X^2.

Best (1994) used a partitioning of chi-squared into components analogous to linear and quadratic components in a regression type analysis of variance. He was specifically interested in looking at centrality and dispersion differences for data obtained in tasting trials, but appropriate use of this partitioning extends to other types of data and it can also be generalized to look at other features such as skewness. Best's approach uses a partitioning proposed by Lancaster (1953). An account together with a comparison with some alternatives, is given by Sprent (1998, Section 13.2). Partitioning of chi-squared is the main tool used in a wide range of nonparametric techniques discussed by Rayner and Best (2001).

13.3.3 Partitioning in a Different Kind of Contingency Table

To compare preferences for competing brands of a product like tomato soup, or for similar types of wine made from the same grape in different countries, or for different varieties of peas or beans, a common procedure is to ask a number of potential consumers, N, to rank each of k brands of the product from 1 to k in order of preference. In practice, the ranking is usually in descending order of preference, the brand ranked 1 being the first choice. The results may be expressed in a $k \times N$ table like Table 10.7.

If there are no tied rankings the results may also be expressed in a $k \times k$ table like Table 13.11 with each row corresponding to one of the k brands and each column to a rank, the ith column corresponding to rank i. Here the rows have no natural ordering, but the columns are ordered. The entry in cell (i, j) is the number among the N testers who gave rank j to brand i. A classic paper on this topic by Anderson (1959) gave data for a study in which 123 consumers were each asked to rank three varieties of snap beans in order of preference. Table 13.11 is a similar table for rankings by 90 potential purchasers for three models of broadly similar cars.

The basic difference between this contingency table and those considered earlier in the chapter is that the cell entries are not sampled independently from 270 units. Instead, each of the 90 would-be purchasers, independently of each other, allocates the ranks 1, 2, 3 to each of the three models. Since each purchaser uses each rank he or she is represented once in each column. This means one may validly carry out a goodness-of-fit test to see, for example, if the hypothesis that there is the same probability of each variety being ranked 1, is supported by considering only the data in column 1. This is because

Table 13.11 *Numbers of potential purchasers giving various rankings to each of three models of car.*

	Rank			
Model	1	2	3	Total
I	27	47	16	90
II	35	31	24	90
III	28	12	50	90
Total	90	90	90	270

what happens in the allocations in the remaining columns is irrelevant to that question. The appropriate test is one for a uniform distribution, using the methods of Section 12.4 leading to an X^2 statistic that has asymptotically a chi-squared distribution with 2 degrees of freedom.

If there is no preferred first choice, the expected numbers in column 1 are each $90/3 = 30$. A similar argument could be applied to column 2, or column 3, but these tests would not be independent of one another, because, if they were we could add the components and get a chi-squared with 6 degrees of freedom for the whole table. However, with fixed marginal totals there are only 4 degrees of freedom associated with Table 13.11. A moment's reflection shows that if we performed a conventional Pearson X^2 test with these fixed marginal totals the value of X^2 (with 4 degrees of freedom) would be the same as that obtained by adding the goodness-of-fit statistics for the no-preference goodness-of-fit test applied to each row with 6 associated degrees of freedom.

Anderson conjectured that if we reduced the usual X^2 statistic by a factor $2/3$ (i.e., the ratio 4 to 6 of the degrees of freedom) we might get a statistic that had asymptotically a chi-squared distribution with 4 degrees of freedom under a hypothesis of independence. He proved this was so, and that the result extended to the case of k items to be ranked where we have a $k \times k$ table, where our conventional Pearson X^2 statistic is reduced by a factor $(k-1)/k$. The modified statistic is $X'^2 = (k-1)X^2/k$.

For the data in Table 13.11 we find $X^2 = 42.8$. Since $k = 3$, the chi-squared statistic for the overall test is $X'^2 = 2 \times 42.8/3 = 28.53$ with 4 degrees of freedom. It is easily verified that $P < 0.0001$ indicating, not surprisingly in view of the data pattern, strong evidence of association between models and purchaser preferences.

Anderson discusses partitioning X'^2 into single additive degrees of freedom but, as in the previous section, these are not unique. The important thing is to choose a partitioning that has a sensible interpretation. For a 3×3 table one partitioning into contrasts proposed by Anderson, and the interpretation of each contrast, is:

- $C_1 : n_{21} - n_{23}$ Linear ranking effect for model II.
- $C_2 : n_{11} - n_{13} - (n_{31} - n_{33})$ Difference between linear rankings for models I and III.
- $C_3 : n_{21} - 2n_{22} + n_{23}$ Quadratic ranking effect for model II.
- $C_4 : n_{32} - n_{12}$ Difference between quadratic rankings for models I and III.

That this is the interpretation of C_4 becomes clear if we write it in an alternative form:

$$[(n_{11} - 2n_{12} + n_{13}) - (n_{31} - 2n_{32} + n_{33})]/3.$$

Equivalence is established by noting that $n_{i1} + n_{i2} + n_{i3} = N, (i = 1, 3)$.

Anderson also establishes that each single degree of freedom component of X'^2 takes the form $C_i^2/\mathrm{Var}(C_i)$. In determining $\mathrm{Var}(C_i)$ one must take into account correlations between the various n_{ij} that arise because the same observers are involved in all counts. The calculations of the $\mathrm{Var}(C_i)$ are explained in Anderson's original paper and outlined in Sprent (1998, Section 13.3). We only quote the result, viz. $\mathrm{Var}(C_1) = \mathrm{Var}(C_4) = 2N/3$ and $\mathrm{Var}(C_2) = \mathrm{Var}(C_3) = 2N$.

An interesting feature of these contrasts is that if we add the components X'^2 for the first two contrasts, their sum is equal to the Friedman statistic T given by (7.5) based on the rank sums for each variety given by all 90 purchasers, equivalent to Kendall's coefficient of concordance. We illustrate these points in an example.

Example 13.12

The problem. For the data in Table 13.11 partition the statistic X'^2 into components corresponding to the contrasts C_1 to C_4 described above. Verify the assertion that addition of the first two components gives the Friedman statistic T in this case.

Formulation and assumptions. The necessary computations are based on the description of components given above.

Procedure. We see immediately from Table 13.11 that $C_1 = 35 - 24 = 11$, and since $\mathrm{var}(C_1) = 2N/3$ and $N = 90$ it follows that the corresponding component of X'^2 is $11^2/[2 \times 90/3] = 2.02$ and in like manner the remaining components are found to be 6.05, 0.05 and 20.42, summing to $X'^2 = 28.54$, the value quoted above apart from rounding error. To calculate the Friedman statistic for these data we note that the sum of the ranks given by all 90 purchasers for model I is $27 + 2 \times 47 + 3 \times 16 = 169$. Similarly the sums for models II and III are 169 and 202, whence the Friedman statistic is easily found to be $T = 8.07$, equal to the sum $2.02 + 6.05 = 8.07$ of the first two components above of X'^2.

Conclusion. If an asymptotic chi-squared test is applied to each component there is strong evidence of association between models and consumer preference for components C_2 and C_4, but not for C_1 or C_3.

Comments. 1. Components C_3, C_4 are associated with departures from linearity of ranking. Linearity, broadly speaking, represents average rankings and departures therefrom represent differences in dispersion. For the data in Table 13.11 it is clear that there is a monotonic fall-off for rank preferences for model II, but that the pattern is different for models I and III. Model I is most frequently ranked 2, and least frequently ranked 3. On the other hand model III is seldom ranked 2; a large number rank it 3, but it gets one more first ranking than does model I. Thus model III shows a greater dispersion of ranks than do the other models.

2. Anderson (1959) suggests that in many practical situations the experimenter may be more interested in model or varietal contrasts at the more favourable ranks, 1 and 2, in which case a more useful set of contrasts would be

$$D_1 : n_{31} - n_{11}, \quad D_2 : n_{31} - 2n_{21} + n_{11},$$
$$D_3 : n_{12} - n_{13} - (n_{32} - n_{33}), \quad D_4 : n_{23} - n_{22}.$$

The components D_1 and D_2 are linear and quadratic comparisons for the first rankings — of main interest in this approach — while the remaining two examine differences in the second and third rankings.

Problems similar to those briefly discussed in this section have been considered by Scholz and Stephens (1987), Best (1994), Best and Rayner (1996) and others.

13.4 A Legal Dilemma

A relatively new field of application where statistical notions such as P-values and power of a test may be especially relevant is law and jurisprudence. We illustrate some difficulties that may arise in the use of statistical notions in court proceedings, and also in some other fields, by a simple example using hypothetical data given by Gastwirth and Greenhouse (1995).

They give two sets of data for comparable groups of 100 employees for each of two companies, A and B where employees are drawn from the same population of potential and eligible employees. For each company the numbers are broken down by majority and minority ethnic groups and within each group by those promoted and those not promoted as follows:

	Company A		Company B	
	Promoted	Not promoted	Promoted	Not promoted
Minority	1	31	2	46
Majority	10	58	9	43

Gastwirth and Greenhouse indicate that if these data were presented in a U.S. legal action alleging discrimination on ethnic grounds in promotion policy the courts may require a further explanation from a company to justify a claim that its policy is not discriminatory on those grounds if the statistical evidence indicates rejection at the 5 percent significance level of the hypothesis of no association between promotion and ethnic grouping. This further explanation might be that at the time of an offer for engagement employees were notified that promotion would only be granted to those willing to work unsocial hours or to work for long periods away from home and that relatively few of the minority group who accepted employment met that requirement.

For the Company A sample it is easily seen that for the minority group the promotion rate is $1/32$, i.e., 3.125 percent compared to an overall rate of 11 percent for all 100 employees. For Company B the corresponding percentages are 4.167 and again 11 percent. It appears then that Company A show slightly greater discrimination, but is the evidence sufficiently strong to support the 5 percent significance criterion in a Fisher exact test which is appropriate here. Even if it is accepted that a one-tail test is approriate the critical significance level is not achieved since the exact $P = 0.0765$.

Intuitively one feels this lets Company B off the hook because they promote 4.167 percent from the minority group compared to 3.125 for Company A. However, we have met situations before where intuition leads us astray, e.g., that arising with Simpson's paradox, so we should apply the Fisher test to the Company B data. Here the relevant one-tail test gives $P = 0.0351$.

Given these data, although Company B promotes a higher percentage of minority group employees, because of the court ruling that a significance level of 5 percent is to be used to indicate acceptable evidence of discrimination, it is the one that is going to have to do some explaining! This may appear to the layman, and to the legal profession, somewhat rough justice. The court may well want an explanation from a statistician as to how the apparent anomaly arises. One contributing factor is that Company A is promoting from a smaller pool of minority employees — 32 percent as opposed to 48 percent for Company B. Sample sizes do affect the behaviour of the Fisher exact test not only with regard to power, but they may also affect discontinuities in possible P-values. It may be appropriate for a statistician to draw the attention of the court to this difficulty. The difference in proportion of minority group employees in the two companies — 32 percent for Company A and 48 percent for Company B — is also of relevance. As Gastwirth and Greenhouse point out, this difference probably arises because each company has its own hiring policy. They suggest that in this example in the workforce available to either company for hire there may be 40 percent belonging to the minority group. A question that then arises is whether either company discriminates between ethnic groups in their hiring policy. There is not acceptable evidence that either does, for assuming random hiring the number of minority hirings per 100 employees has a $B(100, 0.4)$ distribution and using an exact test as, for example, that in Minitab, a hiring of 48 then has an associated two-tail $P = 0.125$ while a hiring of 32 has a two tail $P = 0.104$. However, if we assume simply that each company is hiring from a pool with some fixed, but not necessarily known, proportion in the minority ethic group, hirings of 32 and 48 from that group do indicate a different hiring policy, the relevant P being 0.019.

The effect of this is that for the company with the smaller number of minority group hirings the Fisher exact test has lower power than it has when applied to data for the firm with the greater number of minority group hirings. There is a timely warning here about the need for care when using conditional tests when the marginal totals used for conditioning may themselves be conditional on another factor — such as hiring policy in this example. This again, for statisticians, serves as a warning against placing too much faith in formal significance levels such as 5 percent.

Difficulties of the type outlined above in a legal context arise in other fields, especially medical statistics. For example, in a study of lung cancer in relation to smoking, the difficulty arises if one conditions upon events in the causal path such as patient status for other disorders such as emphysema or hypertension which are often related to smoking.

Use of statistical arguments in courts and in related fields such as forensic medicine is a fast growing area. Courts on the whole prefer straightforward and not too complicated statistical arguments providing these are sound. There are many recorded cases where there have been miscarriages of justice because of crucial weaknesses in statistical evidence — evidence not necessarily given by statisticians. In one recent case in Great Britain there was a substantial miscarriage of justice because a nonstatistician gave evidence quoting probabilities based on an assumption that two events were independent when there was a wealth of empirical evidence that in the relevant circumstances they were not.

Statisticians often complain of a reluctance by lawyers and the judiciary to consider statistical or probabilistic arguments, finding this strange in view of the standard requirement of proof beyond reasonable doubt in criminal cases and the tendency to resolve civil disputes on a balance of probabilities. This reluctance may stem from an inbred hostility to attempts to quantify uncertainty coupled to an often nonunique nature of such quantifications. Another factor may be ambiguity in interpretation as indicated in our example on promotion rates in minority groups. Such difficulties are most apparent in adversarial systems when prosecution and defence experts may often present plausible analyses of the same data that lead to different conclusions.

Many examples of statistical problems in relation to justice and law are given by Gastwirth (1988) and by Aitken and Taroni (2004).

13.5 Power

Increasing attention has been given in recent years to power and sample size relationship for tests involving contingency tables. Exact power computations in practice are limited to fairly restricted and clearly defined models such as the comparison of binomial populations in $2 \times k$, or $k \times 2$ tables, particularly when there is ordering in the k categories. The subject is a complex one. An account of the practical difficulties involved in a wide range of power calculations, together with examples of applications, is given in the StatXact Manual, Chapter 29 to 32.

We showed in Section 6.6.1, where we considered power calculations for the median test, that exact power calculations were possible if we could determine the binomial probabilities relevant under H_0 and a specific alternative of interest. The latter binomial probability was, in that context, dependent upon what assumptions we made about the population distributions giving rise to our samples. However, exact power computations are feasible once that binomial probability, p_1, is ascertained. The procedure is then one for determining power of the Fisher exact test of $H_0: p = 0.5$ against $H_1: p = p_1$. The approach generalizes easily to any p_0 specified in H_0.

StatXact includes a facility for exact power calculations for $2 \times k$ tables where the columns correspond to k binomial samples, or the rows correspond to two multinomial samples each with k possible outcomes. These cover many potential situations of practical importance providing we can specify meaningful binomial or multinomial probabilities for the alternative hypothesis of interest. When we can, the programs may be applied to tests such as the Cochran–Armitage test, and the extensions to that test with arbitrary scores given in Section 13.2.6, as well as to two independent sample permutation tests including WMW, normal scores, logrank scores or even the Pitman raw data test. The reader should refer to the StatXact manual for technical details, and to see examples of what can be achieved.

An example of how relevant binomial probabilities may be assigned for power calculations for a Cochran–Armitage test in a clinical trial context using a logistic regression model is given in Mehta, Patel and Senchaudhuri (1998).

13.6 Fields of Application

Tests of association may be relevant in any situation where independence may not be an acceptable hypothesis. In three-way, or higher, dimensional tables especially, the nature of association is often of special interest.

Drug Testing

Situations like that in Example 13.7 where side-effects of drugs are of interest are common in clinical trials. In such cases the nature of association between dose level and side-effects is often the main interest in the study. An appropriate linear-by-linear association model with suitable choice of row and column scores adequately describes many such associations.

Medicine

In medical research it is often felt that physiological abnormalities may produce undesirable responses. The seriousness of these may increase with the severity of the disorder. An example of such a trend occurred with the spleen size/blood platelet count data considered in Example 12.8. Similarly, responses to environmental factors such as different levels of a known carcinogen nearly always show an ordered response. Again linear-by-linear association models may be appropriate.

Administration of Justice

There has been considerable interest in recent years in differences in criminal trends and in the way the courts treat offenders in different ethnic groups. Agresti (1984, p. 32) considers the analysis of counts of death penalty verdicts based on ethnic grouping of victims and defendants using data given by Radelet (1981). In civil proceedings relating to redundancy there have often been claims that company policy on laying off staff differs between age groups or between ethnic groups, or perhaps gender, even when staff are performing similar functions. Many examples illustrating the use of nonparametric methods in a legal context are given by Gastwirth (1988).

Sociology

Many studies have been made of association between socio-economic background and educational or career achievement, and between that background and attitudes towards social problems. Loglinear models are often relevant to studies of such associations.

Ornithology

To obtain information on the number of birds of a particular species in a defined area, samples of birds can be caught, ringed and released on several occasions. A study of the overlap between the samples can be used to obtain capture-recapture estimates of the population size.

Consumer Preference

While the Kendall coefficient of concordance is adequate for a study of many linear (averaging) aspects of consumer preferences, more sophisticated aspects such as dispersion often require an approach based on partitioning into contrasts that give rise to additive components of the X^2 or G^2 statistics. For example, one might be interested in ascertaining whether the degree of sweetness that consumers of soft drinks prefer shows different patterns in different countries, or whether men and women show differing preferences.

13.7 Summary

The **loglinear model** (Section 13.2.1) is used for the study of many aspects of association. It has close parallels to the linear model for treatment comparisons such as those arising in the analysis of factorial treatment structures in the analysis of variance. A no-interaction model corresponds to independence and a model with interactions to patterns of association. These can be applied to **capture-recapture analysis** (Section 13.2.7) in which one of the cell values is missing (the unobserved cases).

Sets of k tables each 2×2 (Sections 13.2.2 to 13.2.5) are often associated with comparisons of binomial responses to each of two treatments at each of k levels of an explanatory variable (e.g., different age groups or different sources of a raw material). This explanatory variable is often called a *covariate*. Both asymptotic and exact test procedures are available to test whether the **odds ratios** may be supposed equal for all 2×2 tables, and if they are, whether or not they all take the common value unity. The latter implies independence. Equality at a value other than unity implies a first-order interaction.

The **linear-by-linear association model** (Section 13.2.6) is a loglinear model of special importance in nonparametric methods. Many well-known tests may be formulated as special cases and may also be extended by modification of the scoring system used in the original test. The **Cochran–Armitage test** for monotonic trends provides a good example of the possibility for such modifications in binomial probability models with the parameter p changing monotonically with covariate values.

Combining and partitioning of contingency tables (Section 13.3) is often a useful tool. The former must be used with caution, as is illustrated by **Simpson's paradox** (Section 13.3.1). **Partitioning into subtables** each corresponding to a single degree of freedom in a statistic having an asymptotic chi-squared distribution is often useful in elucidating the structure of associations. Most partitionings are not unique and certain necessary conditions must be met to ensure additivity of components (Section 13.3.2).

In discrimination tests where panels of N testers are each required to rank in order of preference a range of similar products such as k different brands of tomato soup or varieties of apple or plum or makes of washing machine the numbers allocating each rank to each product may be presented in a $k \times k$ contingency table. Cell entries are the numbers of tasters allocating each designated rank to each product. The appropriate analysis for association, i.e., consistency between tasters, or no particular preferences (independence) requires a modified analysis that takes account of the fact that although there are only N testers the total number of counts (entries in all k^2 cells) is Nk. Partitioning of a chi-squared statistic associated with such tables is often informative about the nature of any association (Section 13.3.3).

13.8 Exercises

13.1 Prepare cross-sectional tables for the data in Table 13.1 to show
(i) separately for each cholesterol level, presence or absence of CHD at each blood pressure level;
(ii) separately for each blood pressure level, presence or absence of CHD for each cholesterol level.

13.2 Verify the value $G^2 = 2.30$ obtained in Example 13.1.

13.3 Verify that an analysis similar to that in Example 13.1 applied to Table 13.3 in place of Table 13.2 leads to identical conclusions about the nature of any association.

13.4 Use an asymptotic test based on $\phi = \log \theta$, to show that for each of the 2×2 subtables in Table 13.2 there is strong evidence against the hypothesis of no association.

13.5 Verify the numerical results stated in Example 13.3.

13.6 Perform the computations requested in Comment 2 on Example 13.4.

13.7 Confirm the result stated in the comments on Example 13.8 that with column scores 1, 2, 20, 200 the linear-by-linear test gives an asymptotic $P = 0.0101$.

13.8 O'Muircheartaigh and Sheil (1983) gave the following data for the numbers of players with scores (i) par or better (P_1), (ii) over par (P_2) for low and high handicap golfers under two wind conditions, W_1 and W_2. Does a first-order interaction model adequately describe the data?

	Low handicap		High handicap	
	P_1	P_2	P_1	P_2
W_1	9	35	1	49
W_2	37	51	12	115

Use at least two different methods of analysis and compare your results. Also obtain a 95 percent confidence interval for the common odds ratio if you accept the hypothesis that they are the same for each 2×2 table.

*__13.9__ Howarth and Curthoys (1987) give numbers of males and female students in English and Scottish universities in the years 1900–01 and 1910–11. Are proportions in the genders independent between countries for each year? Is a first-order interaction model (i) necessary and (ii) sufficient to explain the observations?

	1900–01		1910–11	
	M	F	M	F
England	11755	2080	16038	3579
Scotland	4432	719	5137	1599

13.10 Considering only the data for coronary heart disease (CHD) in Table 13.1, and assuming that blood pressure levels and cholesterol levels are both in increasing orders, use an appropriate test to decide if there is evidence of an association between blood pressure and cholesterol levels. If you were told also that the blood pressure levels were I = normal or below normal; II = between 1 and 10 percent above normal; III = 11 to 20 percent above normal; IV = 21 − 50 percent above normal; V = more than 50 percent above normal, and that the cholesterol levels are A = normal or less; B = up to 50 percent above normal; C = over 50 percent above normal and including some patients as much as 200 percent above normal, how might you take this information into account in your analysis?

*__13.11__ In an English parliamentary electoral constituency a random sample of 400 voters are classified by age and political party affiliation as follows:

	Age group			
Party	30 or under	31–40	41–55	56 or over
Conservative	31	32	39	34
Liberal Democrat	16	19	25	31
Labour	36	27	58	52

Is there evidence of an association between political affiliation and age? It is generally, though not universally, accepted that the Conservative, Liberal Democrat and Labour parties represent an ordering of right, middle and left in the political spectrum.

13.12 Agresti (1984) quotes the following data on cross-classification of attitudes towards abortion and amounts of schooling based on the U.S. General Social Survey, 1972. Test these data for evidence of association between attitudes and educational background.

	Attitude toward abortion		
	Disapprove	Neutral	Approve
Schooling			
Less than high school	209	101	237
High school	151	126	426
More than high school	16	21	138

*13.13 Use a Pearson chi-squared test or a likelihood ratio test to determine whether either of the 2×2 subtables in Table 13.10 indicates association, and if so whether it is reasonable to suppose a first-order association model is adequate.

13.14 Verify that the partitioning given in Section 13.3.2 for a 3×3 table into four 2×2 tables satisfies the necessary conditions quoted in that section.

13.15 Verify the value of G^2 and each component thereof in Example 13.11.

13.16 Thirty adults are each asked to indicate their preferences for each of three brands of competing detergents by ranking them 1, 2, 3. The results are presented in the following tables of numbers giving each rank:

		Pref. rank		
		1	2	3
	A	12	2	16
Brand	B	9	12	9
	C	9	16	5

Partition the Anderson statistic into components C_1 to C_4 similar to those given in Section 13.3.3. Note that if we had four type of detergent we would get a 4×4 table with an adjusted chi-squared statistic with 9 degrees of freedom. This could be split into 9 contrasts. Choice of appropriate contrasts and calculation of the variance of each is appreciably more complicated than it is for a 3×3 table.

*13.17 The Scottish Office Statistical Bulletin, Nov. 1995, published by the UK Government Statistical Service gave for the academic years indicated the number of first-year undergraduate student awards in Scotland for students aged under 21 and those aged 21 or over. Do these data indicate an increasing trend in the proportion of adult students who received awards?

Year	1990–1	1991–2	1992–3	1993–4	1994–5
Under 21	18349	21379	23412	25325	26606
21 and over	5166	8079	10269	12535	13355

13.18 When age discrimination is alleged in compulsory layoff of employees courts often compare layoff rates among those aged less than 40 and those aged 40 or over. Gastwirth (1997) for a case settled out of court before trial gives these data for layoffs.

	$Age < 40$	$Age \ 40+$
Not laid off	787	1113
Laid off	18	50

If the court consider there is sufficient evidence of discrimination only if significance is established at the 5 percent level, using an appropriate test what conclusion should they reach on this evidence?

*13.19 Gastwirth report that for the case considered in Exercise 13.18 a further breakdown of the data by age became available as follows:

Age	< 40	$40 - 49$	$50 - 59$	$60+$
Not laid off	787	632	374	107
Laid off	18	14	18	18

Should the court reach a different decision using this evidence than that reached using the information given in Exercise 13.18? If a different conclusion is reached, or the evidence is substantially stronger, what are the statistical factor or factors that explain that difference?

13.20 Town A is the only large community in an otherwise thinly populated area. During a particular year, of those who received medical attention for an injury from using a kitchen knife, 40 visited a general practice and 90 were seen at the hospital casualty department. In 15 cases the individual attended both a general practice and hospital casualty. Use this information to estimate the total number of kitchen knife injury cases in Town A during that year, giving an approximate 95 percent confidence interval. Are the assumptions that you made appropriate?

CHAPTER 14

ROBUST ESTIMATION

14.1 When Assumptions Break Down

We have emphasized that distribution-free does not mean assumption free. Sometimes we assumed samples were from populations whose cumulative distribution functions differed, if at all, only in their centrality measures, e.g., in means or in medians. In other cases we assumed only that the population distributions were symmetric, in yet others that samples came from populations where any differences were, in some sense, ordered. Independence, both between observations within a sample, and between different samples, was often of major importance.

Our most relaxed assumption was that associated with the sign test applied to a single sample where the null hypothesis required only that each observation came from unspecified distributions (not necessarily all the same) providing they all had the same median under H_0. We saw that some procedures were more sensitive than others to breakdowns in assumptions.

The breakdown often took the form of a few sample values being inconsistent with the pattern associated with the bulk of the observations. Such abberant observations are often referred to as *outliers*. Outliers may arise in several ways. They may be:

- Values that are incorrectly measured or recorded.
- Measurements made on units that are atypical of the bulk of those in the population under study.
- Measurements made with less accuracy or precision than the bulk of the observations.

There are many causes of incorrect data. A measurement of 3.6 cm may be recorded as 33.6 cm or as 36 cm. Repeating a digit is a common error with keyboard data entry, as is omission of a decimal point. A temperature measurement might be incorrectly reported in degrees Fahrenheit when it should have been in degrees Centigrade. An observer who forgets to take a measurement at a critical time and fears a reprimand might cheat and insert a faked or guessed value. Such incorrectly reported data are not always easy to detect. For temperature measurements in a hot climate a reading given in degrees Fahrenheit when it should have been in degrees Centigrade is readily picked up. This may not be the case in polar climates, however, where readings on either scale can be plausible.

It is sometimes hard to find the cause of correctly recorded but atypical data. One machine among many may produce poor quality goods because it is operating at an abnormal temperature, a factor that may not be observed, or is thought to be irrelevant, by the person collecting the data. The milk production of one cow in a herd may be exceptionally low because it suffers from a yet-to-be-diagnosed illness that affects yield.

A frequent complication when data from several sources are combined is that those from some sources may be less reliable than those from others. For example, if the calcium content of milk samples is measured in several laboratories, most may make very precise and accurate determinations while one or two may make less precise measurements (increasing dispersion about the correct values). Another possibility is that one laboratory may return values that are consistently either too high, or too low.

Higher dispersion around a similar mean could indicate the use of an incorrect scale. For instance, at polar research stations temperature measurements may be recorded using the Centigrade scale at all stations apart from one where the Fahrenheit scale is used. The Fahrenheit readings will have a greater spread of values, a factor that might draw attention to the different scales in use.

How to proceed in such circumstances depends to some extent on one's objectives. Obvious measurement or recording errors should be corrected if possible, or else the observation should be discarded. Some such errors nearly always sneak through, especially in large routine data collections. Although one might also try and reduce the other sources of outliers such as atypical experimental units, or poor laboratory performance, one often has to live at least in part with these. It is important to draw attention to anomolies and to state clearly what action, if any, one takes regarding them.

Recognition of these problems has led to several major statistical developments especially since the 1970s. These include the use of *robust methods*. These aim to minimize the influence of outliers or other anomalous observations when these lead to a breakdown in basic assumptions, while at the same time performing almost as well as the optimum methods when relevant assumptions hold. Robust methods may also be highly efficient while requiring only minimal assumptions.

We give an introduction to two approaches that have proved valuable in the context of possible outliers and refer briefly to some others in passing.

The first of these — known as *the bootstrap* — has the added benefit that it is also valuable in situations where existing theory is intractable or difficult to apply. It has both a parametric and a nonparametric form. We consider only the latter here. An interesting feature of the nonparametric version is that it often comes close to being assumption free. The one key assumption always needed is that the sample cumulative distribution function, which we met in Section 2.3, is a reasonable approximation to the population cumulative distribution function for samples that are not too small.

Compared to some longer-established nonparametric methods, there is a price to pay for this relaxation of assumptions, but that price is often less than any that may arise from making unjustifiable assumptions.

The second main approach we consider is known as *M-estimation* with the key property that the resulting estimators behave like conventional maximum likelihood estimators when the latter are appropriate, while protecting against certain types of outliers by reducing the influence of these in determining the outcome of the analysis.

The basic ideas behind both the bootstrap and M-estimation are straightforward, but their practical application often requires care if some not-always-obvious pitfalls are to be avoided. We outline a few basic principles here, referring the reader to more detailed texts for practical details.

As a preliminary, we give a brief description of a characteristric called *influence* which helps to clarify the relationship between outliers and estimation, and we also say something about detecting outliers.

14.2 Outliers and Influence

14.2.1 Nature and Detection of Outliers

This topic is discussed fully by Barnett and Lewis (1994). In broad terms an outlier is an observation so remote from, or out of line with, other observations as to cause surprise. Whether an observation surprises us is subjective. It may depend upon what we know about the source of the data. For example, given the data set 0, 5, 9, 8, 3, 0, 125, 9, 17 without further information most people would say the observation 125 was sufficiently remote from the others to cause surprise. However, an entomologist who knew these were counts of numbers of aphids on each of nine plants of the same rose species may not consider 125 to be odd, for in many situations involving insect populations heavy infestation on just one or two among a large number of plants is common. Barnett and Lewis (p. 16) quote even more extreme data from Fisher, Corbet and Williams (1943) for numbers of moths caught in a light trap, viz.:

$$11 \quad 54 \quad 5 \quad 7 \quad 4 \quad 15 \quad 560 \quad 18 \quad 120 \quad 24 \quad 3 \quad 51 \quad 3 \quad 12 \quad 84$$

Given the observations 2.7, 3.3, 3.5, 2.8, 4.1, 4.3 and no other information none may seem surprising. But if we are told that these are weights in kg for one growing animal recorded at fortnightly intervals, a zoologist would have doubts about the validity of the fourth observation 2.8. A growing animal may suffer weight loss at some growth stage but the decrease is unlikely to be as large as this. If a loss of this magnitude were recorded the animal would probably be dead a fortnight later; it would certainly be unlikely to return to a weight consistent with a normal growth pattern within a further two weeks. This observation is not an outlier in the sense of being remote, but it is probably a contaminated observation, the contamination being a measurement error. One possibility is that a true weight of 3.8 was recorded as 2.8. This might happen if, with the balance used for weighing, the 1 kg and 2 kg weights placed in the scale pan were similar in size and design (differing mainly in density), or if the weights were read from a digital display it could be a careless misreading.

Outliers or contaminated observations are a nuisance that must, or should, be taken into account when making inferences. Criteria have been evolved that give a basis for excluding outliers from an analysis. The aim here is often to make the remaining data more consistent with some parametric inference model.

Barnett and Lewis (1994, Section 6.3) list 48 tests for outliers in normal distributions alone, most of these being optimal only for fairly specific alternatives to the null hypothesis that all data belong to the same normal distribution where neither or one or both parameters may be known. Difficulties are often caused by a masking effect, whereby the power of a test for one outlier is reduced by the presence of other outliers.

We have already mentioned that if an outlier can be shown to be an error it should be corrected if possible; if it cannot be corrected it should be rejected. When there is no clear indication that an outlier is a measuring or recording error the appropriate action is less clear. It depends upon the population of interest and on what questions are being asked about that population. For example, in an experiment to test a drug for reducing blood pressure it may be known that it is ineffective if the recipient drinks alcohol. All recipients might be instructed not to drink alcohol while taking the drug. If, in a group of 25 patients receiving the drug the decreases in systolic blood pressure in mm Hg are

$$-3 \quad 0 \quad 2 \quad 5 \quad 6 \quad 21 \quad 23 \quad 23 \quad 27 \quad 30 \quad 32 \quad 35 \quad 37$$
$$39 \quad 41 \quad 43 \quad 47 \quad 47 \quad 49 \quad 52 \quad 54 \quad 57 \quad 59 \quad 60 \quad 64$$

a clinician might suspect that the first five readings were for patients who had almost certainly ignored the alcohol ban. How firmly the clinician held that view would be a matter of experience. If it were known, for instance, that the drug is ineffective for about 1 person in 50 whether or not they drink alcohol, it would be reasonable to suppose that one or two of the first five readings might be for such cases, but rather unlikely that all five would. If, on the other hand, there were strong grounds for believing it was almost certain that a substantial reduction in blood pressure would take place if alcohol was not consumed, and what was of interest was the mean or median reduction in blood pressure in such cases, it makes sense to omit the five lowest readings. To assume these five had all ignored the alcohol ban might be unfair. It would be a matter of clinical judgement whether to ask those, or perhaps all, participants if they had taken alcohol, and if so how much. A dose-response relationship may mean that a little alcohol does not have the same impact as a larger intake. There may of course be doubts about the truth of the answers given if such a question were asked!

If there were indications that between 5 and 10 percent of the population might not respond positively to the drug (perhaps for some genetic reason) the above results are consistent with that hypothesis. In that case, if one were interested in the whole population one should not reject any observations when making inferences. If one were interested only in the population other than those who clearly did not respond positively, the five observations should be rejected. The purpose of this hypothetical example is to show that there may be no easy answers to dealing with outliers. It is of course important, and ethically sound, to indicate clearly what has been done — and why — if outliers are present or suspected.

14.2.2 A Test for Outliers

Many tests for outliers lack robustness. Some are notoriously bad at detecting more than one outlier in the same tail; others tend to miss a pair of outliers in opposite tails. A simple and reasonably robust test is to classify any observation x^* as an outlier if

$$\frac{|x^* - \text{med}(x_i)|}{\text{med}[|x_i - \text{med}(x_i)|]} > 5.$$

Here $\text{med}(x_i)$ is the median of all observations in the sample. The denominator is a measure of spread called the *median absolute deviation*, often abbreviated to MAD. The choice of 5 as a critical value is motivated by the reasoning that if the observations, other than outliers, have an approximately normal distribution, the test then picks up as an outlier any observations more than about three standard deviations from the mean.

Example 14.1

The problem. Use the above test to detect any outliers in the data set

$$8.9 \quad 6.2 \quad 7.2 \quad 5.4 \quad 3.7 \quad 2.8 \quad 22.2 \quad 12.7 \quad 6.9 \quad 3.1 \quad 29.8$$

Formulation and assumptions. It is easiest to determine the median and MAD after ordering the observations. We first test the observation furthest from the median. We stop if this is not an outlier. If it is an outlier we test the next most extreme in either tail, proceeding until we find an observation that is not an outlier.

Procedure. The ordered observations are

$$2.8 \quad 3.1 \quad 3.7 \quad 5.4 \quad 6.2 \quad 6.9 \quad 7.2 \quad 8.9 \quad 12.7 \quad 22.2 \quad 29.8.$$

The median is 6.9 and the absolute deviation of the observation 2.8 from this median is $|2.8 - 6.9| = 4.1$. Similarly, the remaining absolute deviations are

$$3.8, 3.2, 1.5, 0.7, 0.0, 0.3, 2.0, 5.8, 15.3, 22.9.$$

Ordering these we easily find that their median, the MAD, is 3.2. Setting $x^* = 29.8$, the left-hand side of our statistic is $22.9/3.2 = 7.16$ so we class 29.8 as an outlier. Setting $x^* = 22.2$ we find $15.3/3.2 = 4.78$ so we do not class this, or any other, observation as an outlier.

Conclusion. The data set contains one outlier, $x^* = 29.8$.

Comments. 1. Having classed 29.8 as an outlier we still have to decide what to do about it. An obvious line to follow is to check (a) whether it may be an error, and if it is whether it can be corrected. If it is not obviously an error then (b) is there anything peculiar about the experimental unit giving rise to that value?

2. The data look likely to have come from a skew rather than a normal distribution. The test we have given is not a test for normality, however. A test such as Lilliefors' test should be used to assess normality if that is relevant.

14.2.3 Influence and Robustness

In robustness studies a key role is played by *influence functions*. The idea is due to Hampel who describes their role in robust estimation in Hampel (1974). We consider only briefly two simple influence functions. A detailed treatment of these functions including applications in various fields is given in Barnett and Lewis (1994, Section 3.1.3).

We consider first the effect of a single outlier in relation to the mean. Suppose we have n uncontaminated observations $x_1, x_2, \ldots, x_n$ which we call *good* observations. There is also one contaminated observation, z, which is an outlier in the sense that it takes a value greater than any good observation. Let $\bar{x} = (\sum_{i=1}^{n} x_i)/n$ be the mean of the good observations and $\bar{x}_a$ the mean of the augmented set that includes z. It is easily seen that

$$\bar{x}_a - \bar{x} = \frac{n\bar{x} + z}{n+1} - \bar{x} = \frac{z - \bar{x}}{n+1}, \tag{14.1}$$

whence it follows that the effect of z on the sample mean $\bar{x}$ for the good observations is a linear function of z which tends to infinity as $z \to \infty$. In other words the effect of just one contaminated observation on the sample mean may be infinite! While (14.1) is a measure of the influence of one contaminated observation the influence function $I_n(z)$ for a sample of n is obtained from (14.1) by multiplying by $n + 1$, i.e.,

$$I_n(z) = (n+1)(\bar{x}_a - \bar{x}) = z - \bar{x}.$$

As $n \to \infty$, $\bar{x} \to \mu$, the population mean for the good observations. The limiting function

$$I(z) = z - \mu \tag{14.2}$$

is the *asymptotic influence function*, but it is often referred to simply as the *influence function* as it is the form of greatest interest. Broadly similar and easily modified arguments apply if z is an outlier in the lower tail, i.e., has a value markedly less than all good observations.

In contrast to the situation with the mean where one outlier may have an appreciable, indeed unbounded, influence the effect of a single outlier upon the median is usually small. It is indeed bounded providing $n \geq 2$. The situation is slightly, but not essentially different, depending upon whether the number of good data is odd or even.

Consider first the case where we have $n = 2m$ good data. If the observations are arranged in ascending order and we denote by $x_{(i)}$ the ith largest sample value then the median of the good observations is $(x_{(m)} + x_{(m+1)})/2$. If an outlier, z has a value greater than $x_{(2m)}$ the effect is to shift the median of the combined sample only to $x_{(m+1)}$. Similarly, an outlier below $x_{(1)}$ only shifts the median to $x_{(m)}$.

If the number of good observations is odd and $n = 2m + 1$, giving a median $x_{(m+1)}$ it is easy to see that an outlier $z > x_{(2m+1)}$ only shifts the median to $(x_{(m+1)} + x_{(m+2)})/2$. Similarly, any $z < x_{(1)}$ shifts the median to $(x_{(m)} + x_{(m+1)})/2$.

Because the underlying population distribution influences the values of the order statistics a different approach to that used for the mean is needed to determine the asymptotic influence function. Details are given in Barnett and Lewis (1994, Chapter 4). They show that an appropriate expression is

$$I(z) = \frac{\text{sgn}(z - m)}{2f(m)} \qquad (14.3)$$

where m is the median for the good observations that are distributed with a frequency function $f(x)$ and $\text{sgn}(z - m) = +1, 0$ or -1 depending upon whether z is greater than, equal to, or less than m. Clearly $I(z)$ given by (14.3) is bounded for a continuous distribution and the *supremum* or greatest value of $|I(z)|$ is $1/[2f(m)]$. This supremum is called the *gross error sensitivity*. If we have a large sample from a $N(\mu, \sigma^2)$ distribution and just one contaminated observation is added then the possible effect upon the sample mean is unbounded, while the maximum effect upon the sample median cannot exceed $1/[2f(\mu)] = \sigma\sqrt{(2\pi)}/2 = 1.253\sigma$, since the median of the normal distribution of the good values is μ.

If there are $n = 2m + 1$ good observations and two contaminated observations are added then, whatever their magnitude, the greatest effect upon the adulterated sample median is easily seen to shift it from $x_{(m+1)}$ to either $x_{(m)}$ or $x_{(m+2)}$. The argument easily extends to 4 contaminants with bounds at most $x_{(m-1)}$ or $x_{(m+3)}$. Proceeding in this way, when $2m$ contaminated observations are added the bounds become $x_{(1)}$ and $x_{(2m+1)}$. Only when there are at least $2m + 1$ contaminants is it possible for the median to include some of the contaminated observations and thus become unbounded if some of the contaminated values are unbounded. Thus, at least 50 percent of the observations must be contaminated in a way that makes them outliers before the median becomes unbounded. A broadly similar argument may be used for an even number of good observations.

The number of contaminated observations needed to make a mean or median unbounded determines what is called the *breakdown point*. This is sometimes expressed as a percentage, i.e., 50 percent for the median, but more usefully as a fraction, i.e., $1/2$ for the median. As we have seen, the breakdown point is $1/(n + 1)$ for the mean, since only one contaminated observation need be added to a sample of n to make the sample mean potentially unbounded. The breakdown point is an important measure of robustness because, in many practical situations there is a strong suspicion, or even direct evidence, that an appreciable proportion of observations may be contaminated.

It should now be intuitively clear why in earlier chapters we often found that estimators based on medians are more robust against outliers than are those based on means.

14.3 The Bootstrap

14.3.1 Motivation

We indicated that the motivation for the bootstrap is that the sample cumulative distribution function $S(x)$ introduced in Section 2.3 should, for all but small samples, reflect many characteristics of the population cumulative distribution function, $F(x)$, for the population from which the random sample was obtained. This was implicit in the Kolmogorov test introduced in that section where we used $S(x)$ in deriving a statistic to determine whether a sample was consistent with some given $F(x)$. For the bootstrap we are often less concerned with hypotheses about any particular $F(x)$ but more often with making inferences simply on the assumption that $S(x)$ is a good approximator to some $F(x)$ without being specific about what that $F(x)$ is. The more vague our assumptions about a population distribution, the more useful the bootstrap becomes.

The method is based on repeated resampling of data to tell us more about characteristics of the population from which the data are a random sample. Resampling, as we saw as early as Example 2.1, and in numerous other examples of permutation test procedures throughout this book, is at the heart of permutation or randomization test theory. Permutation tests are based on resampling *without replacement* and where appropriate lead to exact tests and estimation procedures. Bootstrapping uses sampling *with replacement* and leads to approximate tests and estimation procedures; however, it is also applicable in many situations where no permutation test is available or appropriate.

Bootstrapping can be used without prior specification of any distributional model, but the procedure itself can only be justified by fairly complicated mathematics. The main practical requirement for its application is suitable computing facilities. Although bootstrapping results are by their nature usually only approximate, they are often more reliable and informative than those obtained by fitting a wrong model, e.g., assuming normality when that is not valid, or when it is not justified by some asymptotic result like the central limit theorem.

There is seldom a unique or best solution to a bootstrap problem. In practice the method is used mainly either because there is no tractable analytic solution, when a permutation test or the facility to carry it out is not available, or when there is doubt about whether conditions needed for a particular analytic solution or permutation test actually hold.

The approach is intuitively reasonable in many applications and confidence in the method is enhanced because when it is correctly used inferences are usually similar to those given by analytic or permutation solutions when these exist or are relevant. Important applications include those to complicated data structures or ones that involve inferences about concepts such as correlation or ratios of variables where analytic results are not readily available except under very restrictive distributional assumptions.

Use of the bootstrap stems largely from work by Efron (1979). A straightforward introduction to it and related techniques is given by Efron and Gong (1983). Full treatments at the elementary and intermediate level, with many examples are given by Efron and Tibshirani (1993), Davison and Hinkley (1997) and Chernick (1999).

14.3.2 Bootstrap Samples

Given a random sample of n observations $x_1, x_2, \ldots, x_n$ from some population a bootstrap sample is a random sample of size n obtained from these data by sampling with replacement. This means some x_i may occur more than once, others not at all, in a bootstrap sample.

It is notionally possible to determine the distribution of all possible bootstrap samples and the distribution for many associated statistics such as the means or medians of the bootstrap samples. Such distributions are called *true bootstrap distributions* to distinguish them from estimates based on a random subset of all possible bootstrap samples. The latter are important in practice because determining true bootstrap distributions analytically is a formidable task for all but small n except for a few statistics where some general analytic results hold. In this latter case inferences can usually be made easily without bootstrapping.

We derive here the complete bootstrap distribution for the mean and median for a sample of three observations. This is an interesting illustration, but trivial in practice because so small a sample is not very informative about the distribution associated with a large population. We consider all possible samples of three formed with replacement from three distinct data x_1, x_2, x_3. Because of replacement the probability of any x_i being selected is always $1/3$. In a bootstrap sample of three any x_i may appear r_i times where $r_i = 0, 1, 2$ or 3 subject to the constraint $r_1 + r_2 + r_3 = 3$, where the probability of each possible sample is given by the multinomial distribution. If we write $n(x_i)$ for the number of times each x_i occurs the probability of a sample where x_1 occurs r_1 times, x_2 occurs r_2 times and x_3 occurs r_3 times is

$$\Pr[n(x_1) = r_1, n(x_2) = r_2, n(x_3) = r_3] = \frac{3!}{r_1! \times r_2! \times r_3!} \times \left(\frac{1}{3}\right)^3 \quad (14.4)$$

It is easy to record all possible bootstrap samples and the associated probabilities given by (14.4). In Table 14.1 we give these together with each sample mean and median where we assume, without loss of generality, that $x_1 < x_2 < x_3$.

For each sample there is an associated probability p_i given by (14.4). These are shown in the second column of Table 14.1 so that we know the probability distribution of the bootstrap samples. These probabilities also apply to the

Table 14.1 *Bootstrap samples for $n + 3$ and sample means and medians.*

Sample	Probability	Sample mean	Sample median
x_1, x_1, x_1	$1/27$	x_1	x_1
x_1, x_1, x_2	$3/27$	$(2x_1 + x_2)/3$	x_1
x_1, x_1, x_3	$3/27$	$(2x_1 + x_3)/3$	x_1
x_1, x_2, x_2	$3/27$	$(x_1 + 2x_2)/3$	x_2
x_1, x_2, x_3	$6/27$	$(x_1 + x_2 + x_3)/3$	x_2
x_1, x_3, x_3	$3/27$	$(x_1 + 2x_3)/3$	x_3
x_2, x_2, x_2	$1/27$	$(x_2$	x_2
x_2, x_2, x_3	$3/27$	$(2x_2 + x_3)/3$	x_2
x_2, x_3, x_3	$3/27$	$(x_2 + 2x_3)/3$	x_3
x_3, x_3, x_3	$1/27$	x_3	x_3

sample means y_i in column 3. Using the standard formulae $E(Y) = \sum p_i y_i$ and $\text{Var}(Y) = \sum p_i y_i^2 - (\sum p_i y_i)^2$ we leave it as an exercise for the reader to show that $E(Y) = \bar{x} = (x_1 + x_2 + x_3)/3$, the mean of the original sample of 3. Rather more tedious algebra shows that the variance of the bootstrap sample means is given by $\text{Var}(Y) = \sum (x_i - \bar{x})^2/9$.

These results generalize to samples of size n where once again the mean of the bootsrap means is the overall sample mean and now the variance of the sample means is s^2/n where $s^2 = \sum (x_i - \bar{x})^2/n$. This is similar to the classical estimator of variance of the mean and is indeed equal to the maximum likelihood estimator for samples from a normal distribution. Thus bootstrapping here as a method of determining standard error of the mean shows no advantage over classical methods. Any advantage shows for estimators other than the mean where standard errors are not easy to obtain analytically except under restrictive assumptions about the population distribution.

For the small sample of three bootstrap estimation of the standard error of the median is easy to illustrate. The bootstrap medians in column 4 of Table 14.1 take only three possible values x_1, x_2, x_3. For example, x_1 is the median of each of the three samples (a) x_1, x_1, x_1, (b) x_1, x_1, x_2, (c) x_1, x_1, x_3 and the total associated probability calculated from column 2 is $1/27 + 3/27 + 3/27 = 7/27$. Similarly, the median x_2 has total probability $13/27$ and the median x_3 total probability $7/27$, whence, if Y represents the bootstrap median

$$E(Y) = \frac{7x_1 + 13x_2 + 7x_3}{27}$$

and

$$\text{Var}(Y) = \frac{7x_1^2 + 13x_2^2 + 7x_3^2}{27} - \frac{(7x_1 + 13x_2 + 7x_3)^2}{729}.$$

The bootstrap standard error of the median is the square root of this variance.

One must not expect to make useful inferences about characteristics such as the mean or median of a large population from a sample of three by bootstrapping, or by any other method. Even with realistic but still relatively small samples of, say, 10 or 12, there is only limited evidence about the nature of the population from such scant data. In particular, when sampling from a large population the sample cdf of a small sample may not be a good approximation to the population cdf.

The number of possible different bootstrap samples when $N = 3$ is only the 10 given in Table 14.1. The number of possible samples increases rapidly as n increases. When $n = 4$ it is 35, with the probabilities associated with each ranging from $1/256$ to $24/256$, while for n $= 10$ there are 92,378 distinct bootstrap samples.

In practice bootstrap inferences are nearly always based on Monte Carlo sampling to generate a predetermined fixed number, B, say, of bootstrap samples. We denote a typical bootstrap sample by

$$x_1^*, x_2^*, \ldots, x_n^*$$

where each x_i^* is equal to one of the original observations. As already indicated, because sampling is with replacement some of the original sample values (the x_j) may not appear while others appear more than once.

For example, if $n = 9$ and the observations are

$$2.5, 3.1, 4.2, 5.1, 5.3, 5.9, 6.7, 7.2, 10.5$$

a typical bootstrap sample might be

$$x_1^* = 4.2 \quad x_2^* = 6.7 \quad x_3^* = 5.1 \quad x_4^* = 6.7 \quad x_5^* = 7.2$$
$$x_6^* = 6.7 \quad x_7^* = 2.5 \quad x_8^* = 10.5 \quad x_9^* = 5.1.$$

If B bootstrap samples are generated the bth may be written

$$x_1^{*b}, x_2^{*b}, \ldots, x_n^{*b}.$$

Vector notation provides a convenient shorthand if we write $\mathbf{x}^*$ for any bootstrap sample and $\mathbf{x}^{*b}$ for the bth sample. For the numerical example above

$$\mathbf{x}^* = (4.2, 6.7, 5.1, 6.7, 7.2, 6.7, 2.5, 10.5, 5.1).$$

In bivariate or multivariate situations such as those in correlation or regression each x_i may itself be a vector. For each bootstrap sample $\mathbf{x}^*$ we often compute statistics such as the sample mean or median or sample variance and use these to estimate the corresponding population distribution characteristic. We denote the parameter or other population characteristic we are interested in by θ and the statistic used to estimate it from the bootstrap sample $\mathbf{x}^*$ by $s(\mathbf{x}^*)$. This gives an estimate $\theta^* = s(\mathbf{x}^*)$ for θ.

Because we are sampling with replacement the numerical value of $s(\mathbf{x}^*)$ changes from sample to sample and so the statistic has a distribution. As the number of bootstrap samples, B, tends to infinity the mean of any statistic $s(\mathbf{x}^*)$ will tend to the mean computed for the true bootstrap sampling distribution. However, as pointed out above, this is only known in a few special cases, or can only be worked out for general cases for small values of n, as in the illustration given earlier for $n = 3$.

Practical experience shows that in many, but not in all, situations even for small n the mean of B bootstrap samples converges rapidly to the limiting value that holds when $B \to \infty$.

If we generate B bootstrap samples and denote the mean of the $s(\mathbf{x}^{*b})$ by $s(.^*)$, i.e., $s(.^*) = \sum_b [s(\mathbf{x}^{*b})]/B$, then the appropriate estimator of the true bootstrap standard error of θ^*, which we denote by $\mathrm{se}_B(s)$, is

$$\mathrm{se}_B(s) = \left\{ \frac{\sum_b [s(\mathbf{x}^{*b}) - s(.^*)]^2}{(B-1)} \right\}^{1/2}. \tag{14.5}$$

This is the usual estimator of a population standard deviation based upon a random sample of B from some population and tends to the true bootstrap standard deviation for the statistic s as $B \to \infty$. In practice the approximation is often good for B as low as 20, providing n is not too small. Even for small n reasonable estimates may be obtained with $B = 100$, although again we caution against indiscriminate use of the bootstrap with very small samples because then the sample distribution may not truly reflect a population characteristic of interest.

Pitfalls to watch out for when using the bootstrap with small samples include difficulties with bias. This may carry over to moderate or large samples in some contexts. Bias is discussed for a range of applications by Efron and Tibshirani (1993). It is important to realise that in applications using only a finite number, B, of bootstrap samples there are two sources of error.

The first is the usual *sampling error* applicable to all sampling based inferences about a population no matter what method of inference is used. For example, in parametric inference about the mean of a normal population based on a sample of n the sample mean $\bar{x}$ will in general not equal the population mean μ. The sampling 'error' in taking $\bar{x}$ as an estimator of μ is measured by the standard error.

The second source of error specific to bootstrap sampling is that made when we approximate to the true bootstrap standard error using only a finite number, B, of bootstrap samples. We do not prove it here, but it can be shown that the true bootstrap standard deviation is approximately equal to the estimated population standard deviation for the corresponding estimator. Example 14.2 sheds some light on the way sampling variation is reflected in bootstrap estimates.

Example 14.2

The problem. For the following sample data use the bootstrap for estimating the population median.

0.04	0.06	0.27	0.32	0.33	0.40	0.50	0.63	0.69	0.92
1.09	1.10	1.35	1.61	1.66	1.69	1.71	1.80	1.98	2.65
2.83	3.50	3.72	3.75	3.99	5.16	5.49	6.31	7.05	16.05

Formulation and assumptions. The data are a random sample of 30 from a known distribution, but they are not unlike some that may arise in practice. They might, for example, be the percentages of some pollutant in various water sources, a situation where it is not uncommon to find small levels of pollution in many of the sources but quite high and wide ranging levels in others. The data are arranged in ascending order to indicate more clearly the skewness with a long upper tail including the value 16.05 that might suggest itself as an outlier. A boxplot would be a useful EDA tool to show these characteristics. The sample median is $(1.66 + 1.69)/2 = 1.675$. We explore the information available by generating $B = 40$ bootstrap samples and obtaining the median of each for use in estimating the bootstrap standard error of this statistic.

Procedure. We used the facility for sampling with replacement in Minitab to generate the 40 bootstrap samples. The median was computed for each sample and ranged from 0.805 to 3.17. The estimated bootstrap standard error given by (14.5) was 0.4116.

Conclusions. To help interpret these results we now disclose that the data are a computer generated random sample from an exponential distribution with mean $\mu = 3$. The theoretical median of that distribution is $\theta = 2.079$. A crude rule-of-thumb indication used in bootstrapping is that we should accept a hypothetical value of a parameter θ if the sample equivalent estimator (here the sample median) lies within two standard deviations of that hypothesized value. Taking the estimated bootstrap standard error as a reasonable estimate of this standard deviation, in this example the difference $2.079 - 1.675 = 0.404$ is less than one estimated standard error, which here is 0.4116.

Comments. 1. There may be some unease as to whether 40 bootstrap samples give a reliable estimate of the true bootstrap standard error. That it does so is also an implicit requirement for validity of the crude rule of thumb used above. This rule is based on the assumption that the true bootstrap standard error is reasonably close to the best estimated standard error where this is known.

In this example, where we sampled from a known distribution, there is an analytic expression for the standard error of a median estimator from a sample of n, namely

$$\text{se(median)} = \frac{1}{\sqrt{4nf^2}}, \qquad (14.6)$$

where f is the ordinate of the probability density function at the median. For an exponential distribution with mean 3 it can be shown that $f = 0.1667$ and when $n = 30$ that se(median) $= 0.5476$. Here we see that our estimate of the bootstrap standard error, 0.4116, underestimates the true standard error. This is a bias introduced by the fact that our chosen sample has a median $m = 1.675$ which is appreciably below the population median $\theta = 2.079$. In fact, 19 of the 30 sample values are less than the population median. This is not an alarming departure from the expected number, 15. The situation is not very different from that with many familiar tests. Recall that when sampling from a normal distribution the usual sample estimator s^2 of σ^2

may differ appreciably from the latter in many samples. That is why, when σ^2 is unknown the t-test is invoked to allow for such uncertainty.

It is interesting to note that if we assumed the population median was in fact equal to the sample median of 1.675, then the standard error of the median estimator, assuming our sample was from an exponential distribution with this median, would now be 0.4412, close to the bootstrap estimate of 0.4416 obtained above.

2. The bootstrap may also be used to obtain an estimated confidence interval for a parameter such as the median. There are a number of practical difficulties in obtaining such intervals but we briefly outline one approximate procedure in Section 14.3.4.

3. In this example we considered a sample from a known distribution and so were able to invoke some theoretical results for the standard error of the median because f was known in (14.6). Had we not known f, or if we suspected the sample was from a mixture of distributions, or that some observations may be outliers, the bootstrap might be a serious competitor to, say, sign test procedures. We explore this point further in Example 14.5.

14.3.3 Bootstrapping versus Permutation Procedures

The bootstrap is particularly useful when no simple analytical procedure exists. It is also useful when no exact test based on permutations is available, because it may still provide an approximation to a permutation procedure. Its application in these circumstances is in general not as straightforward as that of exact procedures, or even that for Monte Carlo approximations in permutation tests. We have pointed out that a key difference between a permutation test and a bootstrap procedure is that the former is based on sampling *without* replacement, while the latter is based on sampling *with* replacement. The consequences of this apparently small difference are far reaching. We illustrate some of them for two-independent-sample problems.

We did not consider the Pitman permutation procedure for two independent samples from continuous distributions in much detail in Chapter 6 because it lacks robustness. For hypothesis testing it is in theory straightforward, and as in the case of most permutation tests there are many equivalent statistics. Given samples

$$x_1, \quad x_2, \quad \ldots, \quad x_m$$
$$y_1, \quad y_2, \quad \ldots, \quad y_n$$

it is well-known, and easily verified, for the Pitman or raw data permutation test that equivalent test statistics for the hypothesis H_0: *the population means are identical* against the usual one- or two-tail alternatives under the assumption that the samples are from populations with otherwise identical distributions include the usual t-statistic, the first sample sum $\sum_i x_i$, the second sample sum $\sum_j y_j$, the first sample mean $\bar{x}$, the second sample mean $\bar{y}$, and the difference between sample means $\bar{x} - \bar{y}$. Equivalence follows because under permutation both the denominator in the usual t-statistic and the sum of the combined sample values remain constant so that all the above statistics are linear functions of one another and have the same ordering and the same

P-value is associated with each corresponding value in that ordering. We must remember that if this permutation *t*-statistic is used it does not have the tabulated *t*-distribution values relevant to the normal theory parametric test.

The above equivalences all assume sampling is without replacement. They do not carry over to the bootstrap where we sample with replacement. Then the sum of all $m + n$ values will change between bootstrap samples depending upon which of the original sample values do not appear at all, appear only once, or appear more than once in a given bootstrap sample. Thus, we will arrive at different conclusions if we base our bootstrap inferences on each of the above possible permutation distribution statistics. We have already pointed out that there is, in general, no unique and best bootstrap for a particular situation. What is important is to use a statistic that is intuitively reasonable in a particular case. If we believe our samples come from identical distributions that differ only in mean it makes sense to base bootstrap inferences on the sample mean difference $\bar{x} - \bar{y}$. How we might proceed is shown in Examples 14.3 and 14.4.

Example 14.3

The problem. In Example 6.1 we considered two data sets giving times to perform some calculations. These were

Group A	23	18	17	25	22	19	31	26	29	33	
Group B	21	28	32	30	41	24	35	34	27	39	36

Assuming the samples are from populations with distributions that differ if at all only in their means μ_1 and μ_2 use a bootstrap analysis to assess the strength of evidence against the hypothesis $H_0: \mu_1 = \mu_2$ when the alternative is $H_1: \mu_1 \neq \mu_2$.

Formulation and assumptions. The analysis in Example 6.1 was based on the Wilcoxon rank sum test. Here we base our analysis on the raw data, so that the bootstrap may be regarded as an approximation to a Pitman permutation test. We consider 1000 bootstrap samples and for each we compute $\bar{x}^* - \bar{y}^*$ in the way described under *Procedure*. We estimate the relevant *P*-value as the proportion of the $\bar{x}^* - \bar{y}^*$ that exceed the observed $\bar{x} - \bar{y}$ in magnitude.

Procedure. In a Pitman permutation test where sampling is without replacement each permutation sample of size m is obtained by drawing a sample of m observations from the combined sample of $m + n$ observations. The n observations not selected automatically form the complementary permutation sample of size n. In this example $m = 10$ and $n = 11$.

To obtain an analogous bootstrap sample we first select a sample of $m + n = 21$ *with replacement* from the combined Group A and Group B data. The first 10 values chosen for this sample are designated as a bootstrap sample of size $m = 10$. The remaining $n = 11$ sample values constitute the bootstrap sample of size 11. For each such bootstrap sample we compute $\bar{x}^* - \bar{y}^*$ and compare its magnitude with that of the observed $\bar{x} - \bar{y} = -7.245$ for the given data.

Using this procedure, one sample of 21 we obtained by sampling with replacement from the given data was

17 22 22 27 31 25 41 34 26 17 30 23 28 21 33 21 27 34 23 33 19

leading to bootstrap samples of 10 and 11 values respectively

$$
\begin{array}{cccccccccc}
17 & 22 & 22 & 27 & 31 & 25 & 41 & 34 & 26 & 17 \\
30 & 23 & 28 & 21 & 33 & 21 & 27 & 34 & 23 & 33 & 19
\end{array}
$$

whence it is easily verified that $\bar{x}^{*} - \bar{y}^{*} = 26.2 - 26.545 = -0.345$.

An estimate of the two-tail test P-value based on B bootstrap samples is then given by $P^{*} = $ (number of samples for which $|\bar{x}^{*} - \bar{y}^{*}| \geq 7.245)/B$. For 1000 such samples we found $|\bar{x}^{*} - \bar{y}^{*}| \geq 7.245$ in 15 cases implying $P^{*} = 0.015$. For a second sampling, again with B = 1000, we found $P^{*} = 0.016$.

Conclusion. There is strong evidence against the hypothesis that the means are equal.

Comments. 1. For these data a Pitman permutation test gives the exact $P = 0.014$, in close agreement with the bootstrap result obtained above.

The Pitman test usually gives similar results to the t-test even when assumptions relevant to the latter do not hold. For the above data few statisticians would have reservations about use of the parametric t-test, and indeed for these data the two-tail P-value given by that test is $P = 0.011$, again in broad agreement. Also in Example 6.1 we showed that using the exact WMW test for these data gave $P = 0.016$, so there is good agreement between all tests considered here.

2. Using the bootstrap statistic $\bar{x}^{*} - \bar{y}^{*}$ is intuitively reasonable under the assumption that the only possible population distributional differences involve the means. A different approach is appropriate if we drop the assumption of identical distributions under H_0. We do not pursue this further here, but this problem is discussed fully, together with some alternatives to the approach used in this example, by Efron and Tibshirani (1993, Chapters 15 and 16). Even in simple parametric situations choice of appropriate procedures depends strongly on relevant assumptions. For instance, validity of the t-test depends not only on an assumption of normality but also upon one of equal population variances.

Computational aspects. Nearly all major statistical packages have a facility for random sampling with replacement and even if no direct bootstrapping program is included for the simple example considered here it is usually easy to write a macro to form many bootstrap samples quickly. We used Minitab. For more sophisticated applications of the bootstrap the packages S-PLUS or R are particularly well suited.

Bootstrapping following the broad pattern in Example 14.3 may also be applied in situations where WMW methods are used. With one possible modification the situation is essentially the same as that for the raw data except that these are replaced by ranks. The one modification that might be considered is that if we proceed in this way we are not strictly using a WMW procedure in the bootstrap. Why this is so is illustrated by a trivial example. We consider the situation in Example 2.1 where patients were ranked 1 to 9 in response to two drugs, four of them having received one drug and the remaining 5 another. We learnt in Chapter 6 that essentially what we did in that example was a WMW test. If we follow the procedure in Example 14.3 given the combined sample raw data 1, 2, 3, 4, 5, 6, 7, 8, 9 we may obtain a combined bootstrap sample, say, 1, 2, 9, 2, 8, 1, 3, 3, 9. Continuing as we did in Example 14.3 we would split this into bootstrap samples 1, 2, 9, 2 and

8, 1, 3, 3, 9 of size 4 and 5. To do so is not unreasonable, but proceeding in this way we are not performing a strict WMW bootstrap because we have not reranked the combined sample bootstrap data to allow for ties induced by the resampling. Doing so using appropriate mid-ranks would replace 1, 2, 9, 2, 8, 1, 3, 3, 9 by 1.5, 3.5, 8.5, 3.5, 7, 1.5, 5.5, 5.5, 8.5 giving bootstrap samples 1.5, 3.5, 8.5, 3.5 and 7, 1.5, 5.5, 5.5, 8.5. In practice the difference tends to be small for samples of reasonable size, although it may be appreciable in small samples. We do not recommend the bootstrap for small samples for a reason already explained — namely that a small sample may not be a good representation of an underlying distribution. This is in itself a limitation in any form of inference, but for bootstrapping it is compounded by the bootstrap sampling error.

Example 14.4

The problem. For the data in Example 14.3 explore the use of a bootstrap in the estimation of a P-value using ranked data.

Formulation and assumptions. Bootstrap sampling may be used to obtain an approximation to the exact $P = 0.016$ for a two-tailed test obtained in Example 6.1 using the WMW procedure. The process used follows broadly the pattern in Example 14.3.

Procedure. When the combined samples are ranked we established in Example 6.1 that the rank sums were $S_m = 76$ and $S_n = 155$. These statistics are inappropriate for bootstrapping. This is because for bootstrap samples the corresponding S_m^* and S_n^* no longer satisfy the condition $S_m^* + S_n^* = (m + n)(m + n + 1)/2$ in general. In an approach like that in Example 14.3, we used the difference between means of our bootstrap samples as the appropriate statistic, proceeding for 1000 samples in the manner described in Example 14.3 using (a) the data obtained by resampling ranks directly and (b) that obtained by modification by reranking the resampled data to allow for ties introduced by resampling with replacement. The respective two-tail P-values obtained in the way described in Example 14.3 were $P = 0.012$ for (a) and $P = 0.008$ for (b). While the result using (a) is close to the exact result that for (b) is disappointing. We took a second sample of 1000 bootstrap estimates using (b) and this time we found $P = 0.022$. Combining the two samples gave an estimated $P = 0.015$ in close agreement with the WMW result (see *Comment* 1 below).

Conclusion. There is strong evidence against H_0.

Comments. 1. The fairly marked disagreement between the two estimates of P from two different samples of 1000 using method (b) suggests that our chosen statistic may be somewhat unstable in its properties. This phenomenon is familiar to serious users of the bootstrap, especially with small samples. It may arise from two causes. We may simply be unlucky with the way the sampling scheme is working for the particular samples that were selected, but in this particular example we conjecture that the effect may in part be due to introducing the extra step of modifying the resampled values by reranking them to allow for ties. It is possible that the statistic we used (difference in means) might not then be the most appropriate statistic; however we have not explored this further.

2. Despite the discrepancy discussed in *Comment* 1 the results lead to essentially the same conclusion as that using an exact WMW test and in practice would be comparable to those obtainable using the same number of Monte Carlo samples for a permutation test.

14.3.4 Bootstrap Confidence Intervals

Bootstrapping is often used to obtain approximate confidence intervals. Simplistic approximations suffer from two defects. The first is one of bias which we cover briefly in Section 14.3.5. The other is that they often tend to provide less than the claimed coverage, e.g., an interval computed in a manner that purports to give a 95 percent coverage may only give about 91 or 92 percent coverage. At a basic level Efron and Tibshirani (1993, Section 14.2) list five different ways to compute confidence intervals using the bootstrap. Each works tolerably well when appropriate, but it is hard to know which is appropriate if little is known about distributional properties of the population from which the data are obtained — a situation when we are particularly likely to want to use the bootstrap. Another approach to bootstrap confidence intervals is given by Hall (1992).

We start our discussion from the well-known result that given a sample of n from a normal distribution with variance σ^2 the 95 percent confidence limits for the population mean μ are $\bar{x} \pm 1.96\sigma/\sqrt{n}$ where $\bar{x}$ is the sample mean. If σ is unknown we replace $\sigma/\sqrt{n}$ by its usual sample estimate $s/\sqrt{n}$, where $s = \sqrt{[\sum_i (x_i - \bar{x})^2]/(n-1)}$, and 1.96 by the appropriate quantile of the t-distribution with $n - 1$ degrees of freedom. For values of $n > 30$, the t-distribution quantiles closely approach those of the standard normal distribution. Modifications for other confidence levels are straightforward. In practice the above limits are widely used even if there is little evidence that the sample is from a normal distribution, faith in the outcome relying on the central limit theorem holding, despite its asymptotic nature, even for moderate n.

This fundamental result for parametric inference stimulated two approaches to forming bootstrap confidence intervals. The first is that for reasonably large samples approximate 95 percent confidence limits for a parameter θ may be based on a sample estimator $\hat{\theta}$ of θ and are given by $\hat{\theta} \pm 2\text{se}^*(\hat{\theta})$ where $\text{se}^*(\hat{\theta})$ is an estimate of the true bootstrap standard error based on B bootstrap samples. For samples from a normal population this works for the mean because it can be shown that in that case the bootstrap standard error is a good approximation to the usual estimated standard error (although it shows a small bias). However, as we have emphasized the bootstrap is most useful when analytic theory is nonexistent or is highly distribution dependent and therefore any parametric theory may not hold for a particular sample. Experience has shown that this bootstrap method does not translate well to such situations. This is partly due to potential bias in bootstrap standard errors, but more importantly because confidence intervals based on symmetry about an estimator $\hat{\theta}$ are inappropriate if that estimator has a skew distribution.

A more fruitful and relatively easy-to-apply approach to bootstrap confidence intervals is based on the distribution of appropriate bootstrap estimators, $s(x^{*b})$ of each of B bootstrap samples. When B is large, then providing the $s(x^{*b})$ are reasonably free from bias the distribution of $s(x^{*b})$ is likely to approach the sampling distribution of the estimator of the parameter θ we are interested in. An obvious way to estimate a 95 percent confidence interval is to take the 0.025 and 0.975 quantiles of the distribution of the B computed $s(x^{*b})$. Thus if $B = 1000$ for a 95 percent confidence interval the limits are the 25th and 975th largest values of the $s(x^{*b})$. More generally, for a $(1 - 2\alpha)100$ percent interval with B bootstrap samples the limits are the αBth and the $(1 - \alpha)B$th largest sample value of $s(x^{*b})$. For nonintegral values of such quantiles it usually suffices to round to the nearest integer. Confidence intervals formed in this way are called *quantile-based intervals*. The main defects of these intervals are that they may be misleading due to bias and they tend to give less than the nominal coverage. The effect of bias is that the true probabilities associated with each tail are not equal to each other, and which is the greater depends upon the direction of bias. Efron and Tibshirani (1993, Chapter 14) present two commonly used corrections to remedy these weaknesses, but we do not discuss these here.

Example 14.5

The problem. For the data in Example 14.2 obtain 95 and 99 percent quantile based confidence intervals for the population median using 2000 bootstrap samples.

Formulation and assumptions. For each of 2000 bootstrap samples we obtain the median. These are arranged in order and the appropriate limits are obtained in the way described under *Procedure*.

Procedure. We obtained 2000 medians of bootstrap samples using Minitab and arranged these in ascending order. The 95 percent limits were the 50th and 1950th largest. For our samples these turned out to be 0.92 and 2.83. Similarly 99 percent limits are given by the 10th and 1990th largest and turned out to be 0.69 to 3.50.

Conclusion. Estimated quantile-based 95 and 99 percent bootstrap confidence intervals for the population median are (0.92, 2.83) and (0.69, 3.50).

Comments. 1. It is interesting to compare these with confidence intervals based on the sign test procedure. Under that procedure the interval (0.92, 2.83) has an exact 95.72 percent coverage and the interval (0.69, 3.50) has an exact 99.48 percent coverage, so there is no evidence here of the potential undercoverage of quantile-based intervals that we referred to above. This may be fortuitous, for another sample of 2000 bootstrap medians would be unlikely to lead to exactly the same intervals.

2. For our 2000 samples the estimated standard error of the bootstrap median turned out to be 0.4213, in close agreement with the estimate 0.4116 obtained in Example 14.2 from only 40 samples. It is common experience with bootstrapping that while samples of 100 or less give good estimates of standard errors, samples of 1000 or more are needed for satisfactory quantile-based or related estimates of confidence intervals.

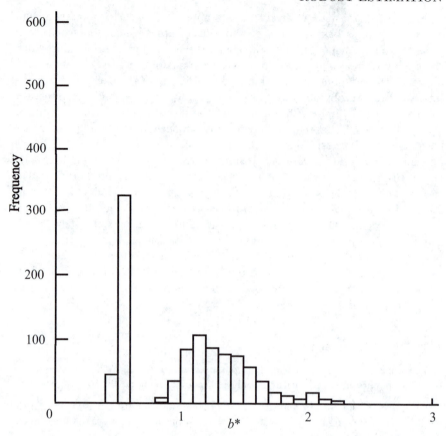

Figure 14.1 *Histogram for 1000 bootstrap least squares slope estimators for data in Example 11.6. Eight values of* b^* *exceeded 2.4.*

3. Because of the asymmetric sample [which we in fact know came from an asymmetric distribution (see Example 14.2)] approximate confidence intervals based on the estimated bootstrap standard error are not appropriate even if that standard error estimate is itself good. In this example using this approach approximate 95 percent confidence limits are $1.675 \pm 2 \times 0.4213$, giving the unsatisfactory interval (0.83, 2.52).

Computational aspects. See remarks under this heading in Example 14.3.

When a statistic of interest is calculated for a large number of bootstrap samples it is often useful to examine the distribution of these bootstrap statistics informally, perhaps presenting them in a histogram or as a box and whisker plot. Although the data set is too small to give bootstrap results likely to be relevant to a larger population we consider some bootstrap samples using the data in Example 11.1 to show what we mean.

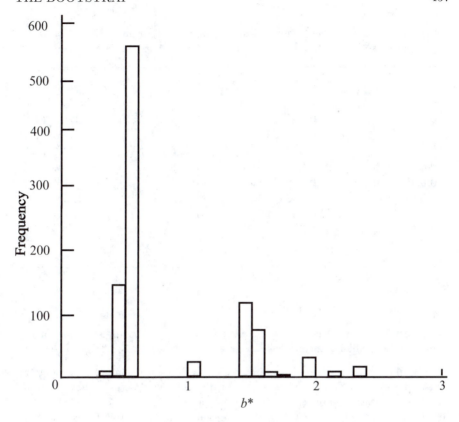

Figure 14.2 *Histogram for 1000 bootstrap Theil–Kendall slope estimators for the data in Example 11.6. Ten values of b* exceeded 2.4.*

Example 14.6

The problem. In Example 11.1 we gave the data set

Hours from start (x)	0	1	2	3	4	5	6
Flow rate (y)		2.5	3.1	3.4	4.0	4.6	5.1	11.1

which was displayed in Figure 11.1. In that example and in Example 11.3 least squares and Theil–Kendall estimates of the regression line slope β were obtained. Use a bootstrap with 1000 samples to obtain approximate 95 percent inter-quantile based confidence intervals for each of these estimates. Use histograms to illustrate the main characteristics of bootstrap estimates of β in each case. Explain the main characteristics of, and differences between, the forms of these histograms.

Formulation and assumptions. One, but not the only, basis for bootstrapping in a bivariate regression problem with n observed data points (x_i, y_i) is to take samples of n of these points with replacement. Using this approach for least squares bootstrap estimates b^* of β we obtain the usual least squares estimate for each bootstrap

Table 14.2 *Values of 1000 bootstrap slope estimates b^* and number of times r that each occurs using the Theil–Kendall estimator for Example 14.6 data.*

b^*	r	b^*	r	b^*	r	b^*	r
0.3	4	0.535	1	0.583	12	1.762	3
0.45	17	0.537	17	0.6	183	1.9	1
0.5	132	0.546	10	1.017	18	1.925	29
0.51	6	0.55	63	1.1	3	1.983	1
0.512	4	0.558	18	1.433	125	2.146	3
0.52	90	0.562	4	1.517	7	2.367	6
0.522	13	0.567	86	1.6	64	3.25	5
0.525	52	0.575	3	1.679	5	6.0	5

sample. Similarly, for the Theil–Kendall method our bootstrap estimator for each sample is the median of the pairwise b_{ij} (as defined in Section 11.1.3 for that sample).

Procedure. For each bootstrap sample of data points we compute the bootstrap least-squares or Theil–Kendall estimator of β in the way described above. For 1000 samples the least squares estimators in one run gave a quantile-based 95 percent confidence interval for β as (0.49, 2.01), in reasonable agreement with the classic least squares interval (0.23, 1.98). The Theil–Kendall estimator for a different sample of 1000 gave an interval (0.50, 2.37), rather longer than the theory-based interval (0.50, 1.93) obtained in Example 11.3. Figures 14.1 and 14.2 show histograms of the 1000 bootstrap estimators of β obtained for each method using a class interval of 0.1. The class interval for 0.8 to 0.9, for example, contains all b^* such that $0.8 < b^* \leq 0.9$, with similar inequalities applying to other intervals.

Conclusions. Owing to appreciable discontinuities in possible values of the b^* (especially in the case of the Theil–Kendall estimator) not too much reliance should be placed in confidence intervals for so small a value of n. The most interesting feature of the histograms is the evident bimodality (or in the case of the Theil–Kendall estimator perhaps multimodality). A moment's reflection would suggest that nearly all bootstrap samples that do not include the aberrant point $(6, 11.1)$ give $b^* \leq 0.6$ and the breaks in the histograms between 0.6 and 0.8 in both Figures 14.1 and 14.2 suggest that values of $b^* > 0.8$ arise from samples where the point $(6, 11.1)$ occurs at least once. The fact that $b^* \leq 0.6$ for only 36.2 percent of the samples for least squares, while $b^* \leq 0.6$ for 71.5 percent of all samples using Theil–Kendall reflects the more robust nature of the Theil–Kendall estimator. Most samples in which the point $(6, 11.1)$ occurs only once are likely to have little influence upon the median of the b_{ij}, whereas only one occurrence of this point in a sample will influence the least squares estimator appreciably. In Section 11.1.4 we pointed out that the least squares estimator can be expressed as a weighted mean of the b_{ij}

and the robustness properties noted here are broadly in line with our findings on influence in Section 14.2.2.

Comments. 1. The number of b_{ij} in many bootstrap samples will be less than the total of 21 for the original sample because of repeated values and also many of the pairwise estimates will be identical if there are several repeated values. These factors may also influence robustness in a rather complicated manner. With either estimator in an extreme case the bootstrap sample may consist of seven replicates of the same point in which case the estimate of β is undefined for both least squares and the Theil–Kendall method.

2. The results discussed here are based on one sample of 1000 for least squares estimators and on another sample of 1000 for Theil–Kendall estimators. It would of course be feasible, and indeed interesting, to compare both estimators for the same sample. We referred above to discontinuities in the Theil–Kendall estimators. Table 14.2 indicates the extent of these for the particular sample we used. These arise because a variety of samples give the same sampling median (which must be an observed b_{ij} or the mean of two observed b_{ij}).

Our simple illustrations of bootstrapping convey only the flavour of this approach and indicate in an intuitive way some of its strengths and weaknesses. As we have pointed out the method is most useful in situations where there is no simple analytic method or we are dealing with data that are samples from essentially unknown distributions. Correlation problems and inferences based on ratios of two variables are situations often met where theory is generally intractable unless very simple distributional assumptions can be made. Several detailed examples of its use in problems involving both correlation coefficients and ratios are given by Efron and Tibshirani (1993) and by Davison and Hinkley (1997). Efron (1981) and Robinson (1983) applied the bootstrap to censored data while Hall and Hart (1990) used the bootstrap to test for differences between means expressed as very general regression functions.

14.3.5 Related Techniques

Although it was not apparent in our examples, the bootstrap frequently leads to biased estimators. In a few cases theoretical adjustments may be made to remove bias and bootstrap methods exist for estimation of bias. In many practical applications bias is more easily estimated by a related but older technique called the *jackknife*. The idea was introduced specifically for bias estimation by Quenouille (1949), but the name jackknife is due to Tukey (1958). The jackknife will not always work; in particular it will not work well when the median is used as an estimator.

Many commonly used estimators that are sample analogues of their population counterparts are biased except perhaps when rather strong distributional assumptions are made. Indeed one of the few universally unbiased estimators in common use is the sample mean, which is unbiased when used as an estimator of the population mean. A simple example of a biased estimator is the sample variance $s^2 = \sum_i (x_i - \overline{x})^2/n$ as an estimator of a population variance σ^2. Here the bias is easily estimated, it being well-known that $E(s^2) = (n-1)\sigma^2/n = \sigma^2 - \sigma^2/n$, so the bias is $-\sigma^2/n$.

The jackknife works this way. Given a sample of n, for any parameter θ for which an estimator is the sample analogue of the parameter being estimated, and is denoted by $\hat{\theta}$, we form n further estimators by replacing the original sample by a set of samples each identical with the original except that one observation is omitted from each in turn. These are called *jackknife samples*. Thus the ith jackknife sample is

$$x_1, x_2, \ldots, x_{i-1}, x_{i+1}, \ldots, x_n, \quad i = 1, 2, \ldots, n.$$

We denote the estimator of θ based on the sample with x_i omitted by $\hat{\theta}_{(i)}$ and the mean of the $\hat{\theta}_{(i)}$ by $\hat{\theta}_{(.)}$, i.e., $\hat{\theta}_{(.)} = \sum_i \hat{\theta}_{(i)}/n$. Finally, the jackknife estimator, $\theta^\dagger$, of θ is

$$\theta^\dagger = n\hat{\theta} - (n-1)\hat{\theta}_{(.)} \tag{14.7}$$

and the jackknife estimator of bias is

$$bias(\hat{\theta}) = (n-1)(\hat{\theta}_{(.)} - \hat{\theta}). \tag{14.8}$$

The motivation for (14.7) and (14.8) is that for both the sample mean and the sample variance as estimators of the corresponding population parameters (14.7) gives unbiased estimators of these parameters and (14.8) correctly estimates the bias of the sample mean as estimator of the population mean as zero and that of the sample variance as estimator of the population variance as $-\sigma^2/n$. These results are easily verified by calculating expectations. In general (14.8) does not reduce the bias of every estimator to zero, nor does (14.7) always provide an unbiased estimator of any parameter. However, the latter will usually reduce any bias in $\hat{\theta}$.

The jackknife and its use in association with the bootstrap to reduce bias and make other improvements to bootstrap estimation is discussed in detail both by Efron and Tibshirani (1993, especially Chapter 11) and by Davison and Hinkley (1997).

Another widely used technique that bears some resemblance to the bootstrap is cross-validation. In fields like regression and classification methods, parameter estimation is often a part of the model building process to produce a model that will be used to make predictions when new data become available. One source of concern is how good these predictions will be when applied to the new data. Ideally we should aim to use all available data to fit a model and then take further samples from the original population and see how well our fitted model acts as a predictor. This may be costly or impossible, in which case an alternative is cross-validation. For this the data are divided into two or more portions and a model is fitted in turn to all but one of these portions. The prediction error of the fitted model when it is fitted to the omitted portion is then calculated. Each portion is omitted in turn and a combined estimated prediction error is obtained.

An extreme case especially suited to relatively small data sets is the *leave-one-out* cross-validation procedure, where each of n observations is omitted in turn and a model is fitted to the remaining $n-1$ data. A predicted value is then obtained for the omitted observation using that model. For example, in a regression context if, when the ith observation is omitted, the predicted value of y_i is $y^*_{(i)}$ the cross-validation prediction error is defined as $\sum_i (y_i - y^*_{(i)})/n$. The computational procedure is similar to that for the jackknife but the objective is different. A more detailed description of cross-validation and its relationship to the bootstrap is given by Efron and Tibshirani (1993, Chapter 17).

14.4 M-Estimators and Other Robust Estimators

Throughout this book we have met estimators that show varying degrees of tolerance to breakdowns in assumptions, especially those associated with a few observations being out of line with the general patterns. We indicated in Section 14.2, and elsewhere, that the median is generally more robust than the mean in the presence of a few rogue observations – behaviour explained by the *influence functions*. The price to be paid for this robustness may be an increase in the standard error of our estimator, or perhaps the introduction of bias in procedures such as the bootstrap. There are several compromise estimators that retain some of the desirable properties of the mean as an estimator, but either down-weigh, or eliminate, the effect of more extreme observations.

One such estimator is the *trimmed mean* in which we eliminate a proportion (typically the 10 percent most extreme observations) and base inferences on the mean of the remaining observations. Another is the *Winsorized mean* where extreme observations are shrunk to the value of the largest remaining observation. We do not discuss these methods in detail but these and related ones are fully covered in Barnett and Lewis (1994) in a range of contexts.

The mean, median, trimmed mean and Winsorized mean are members of a class of estimators sometimes called *L-estimators*. Here L stands for *linear*, the name being given because if we write the ith ordered sample value $x_{(i)}$ then any of these estimators take the form

$$\mu^* = \sum_i w_i x_{(i)}$$

where the w_i are weights.

Putting $w_i = 1/n$ for all i gives μ^* as the sample mean. If n is odd, say $n = 2m + 1$, and we put $w_{m+1} = 1$ and $w_i = 0$ if $i \neq m + 1$ then μ^* is the sample median. For $n = 2m$ the median is the mean of the middle pairs of values, so the weights are $w_m = w_{m+1} = 0.5$ and $w_i = 0$ otherwise. For a trimmed mean where the k largest and k smallest observations are trimmed the weights are $w_i = 0$ if $i \leq k$ or if $i \geq n - k + 1$, and $w_i = 1/(n - 2k)$ otherwise. For Winsorization, if k values in each tail are shrunk to $x_{(k+1)}$ and $x_{(n-k)}$ respectively, then $w_i = 0$ if $i \leq k$, or if $i \geq n - k + 1$, while $w_{k+1} = w_{n-k} = (k + 1)/n$, otherwise $w_i = 1/n$.

Since they include the mean, L-estimators are not necessarily robust. They tend to gain robustness when extreme order statistics are downweighted. In passing, we note that the median is a special case of the trimmed mean in which all but one observation is trimmed if n is odd, and all but two are trimmed if n is even.

Inevitably there are often doubts about whether certain observations are or are not in some way distorted relative to the bulk of the observations. Even if there are reasonable models to cover such discrepancies there may still be either one or two gross departures from the general pattern, or alternatively a large number of small perturbations, these often arising as a consequence of rounding or using measuring devices of limited accuracy. This has led to the development of procedures that are almost as good as the best available procedures when there are no such disturbances, and are little affected by just a few major or many minor disturbances.

The Huber-Hampel *M-estimators* are one such class. The M indicates that they are like optimal maximum likelihood estimators when these are appropriate. They are little disturbed by either a few grossly aberrant observations or by small perturbations in many observations. For all but trivial problems their use requires adequate computer programs. We demonstrate the basic ideas for estimation of the mean of a symmetric distribution where the maximum likelihood estimator is the sample mean. In particular, given a sample of n observations $x_1, x_2, \ldots, x_n$ from a normal distribution this is equivalent to least squares estimation where the estimator $\hat{\mu}$ is the value of μ that minimizes

$$U(x, \mu) = \sum_i (x_i - \mu)^2, \tag{14.9}$$

i.e., the sum of squares of deviations of the x_i from μ. The influence function (14.2) shows that for any one observation $\hat{\mu}$ is linearly dependent on the distance of that observation from the true mean. Thus, an outlier z lying six standard deviations from the mean will have twice the influence of one lying three standard deviations from the mean.

Huber (1972) and others developed M-estimators to cope with the possible presence of a few outliers where one really wanted inferences applicable to the remaining data in situations where specific maximum likelihood estimators were available for that remaining data. The proposed estimators have a built-in mechanism for reducing the effect of any outliers. Apart from this they are exactly, or almost, identical to maximum likelihood estimators.

The quantity $(x_i - \mu)^2$ appearing in (14.9) is an example of what is called a *distance measure* because it measures (the square) of the distances of the x_i from μ. The complete function in (14.9) is called a *distance function*. Another example of a distance function is the function $V(x, \mu) = \sum_i |x_i - \mu|$, i.e., the sum of the absolute deviations. This is minimized by setting $\mu^* = \text{med}(x_i)$. It is well-known that in the case of U given by (14.9) we find the minimum by differentiating with respect to μ and equating that derivative to zero leading to the normal, or estimating, equation

$$\sum_i [-2(x_i - \mu)] = 0$$

with solution $\hat{\mu} = (\sum_i x_i)/n = \bar{x}$, the sample mean.

A distance measure is defined as a function $d(t)$ such that for any t

- (i) $d(t) \geq 0$,
- (ii) $d(t) = d(-t)$,
- (iii) the derivative $\psi(t) = d'(t)$ is a nondecreasing function of t for all t.

For robust estimation of the mean, μ, of a symmetric distribution Huber (1972) proposed a function $d(t)$ such that for some fixed $k > 0$,

$$\begin{aligned} d(t) \quad &= \quad t^2/2 \qquad\qquad \text{if } |t| \leq k, \\ &= \quad k|t| - k^2/2 \quad \text{if } |t| > k. \end{aligned}$$

The derivative $\psi(t) = d'(t)$ is

$$\begin{aligned} \psi(t) \quad &= -k \text{ if } t < -k, \\ &= t \text{ if } |t| \leq k, \\ &= k \text{ if } t > k. \end{aligned}$$

We estimate μ by the value μ^* that satisfies the normal equation

$$\sum_i \psi(x_i - \mu) = 0. \tag{14.10}$$

If $k \to \infty$, then for all x, $\psi(x_i - \mu) = x_i - \mu$ and (14.10) gives the sample mean as the appropriate estimator, equivalent to the maximum likelihood estimator in the normal case. For finite k the form of $d(t)$ implies that for all x_i for which $|x_i - \mu| \leq k$ we minimize a function equivalent to that for least squares. For $|x_i - \mu| > k$ we minimize a linear function of absolute differences. We need to choose a suitable value for k. In this simple problem practical experience has indicated that a useful choice is one such that the interval with end points given by $\text{med}(x_i) \pm k$ contains between 70 and 90 percent of all observations. In general (14.10) must be solved iteratively. When k is chosen we proceed by first rewriting that equation in the form

$$\sum_i \frac{\psi(x_i - \mu)}{x_i - \mu}(x_i - \mu) = 0 \tag{14.11}$$

Setting $w_i = \psi(x_i - \mu)/(xi - \mu)$, (14.11) becomes $\sum_i w_i(x_i - \mu) = 0$ with solution

$$\mu^* = \frac{\sum_i (w_i x_i)}{(\sum_i w_i)}. \tag{14.12}$$

The weights are functions of the unknown μ^* so we need an estimate μ_0 of μ^* to calculate initial weights w_0 and use these in (14.12) to calculate a new estimate μ_1. We repeat this procedure until convergence, which is usually achieved in a few iterations. For even moderate sample sizes computation is tedious without a suitable computer program, but we illustrate the steps for three small data sets.

Example 14.7

The problem. Obtain and compare Huber M-estimators of the mean for the three data sets

$$
\begin{array}{cccccccc}
0, & 1.2, & 2.3, & 3.8, & 5.2, & 7.1, & 7.9, & 8.5 \\
0, & 1.2, & 2.3, & 3.8, & 5.2, & 7.1, & 7.9, & 19.7 \\
0, & 1.2, & 2.3, & 3.8, & 5.2, & 7.1, & 7.9, & 115.5
\end{array}
$$

Formulation and assumptions. A suitable choice of k is required, after which the iterative procedure outlined above is carried out for each data set.

Procedure. We illustrate the procedure for the second data set, leaving the reader to follow through the similar steps for the other sets (Exercise 14.7). The median of the second set is $(3.8 + 5.2)/2 = 4.5$. If we take this as our first estimate of μ and choose $k = 4$ the interval 4.5 ± 4 includes 75 percent of all observations. If we choose $k = 5$ the interval 4.5 ± 5 includes 87.5 percent. Both are within the suggested range 70 to 90 percent. We illustrate the procedure with $k = 5$. Only the last point falls outside the interval 4.5 ± 5 (i.e., -0.5 to 9.5). Since we set $\psi(x_i - \mu) = (x_i - \mu)$ for observations in the interval $(-0.5, 9.5)$ it follows that $w_i = 1$ for all but the observation $x_8 = 19.7$. For that observation $\psi(x_i - \mu) = k = 5$, so that $w_8 = 5/(19.7 - 4.5) = 0.3289$, whence our new estimate, μ_1, is

$$
\mu_1 = \frac{0 + 1.2 + 2.3 + 3.8 + 5.2 + 7.1 + 7.9 + 0.3289 \times 19.7}{1 + 1 + 1 + 1 + 1 + 1 + 1 + 0.3289}
$$

$$
= \frac{33.3733}{7.3289} = 4.5554.
$$

For the next iteration we retain the same $k = 5$. This gives full weight $w = 1$ to all observations in the interval 4.554 ± 5, i.e., $(-0.446, 9.554)$. The weight for $x_8 = 19.7$ is now $w_8 = 5/(19.7 - 4.554) = 0.3301$ whence a similar calculation to that above gives a new estimate $\mu_2 = 4.639$. For the next iteration it is again clear that only x_8 has a weight less than 1 and this weight is easily seen to be $w_8 = 5/(19.7 - 4.639) = 0.3320$ and with this weight our next estimate $\mu_3 = 4.643$. The new adjusted weight for x_8 is $w_8 = 5/(19.7 - 4.643) = 0.3321$, leading to the new estimate $\mu_4 = 4.643$. To this degree of accuracy there is no improvement on the previous iteration so we conclude $\mu^* = 4.643$.

Conclusion. The M-estimator of the population mean based on the second data set is 4.643.

Comments. 1. With the same choice of sample median and k for the first data set it is immediately clear that there are no observations for which $w_i \neq 1$ and that our estimator is thus the sample mean, i.e., $\mu^* = 4.5$. For the third data set similar calculations to that for the second set with x_8 there replaced by $x_8 = 115.5$ lead after several iterations to the estimate $\mu^* = 4.643$.

2. As indicated under *Procedure* a choice $k = 4$ may not be unreasonable. In Exercise 14.7 we ask that this choice be used.

3. For the three data sets here sample means are respectively 4.5, 5.9 and 17.875 indicating the strong influence of the outliers. The above M-estimation procedure

leads to 4.5, 4.643, 4.643, as estimates of the population mean indicating a dramatic effect of downweighting the outliers.

4. Other possible estimators in this case include trimmed means, Winsorized means or the median. The sample median for each set is 4.5, while a trimmed mean excluding k = 1 observation from each tail is easily shown to lead to estimates 4.58 in each case. Similarly, in each case the corresponding Winsorized mean estimates are 4.575 (Exercise 14.8).

Different choices of k or different amounts of trimming or Winsorization will give different estimates. In practice if estimators are robust against the outliers in the data there should be little difference between rational choices, i.e., choices that downweigh rogue observations substantially or even eliminate them from the computation, while directly or indirectly giving strong support to good observations. We might at first thought conclude that the median does not obey these criteria, its computation taking into account at most two observations. However, all observations do have a role in determining the median because the ordering of the observations determines which observation, or pair of observations, make a non-zero contribution to its determination.

Only if the observations are sampled from a symmetric distribution, however, will the sample median be a sensible estimator of the population mean (and then only if the latter exists). In many practical situations, even in the absence of any rogue observations the median estimator will, however, have a greater standard error than that for the mean.

Several practical considerations influence the choice of k when using M-estimators. In particular, if k is too small only a few observations get full weight and the estimator may be strongly influenced by rounding or grouping effects in those fully weighted observations. Also, there may be unsatisfactory features relating to change of scale that can be overcome by using different weights to those used above. Numerous alternative distance functions have been proposed, but we omit details here. Many aspects of these procedures are discussed by Andrews et al. (1972).

Andrews (1974) introduced an M-estimator for robust linear multiple regression and Härdle and Gasser (1984) apply a similar type of estimator for fitting nonlinear curves.

Thomas (2000) describes the use of the bootstrap to determine the precision of robust estimates of centrality, illustrating the use of appropriate methods for ten data sets that are either highly skewed or contain outliers.

14.5 Fields of Application

Efron and Tibshirani (1993) and Davison and Hinkley (1997) include examples on many and varied applications of the bootstrap.

Ratio Estimates

Efron and Tibshirani (1993, Section 10.3) discuss use of the bootstrap in a problem involving a test criterion for bioequivalence used by the U.S. Food and Drug Administration that is based on ratios. The procedure is useful here because there are virtually no analytic results for the properties of the ratios that are of interest.

All statisticians and experimenters handling more than a few data sets come face to face with data that well may be in some way contaminated in the sense that some of the observations do not appear to fit a model that is reasonable for the rest of the data. Robust methods may then have a valuable role in making inferences that are relevant to the uncontaminated data. Again, we just indicate one situation where this may be the case.

Laboratory Determinations

Very often two or more laboratories are asked to carry out the same chemical determination on samples from the same population, e.g., to determine the percentage of fat in samples of milk or the amount of some contaminant, e.g., lead in 100 gm samples from zinc ingots. Precision may vary from laboratory to laboratory or some laboratories may make consistent errors (bias) in their measurements. Robust methods of estimation such as those based on M-estimators might be used to reduce the influence of extreme results from laboratories producing readings with low precision. Conventional parametric or nonparametric tests comparing results from different laboratories for a shift in mean or median may be useful to detect biases.

14.6 Summary

Observations that cause surprise in the sense that they appear far removed from the bulk of the data or seem to be inconsistent with some specified model are generally called **outliers**. For a given data set what is classed as an outlier may be context dependent (Section 14.2.1). The appropriate action to be taken to deal with outliers depends both upon their nature and the objectives of the investigation in which they occur. Erroneous observations should be corrected where possible or else rejected. If it is believed outliers represent a small proportion of units present in the population the action to be taken may depend on whether one wants to make inferences relevant to the complete population or only to the population ignoring outliers.

Influence studies (Section 14.2) indicate both the extent to which outliers may influence sample characteristics such as the mean or median and also explain why some estimators (e.g., the median) will be less affected by substantial numbers of outliers than others (e.g., the mean).

The bootstrap (Section 14.3) is a resampling method of inference that relies for its usefulness on the assumption that for not-too-small samples the sample cdf is a reasonable approximation to the population cdf. It differs from randomization or permutation distribution theory in that it involves sampling with replacement whereas the latter involves sampling without replacement. It is especially useful in situations where there are few analytic results for exact inference or where formulation of a precise model is difficult. Some of the problems with bias that are associated with certain bootstrap estimators may be eased by use of the **jackknife** (Section 14.3.5). **Cross-validation** is useful for exploring the predictive ability of a fitted model.

Two important categories of **robust estimators** (Section 14.4) are certain L-estimators such as trimmed or Winsorized means and the median and M-estimators which behave very like the relevant maximum likelihood estimators when applied to uncontaminated data, but reduce the influence of outliers when these are present. This makes the method particularly suitable when the status of outliers is uncertain, but we want to make inferences applicable to the population from which the bulk of the data was sampled.

14.7 Exercises

*14.1 Use the test given in Section 14.2.2 to test for outliers in each of the following data sets:

Set A	3.2	4.9	1.3	9.7	12.9	−6.3	4.2	22.5	0.3	−7.1	
Set B	1	35	71	13	45	18	91	34	777	29	452

14.2 Use the test given in Section 14.2.2 to test for outliers in the data set used in Example 14.2.

*14.3 Smith and Naylor (1987) give the following data for strengths of 15 cm lengths of glass fibre and suggest that the two smallest observations may be outliers. Does the test proposed in Section 14.2.2 confirm this?

0.37	0.40	0.70	0.75	0.80	0.81	0.83	0.86	0.92	0.92	0.94	0.95
0.98	1.03	1.06	1.06	1.08	1.09	1.10	1.10	1.13	1.14	1.15	1.17
1.20	1.20	1.21	1.22	1.25	1.28	1.28	1.29	1.29	1.30	1.35	1.35
1.37	1.37	1.38	1.40	1.40	1.42	1.43	1.51	1.53	1.61		

14.4 A sample of paired observations of X, Y are

$$
\begin{array}{cccccc}
X & 1 & 2 & 3 & 5 & 8 \\
Y & 3 & 7 & 9 & 7 & 12
\end{array}
$$

Use a bootstrap based on at least 40 samples to obtain an estimate of the bootstrap standard error of the estimator $\bar{x}/\bar{y}$, for the ratio of population means. In addition use 1000 bootstrap samples to obtain 95 percent quantile-based confidence intervals for the population ratio. Appropriate computer software will be needed for the latter computation.

*14.5 For the data in Exercise 14.4 use the jackknife to estimate the bias in the estimator $\bar{x}/\bar{y}$.

14.6 Calculate the mean of the data in Example 14.2 and also the trimmed means trimming the top and bottom 10 percent (deciles) and 25 percent (quartiles) of all observations. Explain any difference between these estimates of centrality. Which do you prefer and why?

14.7 For the first and third data sets in Example 14.7 verify the values for the M-estimators quoted in that example when $k = 5$. Also obtain the corresponding estimates when $k = 4$.

14.8 Verify the values for the median, trimmed mean and Winsorized estimates given in *Comment* 4 in Example 14.7.

*14.9 The data below give the age of first stroke (years) for 10 patients in a hospital:

$$0 \quad 34 \quad 52 \quad 57 \quad 61 \quad 65 \quad 68 \quad 76 \quad 81 \quad 85$$

Use the method described in Section 14.2.2 to test for outliers. Can you think of possible reasons why outliers might arise in this type of data?

14.10 Records at a medical centre indicate for nine patients the following numbers of doctor consultations in a twelve-month period:

$$0 \quad 1 \quad 1 \quad 1 \quad 1 \quad 2 \quad 3 \quad 6 \quad 12$$

Calculate the mean and median, and compare these values with the trimmed mean and Winsorized mean based on an assumption that the smallest and largest values may be outliers.

CHAPTER 15

MODERN NONPARAMETRICS

15.1 A Change in Emphasis

To describe topics covered so far in this book as *classic nonparametrics*, and those in this chapter as *modern nonparametrics*, is an oversimplification. That classification is appropriate only in the sense that many of the methods in earlier chapters date back to the first half of the twentieth century, while those in this chapter are either more recent developments or are of a more specialized nature.

Since their introduction most classic methods have evolved gradually to the form in which they are now widely used. For example, while the concept of permutation tests as the basis for exact tests dates back at least to the 1930s, in all but a few situations their widespread use only became possible with the unprecedented growth of computer power from about the 1970s onward.

Full appreciation of the properties of many procedures evolved over time thanks to numerous studies of Pitman efficiency and power. Monte Carlo sampling provides, for samples of moderate size, a bridge between exact permutation tests and asymptotic approximations. It gives a better understanding of when it is appropriate to use asymptotic results, or when they may be misleading.

The classic procedures evolved in an era when using statistics to make inferences from samples of small to moderate sizes dominated developments in theory and application. This is still an important area of application, especially in the early stages of many research and other studies.

From about the 1980s onward interest has increasingly turned to methodologies appropriate to large data sets, often involving several, or even a large number, of variables. Here nonparametric methods of a rather different type come into play. These are largely aimed at distilling out key characteristics of such data. For example, given a univariate data set with hundreds, or even thousands of observations, obtained from some population with an unknown distribution we may wish to estimate the population probability density function making no more than an assumption that it is some continuous function $f(x)$ with the property that $\int_{-\infty}^{\infty} f(x)dx = 1$ or perhaps of the form $\int_a^b f(x)dx = 1$, where both, only one, or neither, of a, b are known. With no restriction on the form of $f(x)$ this is a completely nonparametric and distribution-free specification.

Given large sets of paired observations (x_i, y_i) we may wish to fit a curve of the form $y = g(x)$ that provides a summary of the main data characteristics.

One approach to this is to use nonparametric regression methods with $g(x)$ unspecified.

Another growth area in recent years has been multivariate nonparametric methods, including studies of relationships between different variables. One such development is an extension of the ideas of correlation and linear regression from the bivariate situations considered in Chapters 10 and 11 to more than two explanatory variables. Further extensions drop the linearity assumption, while still allowing some parameters in the model, the corresponding methods then often being described as *semiparametric*. In this chapter we give a few examples of methods evolved to cover some of the situations outlined in this section. The treatment is indicative and only hints at what is happening in a rapidly expanding area of nonparametrics.

15.2 Density Estimation

We consider first the density estimation problem. A histogram (or in some cases as in Figure 3.1 a bar chart) is a simple form of graphical approximation to an unknown $f(x)$. In Figure 3.1 this worked well to indicate how the Wilcoxon statistic distribution approached normality even when the sample size n was not large.

Histograms have major practical limitations. In particular:

- they are not smooth,

- they are sensitive to the choice of class interval,

- they depend on the choice of end points of the intervals,

- they are, in essence, discontinuous.

An example highlights some implications of these shortcomings.

Example 15.1

The problem. Explore the limitations of histograms to indicate the shape of the population probability density function using the following sample of observations

6.1	4.7	4.1	4.6	6.0	5.5	4.6	5.8	5.6	5.0
5.9	4.9	3.9	4.9	4.0	3.1	4.9	6.1	3.1	4.3
4.9	6.0	1.4	3.9	6.1	5.3	2.7	6.9	5.2	3.8
8.2	8.5	8.1	4.9	8.4	8.4	7.3	3.4	10.7	4.7

Formulation and assumptions. Most statistical software packages will produce histograms with a wide choice of equal or unequal class intervals. For density estimation purposes equal intervals are usually appropriate. In this context the subintervals are often referred to as *bins* and the width of the interval as the *binwidth*.

Procedure. The histograms in Figures 15.1, 15.2 and 15.3 are each based on the above data. In the first two the binwidth is 1 unit, but the end points differ. For these data the choice of end points does not affect the histograms so much as it sometimes does. Nevertheless, Figure 15.2 gives a slightly stronger impression that there may

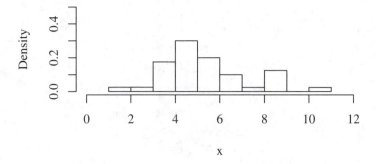

Figure 15.1 *Histogram for Example 15.1 data, unit binwidth, integer endpoints.*

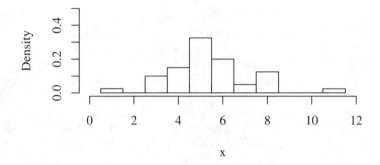

Figure 15.2 *Histogram for Example 15.1 data, unit binwidth, noninteger endpoints.*

be two outliers, and reduces the evidence of bimodality compared to the impression given by Figure 15.1. In Figure 15.3 the binwidth is 0.5 and there is a strong suggestion both of outliers and multimodality. Intuition rightly suggests that the reduced binwidth picks up local detail rather than highlighting key characteristics of the data.

Conclusion. Histograms, except perhaps for very large data sets (many hundreds or thousands of observations) are, for the reasons itemized before this example, of limited value for estimating density functions for a continuous distribution.

Comment. Choice of binwidth and number of bins has a strong influence on the general appearance of histograms. Histograms are best regarded as no more than an EDA tool. In this example we have shown that whereas one may be reasonably comfortable with binwidth 1, halving the binwidth is less satisfactory. So too is increasing the binwidth and thus effectively reducing the number of bins. This tends to obscure the broad picture. The reader should check this by forming histograms for the data in this example with binwidths of, say, 2 and 2.5, and several choices of endpoints.

Computational aspects. Appropriate computer software for histograms makes it easy to compare the results for various choices for a range of bin positions and binwidths.

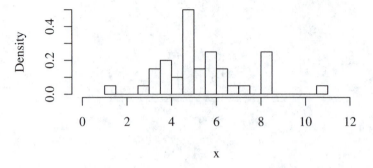

Figure 15.3 *Histogram for Example 15.1 data with half unit binwidth.*

An alternative form of estimation, *Kernel Density Estimation*, remedies many of the deficiencies of the histogram. In particular, it provides a continuous estimate and at the same time removes the dependence on end points of bins implicit in fitting histograms. The problem of choosing something equivalent to binwidth remains, but criteria exist to determine an appropriate choice.

For any x in a local neighbourhood of each data value x_i, fitting is controlled in a way that depends upon the distance of x from x_i. If we denote the estimate of $f(x)$ by $\hat{f}(x)$ the contribution at x depends upon a function called the *kernel function* and a notion called the *bandwidth* we associate with it. If we denote the kernel function by $k(u)$ and its bandwidth by Δ and there are n data points x_i the estimated density function at any point x is

$$\hat{f}(x) = \frac{1}{n\Delta} \sum_{i=1}^{n} k\left(\frac{x - x_i}{\Delta}\right) \qquad (15.1)$$

How do we choose the function $k(u)$ and Δ? There are a number of possible choices for $k(u)$. Practical experience has shown that some half-dozen widely used ones tend to give similar results for a chosen Δ. However, the choice of bandwidth, Δ, often has a marked influence, as does the choice of binwidth for a histogram.

A common choice for $k(u)$ is the normal or *Gaussian kernel*

$$k(u) = \frac{1}{\sqrt{2\pi}} e^{-u^2/2}.$$

Several other common choices are listed and discussed by Wasserman (2006, Section 4.2) in a broader context. Modern software, such as the graphics package in R, provides easily used facilities for fitting kernel density estimates for a range of bandwidths. Figures 15.4 and 15.5 show kernel density functions fitted to the data in Example 15.1 using the Gaussian kernel with two different bandwidths, one twice as wide as the other. In each case the histogram in Figure 15.1 is superimposed.

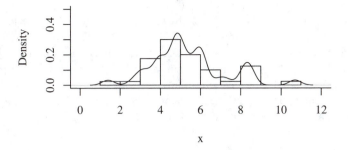

Figure 15.4 *Kernel density with bandwidth 0.2971.*

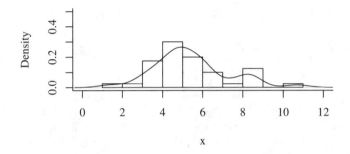

Figure 15.5 *Kernel density with bandwidth 0.5942.*

Which fit do you prefer? That with the wider bandwidth, like the superimposed histogram, gives a reasonable picture of the general data pattern. As mentioned above there are certain criteria that allow an objective choice of bandwidth. One widely used is based on what is called the *estimated mean integrated square error*. This may be computed for a range of bandwidths and the optimum choice is the width that minimizes this error. For a detailed discussion see Wasserman (2006, Sections 6.1 and 6.3). Wasserman suggests using cross-validation techniques for the required estimation.

There are also some rule-of-thumb guides for choosing bandwidth. One often used is to choose Δ so that

$$\Delta = \frac{1.06s}{n^{1/5}}$$

where s is the sample standard deviation. It is sometimes suggested that rather than using s one should insert the minimum of s and $Q/1.34$ where Q is the interquartile range for the data. Using this minimum is more appropriate if there are outliers, or gross departures from normality, in the data.

For the record, the data in Example 15.1 were obtained by sampling from two normal distributions, the first having mean 5 and standard deviation 1, the second having mean 7 and standard deviation 3. The first 30 readings are from the former distribution, the remaining 10 from the latter. It is not surprising, therefore, that the kernel density estimate shows bimodality. In Figure 15.5 the more pronounced mode is close to 5, as one might expect when the source of the data is known, but it is perhaps surprising that the secondary mode is close to 8, rather than near 7. We ask you to explore further why this might not be unreasonable in Exercise 15.1.

For a detailed discussion of kernel density estimation and an account of alternative approaches to density estimation see Wasserman (2006, Chapter 6).

15.3 Regression

There are many bivariate situations where it is immediately evident that a model

$$y_i = \beta_0 + \beta_1 x_i + \varepsilon_i$$

is inadequate to describe the data generating process. A very general model is one where

$$y_i = g(x_i) + \epsilon_i, i = 1, 2, \ldots, n$$

where the function g may be any function and the ε_i have median zero for all i. We now consider one completely distribution-free method for estimating a curve $\hat{g}(x)$ descriptive of the unknown $g(x)$, and briefly mention an alternative approach. These methods are commonly referred to as *data smoothing procedures*. They determine a curve for a function $g(x)$ that gives a good representation of the trend in a scatter plot in a manner somewhat reminiscent of the way kernel density estimation gives a function representative of a population probability density function.

15.3.1 Lowess Curve Smoothing

The procedure in this section was proposed by Cleveland (1979). It is not only capable of providing sensible estimates of trend, but is robust against the presence of a few outlying, or rogue, observations. Its successful application requires care in the choice of certain quantities that have to be user-specfied. One of these is reminiscent of the choice of bandwidth in kernel density estimation. It often helps to try various choices of these quantities in what amounts to a sophisticated exploratory data analysis. This may be especially useful with large data sets.

The method proposed by Cleveland is usually referred to as *lowess*, this being derived as an acronym from **LO**cally **WE**ighted **S**moothing **S**catterplots. The name *loess* also appears in the literature, sometimes being used for a modification of the scheme described here, and sometimes simply as an alternative name for lowess. Regrettably, both names are sometimes used for a partial procedure that fits locally weighted regressions, but does not carry out the iterative reweighting procedure that gives the full method its robustness.

The basic idea behind locally weighted regression predates the paper by Cleveland. It extends the concept of a *weighted moving average* widely used in the analysis of time series. In effect a linear, or sometimes a second degree, polynomial is fitted at each data point usuing weighted least squares with the weights so chosen that they decrease as x moves away from the x_i value at which the fit is being made. The weights are zero for points more than a certain distance away.

The weight function is chosen to have a maximum at the x_i at which the fit is being made, and is symmeric about that point.

First, we decide the degree of the polynomial to be fitted at each point. Then we decide how many neigbouring points are to be given nonzero weight. Next, we need a non-negative weight function w meeting these stipulations. Once the choices of degree of polynomial, d, of the weighting function, $w(x)$, and the local range have been made a localized weighted least squares procedure is used to determine a fitted estimate of $g(x)$, denoted by $\hat{y}_i$, at x_i.

Before proceeding we describe the extension to robustness that is the key part of the lowess process if there are, or may be, outliers. This is an iterative procedure using a different set of weights. Once we have initial estimates $\hat{y}_i$ for all data points based on localised weighted regressions we calculate residuals $r_i = y_i - \hat{y}_i$. To reduce the impact of outliers, a further iteration is performed with new weights that take account of the magnitude of the residuals. Points with large values of $r_i = |y_i - \hat{y}_i|$ are now given reduced weights. This iterative process may be repeated any number of times using the new $\hat{y}_i$ determined at the previous iteration.

The practical problems now are:

- What are approriate values for d, the degree of the locally fitted polynomial?

- What is an approriate form for the weighting function, $w(u)$?

- How many neigbouring points should be used (i.e., given nonzero weight in the fitting process)?

- How many iterations, i, are needed to ensure robustness?

These questions are considered in some detail in Cleveland's original paper and have been widely discussed in the literature since. We give here some answers that have been found to work well in practice, commenting briefly on the rationale behind some of the choices.

The common objective of data smoothing is to detect general patterns rather than detailed local variation, so it usually suffices to choose the degree of the locally fitted polynomial to be $d = 1$ or 2, i.e., to fit straight lines, or quadratics, locally. To avoid an undue influence of a few near neighbours, some of which may be outliers, the range of neighbouring points should not be too restricted. This range is often determined as a fraction of the total number of points. When n is large this fraction might be relatively small, but for small to moderate n a fraction, or proportion, between $f = 0.5$ and $f = 0.7$ is likely to be satisfactory unless the data indicate some abrupt changes in the relationship between x and y. For larger data sets smaller values of f may give satisfactory fits if a suitable weighting function is chosen. If, as is often the case, the choice of f gives a nonintegral number of points for inclusion in the local interval, this may be rounded to the next integer above.

The most widely used weighting function has the form

$$w(u) = (1 - |u|^3)^3$$

where $|u| \leq 1$. This function is symmetric about $u = 0$. We also set $w(u) = 0$ if $|u| > 1$. If the fraction of points to form the predictor at the point with abscissa x_i is such that the most distant point from that x_i to be included in estimating that predictor is at distance Δ from it, then the weight associated with any point x_j in the predicting equation for the point with abscissa x_i is

$$w_j = \left(1 - \left|\frac{x_j - x_i}{\Delta}\right|^3\right)^3.$$

These weights are used in a weighted least squares regression to obtain an estimator $\hat{y}_i$ corresponding to x_i. The process is repeated for all n observed x_i. That is, we fit a total of n local weighted regressions. The method is computer intensive for all but small n even if fitting only low degree poynomials locally.

For the robust smoothing procedure we have indicated that we use weights depending on the magnitude of the residuals $r_i = |y_i - \hat{y}_i|$ at each observed abscissa x_i. Points with large residuals are given low weight. The first step in determining weights with this property is to compute the median of the absolute residuals — or median absolute deviation — often abbreviated to MAD. We have already met this concept in Section 14.2.2.

A suitable weighting scheme for the iterations uses a weight suggested by Cleveland, namely

$$w_i = \left(1 - \left(\frac{r_i}{6MAD}\right)^2\right)^2 \quad \text{if } |r_i| < 6MAD,$$

and zero otherwise. At each iteration the r_i are residuals for the current fit.

If there are no outliers there will generally be little change in the fit betweeen iterations. Even for a few outliers two or three iterations generally suffice to give a reasonable fit.

Cleveland discusses estimation of error variance and relevant standard errors for fitted values under a normality assumption for the distribution of the ϵ_i. This may seem very restrictive, but remember we make no assumption about the form of the deterministic function $g(x)$.

Example 15.2

The problem. Explore the lowess procedure for smoothing the scatter for the following data:

x	1.0	4.7	5.1	8.2	9.1	16.4	20.1	25.2	29.1	35.4	42.1	49.3
y	3.2	3.9	4.7	4.8	5.4	6.7	16.5	7.3	7.1	7.9	8.7	9.3

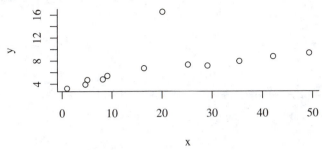

Figure 15.6 *Scatter plot of Example 15.2 data.*

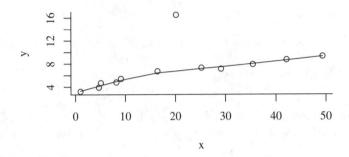

Figure 15.7 *Lowess fit with f = 2/3, after 3 iterations.*

Formulation and assumptions. A scatter plot of the data will show whether some simple function such as a straight line or low degree polynomial might provide a reasonable fit. It will also indicate possible outliers. If there is no obvious simple polynomial fit, or there are suspect observations, a lowess fit might be attempted. When using a program for lowess it is best to use at first reasonable default values for the degree of polynomial, usually starting with a linear local weighted regression fit. Similarly, a default choice of the fraction f of the points to be included might for this small data set be given a value $f = 2/3$. The number of iterations might be fixed at 3. If the smoothing appears unsatisfactory, other values should be tried. Even if the smoothing is deemed satisfactory, it is often salutary to look at alternative fits and compare them with a view to understanding why they take the form they do.

Procedure. Figure 15.6 gives a scatter plot of the data. There is a clear sugges-tion of a reasonably smooth but non-linear trend in the bulk of the data, with the observation (20.1, 16.5) a prominent outlier.

Figure 15.7 shows a curve obtained by joining up by straight lines the lowess $\hat{y}_i$ corresponding to the observed x_i where we took $f = 2/3$. We carried out three robust iterations. It is clear that the gross outlier has virtually no effect on the smoothed curve, which passes close to the other points on the scatter diagram.

Figure 15.8 shows the smoothing effect when no robustness iterations are per-formed, only the local weighted regression being fitted. The smoothing curve obtained by joining the predicted $\hat{y}_i$ shows the strong influence of the outlier, giving a curve

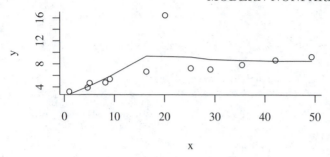

Figure 15.8 *Local weighted regressions fit with f = 2/3 and no iterations.*

that, by no stretch of the imagination could be called either a useful representation of the data as a whole, or of the data excluding the outlier.

If the value of f is set low so that only a few neighbouring points are used in the local weighting one might expect local variation to play a key role. This is confirmed by Figure 15.9 where we have reduced f from 2/3 to 0.2, while again returning to 3 robustness iterations. Because the fitting is so localized, rather than a smooth curve we get a series of lines that effectively join neighbouring points (including the outlier).

Finally, to illustrate the bizarre situation that may arise with local weighted fits that involve a moderate number of points, Figure 15.10 shows a fit with $f = 0.4$ again after 3 iterations for robustness.

Conclusions. Lowess is capable of providing useful smoothing of scatter plots, but for meaningful results care is needed in choosing the parameters, especially the fraction f.

Comment. The data used in this example is a small set generated, apart from the outlier, by adding small errors to the y_i corresponding to each given x_i determined by the relationship $y = 2 + \sqrt{x}$. In most practical applications the number of data points is usually considerably greater, and the scatter about any potentially reasonable relationship is more pronounced.

Computational aspects. Software for lowess and related calculations is fairly widely available. There is a program in Minitab. Computation in this example was done using the comprehensive graphing facilities in R.

Example 15.3

The points plotted in Figures 15.11 and 15.12 are for a data set of 200 observations. Ignoring for the moment the curves in these figures, one still detects an underlying pattern in the data on which is superimposed random variation. The curve in Figure 15.11 is composed of straight line joins of adjacent $\hat{y}_i$ for a lowess fit with $f = 2/3$. This does not pick up the underlying data pattern. This is because taking such wide windows for the local fits oversmoothes the data. The smoothing curve in Figure 15.12 on the other hand where $f = 0.1$ picks up the underlying trend very

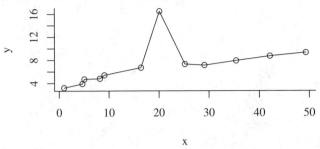

Figure 15.9 *Lowess fit with f = 0.2 and three iterations.*

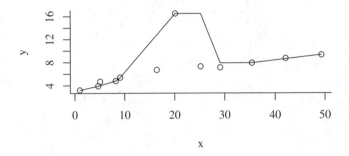

Figure 15.10 *Lowess fit with f = 0.4 and three iterations.*

well, suggesting a wavelike pattern. If suitable software is available the reader shoud experiment with other values of f.

How good the lowess approximation is becomes evident when we explain how the data were produced. For 200 equally spaced x values between 0.1 and 20, random iid N(0, 0.25) ε_i were added to

$$y_i = \sin x_i.$$

Apart from wandering a bit at the last data point, the sinusoidal form is closely reproduced by the lowess smoothing in Figure 15.12.

15.3.2 Some Other Methods of Smoothing

There are alternatives to lowess for scatter plot smoothing. We describe one briefly in this section without giving computational details.

This is a nonparametric regression method using *roughness penalties*. It is especially appropriate for data like those in Figure 15.12. Visual examination suggests some sort of wavelike curve might be suitable to indicate a relationship between the mean or median of Y for each x. If there is little empirical or theoretical knowledge to suggest a specific form of $g(x)$ a nonparametric fit is appropriate.

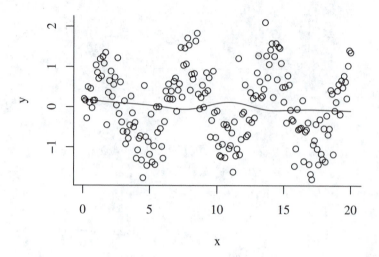

Figure 15.11 *Lowess fit to Example 15.3 data with f = 2/3 and three iterations.*

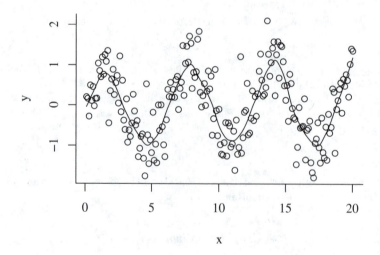

Figure 15.12 *Lowess fit to Example 15.3 data with f = 0.1 and three iterations.*

We have explained how these data were generated, but without such hindsight we might only assume, as we did in the previous section, that

$$y_i = g(x_i) + \epsilon_i, i = 1, 2, \ldots, n$$

where the function g may be any function and the ε_i have median zero for all i.

A full treatment of the theory and practical implications of the technique using penalty functions is given by Green and Silverman (1994). In general, the more parameters that are needed to specify a model, the more flexible that model becomes. At one extreme if a model with sufficient flexibility is chosen the resulting estimation of parameters leads to a curve passing through all points. For example, if there are n distinct points (x_i, y_i) we can always find a polynomial of degree not greater than $n - 1$ to pass through all the points. This is a generalization of the obvious and well-known fact that one can always find a straight line to pass through two points and a quadratic to pass through three points (which reduces to a straight line if the points are collinear). One would usually not be happy to use the resulting polynomial for prediction or interpolation purposes. This is a clear case of oversmoothing.

The key idea behind curve fitting with a smoothness penalty — sometimes called *penalized least squares regression* — is that a term is added to the sum of squares $\sum_{i=1}^{n} [y_i - g(x_i)]^2$ that is minimized in least squares. This additional term has a form that reduces the effect of roughness while avoiding the difficulties of oversmoothing.

For a reasonably smooth curve a measure of such roughness in a broad sense is the amount of change in the second derivative over the interval (a, b) where typically $a = x_1$ and $b = x_n$ for a data set with the x_i in ascending order. This is intuitively obvious if $g(x)$ happens to be a polynomial because, for a straight line the second derivative is zero, so the roughness is always zero. There is no smoother curve than a straight line. For a quadratic the second derivative is constant. For a polynomial of higher degree, n, the second derivative is a polynomial of degree $n - 2$ at most and in general this means the measure of roughness acts as a penalty because it tends to increase as the degree of the polynomial increases.

In a more general context, where $g(x)$ need not be a polynomial, a closer fit to the data still tends to produce more roughness. Thus the idea of a roughness penalty is that there is a payoff between the closeness of fit at individual points and the smoothness of the fitted curve overall. The function to be minimized takes the form

$$U(g) = \sum_{i=1}^{n} [y_i - g(x_i)]^2 + \alpha \int_a^b [g''(x)]^2 \, dx$$

where α is a positive constant. We denote by g^* the function that minimizes $U(g)$. The value of α has to be chosen, and clearly the greater the value of α, the greater the influence of the roughness penalty $\int_a^b [g''(x)]^2 \, dx$ on the solution. When $\alpha \to 0$ the solution approaches least squares but when $\alpha \to \infty$ for minimization of $U(g)$ the second derivative $g''(x)$ is forced towards zero, implying that the fitted curve does not change in slope, and thus approaches a straight-line regression no matter what g is chosen.

A remarkable property of the method is that once α is chosen we do not need to specify the family g to estimate g^*. This is because the latter can always be expressed in terms of *cubic splines*. A description of these, or the fitting procedures, is beyond the scope of this book. A detailed account of both the theory and application of nonparametric regression using penalty functions is given by Green and Silverman (1994), their Chapters 2 and 3 being devoted to practical aspects of the use of cubic splines in fitting.

15.4 Logistic Regression

We develop the key ideas in this section by discussing the data in Table 15.1. This classic data set was given by Millicer and Szczotka (1966) and is reproduced in Hand *et al* (1994, Set 476). It has been analysed by Finney (1971) and again by Morgan (1992) and is also discussed by Sprent (1998).

Table 15.1 indicates an increasing trend in the proportion, p, who have menstruated as age increases. Figure 15.13 is a plot of p against age. The basic observation needed to form Table 15.1 is whether or not each individual in the study has menstruated at a particular age. This is coded as an observed value of a *dichotomous* or *binary* variable, Y, which may take the value 0 or 1. We record a value 0 for an individual who has not menstruated, and a value 1 for an indivudal who has menstruated. The prime interest in a study such as this is centred on the proportion $p_i = r_i/n_i$, $i = 1, 2, \ldots, 25$ who have menstruated in each age group, and how that proportion changes with age. This proportion is an estimate of $\mathrm{E}(Y|x = x_i) = \pi_i$.

One might want to estimate, for example, the age at which 50 percent of girls would be expected to have menstruated. This is the *median menstrual age*. We might also be interested in the quartile ages, or the age at which one might expect 95 percent of girls to have menstruated.

To summarize, we look upon the observed p_i as an estimate of an unknown population probability π_i that a girl of age x_i will have menstruated. Regression methods would seem an obvious way to estimate π for any given age, x.

It is clear from Figure 15.13 that a straight line regression would not provide an adequate, nor indeed a meaningful fit, because for certain ages it would give impossible estimates of π lying outside the permissible range $0 \leq \pi \leq 1$. Choice of an appropriate function to fit to data like those in Figure 15.13 is no easy task and a fruitful approach is to consider some transformation of the data that allows us to work with a straight line regression. A variable such as age, x, is called an *explanatory variable*. It is, in essence, a covariate that helps explain changes in the p_i.

A versatile transformation is one associated with a family of distributions called *logistic distributions*. For an explanatory variable x, suppose that when $x = x_i$ the probability an event of interest (conveniently called a *success*) is π_i. If we assume that

Table 15.1 *Number (r) and proportion (p = r/n) of Warsaw girls in 1965 in each of 25 age groups of size (n) who had menstruated.*

Mean age	n	r	p
9.21	376	0	0
10.21	200	0	0
10.58	93	0	0
10.83	120	2	0.02
11.08	90	2	0.02
11.33	88	5	0.06
11.58	105	10	0.10
11.83	111	17	0.15
12.08	100	16	0.16
12.33	93	29	0.32
12.58	100	39	0.39
12.83	108	51	0.47
13.08	99	47	0.47
13.33	106	67	0.63
13.58	105	81	0.77
13.83	117	88	0.75
14.08	98	79	0.81
14.33	97	90	0.93
14.58	120	113	0.94
14.83	102	95	0.93
15.08	122	117	0.96
15.33	111	107	0.96
15.58	94	92	0.98
15.83	114	112	0.98
17.58	1049	1049	1.00

$$\pi_i = \frac{exp(\alpha + \beta x_i)}{1 + exp(\alpha + \beta x_i)} \tag{15.2}$$

where α and β are constants, we find by straightforward algebra that

$$\lambda_i = \log\left[\frac{\pi_i}{1 - \pi_i}\right] = \alpha + \beta x_i, \tag{15.3}$$

i.e., that the logarithm of the odds, often called the *logit*, is linearly related to x.

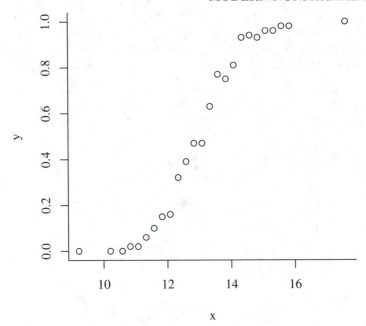

Figure 15.13 *Proportions of Warsaw girls having menstruated in different mean age groups.*

If we replace the π_i by the observed p_i and the model (15.2) is appropriate, then there should be a near linear relationship between the estimated log odds $\log[p_i/(1 - p_i)]$ and the corresponding x_i.

We may use the data set derived from the dichotomous variable Y to estimate $\mathrm{E}(Y|x) = p_i$ by replacing π_i by p_i in (15.2). The usual method of estimation is by maximum likelihood. We omit details, but the estimating equations turn out to be nonlinear and iterative solutions are required. Computer software to obtain these iterative solutions is widely available. These programs usually compute P-values for relevant hypothesis tests and often also relevant confidence intervals. For all but very small samples asymptotic results are usually adequate. What inferences one wishes to make depend on the problem being tackled.

This approach is often applied to toxicity studies. Basically these are studies where a series of increasing doses of some substance such as a drug are given to independent groups of subjects. Each subject in a group might be an animal or insect of a given species. At each dose level the number of subjects that show a specific response in the group given that dose is recorded. For a toxin the response of interest is often death, and interest attaches to the proportion showing this response at each dose level. The experimenter is usually interested in how the proportion responding increases with increasing dose. The situation here is analogous to that considered in Table 15.1 with the variable *age* replaced by *dose level,* the numbers of girls in each group being replaced by the number of experimental units in each group, the number and proportion having menstruated being replaced by number and proportion exhibiting the response under study such as *death, losing consciousness,* or whatever is of interest. Such *all or nothing* responses are called *quantal responses.*

In toxicity experiments the *median lethal dose,* i.e., the dose that will produce the response in half the subjects exposed to it is often of interest.

It follows from (15.3) that the median lethal dose is determined by setting $\lambda = 0$, since this implies $\pi = 0.5$. If the estimates of α, β are $\hat{\alpha}$, $\hat{\beta}$ an appropriate estimator of the median lethal dose is $x = -\hat{\alpha}/\hat{\beta}$.

In other situations we may want to test hypotheses about β only. For example, acceptance of the hypothesis $\beta = 0$ would imply insufficient evidence to conclude that the proportions responding changed weith x.

Maximum likelihood estimation using iterative methods is usually adequate, but conditional inferences based on permutation theory are described by Cox and Snell (1989, Sections 2.1.3 and 2.1.4). The method may be extended to include more than one explanatory variable. Statistical packages having programs for logisitic regression include Minitab and R. LogXact is a specialized package that provides exact permutation tests for small samples, and asymptotic tests based on maximum likelihood for larger samples. These programs also produce a host of supplementary output including relevant standard errors, confidence limits for α, β, as well as goodness of fit tests, the user being allowed some choice of precisely what is output.

The method is reasonably robust against certain small departures from the model, but may prove unsatisfactory if, for example, some explanatory variables, which may be continuous or discrete, that influence the binary response are omitted. In Table 15.1 the only explanatory variable is *age.* One can envisage other factors that may influence the onset of menstruation such as levels of nutrition, individual health status, amount of exercise. There is no evidence that such factors are relevant for this data set. This does not mean that they may not affect the age at which menstruation first occurs, but if they do, then the effect appears to be proportionate over age groups.

As we have omitted details of the maximum likelihood method of estimating parameters in logistic regression, we consider only briefly the nature of output from standard statistical software and its interpretation.

Example 15.4

The problem. For the data in Table 15.1 fit a logistic regression and estimate the median age of menarche for Warsaw girls in 1965.

Formulation and assumptions. We assume a logistic regression is appropriate and that age, x is the only relevant explanatory variable.

Procedure. We used both Minitab and LogXact with default settings to produce relevant output. The key estimates for each program differed only slightly. Such differences were attributable to iterative procedure details such as the number of iterations, amount of rounding, etc. The logistic equation obtained was

$$\lambda = -21.227 + 1.632x$$

with a 95 percent confidence interval for β of (1.517, 1.748).

Conclusions. There is clear evidence from the estimator $\hat{\beta}$ and the associated confidence interval that π increases with age (as of course is evident from inspection of the data). The estimate, x_m of the median age of menarche at which $\pi = 0.5$ is given by setting $\lambda = 0$, whence $x_m = 21.227/1.632 = 13.01$.

Comment. We have not discussed other aspects of the computer output, the exact nature of which depends on the software used, but for this example there is every indication that the model is adequate.

Logistic regression is not in a strict sense a nonparametric procedure, but rather a mathematical model that requires only limited assumptions for its validity. It is only distribution-free in the sense that we do not need to be specific about the distribution of the p_i except in very broad terms. The logit transformation is a competitor for at least one other transformation, the *probit* transformation, which very often leads to similar inferences. Both are members of a wider class of models known as *generalised linear models*. We do not discuss these further here. They are important models, but a detailed study would take us beyond the field of nonparametics.

To indicate the versatility of logistic regression we look briefly at another example where logistic regression leads to similar conclusions to those reached earlier by other means.

Example 15.5.

In Table 13.8 we reproduced data given by Graubard and Korn (1987) on abnormalities in offspring and were interested in whether these were related to alcohol consumption by mothers during pregnanacy.

In Example 13.9 we compared methods using different scoring systems to see if we could detect evidence that the probability of abnormalities depended upon the level of alcohol consumption. We found that results of our analysis were strongly influenced by the scoring system used. It would seem appropriate to use such a scoring system as an explanatory variable in a logistic regression analysis. Referring to Table 13.8, and taking a malformation as the event of interest, if we use the scores 1, 2, 3 4, 5 for alcohol consumption as an explanatory variable Minitab gives

$$\lambda = -6.209 + 0.228x,$$

and the P-value for a two-tail test with H_0: $\beta = 0$ is $P = 0.176$, suggesting no evidence of a trend with increasing alcohol consumption. This is consistent with our finding using these scores in Example 13.9.

Using the alternative scores suggested in that example, namely 0, 0.5, 1.5, 4 and 7, we find logistic regression gives

$$\lambda = -5.961 + 0.316x$$

and the P-value for a two-tail test with H_0: $\beta = 0$ is now $P = 0.012$, suggesting there may be a trend. This is again consistent with our finding in Example 13.9.

In this example the estimate of the median lethal dose of alcohol, i.e., the dose that would result in 50 percent of births showing malformations is not a useful concept. The estimate is $5.961/0.316 = 18.86$. This involves extrapolation well beyond the data set to a situation where the model is almost certainly irrelevant. It is doubtful whether a mother with that alcohol consumption would conceive, or even if she did, whether she would survive to give birth.

15.5 Multivariate Data

Often more than one variable is measured on each experimental or sample unit. We might measure the height, weight and waist girth of a group of 10 men aged 30 years and corresponding measurements on a group of 8 men aged 40 years. We could perform separate univariate analyses on each of the variables *height*, *weight* and *waist girth* to test for differences between *age*. If we were interested in mean, or median, differences we might use an appropriate parametric or nonparametric procedure to decide whether there were significant differences attributable to age. The data may give a general impression that the height distribution does not differ much between ages, but that both weight and waist girth looked to have a greater median for the older group. However, because of the relatively small samples the appropriate test — perhaps a t-test or a WMW test — may return a $P > 0.05$ for both these measurements. A question that then arises is whether there is some overall test that combines the information on the different variables. It is not unreasonable to hope that combining the information in some appropriate way may give a more powerful test that using a test that only looks at one variable at a time.

Parametric methods of *multivariate analysis* are well developed. When there are two treatments and k variables are measured on each experimental unit the t-test generalizes to Hotelling's T^2-test. This uses a statistic inspired by the univariate t statistic, but generalized to take into account the pooled sample covariances. For k variables observed on each unit in samples of size m, n the statistic, T^2 used in the test has an F-distribution with k and $m + n - k - 1$ degrees of freedom. When $k = 1$ this, not surprisingly, reduces to the square of the usual t-statistic. For situations with more than two treatments the classic one-variable ANOVA has been extended to *Multivariate Analysis of Variance*, often abbreviated to MANOVA.

We do not consider these parametric approaches further here, except to remark that in a nonparametric context, at least in some simple cases, we might use multivariate permutation procedures. For example, for the T^2 statistic these are analogous to the Pitman or raw data permutation tests for univariate data. Higgins (2004, Section 6.1) gives a simple example to show how this proceeds. Although we have not investigated this, and do not know of any other reports, we suspect that, like its univariate counterpart, the raw-data permutation procedure may lack robustness.

Several tests based on ranks have been suggested, making use of statistics such as Wilcoxon rank sums. What test of this type are appropriate depends largely on the hypotheses to be tested. We consider one example in this category. A more extensive introductory treatment is given by Higgins (2004, Chapter 6) but serious users of multivariate nonparametrics will need to refer to specialist books or papers on specific aspects such as multivariate density esitmation, multiple regression and so on. A classic overview at a fairly advanced level of multivariate nonparametric methods is given by Puri and Sen (1971), although there have been many practical developments since then. Some of these are discussed by Oja and Randles (2004).

15.5.1 A Multivariate Rank-Based Test for Observational Data

The test we illustrate in Example 15.6 is only relevant for a two independent sample problem where we believe any shift in location is in the same direction in each consitituent variable.

Example 15.6

The Problem. The data below are parts per million of 3 toxins, *A, B, C* in the discharge from 11 sewage tanks. For the first five tanks (*Group I*) a chemical has been added which it is hoped might decrease the level of all three toxins, while no chemical has been added in the last six tanks (*Group II*). Do the data indicate that the added chemical may have reduced the mean toxin levels?

	A	B	C
	32	44	207
	24	46	309
Group I	16	92	222
	11	29	415
	15	84	215

	A	B	C
	17	112	412
	23	83	478
Group II	29	91	429
	31	47	307
	37	89	225
	39	63	302

Formulation and Assumptions. A preliminary inspection of the data indicates that levels may be lower for all three toxins in *Group I*. On the assumption that the added chemical could not increase levels, one possibility is to use a one-tail WMW test for each chemical separately. However, the relatively small data sets may make such tests somewhat inconclusive. Because of the wide variation in toxin levels, both in and between tanks, it makes little sense to add the 3 toxin levels within each tank and to base an analysis on such totals. A more logical procedure is to rank the combined group data separately for each chemical and to analyse either the sum of these ranks as scores for a permutation test or else rank the sums and perform a WMW test on those ranks.

Procedure. The ranks corresponding to the above toxin levels, together with the total ranks over all three toxins for each tank are set out below. It is easy using any standard program to calculate the WMW one-tail P for each chemical using the *Group II* sums as the test statistic. The relevant P-values turn out to be 0.063, 0.123 and 0.089 for toxins A, B and C. For either the sum of the data as scores, or the rank of those sums, the one-tail $P = 0.002$.

	A	B	C	Total
	9	2	1	12
	6	3	7	16
Group I	3	10	3	16
	1	1	9	11
	2	7	2	11

	A	B	C	Total
	4	11	8	23
	5	6	11	22
Group II	7	9	10	26
	8	4	6	18
	10	8	4	22
	11	5	5	21

Conclusion. Although all three P-values for the univariate analysis for each toxin give P slightly greater than 0.05, the P-value for the summed ranks provides strong evidence that the added chemical reduces levels of toxins overall.

Comment. The procedure described in this example will only work if the order of the effect is the same for each of the k (here $k = 3$) variables. There is a similar concept here to the ordering implicit in the Jonckheere–Terpstra test.

Multivariate data sets may take a wide variety of forms, involving data sets for large numbers of treatments and complex covariance patterns. We have considered in this example perhaps the simplest possible situation with continuous experimental data. The general application of MANOVA requires adequate computer facilities and few exact permutation tests are possible. Where asymptotic results are of doubtful validity, Monte Carlo simulations may be valuable.

15.5.2 Partial Correlation

Our second example of nonparametric multivariate methods covers partial correlation. The concept is well-developed for Pearson correlation coefficients in the parametric field.

Partial correlation is relevant to determining whether the correlation between two variables X and Y reflects only that each is correlated with a third variable Z, or whether correlation between X and Y would still exist if Z is fixed.

For example, a group of students might sit examinations in mathematics, biology and also be scored in a general intelligence test. If their marks in mathematics, X, and biology, Y are highly correlated we may ask whether this simply reflects general intelligence levels, Z, or whether this correlation would still be apparent if all students had a fixed level of general intelligence.

The classic approach for assessing this situation is to calculate the Pearson correlation coefficient between the conditional distributions of X and Y with Z held fixed. A commonly used notation in this context is to denote the ordinary Pearson coefficients between X and Y by r_{XY}, that between X and Z by r_{XZ} and that between Y and Z by r_{YZ}. The partial correlation between X and Y conditional on fixed Z, denoted by $r_{XY.Z}$, is well-known to be

$$r_{XY.Z} = \frac{r_{XY} - r_{XZ} r_{YZ}}{\sqrt{(1 - r_{XZ}^2)(1 - r_{YZ}^2)}}. \tag{15.4}$$

This result is derived in standard textbooks that deal with multiple and partial correlation. It may be extended to deal with partial correlations conditional on fixed values of several varianbles $Z_1, Z_2, \ldots$.

If data consist of ranks a partial Spearman correlation coefficient may be computed using (15.4). A partial coefficient may also be computed for Kendall's τ_k. These distribution-free coefficients are discussed in detail in Kendall and Gibbons (1990, Chapter 8) and by Siegel and Castllan (1988, Section 9.5).

We use an example to indicate the notions behind a partial correlation coefficient for Kendall's τ_k.

Example 15.7

The problem. For three variable X, Y, Z for 8 subjects the rankings are

Subject	A	B	C	D	E	F	G	H
X	3	1	2	4	7	5	6	8
Y	2	4	1	3	5	8	7	6
Z	1	2	3	4	5	6	7	8

Obtain the partial Kendall correlation coefficient $t_{XY.Z}$ for these data and consider how these are generalized from 8 to n observations on each variable.

Formulation and assumptions. We first obtain the Kendall coefficients t_{XY}, t_{XZ} and t_{YZ} using counts of concordances and discordances as in Section 10.1.4. We

Table 15.2 *Concordances* (+) *and discordances* (−) *among pairs of ranks for Kendall correlation for each variable with* Z.

Pair	X	Y	Z	Pair	X	Y	Z
A, B	−	+	+	C, E	+	+	+
A, C	−	−	+	C, F	+	+	+
A, D	+	+	+	C, G	+	+	+
A, E	+	+	+	C, H	+	+	+
A, F	+	+	+	D, E	+	+	+
A, G	+	+	+	D, F	+	+	+
A, H	+	+	+	D, G	+	+	+
B, C	+	−	+	D, H	+	+	+
B, D	+	−	+	E, F	−	+	+
B, E	+	+	+	E, G	−	+	+
B, F	+	+	+	E, H	+	+	+
B, G	+	+	+	F, G	+	−	+
B, H	+	+	+	F, H	+	−	+
C, D	+	+	+	G, H	+	−	+

then develop procedures to obtain $t_{XY.Z}$ for this particular data set and consider its generalization.

Procedure. The data above are so arranged that the Z-ranks are in ascending order. This simplifies some computations, but results in no loss of generality. Using the methods outlined in Section 10.1.4 and counting the numbers of concordances and discordances you should confirm that

$$t_{XY} = \frac{20 - 8}{28}, \qquad t_{XZ} = \frac{24 - 4}{28}, \qquad t_{YZ} = \frac{22 - 6}{28}.$$

In Table 15.2 we record for each pair of subjects whether they are concordant (+) or discordant (−) for each variable relative to the ordered Z. For example, for the pair A, C since the X ranks are respectively 3, 2 we record a *minus* for X since $2 - 3$ is negative. Similarly, for the pair E, G since the Y ranks are respectively 5, 7 we record a *plus* for Y since $7 - 5$ is positive. Since the Z are in ascending order for all pairs we record a *plus*.

The data in Table 15.2 indicates that the signs associated with X, Y are sometimes a pair of plus, sometimes a plus and a minus, and in one case both minus. The numbers in the four possible categories ++, +−, −+, −− are summarized in Table 15.3. Inspecting that table it is clear that the row totals are the numbers of concordances and discordances relevant to calculating t_{XZ} and the column totals are those relevant to calculating t_{YZ}. The grand total of 28 is the total number of concordances and discordances providing the denominator $N = n_c + n_d$ in the notation used in Section 10.1.4. The sum $19 + 1$ on the main diagonal are the number of

Table 15.3 *Numbers of pairs where* X, Y *rankings relative to corresponding* Z *ranking are concordant* (+) *or discordant* (−).

		Y		
		+	−	Total
X	+	19	5	24
	−	3	1	4
	Total	22	6	28

Table 15.4 *Numbers of pairs among* N *subjects where* X, Y *rankings relative to corresponding* Z *ranking are concordant* (+) *or discordant* (−).

		Y		
		+	−	Total
X	+	a	b	$a+b$
	−	c	d	$c+d$
	Total	$a+c$	$b+d$	N

concordances relevant to calculating t_{XY}, while the number of discordances is given by the other diagonal sum $3 + 5$.

These results generalize to the ranking of n subjects or rankings for variables X, Y, Z giving rise to the structure in Table 15.4.

For this generalization we easily see that

$$
\begin{aligned}
t_{XZ} &= [(a+b) - (c+d)]/N \\
t_{YZ} &= [(a+c) - (b+d)]/N \\
t_{XY} &= [(a+d) - (b+c)]/N
\end{aligned}
\tag{15.5}
$$

The Kendall partial tau is defined as

$$
t_{XY.Z} = \frac{ad - bc}{\sqrt{(a+b)(c+d)(a+c)(b+d)}}
\tag{15.6}
$$

For the data in this example we find

$$
t_{XY.Z} = \frac{19 - 15}{\sqrt{24 \times 4 \times 22 \times 6}} = 0.0355
$$

Conclusion. Since the partial correlation coefficient $t_{XY.Z}$ is appreciably less than t_{XY} we conclude that the correlation betweeen X and Y is largely a reflection of correlations of each with Z.

Comment. The 2×2 tables of sign pairings relative to concordances with Z indicate complete independence if the cross product $ad - bc = 0$. It is not difficult to show that (15.6) only takes values in the interval$[-1, 1]$ and that $t_{XY.Z} = \pm 1$ if and only if

$$(ad - bc)^2 = (a + b)(c + d)(a + c)(b + d).$$

This implies

$$4abcd + a^2(bc + cd + bd) + b^2(ac + ad + cd)+$$

$$c^2(ab + ad + bd) + d^2(ac + ab + bc) = 0.$$

Since all of a, b, c, d are nonnegative a necessary and sufficient condition for this is that two of a, b, c, d must be zero. If two in the same row or column of the 2×2 array are zero this implies that either X or Y are in complete agreement, or complete disagreement with Z. Cases of interest are both a and d zero, or both b and c zero. The former implies X and Y disagree completely in concordances with Z, giving $T_{XY.Z} = -1$, the latter that they agree completely giving $T_{XY.Z} = +1$.

It is not difficult to show using tedious elementary algebra (see Exercise 15.3) that

$$t_{XY.Z} = \frac{t_{XY} - t_{XZ}t_{YZ}}{\sqrt{(1 - t_{XZ}^2)(1 - t_{YZ}^2)}},$$

a form similar to that given in (15.4) for the Pearson coefficient. This form is convenient for computation of the partial coefficient. Permutation tests for a departure of the partial coefficient from zero are laborious and not widely available in standard statistical software. Tables for commonly occuring situations are given by Siegel and Castellan (1988, Table S).

15.6 New Methods for Large Data Sets

In Section 8.6 we cautioned against indiscriminate use of multiple comparisons even when the number of possible comparisons was relatively small. Modern genetics, and other branches of science, often produced situations where an obvious approach to a statistical analysis is to test the same hypothesis for data sets from each of many hundreds or even thousands of samples with a view to selecting, perhaps for further study, only a small proportion of the corresponding populations that show marked departures from the null hypothesis.

Bearing in mind the philosophy behind the conventional hypothesis test, with significance level set at 0.05, this means that even if the null hypothesis holds for every one of, say, three thousand samples, and we assume these are independent we would expect approximately $3000/20 = 150$ of these to produce a test statistic indicating significant departure from the null hypothesis. These significant results are usually described as *false positives*.

In genetics there is often a subpopulation of just a few genes among many for which for biological reasons the null hypothesis may be expected not to hold. For that subpopulation we expect, if our test is reasonably powerful, that most will also yield a significant result, i.e., a *true positive*.

Efron (2005) suggests that situations like that outlined above call for new approaches to deal sensibly with the problem of distinguishing between true and false positives. In an earlier paper, Efron (2004) had already dealt with some of the technicalities of the new approaches he favours.

In this brief discussion we do not present these technical details, but concentrate on the motivation for some of these ideas that are set out in Efron (2005). There he gives what is essentially a nontechnical account of principles, and the paper is recommended reading for anyone interested in this increasingly important area of application.

Efron uses several examples to illustrate key points. We concentrate on one of these. It is a study reported originally by Hedenfalk et al (2001). It involved an investigation where two mutations each led to an increased risk of breast cancer, called BRCA1 and BRCA2. These are different mutations on different chromosomes. Heldenfalk and her coworkers wanted to assess whether the tumours resulting from the two different mutations were themselves genetically different.

They took tumours from 15 breast cancer patients — 7 from subjects with the BRCA1 mutation, 8 from subjects with the BRCA2 mutation. A separate micro-array was developed for each tumour and each micro-array involved the same 3226 genes. The data recorded for each tumour and for each gene was a *genetic activity number*. Efron reports that these numbers have little microbiological meaning individually, but that they can be compared with each other statistically. The data for the first gene for the 15 tumours were

BRCA1	-1.29	-1.41	-0.55	-1.04	1.28	-0.27	-0.57	
BRCA2	-0.70	1.33	1.14	4.67	0.21	0.65	1.02	0.16

Intuitively, we feel that these data probably will lead to rejection of the null hypothesis that the samples come from the same distribution in favour of a hypothesis specifying a location shift. This is suggested by the fact that all but one of the BRCA1 datum are negative, while only one BRCA2 datum is negative. Obvious tests here are the t-test, or perhaps, more appropriately, because of some vagueness about the physical meaning of, and even the statistical distribution of, these activity numbers, the WMW test. We consider the latter test as in Efron (2005)

Using the sum of the ranks for the BRCA2 data as the test statistic it is easy to see that the sum for any sample of 8 from 15 must lie between 36 and 92. For the above data you should verify that the sum of the ranks $S_8 = 83$ corresponding to a two-tail $P = 0.029$.

This test involves only the first of 3226 genes. Efron computed this statistic for the data for all 3226 genes. If the null hypothesis of zero difference between the median gene activity numbers held for all 3226 genes the expected number of occcurences for each rank sum of 8 from 15 corresponding to BRCA2 is obtained by multiplying the probabilities of each rank sum under the null hypothesis by 3226. These expected numbers are shown on a bar chart in Figure 15.14.

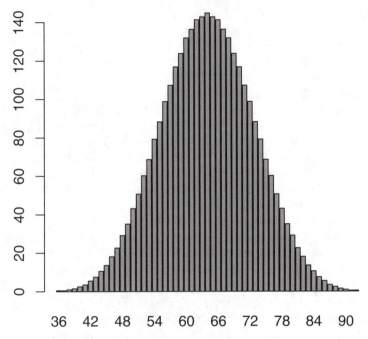

Figure 15.14 *Expected number of occurences of each score for 3226 genes in a WMW test for breast cancer study data.*

From Fig 15.14 it is apparent that scores between 62 and 66 have the highest probabilities and that we might expect each to occur between about 140 and 145 times. In fact none of these (or any other relevant rank sum) occurred more than 110 times among the 3226 genes. Efron gives a bar chart showing the number of times each score between 36 and 92 was obtained. This is appreciably flatter, but reasonably symmetric, having much higher tail numbers than those in Figure 15.14. For example, the score of 83 observed for the first gene was also obtained for 35 other genes. Under H_0 the expected number of occurences of a score 83 in 3226 tests is 13.55, meaning we might expect about 13 or 14 false positives. Therefore, among the 36 observed it is likely that something like 40 percent would represent *false positives*. This percentage is sometimes called a *false discovery rate*. This means that if we decide to explore further the nature of any genetic differences in each of these genes we might be wasting our time for something like 40 percent of the genes which are in reality *false positives*.

There is no reason to confine our attention to genes giving the score 83. Efron reports that 580 genes showed significance at the conventional 5 percent level, compared to an expected $3226/20 = 161.3$. Among all *significant* genes the expected false discovery rate is about 28 percent.

If we want to be fairly sure of selecting genes that are not likely to be *false positives* we might do better to look at scores where false positives are a relative rarity. For example, Efron reports 13 genes for which the score is 90 compared to an expected 1 under H_0, giving a false discovery rate of about 8 percent.

The above account of the issues involved is superficial and Efron suggests that the question of false detection rates lends itself nicely to what is known as an *empirical Bayes solution*. This is discussed in detail in Efron (2004).

Efron raises another important point. He suggests that the calculations of false discovery rate on the basis of the null hypothesis of no distributional difference might be wrong. We have assumed that for all genes the same H_0 is relevant, and that samples are independent. Both may be false. What happens in one gene may be correlated with what happens in another in a way not related to the mutations associated with breast cancer. Also, some genetic differences may be irrelevant to the onset of breast cancer, but be associated with other characteristics of individuals falling into each disease-type category. In the large set of 3226 it may be reasonable to suppose that only a few exhibit score differences attributable to the type of breast cancer

Therefore Efron reasons that the observed differences give us the basis for a better representation of the null distribution than does the WMW null distribution pictured in Figure 15.14. Efron (2004) explains both how we might estimate that null distribution, and also indicates why the observed distribution may differ from the null WMW distribution, or for that matter from the very similar normal-based t-distribution. The reader should refer to Efron's two papers cited in this section for details of this approach. He illustrates the procedure using the breast cancer data and some other large data sets.

The approach brings together, not only a link between frequentist and Bayesian methods through the notion of *empirical Bayes*, but also illustrates the importance of, and the need to consider along side each other, concepts such as those of EDA, classic nonparametric methods, and modern developments like bootstrapping applied to processes such as a lowess fit. An application of lowess in this context is illustrated in Efron (2005) using data for a study of the relation between liver function and age.

15.7 Correlations within Clusters

Situations often arise where a binary response is measured on a number of units that are subject to some form of clustering. For instance, in animal experiments each litter may form a cluster. In many areas of biomedical research responses of individuals within each cluster may no longer be independent. This is especially true for genetically transmitted characteristics where, if one sib exhibits a certain characteristic this may increase the probability that it will be exhibited in other sibs.

For binary data this implies that the probability of a positive response may be conditional upon whether other units in a cluster such as a litter show a positive response. If the positive response is rare there may be only a few such responses observed among a large number of individuals, a situation where, as we have seen elsewhere, asymptotic results are often misleading. Here we have the added complication that responses within clusters may no longer be independent.

In these circumstances a simple binomial model that applies to individual units is no longer appropriate and parametric modifications are available to cover this situation. In that context the phenomenon is often described as *overdispersion*. However, these parametric models often prove unsatisfactory when the probability of the event of interest is small in the population at large, but may sometimes be appreciable within some clusters, while in others it may be effectively zero. The reason for this breakdown is partly that parametric tests often rely on large sample theory but this theory is not reliable if the overall probability of the event of interest is small, even when the number of units observed is large.

Corocoran et al (2001) have developed exact tests for trends such as dose-related responses when the binary data are associated with clusters in which there is correlation in the sense described above. That is, in some of these clusters a greater number of individuals show the response than one would expect on the basis of the overall binomial p associated with a given dose. Here we only outline in terms of an example given in their paper the kind of situation where the test is useful. Their test is available in StatXact 7, though not in earlier versions of this software.

The data they discuss were given originally by Bradstreet and List (1995). One hundred female mice were randomly allocated 25 each to a control and three increasing dose levels of a potentially harmful drug. They were mated with untreated control mice and all but four were impregnated, so there were 96 litters in the study. The mice were killed after 17 days gestation and the fetuses examined for certain malformations. The total number of fetuses observed over all four treatments was 1248, but only 28 of these showed one or more malformations. This gives an overall estimate of the probability of a malformation of $p = 28/1248 \approx 0.0224$. If we assume there is no treatment (drug) effect and that all fetuses have the same probability of showing an effect irrespective of litter, i.e., we assume independence, the probability that in a litter of size 13 we observe r malformed fetuses, p_r, is

r	0	1	2	3	> 3
p_r	0.745	0.222	0.030	0.003	0.000

If the 96 litters were all of size 13, the average size of litters in this experiment, it is easily established that among 96 litters the expected number of litters showing 0, 1 , 2 or 3 malfunctions would be approximately 72, 21, 3, 0. If, however, in some clusters there was a higher value of p and in some a lower value of p we would expect higher numbers at the extremes, 0 or 3 or more malformations. The observed number of occurences of 0, 1, 2 and 3 malformations were 74, 17, 4 and 1, suggesting a tendency for nonindependence or correlation within clusters, or what is often described as overdispersion.

In the above arguments we have made a simplifying assumption that all clusters were of size 13, the average size. In fact all but 17 were between 11 and 15. The smallest litter was of size 7, the largest of size 17. The effect of ignoring this variation in litter size in this example was to underestimate the degree of correlation because the one cluster with 3 malformations occurred in a litter of size only 10, and 2 of the cases of 2 malformations were in litters of size less than 13. Only one malformation was recorded among the 10 litters of size 16 or over, compared with an expected number 2.6.

A second simplification was that we ignored a possible treatment effect with increasing dose levels of the potentially harmful drug. However, any such effect would be that the value $p = 0.0224$ used in our calculation would overestimate the expected number of malformations in each cluster under an independence assumption in the control groups and underestimate these expected numbers in the clusters associated with higher doses of the drug. These opposing trends would tend to have a neutralizing effect on our simplified argument used above.

As Cocoran et al point out, if we ignore the clustering in litters and assume that there is a monotonic treatment effect with increasing dose level we might apply a Jonckheere–Terpstra test or alternatively on the assumption that p, the probability of a malformation may increase with dose of drug, a Cochran–Armitage test. They report exact P-values for these to be 0.144 and 0.103 repespectively for the data they give. They have developed an exact test that takes intra-cluster correlation or over dispersion into account, allowing also for differences in litter sizes. This gives $P = 0.059$. They also quote results for some parametric tests developed for such data that take correlation within clusters into account. All such tests for these data give P-values ranging from 0.129 to 0.145.

For small p, even in large samples, the number of positive responses is likely to be sparsely distributed, and this is a typical situation where asymptotic tests associated with parametric models (or even nonparametric ones) may prove unreliable. A description of how to implement the exact test in this and certain closely related situations is given in the StatXact 7 manual, Chapter 17.

15.8 Summary

For a large data set a **histogram** may give a broad indication of the nature of the probability density function of the population from which the sample is derived. Some of the disadvantages of the histogram such as discontinuity and its dependence on choice of bin position and width are overcome by **kernel density estimation** described in Section 15.2.

A robust nonparametric and distribution-free approach to **curve smoothing** given large data sets is provided by **lowess** (Section 15.3.1). An alternative smoothing procedure is based on least squares associated with a **penalty**

function. This is designed to strike a balance between indicating the general trend and any overemphasis on local irregularities. A brief outline is given in Section 15.3.2.

Logistic regression (Section 15.4) uses an iterative least squares method to fit a linear regression on the logit transformation of proportions showing a specified response when these proportions are dependent on one or more covariates such as age or dose levels of a toxin. Interest often focuses on specific features of the data, such as estimation of a **median lethal dose** in toxicity studies.

Nonparametric **multivariate analysis** (Section 15.5) is a rapidly expanding field. We illustrate one aspect by a simple application based on ranking within each variable for a two-independent sample model.

Partial correlation coefficients may be calculated for the Pearson, Spearman or Kendall coefficients.

Large data sets arising in fields such as studies of DNA sequences or large numbers of genes require innovative approaches. A possible approach is indicated in Section 15.6.

Exact tests have an important role where positive binary responses have a low overall probability and units are grouped into clusters (e.g., litters) where responses within clusters may be correlated. A simple example is given in Section 15.7.

15.9 Exercises

*15.1 In Section 15.2 we asserted that the final one-quarter of the data in Example 15.1 represented a random sample from a normal distribution with $\mu = 7$ and $\sigma = 3$. We pointed out that there might be some surprise that the kernel density function exhibited a secondary maximum near 8 rather than 7. How would you produce evidence to convince a doubter that the assertion about the source of the data is not contradicted by the position of this maximum?

15.2 Use any available software that provides a random sample generator for normal distributions and a lowess program to (i) generate values of x, y for 121 equally spaced x in the interval $[-2, \ 2]$ where

$$y = f(x) = 4x^3 - x^2 - 4x + 1,$$

(ii) use the random sample program to generate a sample of 121 from a normal distribution with mean zero and standard deviation 0.3. Denote the ith sample value so generated by e_i, (iii) calculate for each of the data points generated in (i) $u_i = y_i + e_i$, (effectively adding a random 'error') and (iv) fit a lowess curve to the points $(x_i, \ u_i)$. Does your lowess fit broadly represent the specified $y = f(x)$. If not, explain why it does not.

15.3 Show that the expression for Kendall's partial tau, $t_{XY.Z}$ in (15.6) implies a form for $t_{XY.Z}$ analogous to the corresponding standard form of Pearson's $r_{XY.Z}$.

15.4   Verify the statement in Section 15.7 that in the absence of correlation within clusters (i.e., litters) of size 13 the expected numbers of clusters showing 0, 1, 2, 3 malformations in a total of 96 clusters are respectively 72, 21, 3, 0.

*__15.5__ Twelve students were given a test in *general science* before commencing one semester courses in *physics* and *biology*. Their test scores in the preliminary test and in tests on completion of their courses are given below. Compute appropriate Kendall correlation and partial correlation coefficients to assess the evidence that any correlation between the scores in the final tests in physics and biology reflects only differences in general scientific ability as measured at the entrance test.

Entry test	Biology	Physics
73	67	80
71	72	79
69	60	82
68	63	64
66	58	77
65	55	56
62	53	52
61	47	50
59	49	48
57	46	45
74	43	44
51	44	42

*__15.6__. In a situation analogous to that described in Section 15.6 for the breast cancer data measurements are recorded for samples of 6 and 10 respectively on each of 4126 genes. If a WMW test is performed on the data for each gene using the sum of the ranks for the sample of 6 as the test statistic and the null hypothesis of no distributional differences is true for all genes what is the expected number of times a score of 33 should be observed among the 4126 genes? If the null hypothesis may not hold for all genes and among the 4126 genes a score of 33 is observed for 36 genes, what is approximately the expected rate of false positives at this score?

15.7   In a laboratory experiment to test the effect of fungicide on growth of cabbages single plants are put in each of eight pots. Three of these are treated with fungicide and the remaining five are untreated. Plants are harvested and for each plant records are taken of the root weight in grams (x_1), the length of the longest root in millimetres (x_2), the top freshweight in grams (x_3) and the area of the largest leaf in square millimetres (x_4). The results are given below. Use a method similar to that in Example 15.6 to test the hypothesis that the fungicide does not influence growth. Perform any other tests you consider appropriate.

	x_1	x_2	x_3	x_4
Fungicide				
I	3.2	42	2.7	1160
II	3.6	38	3.1	1325
III	3.4	46	2.9	1470
No fungicide				
IV	3.0	39	2.4	1180
V	3.1	41	2.8	1220
VI	3.5	37	2.9	1270
VII	3.3	44	3.1	1340
VIII	2.9	43	2.6	1280

Appendix 1

BADENSCALLIE BURIAL DATA

Several examples in this book use data on the age at death of male members of four Scottish clans in the burial ground at Badenscallie in the Coigach district of Wester Ross, Scotland. The data were collected in June 1987. Clan names have been changed but the records are as complete as possible for four clans. There were a few missing values because names or dates were unreadable on a few headstones and several headstones appeared to be missing.

Minor spelling variations, especially those of M', Mc and Mac, were ignored. Ages are given for complete years, e.g., 0 means before first birthday and 79 means on or after 79th but before 80th birthday, according to the information on the tombstone. Ages are given in ascending order within each clan.

McAlpha (59 members)

0	0	1	2	3	9	14	22	23	29	33	41	41	42	44
52	56	57	58	58	60	62	63	64	65	69	72	72	73	74
74	75	75	75	77	77	78	78	79	79	80	81	81	81	81
82	82	83	84	84	85	86	87	87	88	90	92	93	95	

McBeta (24 members)

0	19	22	30	31	37	55	56	66	66	67	67	68	71	73
75	75	78	79	82	83	83	88	96						

McGamma (21 members)

13	13	22	26	33	33	59	72	72	72	77	78	78	80	81
82	85	85	85	86	88									

McDelta (13 members)

1	11	13	13	16	34	65	68	74	77	83	83	87

Appendix 2

SOME SOLUTIONS AND COMMENTS

Chapter 1

1.4 Six tosses suffice since $\Pr(6 \ \textit{heads}) = \Pr(6 \ \textit{tails}) = 1/64$. Critical region composed of both these outcomes gives $P = 1/32 \approx 0.031$ **1.6** (i) 0.119, (ii) 0.5, (iii) 0.9908.   **1.7** For Set I t-test 95 percent confidence interval is $(-1.6, 13.6)$. For Set II interval is $(3.3, 8.7)$. Null hypothesis of mean 10 for second set is only case where we would reject with a $P < 0.05$ (actual $P = 0.014$). For both sets sample mean is 6. Difficult to assess nature of population with such small samples but if sample were from a normal distribution it would be rather surprising if values in first sample fell into two groups, one well below mean and one well above mean. **1.8** Largest attainable P-value is 0.0386. Critical region is 0, 1, 2, 10, 11 12 successes. When $p = 0.75$ power is probability of getting outcome in critical region, which is approximately 0.39.

Chapter 2

2.2 $S = 13$; $P = \Pr(S \le 13) = 7/126 \approx 0.056$, indicating slight evidence against $\mathrm{H_0}$ that suggests a larger experiment might be desirable.   **2.5** Open interval $(156, 454)$. **2.6** Open interval $(-\infty, 454)$ gives approx. 98 percent coverage. In practice since it is impossible to have a negative number of pages in a book the interval is realistically $(0, 454)$. For data where the potential range of possible values is unbounded below and above the lower endpoint $-\infty$ is appropriate.   **2.9** The null hypothesis is that diets are equally effective. An appropriate alternative is that either may be the more beneficial, implying a two tail test. The number of possible rank allocations of 4 from 20 is

$$(20 \times 19 \times 18 \times 17)/(1 \times 2 \times 3 \times 4) = 4845$$

so probability of a rank allocation 1, 2, 3, 4 is $1/4845$. This implies a two-tail $P = 2/4845 \approx 0.0004$.
2.10
$$\Pr(\text{a sample value exceeds } 0.95) = 0.05.$$

Thus
$$\Pr(\text{a sample value does not exceeds } 0.95) = 0.95.$$

So
$$\Pr(\text{no sample value exceeds } 0.95) = (0.95)^{12}.$$

Now
$$\Pr(\text{at least one sample value exceeds } 0.95) =$$
$$1 - \Pr(\text{no sample value exceeds } 0.95) =$$
$$= 1 - (0.95)^{12} = 1 - 0.5404 = 0.4596$$

Chapter 3

3.1 Any s_i is included in S_+ if it has a plus sign, otherwise it makes zero contribution. Probability it has plus sign is 0.5. Thus $E(s_i) = s_i/2$ and $Var(s_i) = s_i^2/4$. Since the s_i are independent it follows that $E(S_+) = \sum_{i=1}^{n}(s_i/2)$ and $Var(S_+) = \sum_{i=1}^{n}(s_i^2/4)$. **3.5** All deviations from 6 are either $+2$ or -2. Thus Wilcoxon test easily seen to be equivalent to sign test. This is not the case if $H_0 : \theta = 7$, for then the signed deviations are of magnitudes 1 and 3. **3.8** $P = 0.13$ (two-tail). 95 percent Wilcoxon interval is (200.5, 443). Normal theory interval is (227.4, 405.9). **3.9** Two-tail $P = 0.3877$. Reflects low power of sign test relative to Wilcoxon signed-rank when no clear evidence of lack of symmetry such as a long right (or left) tail. **3.12** No. Wilcoxon and sign test (exact or asymptotic) all suggest some values above 72 would not be rejected at conventional 5 percent significance level. Sign test probably most appropriate as some evidence of asymmetry. **3.14** Ordered data suggest asymmetry so sign test appropriate. One tail test might be appropriate as experience suggests claims tend to rise as a result of inflation, but this is not certain, so more logical to insist on a two-tail test. Testing $H_0: \theta = 1570$ two-tail $P = 0.0570$. One-tail $P = 0.0287$. 94.3 percent confidence interval based on sign test is (1581, 2713) and 98.7 percent interval is (1327, 3419). Some indication median has risen, but evidence not very strong. Changing currency (a scale change only) would not alter conclusions. **3.15** One-tail test appropriate. For sign test $P = 0.090$. For Wilcoxon test $P = 0.011$. Symmetry assumption not unreasonable, so some evidence claim not justified. **3.17** Sign test interval is (11, 36). Wilcoxon interval is (14, 37). Some reservations about symmetry assumption due to observed time of 145. **3.18** Data median is 0.289. The value 0.2746 is well outside nominal 99 percent confidence limits based on either Wilcoxon or signed rank procedures. Indication of a higher percentage of births per day in September than the annual median. This implies births not uniformly distributed throughout year.

Chapter 4

4.2 $P = 0.165$. No substantial evidence against H_0. **4.3** Both tests suggest strong evidence against normality. $P \approx 0.010$ (Lilliefors'). $P = 0.001$ (Shapiro–Wilk). **4.5** Denoting by r the minimum number of damaged fruit in a batch of n which implies rejection to meet producer's risk $P = 0.10$ if $p = 0.01$ and by pwr the corresponding power if $p = 0.03$, one finds

n	100	200	300	400	500
r	3	5	6	8	9
pwr	0.58	0.72	0.89	0.92	0.96

Because of large discontinuities in exact P-values the exact power fluctuates markedly for small changes in sample size. A power of 0.95 is needed to meet the specified consumer's risk. Depending on available software the reader should experiment with several values of n between 400 and 500. For $n = 440$, for example, StatXact indicates that rejection if there are 8 or more damaged fruit gives an exact producer's risk $P = 0.078$ and a power 0.954 when the alternative is $p = 0.03$. This implies an exact consumer's risk of $P^* = 1 - 0.954 = 0.046$. Asymptotic results show reasonable agreement. **4.7** For smaller sample two-tail $P = 0.24$; for larger sample two-tail $P < 0.001$. Power increased with sample size. The more informative 95 percent confidence intervals for proportions approving are (0.17, 0.59) and (0.27, 0.40). **4.9** Cox-Stuart trend test gives two-tail $P = 0.125$ corresponding to 6 plus and 1 minus. A slight suggestion of a trend but a larger sample would be needed to confirm this. **4.11** Test for runs above and below median gives a two-tail $P = 0.0634$, suggesting slight evidence against the null hypothesis of no effect. Runs above the median dominate for those tested later. Cox–Stuart test may be less powerful as monotonicity assumption is unlikely to be justified (two-tail $P = 0.29$). **4.14** Hodges–Ajne $P = 0.94$. No evidence against hypothesis that all types of question are equally likely. **4.15** Sign test two-tail $P = 0.109$. Wilcoxon $P = 0.049$. Some evidence against H_0. No reservations.

Chapter 5

5.2 Two-tail $P = 0.045$ indicating moderate evidence of median difference. Wilcoxon 95 and 99 percent confidence intervals are $(0.05, 0.85)$ and $(-0.25, 1.05)$ and corresponding normal theory intervals are $(0.03, 0.85)$ and $(-0.15, 1.02)$. **5.4** Sign test appropriate; four M corresponds to 4 plus. Two-tail $P = 0.049$ suggests moderately strong evidence that fathers show better understanding. **5.6** No strong evidence of change in attitudes, two-tail $P = 0.126$. **5.9** For Wilcoxon test exact two-tail $P = 0.125$, so no strong evidence against H_0. t-test $P = 0.063$. **5.11** Wilcoxon two-tail $P = 0.271$. 95 percent confidence interval $(-2.95, 0.95)$. No evidence of one organization consistently returning higher percentages. **5.12** Wilcoxon two-tail $P = 0.031$ so fairly strong evidence against equal loss. **5.14** When $\theta = 2$, $p_1 = 0.8784$. **5.16** For medians, 1, 2 we require $\lambda = 0.6931, 0.3465$ respectively, whence $p_0 = 0.5$ and $p_1 = 0.707$. Sample size of 34 is needed.

Chapter 6

6.2 WMW two-tail $P = 0.0549$; not formally significant at 5 percent level but a slight indication that hard specimens may be associated with lower temperature. Test valid if it is reasonable to suppose any difference is a median shift or dominance by one distribution. **6.4** For WMW two-tail $P = 0.139$; 95 percent confidence interval $(-0.2, 1.9)$; for t-test $P = 0.097$, interval is $(-0.17, 1.83)$. Little reason to doubt validity of normal theory test. Result reflects slightly higher efficiency. **6.6** For two independent sample analysis two-tail $P = 0.272$ using WMW test. Differences between individuals in each field swamps any systematic difference between fields for individuals. Analysis used here is inappropriate for these data since it ignores pairing. **6.8** For asymptotic WMW test allowing for ties two-tail $P = 0.004$ suggesting strong evidence of shorter waiting times in 1983. **6.9** For asymptotic WMW test allowing for ties two-tail $P = 0.280$ so no evidence of difference in average sentence length. Do the data suggest WMW test may not be appropriate? Might there not be distributional differences in features other than average sentence length? **6.13** Exact two-tail $P = 0.208$ ($S_m = 101$). **6.17** Appropriate runs test one-tail $P = 0.72$. This is probability of 11 or less runs under H_0. No evidence of clustering.

Chapter 7

7.1 $F = 6.78$ with 2, 15 degrees of freedom. $P = 0.008$. Similar to Kruskal–Wallis test. **7.3** $Z = 2.85, p = 0.002$. **7.6** Jonckheere-Terpstra $U = 38$. Exact one-tail $P = 0.026$. Evidence supports expert's claim if one accepts one-tail P is justified. This implies he has positive discriminatory powers. Some may argue that it is possible he has negative discriminatory powers in that he might classify the worst author's work as *best*. In that case a two-tail test would be appropriate. **7.7** $T = 6.47$,

$P = 0.089$. Little firm evidence of consistent preferences. **7.8** Cochran Q $=$ 1.435 and exact two-tail $P = 0.855$, asymptotic $P = 0.697$. No evidence that tipster's differ in ability (or in incompetence). **7.11** (i) StatXact Monte Carlo estimated $P \approx 0.267$ (Asymptotic $P = 0.262$). (ii) Exact $P = 0.013$ (Asymptotic $P = 0.020$). **7.13** StatXact Monte Carlo estimated $P \approx 0.108$ (100,000 samples). Asymptotic $P = 0.109$. **7.15** Jonckheere–Terpstra $U = 119$. Exact or asymptotic $P < 0.0001$. Very strong support for experimenter's anticipated response.

Chapter 8

8.2 Setting all $d_{ij}^2 = k^2$ and counting number of terms in each sum gives

$$\nu_1 = \frac{(4k^2)^2}{4k^4 + 12k^4} = 1 \text{ and } \nu_2 = (r-1)\frac{(4k^2)^2}{4k^4} = 4(r-1).$$

8.4. For Wilcoxon test for carryover effect two-tail $P = 0.7337$, for direct effect of treatments, $P = 0.2104$, and for differential residual effect $P = 0.4709$. **8.6** Jonckheere–Terpstra test gives a two-tail $P < 0.0001$ indicating near-certain evidence of a toxicity trend. We now test succeeding higher levels against control until one indicates significance. A simple but ultra-conservative test uses WMW with Bonferroni adjustment. If base level of $\alpha = 0.05$ is chosen, since there are six possible pairwise contrasts critical P-value when comparing *Level O* with *Level 1* is $P = 0.0083$. WMW test for this pair gives $P = 0.0029$. Had this not indicated significance a less conservative approach might have been to use (8.9). **8.8** One-tail Jonckheere–Terpstra $P = 0.2095$ so no evidence of decreasing trend. Not appropriate to test for difference between Dynasties I and IV as no evidence of overall difference.

Chapter 9

9.3 For logrank test two-tail $P = 0.070$. For Gehan–Wilcoxon $P = 0.068$. Some indication that survival times may be longer for Motion A. **9.4** For logrank test StatXact gives exact two-tail $P = 0.0356$ and for Gehan–Wilcoxon with modification $P = 0.0156$. Strong indication here that treatment improves survival prospects. **9.5** StatXact gives $P = 0.0637$ for Gehan–Wilcoxon and $P = 0.0994$ for logrank test. Only slight indication of differences, probably due to small sample sizes.

Chapter 10

10.4 Exact one-tail $P = 0.040$ Same conclusion as that using t_k **10.5** No evidence of trend using Cox–Stuart test. Low power because information used is less than that for other coefficients. **10.7** $r_s = 0$, $t_k = 0.048$. No evidence of association. **10.9** $r_s = 0.90$, exact two-tail $P = 0.009$. $t_b = 0.78$, exact two-tail $P = 0.017$. **10.11** (i) $W = 0.367$, Monte Carlo estimate $P \approx 0.0012$ (Asymptotic $P = 0.0025$). (ii) $W = 0.549$, Monte Carlo estimate $P < 0.0005$ (Asymptotic $P = 0.0005$). Examiner 5 consistently awards higher marks **10.13** Asymptotic one-sided $P = 0.0385$. **10.14** $p_o = 0.76$, $p_e = 0.60107$ whence $\kappa = 0.3984$. The estimated s.e. of κ is 0.06408 giving $Z = 6.22$. implying $P < 0.0001$. Agrees with both exact and asymtotic values of P given by StatXact, although a different approximation for the s.e. is used.

Chapter 11

11.3 $y = -13.2 + 2.375x$. Plot indicates nonlinearity. **11.5** For the B(8, 0.05) distribution $\Pr(X \leq 1) \approx 0.035$, implying we omit the smallest and largest differences before dividing by 9.   **11.7** Plot indicates clear nonlinearity. One possibility would be to fit a monotonic regression, a technique described by Conover (1999, Section 5.6).   **11.9** WMW $U = 5$, exact one-tail $P = 0.075$. Hardly enough evidence to reject equality of slopes, but samples are small for suggested test.

Chapter 12

12.9 The Kruskal–Wallis test is appropriate for testing for differences between countries as the responses are ordered. The asymptotic $P = 0.0504$. For these data the Pearson chi-squared $X^2 = 25.34$, asymptotic $P = 0.0047$. One Monte Carlo estimate of exact P using 10000 samples gave $P = 0.0048$. Likelihood ratio or Fisher–Freeman–Halton test might also be used. It is perhaps surprising that the Kruskal–Wallis test returns the highest P. This may be a consequence of the somewhat arbitrary scoring system into *excellent*, *reasonable* and *poor* and how these scores are reflected as tied ranks.   **12.11** Any valid test (Pearson, Likelihood ratio, Fisher, Jonckheere–Terpstra) indicates overwhelming evidence ($P < 0.001$) that poor make less use of the service.   **12.13** Chi-squared goodness of fit exact $P < 0.0001$. Overwhelming evidence that manufacturer's claim not justified. (Hint: under H_0 the probabilities that 0, 1, 2, 3, 4 parts survive have a binomial distribution with $n = 4$, $p = 0.95$.)   **12.17** Exact Fisher–Freeman–Halton $P = 0.0716$. Assuming ordering indicated, exact Jonckheere–Terpstra $P = 0.0669$ (asymptotic $P = 0.0507$).   **12.21** McNemar $X^2 = 23$. Strong evidence of change in attitudes. Exact $P < 0.0001$.   **12.24** If choices random expected number for each group is 33.33. $X^2 = 38.11$, asymptotic $P = 0.043$. Some evidence selection not random. First choices for A, B, C, D are respectively 248, 181, 198, 173. Expected numbers 200 for each. $X^2 = 16.99$. Asymptotic $P = 0.0007$. Strong evidence for preference for A and some dislike of D.   **12.26** $X^2 = 40.48$, asymptotic $P < 0.0001$; $G^2 = 13.43$, asymptotic $P = 0.266$. In both tests exact $P = 0.0182$. Asymptotic results completely unreliable with so many sparse cells giving rise to low expected frequencies. Different patterns in columns 11, 12 relative to rest of table dominate in determining X^2, G^2; the former is more susceptible to small expected frequencies in row 1.

Chapter 13

13.9 Breslow–Day statistic (13.10) is 52.86 and exact and asymptotic $P < 0.0001$ providing virtually overwhelming evidence that there is not a common odds ratio and thus that a first-order interaction model does not suffice.   **13.11** If we accept ordering of parties we might use a linear-by-linear association model with rank scores. This gives an asymptotic two-tail $P = 0.20$ so no evidence of association. Other tests are possible. One might query whether any association need be monotonic with age.   **13.13** For first table exact $P < 0.0001$ using Pearson, likelihood ratio or Fisher test. For second table exact $P = 0.0017$. Breslow–Day test exact $P = 0.58$. First-order association model appears adequate.   **13.17** Cochran–Armitage test with scores 1, 2, 3, 4, 5 shows overwhelming evidence of increasing trend. One-tail $P < 0.0001$.   **13.19** A Cochran–Armitage style test is appropriate. Using default scores 1, 2, 3, 4 for columns the exact one-tail P reduces to $P < 0.000\,000\,1$. This reflects increasing power available because of the use of more detailed information than that in Exercise 13.18.

Chapter 14

14.1   No outlier in set A. In set B 452 and 777 are outliers.   **14.3**   None detected.   **14.5** -0.0088.   **14.9**   For observation zero, statistic takes value 5.25. No other observation detected as an outlier. Most strokes occur in adults, usually aged 50 or over. Infant strokes occassionally occur in cases of severe heart malformation. This could explain zero (i.e., aged less than 1 year) observation. Another possibility would be a faulty record — a nonzero digit preceding the 0 was omitted.

Chapter 15

15.1   Last 10 from a sample from a N(7, 9) distribution. For this sample 95 per-cent cvonfidence interval for population mean based on t-distribution is (5.67, 8.5). Although the sample mean is only 7.26, the secondary mode is pushed higher by a few high values in the larger sample of 30 more than offsetting three small values in the sample of 10.   **15.5** $t_{XY} = 0.7879$ and $t_{XY.Z} = 0.6749$ suggesting correlation would still be observed even if entry standard were fixed.   **15.6**   Expected number of scores of 33 among 4126 independent tests is approximately 19. If 36 scores of 33 are observed about 52.8 percent are likely to be false positives.

References

Adichie, J.N. (1967) Estimates of regression parameters based on rank tests. *Ann. Math. Statist.*, **38**, 894–904.

Adichie, J.N. (1984) Rank tests in linear models. In *Handbook of Statistics, Vol. 4*, Ed. Krishnaiah, P.B. and Sen, P.K. pp. 229–257. Amsterdam: Elsevier.

Agresti, A. (1984) *Analysis of Ordinal Categorical Data*. New York: John Wiley & Sons.

Agresti, A. (1992) A survey of exact inferences for contingency tables. *Statistical Science*, **7**, 131–153.

Agresti, A. (1996) *An Introduction to Categorical Data Analysis*. New York: John Wiley & Sons.

Agresti, A. (2002) *Categorical Data Analysis, 2nd ed.* New York: John Wiley & Sons.

Aitchison, J.W. and Heal, D.W. (1987) World patterns of fuel consumption; towards diversity and a low cost energy future. *Geography*, **72**, 235–239.

Aitken, C and Taroni, F. (2004) *Statistics and the Evaluation of Evidence for Forensic Scientists*. Chichester: John Wiley & Sons.

Ajne, B. (1968) A simple test for uniformity of a circular distribution. *Biometrika*, **55**, 343–354.

Akritas, M.G. (2004) Nonparametric survival analysis. *Statistical Science*, **19**, 615–623.

Akritas, M.G. and Arnold, S.F. (1994) Fully nonparametric hypotheses for factorial designs. I. Multivariate repeated measures designs. *J. Amer. Statist. Assoc.* **89**, 336–343.

Akritas, M.G. and Brunner, E. (1997) Nonparametric methods for factorial designs with censored data. *J. Amer. Statist. Assoc.*, **92**, 568–576.

Akritas, M.G., Arnold, S.F. and Brunner, E. (1997) Nonparametric hypotheses and rank statistics for unbalanced factorial designs. *J. Amer. Statist. Assoc.* **92**, 256–265.

Altshuler, B. (1970) Theory for the measurement of competing risks in animal experiments. *Math. Biosciences.* **6**, 1–11.

Andersen, P.K., Borgan, O, Gill, R.D. and Keiding, N. (1982) Linear nonparametric tests for the comparison of counting processes with application to censored survival data. *International Statistical Review*, **50**, 219–258.

Anderson, R.L. (1959) Use of contingency tables in the analysis of consumer preference studies. *Biometrics*, **15**, 582–590.

Andrews, D.F. (1974) A robust method for multiple linear regression. *Technometrics*, **16**, 523–531.

Andrews, D.F., Bickel, P.J., Hampel, F.R., Huber, P.J., Rogers, W.H. and Tukey, J.W. (1972) *Robust Estimates of Location. Survey and Advances*. Princeton, NJ: Princeton University Press.

Ansari, A.R. and Bradley, R.A. (1960) Rank sum tests for dispersion. *Ann. Math. Statist.*, **31**, 1174–1189.

Arbuthnot, J. (1710) An argument for Divine Providence, taken from the constant regularity observ'd in the births of both sexes. *Phil. Trans. Roy. Soc.*, **27**, 186–190.

Armitage, P. (1955) Tests for linear trends in proportions and frequencies. *Biometrics*, **11**, 375–386.

Arnold, H.J. (1965) Small sample power for the one sample Wilcoxon test for non-normal shift alternatives. *Ann. Math. Statist.*, **36**, 1767–1778.

Atkinson, A.C. (1985) *Plots, Transformations and Regression. An Introduction to Graphical Methods of Diagnostic Regression Analysis.* Oxford: Clarendon Press.

Bahadur, R.R. (1967) Rates of convergence of estimates and test statistics. *Ann. Math. Statist.*, **38**, 303–324.

Balakrishnan, N. and Koutras, M.V. (2002) *Runs and Scans with Applications.* New York: John Wiley & Sons.

Bardsley, P. and Chambers, R.L. (1984) Multipurpose estimation from unbalanced samples. *Appl. Statist.*, **33**, 290–299.

Barnett, V.D. and Lewis, T. (1994) *Outliers in Statistical Data.* 3rd edn. Chichester: John Wiley & Sons.

Bartlett, M.S. (1935) Contingency table interactions. *J. Roy. Statist. Soc., Suppl.*, **2**, 248–252.

Barton, D.E. and David, F.N. (1957) Multiple runs. *Biometrika*, **44**, 168–178.

Bassendine, M.F., Collins, J.D., Stephenson, J., Saunders, P. and James, O.W.F. (1985) Platelet associated immunoglobulins in primary biliary cirrhosis: a cause of Thrombocytopenia? *Gut*, **26**, 1074–1079.

Berry, D.A. (1987) Logarithmic transformations in ANOVA. *Biometrics*, **43**, 439–456.

Best, D.J. (1994) Nonparametric comparison of two histograms, *Biometrics*, **50**, 538–541.

Best, D.J. and Rayner, J.C.W. (1996) Nonparametric analysis for doubly ordered two-way contingency tables. *Biometrics*, **52**, 1153–1156.

Biggins, J.D., Loynes, R.M. and Walker, A.M. (1987) Combining examination results. *Brit. J. Math. and Statist. Psychol.*, **39**, 150–167.

Bishop, Y.M.M., Fienberg, S.E. and Holland, P. (1975) *Discrete Multivariate Analysis: Theory and Practice.* Cambridge, MA: MIT Press.

Bland J.M and Altman D.G (1994) Correlation, regression, and repeated data. *British Medical Journal*, **308**, 896.

Blomqvist, N. (1950) On a measure of dependence between two random variables. Ann. Math. Statist., **21**, 593–600.

Blomqvist, N. (1951) Some tests based on dichotomization. *Ann. Math. Statist.*, **22**, 362–371.

Bodmer, W.F. (1985) Understanding statistics. *J. Roy. Statist. Soc. A*, **148**, 69–81.

Boos, D.D. (1986) Comparing K populations with linear rank statistics. *J. Amer. Statist. Assoc.*, **81**, 1018–1025.

Borkowf, C.B., Gail, M.H., Carroll, R.J. and Gill, R.D. (1997). Analyzing bivariate continuous data grouped into categories defined by empirical quantiles of marginal distributions. *Biometrics*, **53**, 1054–1069.

Bowker, A.H. (1948) A test for symmetry in contingency tables. *J. Amer. Statist. Assoc.*, textbf43, 572–574.

REFERENCES

Bradley, J.V. (1968) *Distribution-Free Statistical Tests*. Englewood Cliffs, NJ: Prentice-Hall.

Bradstreet, T.E. and Liss, C.L. (1995) Favorite data sets from early (and late) phases of drug research. Part 4. *ASA Proceedings of the Section on Statistical Education*, 335–340.

Breslow, N.E. and Day, N.E. (1980) *Statistical Methods in Cancer Research. I. The Analysis of Case-Control Studies*. Lyons: WHO/IARC Scientific Publications, No. 32.

Brunner, E and Puri M.L. (1996) Nonparametric methods in design and analysis of experiments. *Handbook of Statistics* **13**. Ed. S. Ghosh and C.R. Rao. 631–702. New York: Elsevier.

Brunner, E and Puri M.L. (2001) Nonparametric methods in factoriasl designs. *Statistical Papers*, **42**, 1–52.

Bünning, H. and Kössler, W. (1999) The asymptotic power of Jonckheere-type tests for ordered alternatives. *Australian and New Zealand J. Statist.*, **41**, 67–72.

Burns, K.C. (1984) Motion sickness incidence: distribution of times to first emesis and comparison of some complex motion conditions. *Aviation Space and Environmental Medicine*, **56**, 521–527.

Cannistra, R. (2004) The ethics of early stopping rules: who is protecting whom? *Journal of Clinical Oncology*, **22**, 1542–1545.

Cantor, A.B. (1996) Sample-size calculations for Cohen's kappa. *Psychol. Meth.*, **1**, 150–153.

Capon, J. (1961) Asymptotic efficiency of certain locally most powerful rank tests. *Ann. Math. Statist.*, **32**, 88–100.

Carter, E.M. and Hubert, J.J. (1985) Analysis of parallel line assays with multivariate responses. *Biometrics*, **41**, 703–710.

Chandra, M., Singpurwalla, N.D. and Stephens, M.A. (1981) Kolmogorov statistics for test of fit for the extreme value and Weibull distributions. *J. Amer. Statist. Assoc.*, **76**, 729–731.

Chernick, M.R. (1999) *Bootstrap Methods - A Practitioner's Guide*. New York: John Wiley & Sons.

Christensen, R.A. (1990) *Log Linear Models*. New York: Springer-Verlag.

Cleveland, W.S. (1979) Robust locally weighted regression and smoothing scatterplots. *J. Amer. Statist. Assoc.*, **74**, 829–836.

Cochran, W.G. (1950) The comparison of percentages in matched samples. *Biometrika*, **37**, 256–266.

Cochran, W.G. (1954) Some methods for strengthening the common χ^2 tests. *Biometrics*, **10**, 417–454.

Cohen, A. (1983) Seasonal daily effect on the number of births in Israel. *Appl. Statist.*,**32**, 228–235.

Cohen, J. (1960) A coefficient of agreement fro nominal scales. *Educ. Psychol. Meas.*, **20**, 37–46.

Cohen, J. (1968) Weighted kappa: nominal scale agreement with provision for scaled disagreement or partial credit. *Psychol. Bull*, **70**, 213–220.

Conger, A.J. (1980) Integration and generalization of kappa for multiple raters. *Psychol. Bull.*, **88**, 322–328.

Conover, W.J. (1980) *Practical Nonparametric Statistics*. 2nd edn. New York: John Wiley & Sons.

Conover, W.J. (1999) *Practical Nonparametric Statistics.* 3rd edn. New York: John Wiley & Sons.

Conover, W. J. and Iman, R.L. (1976) On some alternative procedures usung ranks for the analysis of experimental designs. *Commun. Statist.* **A5, 14**, 1349–1368.

Conover, W. J. and Iman, R.L. (1981). Rank transformation as a bridge between parametric and nonparametric statistics. *Amer. Statist.* **35**, 124–133.

Cook, R.D. and Weisberg, S. (1982) *Residuals and Influence in Regression.* London: Chapman & Hall.

Corcoran, C., Ryan, L., Senchaudhuri, P., Mehta, C., Patel, N. and Molenberghs, G. (2001) An exact trend test for correlated binary data. *Biometrics,* **57**, 941–948.

Cormack, R.M. (1989) Log-linear models for capture-recapture. *Biometrics,* **45**, 395–413.

Cox, D.R. (1984) Interaction. *International Statistical Review,* **52**, 1–31.

Cox, D.R. (2006) *Principles of Statistical Inference.* Cambridge: Cambridge University Press.

Cox, D.R. and Snell, E.J. (1989) *Analysis of Binary Data.* 2nd edn. London: Chapman & Hall.

Cox, D.R. and Stuart, A. (1955) Some quick tests for trend in location and dispersion. *Biometrika,* 42, 80–95.

Cox, P.R. (1978) *Demography.* 5th edn. Cambridge: Cambridge University Press.

Cramér, H. (1928) On the composition of elementary errors. *Skand. Aktuarie-tidskrift,* **11**, 13–74, 141–180.

Crane M, Mallol J, Beasley R, Stewart A, Asher M.I. (2003) Agreement between written and video questions for comparing asthma symptoms in ISAAC. *European Respiratory Journal,* **21**, 455–461.

Daniel, W.W. (1990) *Applied Nonparametric Statistics.* 2nd. edn. Boston: PWS–Kent Publishing Company.

Dansie, B.R. (1986) Normal order statistics as permutation probability models. *Appl. Statist.,* **35**, 269–275.

David, F.N. and Barton, D.E. (1958) A test for birth-order effects. *Ann. Hum. Eugenics,* **22**, 250–257.

Davis, T.P. and Lawrance, A.J. (1989) The likelihood for competing risk survival analysis. *Scand. J. Statist.,* **16**, 23–28.

Davison, A.C. and Hinkley, D.V. (1997) *Bootstrap Methods and Their Application.* Cambridge: Cambridge University Press.

Day, S. and Talbot, D.J. (2000) Editorial: Statistical guidelines for clinical trials. *J. Roy. Statist. Soc.* A, **163**, 1–3.

Desu, M.M and and Raghavarao, D. (2004) *Nonparametric Statistical Methods for Complete and Censored Data.* Boca Raton: Chapman & Hall/CRC.

Dietz, E.J. and Killen, T.J. (1981) A nonparametric multivariate test for trend with pharmaceutical applications. *J. Amer. Statist. Assoc.,* **76**, 169–174.

Dinse, G.E. (1982) Nonparametric estimation for partially-complete time and type of failure data. *Biometrics,* **38**, 417–431.

Dobson, A.J. (2001) *An Introduction to Generalized Linear Models.* 2nd edn. Boca Raton: Chapman & Hall/CRC.

Dolkart, R.E., Halperin, B. and Perlman, J. (1971). Comparison of antibody responses in normal and alloxan diabetic mice. *Diabetes* **20**, 162–167.

Donahue, R.M.J. (1999) A note on information seldom reported via the P value. *Am. Statistician,* **53**, 303–306.

Donner A, Eliasziw M, Klar N (1996). Testing the homogeneity of kappa statistics. *Biometrics*, **52**, 176–183.

Durbin, J. (1951) Incomplete blocks in ranking experiments. *Brit. J. Psychol. (Statist. Section)*, **4**, 85–90.

Durbin, J. (1987) Statistics and statistical science. *J. Roy. Statist. Soc. A*, **150**, 177–191.

Easterbrook, P.J., Berlin, J.A., Gopalan, R. and Matthews, D.R. (1991) Publication bias in clinical research. *Lancet*, **337**, 867–872.

Edgington, E.S. (1995) *Randomization Tests*. 3rd edn. New York: Marcel Dekker, Inc.

Efron, B. (1979) Bootstrap methods: another look at the Jackknife. *Ann. Statist.*, **7**, 1–26.

Efron, B. (1981) Censored data and the bootstrap. *J. Amer. Statist. Assoc.*, **76**, 312–319.

Efron, B. (2004) Large-scale simultaneous hypothesis testing: The choice of a null hypothesis. *J. Amer. Statist. Assoc.*, **99**, 96–104.

Efron, B. (2005) Bayesians, frequentists, and scientists. *J. Amer. Statist. Assoc.*, **100**, 1–5.

Efron, B. and Gong, G. (1983) A leisurely look at the bootstrap, the jackknife and cross-validation. *Am. Statistician*, **37**, 36–48.

Efron, B. and Tibshirani, R.J. (1993) *An Introduction to the Bootstrap*. London: Chapman & Hall.

Emerson, J.D. and Hoaglin, D.C. (1983) Analysis of two-way tables by medians. In *Understanding Robust and Exploratory Data Analysis* Eds. D. C. Hoaglin, F. Mosteller and J. W. Tukey., 166–207. New York: John Wiley & Sons.

Eplett, W.J.R. (1982) The distribution of Smirnov type two-sample rank tests for discontinuous distribution functions. *J. Roy. Statist. Soc. B*, **44**, 361–369.

Everitt, B.S. (1992) *The Analysis of Contingency Tables*. 2nd edn. London: Chapman & Hall.

Fienberg, S.E. (1980) *The Analysis of Cross Classified Categorical Data*. 2nd edn. Cambridge, MA: MIT Press.

Finney, D.J. (1971) *Probit Analysis*, 3rd. edn. Cambridge: Cambridge University Press.

Fisher, N.I. (1993) *Statistical Analysis of Circular Data*. Cambridge: Cambridge University Press.

Fisher, N., Turner, S.W., Pugh, R. and Taylor, C. (1994) Estimating numbers of homeless and homeless mentally-ill people in north east Westminster by using capture-recapture analysis. *BMJ*, **308**, 27–30.

Fisher, R.A. (1929) Tests of significance in harmonic analysis. *Proc. Roy. Soc. London*, **A125**, 54–59.

Fisher, R.A. (1935) *The Design of Experiments*. Edinburgh: Oliver & Boyd.

Fisher, R.A. (1948) *Statistical Methods for Research Workers*. 10th edn. Edinburgh: Oliver & Boyd.

Fisher, R.A. and Yates, F (1957) *Statistical Tables for Biological, Agricultural and Medical Research* 5th Edn. Edinburgh: Oliver & Boyd.

Fisher, R.A., Corbet, A.S. and Williams, C.B. (1943) The relation between the number of species and the number of individuals in a random sample of an animal population. *J. Animal Ecol.*, **12**, 42–57.

Fleiss, J.L. (1978) Measuring nominal scale agreement among many raters. *Psychol. Bull.*, **76**, 378–382.

Fleiss, J.L. (1981) *Statistical Methods for Rates and Proportions.* 2nd edn. New York: John Wiley & Sons.

Fleiss, J.L. and Cohen, J. (1973) The equivalence of weighted kappa and the intra-class correlation coefficient as measures of reliability. *Educ. Psychol. Meas.*, **33**, 613–619.

Fleiss, J.L., Lee, J.C.M. and Landis, J.R. (1979) The large sample variance of kappa in the case of different sets of raters. *Psychol. Bull.*, **86**, 974–977.

Fligner, M.A. and Rust, S.W. (1982) A modification of Mood's median test for the generalized Behrens–Fisher problem. *Biometrika*, **69**, 221–226.

Freeman, G.H. and Halton, J.H. (1951) Note on an exact treatment of contingency, goodness of fit and other problems of significance. *Biometrika*, **38**, 141–149.

Freidlin, B. and Gastwirth, J.L. (1999) Unconditional versions of several tests commonly used in the analysis of contingency tables. *Biometrics*, **55**, 264–267.

Freidlin, B. and Gastwirth, J.L. (2000) Should the median test be retired from general use? *Am. Statistician*, **54**, 161–164.

Freidlin, B, Podgor, M.J. and Gastwirth, J.L. (1999) Efficiency robust tests for survival or ordered categorical data. *Biometrics* **55**, 883–886.

Freund, J.E. and Ansari, A.R. (1957) *Two Way Rank Sum Tests for Variance.* Blacksburg VA: Virginia Polytechnic Institute Report to Ordnance Research and NSF, **34**.

Friedman, M. (1937) The use of ranks to avoid the assumptions of normality implicit in the analysis of variance. *J. Amer. Statist. Assoc.*, **32**, 675–701.

Galton, F. (1892) *Finger Prints.* London: Macmillan.

Gart, J. (1970) Point and interval estimation of the common odds ratio in the combination of 2×2 tables with fixed marginals. *Biometrika*, **57**, 471–475.

Gastwirth, J.L. (1965) Percentile modifications of two sample rank sum tests. *J. Amer. Statist. Assoc.*, **60**, 1127–1141.

Gastwirth, J.L. (1968) The first median test: a two-sided version of the control median test. *J. Amer. Statist. Assoc.*, **63**, 692–706.

Gastwirth, J.L. (1985) The use of maximum efficiency robust tests in combining contingency tables and survival analysis. *J. Amer. Statist. Assoc.*, **80**, 380–384.

Gastwirth, J.L. (1988) *Statistical Reasoning in Law and Public Policy.* Orlando: Academic Press.

Gastwirth, J.L. (1997) Statistical evidence in discrimination cases. *J. Roy. Statist. Soc.,A*, **160**, 289–303.

Gastwirth, J.L. and Greenhouse, S.W. (1995) Biostatistical concepts and methods in the legal setting. *Statistics in Medicine*, **14**, 1641–1653.

Gat, J.R. and Nissenbaum, A. (1976) Limnology and ecology of the Dead Sea. *Nat. Geog. Soc. Res. Reports, 1976 Projects*, 413–418.

Geffen, G., Bradshaw, J.L. and Nettleton, N.C. (1973) Attention and hemispheric differences in reaction time during simultaneous audio-visual tasks. *Quart. J. Expt. Psychol.*, **25**, 404–412.

Gehan, E.A. (1965a) A generalized Wilcoxon test for comparing arbitrarily singly censored samples. *Biometrika*, **52**, 203–223.

Gehan, E.A. (1965b) A generalized two-sample Wilcoxon test for doubly censored data. *Biometrika*, **52**, 650–653.

Gibbons, J.D. and Chakraborti, S. (2004) *Nonparametric Statistical Inference.* 4th edn. New York: Marcel Dekker.

Gideon, R.A. and Hollister, R.A. (1987) A rank correlation coefficient resistant to outliers. *J. Amer. Statist. Assoc.*, **82**, 656–666.

Gill, A.N. and Mehta, G.P. (1989) Selection procedures for scale parameters using two-sample statistics. *Sankhya, Series B*, **51**, 149–157.

Gill, A.N. and Mehta, G.P. (1991) Selection procedures for scalar parameters using two-sample *U*-statistics. *Austral. J. Statist.*, **33**, 347–362.

Gillon, R. (1986) *Philosophical Medical Ethics.* Chichester: John Wiley & Sons.

Good, P. (2005) *Permutation , Parametric and Bootstrap Tests of Hypotheses.* New York: Springer.

Graubard, B.I. and Korn, E.L. (1987) Choice of column scores for testing independence in ordered $2 \times k$ contingency tables. *Biometrics*, **43**, 471–476.

Graunt, J. (1662) *Natural and Political Observations made upon the Bills of Mortality.* London.

Green, P.J. and Silverman, B.W. (1994) *Nonparametric Regression and Generalized Linear Models.* London: Chapman & Hall.

Gumpertz, M.L., Graham, J.M. and Ristiano, J.B. (1997) Autologistic model of spatial pattern of Phytophthora epidemic in Bell Pepper: Effects of soil variables on disease presence. *Journal of Agricultural, Biological and Environmental Statistics*, **2**, 131–156.

Hahn, M.E., Haber, SW.B. and Fuller, J.L. (1973). Differential agonistic behaviour in mice selected fro brain weight. *Phys. and Behaviour*, **10**, 759–762.

Hájek, J., Sidák, Z. and Sen, P.K. (1999) *Theory of Rank Tests.* 2nd edn. San Diego: Academic Press.

Hall, P. (1992) *The Bootstrap and Edgeworth Expansion.* New York: Springer–Verlag.

Hall, P. and Hart, J.D. (1990) Bootstrap tests for differences between means in nonparametric regression. *J. Amer. Statist. Assoc.*, **85**, 1039–1049.

Hampel, F.R. (1974) The influence curve and its role in robust estimation. *J. Amer. Statist. Assoc.*, **69**, 383–393.

Hand, D. J., Daly, F., Lunn, A.D., McConway, K.J., and Ostrowski, F. (1994) *Small Data Sets.* London: Chapman & Hall.

Härdle, W. and Gasser, T. (1984) Robust nonparametric function fitting. *J. Roy. Statist. Soc. B*, **46**, 42–51.

Hay, G. (1997) The selection from multiple data sources in epidemiological capture-recapture studies. *The Statistician*, **46**, 515–520.

Hedenfalk, I., Duggen, D., Chen, Y. *et al* (2001) Gene expression profiles in hereditary breast cancer. *New England J. of Medicine*, **344**, 539–548.

Hettmansperger, T.P. and McKean, J.W. (1998) *Robust Nonparametric Statistical Methods.* London: Arnold.

Hettmansperger, T. P. and Elmore, R., (2002) Rank tests for interaction: should they be taught in a nonparametrics nourse? *Proceedings of the 6th International Conference on the Teaching of Statistics, Cape Town, South Africa, July, 2002.*

Hickman M, Hope V, Platt L, Higgins V, Bellis M, Rhodes T, Taylor C, Tilling K. (2006) Estimating prevalence of injecting drug use: a comparison of multiplier and capture-recapture methods in cities in England and Russia. *Drug and Alcohol Review*, **25**, 131–140.

Higgins, J.J. (2004) *Introduction to Modern Nonparametric Statistics.* Belmont: Duxbury.

Hill, N.S. and Padmanabhan, A.L. (1984) Robust comparison of two regression lines and biomedical applications. *Biometrics*, **40**, 985–494.

Hinkley, D.V. (1989) Modified profile likelihood in transformed linear models. *Appl. Statist.*, **38**, 495–506.

Hodges, J.L. Jr. (1955) A bivariate sign test. *Ann. Math. Statist.*, **26**, 523–527.

Hodges, J.L. and Lehmann, E.L. (1963) Estimates of location based on rank tests. *Ann. Math. Statist.*, **34**, 598–611.

Hollander, M. and Wolfe, D.A. (1999) *Nonparametric Statistical Methods*. 2nd edn. New York: John Wiley & Sons.

Hook, E.B. and Regal, R.R. (1982) Validity of Bernoulli census, log-linear, and truncated binomial models for correcting for underestimates in prevalence studies. *Am J. Epidemiol.*, **116**, 168–176.

Hook E.B. and Regal R.R. (1995) Capture-recapture methods in epidemiology: methods and limitations. *Epidemiologic Reviews*, **17**, 243–264.

Hora, S.C. and Conover, W.J. (1984) The *F*-statistic in the two way layout with rank score transformed data. *J. Amer. Statist. Assoc.*, **79**, 668–673.

Hornbrook, M.C., Hurtado, A.V. and Johnson, R.E. (1985) Health care episodes: definition, measurement and use. *Medical Care Review*, **42**, 163–218.

House, D.E. (1986) A nonparametric version of Williams' test for a randomized block design. *Biometrics*, **42**, 187–190.

Howarth, J. and Curthoys, M. (1987) The political economy of women's higher education in late nineteenth and early twentieth century Britain. *Hist. Res.*, **60**, 208–231.

Huber, P.J. (1972) Robust statistics: a review. *Ann. Math. Statist.*, 43, 1041–1067.

Hussain, S.S. and Sprent, P. (1983) Nonparametric regression. *J. Roy. Statist. Soc. A*, **146**, 182–191.

Hutton, J.L. (1995) Statistics is essential for professional ethics. *J. Appl. Philos.*, **12**, 253–261.

Iman, R.L. and Davenport, J.M. (1980) Approximations of the critical region of the Friedman statistic. *Communications in Statistics*, **A9**, 571–595.

Iman, R.L., Hora, S.C. and Conover, W.J. (1984) Comparison of asymptotically distribution-free procedures for the analysis of complete blocks. *J. Amer. Statist. Assoc.*, **79**, 674–685.

Innes, N.P.T., Stirrups D.R., Evans D.J.P., Hall N. and Leggate, M. (2006) A novel technique using preformed metal crowns for managing carious primary molars in general practice — A retrospective analysis. *British Dental Journal*, **200**, 451–454.

Jaeckel, L.A. (1972) Estimating regression equations by minimizing the dispersion of residuals. *Ann. Math. Statist.*, **43**, 144–158.

Jarrett, R.G. (1979) A note on the intervals between coal mining disasters. *Biometrika*, **66**, 191–193.

Jonckheere, A.R. (1954) A distribution free *k*-sample test against ordered alternatives. *Biometrika*, **41**, 133–145.

Kalbfleish, J.D. and Prentice, R.L. (1980) *The Analysis of Failure Time Data*. New York: John Wiley & Sons.

Kaplan, E.L. and Meier, P. (1958) Nonparametric estimation from incomplete observations. *J. Amer. Statist. Assoc.* **53**, 457–81.

Katti, S.K. (1965) Multivariate covariance analysis. *Biometrics*, **21**, 957–974.

Kendall, M.G. (1938) A new measure of rank correlation. *Biometrika*, **30**, 81–93.

Kendall, M.G. and Gibbons, J.D. (1990) *Rank Correlation Methods*. 5th edn. London: Edward Arnold.

Kerridge, D. (1975) The interpretation of rank correlations. *Appl. Statist.*, **24**, 257–258.

Kimber, A.C. (1987) When is a χ^2 not a χ^2? *Teaching Statistics*, **9**, 74–77.

Kimber, A.C. (1990) Exploratory data analysis for possibly censored data from skewed distributions. *Appl. Statist.*, **39**, 21–30.

Kimura, D.K. and Chikuni, S. (1987) Mixtures of empirical distributions: an iterative application of the age-length key. *Biometrics*, **43**, 23–35.

Klotz, J. (1962) Nonparametric tests for scale. *Ann. Math. Statist.*, **33**, 498–512.

Klotz, J. (1963) Small sample power and efficiency for the one-sample Wilcoxon and normal scores tests. *Ann. Math. Statist.*, **34**, 624–632.

Knapp, T.R. (1982) The birthday problem: some empirical data and some approximations. *Teaching Statistics*, **4**, 10–14.

Koch, G.G. (1972) The use of nonparametric methods in the statistical analysis of the two-period change-over design. *Biometrics*, **28**, 577–594.

Kolmogorov, A.N. (1933) Sulla determinazione empirica di una legge di distribuzione. *G. Inst. Ital. Attuari*, **4**, 83–91.

Kolmogorov, A.N. (1941) Confidence limits for an unknown distribution function. *Ann. Math. Statist.*, **12**, 461–463.

Kraemer, H.C. and Thiemann, S. (1987) *How many subjects? Statistical power analysis in research*. Newbury Park: Sage.

Krantz, D.H. (1999) The null hypothesis testing controversy in psychology. *J. Amer. Statist. Assoc.*, **94**, 1372–1381.

Kruskal, W.H. and Wallis, W.A. (1952) Use of ranks in one-criterion variance analysis. *J. Amer. Statist. Assoc.*, **47**, 583–621.

Lancaster, H.O. (1949) The derivation and partitioning of χ^2 in certain discrete distributions. *Biometrika*, **36**, 117–129.

Lancaster, H.O. (1953) A reconciliation of χ^2, considering metrical and enumerative aspects. *Sankhya*, **13**, 1–9.

Landis, J.R. and Koch, G.G. (1977) The measurement of observer agreement for categorical data. *Biometrics*, **33**, 159–174.

Leach, C. (1979) *Introduction to Statistics. A Nonparametric Approach for the Social Sciences*. Chichester: John Wiley & Sons.

Lehmann, E.L. (1975) *Nonparametrics: Statistical Methods Based on Ranks*. San Francisco: Holden Day, Inc.

Lehmann, E.L. (2006) *Nonparametrics: Statistical Methods Based on Ranks*. Revised edn. Berlin: Springer.

Light, R.J. (1971) Measures of response agreement for qualitative data: some generalizations and alternatives. *Psychol. Bull.*, **76**, 89–97.

Lilliefors, H.W. (1967) On the Kolmogorov-Smirnov test for normality with mean and variance unknown. *J. Amer. Statist. Assoc.*, **62**, 399–402.

Lindley, D.V. and Scott, W.F. (1995) *New Cambridge Elementary Statistical Tables*. 2nd edn. Cambridge: Cambridge University Press.

Lindsey, J.C., Herzberg, A.M. and Watts, D.G. (1987) A method of cluster analysis based on projections and quantile–quantile plots. *Biometrics*, **43**, 327–341.

Little, R.J. (2006) Calibrated Bayes: A Bayes/frequentist roadmap. *The American Statistician*, **60**, 213–223.

Lloyd, C.J. (1999) *Statistical Analysis of Categorical Data.* New York: John Wiley & Sons.

Lombard, F. (1987) Rank tests for change-point problems. *Biometrika,* **74**, 615–624.

Lubischew, A.A. (1962) On the use of discriminant functions in taxonomy. *Biometrics,* **18**, 455–477.

McCullagh, P. and Nelder, J.A. (1989) *Generalized Linear Models.* 2nd edn. London: Chapman & Hall.

Mack, G.A. and Wolfe D.A. (1981) K-sample rank test for umbrella alternatives. *J. Amer. Statist. Assoc.,* **76**, 175–181.

McKean, J.W., Sheather, S.J. and Hettsmansperger, T.P. (1990) Regression diagnostics for rank based methods. *J. Amer. Statist. Assoc.,* **85**, 1018–1028.

McNemar, Q. (1947) Note on the sampling error of the difference between correlated proportions or percentages. *Psychometrika,* **12**, 153–157.

Manfredini R., Gallerani M., Caracciolo S., Tomelli A., Calo G. and Fersini C. (1994) Circadian variation in attempted suicide by deliberate self poisoning. *British Medical Journal,* **309**, 774–775.

Manly, B.F.J. (2006) *Randomization, Bootstrapping and Monte Carlo Methods in Biology.* 3rd edn. Boca Raton: Chapman & Hall/CRC.

Mann, H.B. and Whitney, D.R. (1947) On a test of whether one of two random variables is stochastically larger than the other. *Ann. Math. Statist.,* **18**, 50–60.

Mantel, N. and Haenszel, W. (1959) Statistical aspects of the analysis of data from retrospective studies of disease. *J. Nat. Cancer Res. Inst.,* **22**, 719–748.

Marascuilo, L.A. and McSweeney, M. (1977) *Nonparametric and Distribution-free Methods for the Social Sciences* Monterey, CA: Brooks/Cole Publishing Company.

Marascuilo, L.A. and Serlin, R.C. (1979) Tests and contrasts for comparing change parameters for a multiple sample McNemar data model. *Brit. J. Math. and Statist. Psychol.,* **32**, 105–112.

Mardia, K.V. (1972) *Statistics of Directional Data.* London: Academic Press.

Mardia, K.V. and Jupp, P.E. (2000) *Directional Statistics.* Chichester: John Wiley & Sons.

Maritz, J.S. (1995) *Distribution-free Statistical Methods.* 2nd edn. London: Chapman & Hall.

Mathisen, H.C. (1943) A method of testing the hypothesis that two samples are from the same population. *Ann. Math. Statist.,* **14**, 188–194.

Mattingley, P.F. (1987) Pattern of horse devolution and tractor diffusion in Illinois, 1920–82. *Prof. Geographer,* **39**, 298–309.

Mehta, C.R. and Patel, N.R. (1983) A network algorithm for performing Fisher's exact test in $r \times c$ contingency tables. *J. Amer. Statist. Assoc.,* **78**, 427–434.

Mehta, C.R. and Patel, N.R. (1986) A hybrid algorithm for Fisher's exact test on unordered $r \times c$ contingency tables. *Communications in Statistics,* **15**, 387–403.

Mehta, C.R., Patel, N.R. and Gray, R. (1985) On computing an exact common odds ratio in several 2×2 contingency tables. *J. Amer. Statist. Assoc.,* **80**, 969–973.

Mehta, C.R., Patel, N.R. and Senchaudhuri, P. (1988) Importance sampling for estimating exact probabilities in permutational inference. *J. Amer. Statist. Assoc.,* **83**, 999–1005.

Mehta, C.R., Patel, N.R. and Senchaudhuri, P. (1998) Exact power and sample size computations for the Cochran-Armitage trend test. *Biometrics,* **54**, 1615–1621.

Mehta, C.R., Patel, N.R. and Tsiatis, A.A. (1984) Exact significance tests to establish treatment equivalence for ordered categorical data. *Biometrics,* **40**, 819–825.

Millicer, H. and Szczotka, F. (1966) Age at menarche in Warsaw girls in 1965. *Human Biol.*, **38**, 199–203.

Mood, A.M. (1940) The distribution theory of runs. *Ann. Math. Statist.*, **11**, 367–392.

Mood, A.M. (1954) On the asymptotic efficiency of certain nonparametric two-sample tests. *Ann. Math. Statist.*, **25**, 514–522.

Morgan, B. J. T. (1992) *Analysis of Quantal Response Data*. London: Chapman & Hall.

Moses, L.E. (1952) Nonparametric statistics for psychological research. *Psychol. Bull.*, **49**, 122–143.

Neave, H.R. (1981) *Elementary Statistical Tables*. London: George Allen & Unwin.

Nelder, J.A. (1999) From statistics to statistical science (with comment). *The Statistician*, **48**, 257–270.

Noether, G.E. (1984) Nonparametrics: the early years — impressions and recollections. *Am. Statistician*, **38**, 173–178.

Noether, G.E. (1987a) Sample size determination for some common nonparametric tests. *J. Amer. Statist. Assoc*, **82**, 645–647.

Noether, G.E. (1991) *Introduction to Statistics: The Nonparametric Way*. New York: Springer–Verlag.

Oja, H. and Randles, R.H. (2004) Multivariate Nonparametric Tests. *Statistical Science* **19**, 598–605.

O'Muircheartaigh, I.G. and Sheil, J. (1983) Fore or five? The indexing of a golf course. *Appl. Statist.*, **32**, 287–292.

Page, E.B. (1963) Ordered hypotheses for multiple treatments: a significance test for linear ranks. *J. Amer. Statist. Assoc.*, **58**, 216–230.

Paul, S.R. (1979) Models and estimation procedures for the calibration of examiners. *Brit. J. Math. Statist. Psychol.*, **32**, 242–251.

Pearce, S.C. (1965) *Biological Statistics – An Introduction*. New York: McGraw–Hill.

Pearson, J.C.G. and Sprent, P. (1968) Trends in hearing loss associated with age or exposure to noise. *Appl. Statist.*, **17**, 205–215.

Pearson, K. (1900) On a criterion that a given system of deviations from the probable in the case of a correlated system of variables is such that it can reasonably be supposed to have arisen in random sampling. *Phil. Mag.*(5), **50**, 157–175.

Peto, R. and Peto, J. (1972) Asymptotically efficient rank-invariant test procedures. *J. Roy. Statist. Soc. A*, **135**, 185–206.

Peto, R, Pike, M.C., Armitage, P., Breslow, N.ED., Cox, D.R. , Howard, S.V., Mantel, N., McPherson, K., Peto, J. and Smith, P.G. (1977) Design and analysis of randomized clinical trials requiring prolonged observation of each patient. *Brit. J. Cancer*, **35**, 1–39.

Pettitt, A.N. (1979) A nonparametric approach to the change-point problem. *Appl. Statist.*, **28**, 126–135.

Pettitt, A.N. (1981) Posterior probabilities for a change-point using ranks. *Biometrika*, **68**, 443–450.

Pitman, E.J.G. (1937a) Significance tests that may be applied to samples from any population. *J. Roy. Statist. Soc., Suppl.*, **4**, 119–130.

Pitman, E.J.G. (1937b) Significance tests that may be applied to samples from any population, II: The correlation coefficient test. *J. Roy. Statist. Soc., Suppl.*, **4**, 225–232.

Pitman, E.J.G. (1938) Significance tests that may be applied to samples from any population. III. The analysis of variance test. *Biometrika*, **29**, 322–335.

Pitman, E.J.G. (1948) *Mimeographed lecture notes on nonparametric statistics.* Columbia University.

Plackett, R.L. (1981) *The Analysis of Categorical Data.* 2nd edn. London: Charles Griffin & Co.

Pocock, S.J. (1983) *Clinical Trials — A Practical Approach.* Chichester: John Wiley & Sons.

Posner, K.L., Sampson, P.D., Caplan, R.A., Ward, R.J. and Chenly, F.W. (1990) Measuring interrater reliability among multiple raters: An example of methods for nominal data. *Statist. Med.*, **9**, 1103–1116.

Pothoff, R.F. (1963) Use of the Wilcoxon statistic for a generalized Behrens-Fisher problem. *Ann. Math. Statist.*, **34**, 1596–1599.

Prentice, R.L. and Marek, P. (1979) A quantitative discrepancy between censored data rank tests. *Biometrics*, **35**, 861–867.

Puri, M.L. and Sen, P.K. (1971, reprinted 1993) *Nonparametric Methods in Multivariate Analysis.* New York: John Wiley & Sons.

Putt, M.E. and Chinchilli, V.M. (2004) Nonparametric approaches to the analysis of crossover studies. *Statistical Science*, **19**, 712–719.

Quade, D. (1979) Using weighted ranks in the analysis of complete blocks with additive block effects. *J. Amer. Statist. Assoc.*, **74**, 680–683.

Quenouille, M.H. (1949) Approximate tests of correlation in time series. *J. Roy. Statist. Soc. B*, **11**, 18–84.

Radelet, M. (1981) Racial characteristics and the imposition of the death penalty. *Amer. Sociol. Rev.*, **46**, 918–927.

Rae, G. (1988) The equivalence of multirater kappa statistics and intraclass correlation coefficients. *Educ. and Psychol. Meas.*, **48**, 921–933.

Randles, R.H., Fligner, M.A., Policello, G.E. and Wolfe, D.A. (1980) An asymptotically distribution-free test for symmetry versus asymmetry. *J. Amer. Statist. Assoc.*, **75**, 168–172.

Randles, R.H. and Wolfe, D.A. (1979) *Introduction to the Theory of Nonparametric Statistics.* New York: John Wiley & Sons.

Rayner and Best (2001) *A Contingency Table Approach to Nonparametric Testing.* Boca Raton: Chapman & Hall/CRC.

Rees, D.G. (2000) *Essential Statistics.* 3rd edn. Boca Raton: Chapman & Hall/CRC.

Rees, J.H., Thompson, R.D., Smeeton, N.C. and Hughes, R.A.C. (1998) Epidemiological study of Guillain-Barr syndrome in south east England. *J. Neurol. Neurosurg. and Psychiatry*, **64**, 74–77.

Regal, R.R. and Hook, E.B. (1984) Goodness-of-fit based confidence intervals for estimates of the size of a closed population. *Statist. Med.*, **3**, 287–291.

Regal, R.R. and Hook, E.B. (1999) An exact test for all-way interactions in a 2^M contingency table: application to interval capture-recapture estimation of population size. *Biometrics*, **55**, 1241–1246.

Robins, J., Breslow, N. and Greenland, S. (1986) Estimation of the Mantel–Haenszel variance consistent in both sparse data and large-strata limiting models. *Biometrics*, **42**, 311–323.

Robinson, J.A. (1983) Bootstrap confidence intervals in location–scale models with progressive censoring. *Technometrics*, **25**, 179–187.

Rogerson, P.A. (1987) Changes in U.S. national mobility levels. *Prof. Geographer*, **39**, 344–351.

Rosenthal, I. and Ferguson, T.S. (1965) An asymptotic distribution-free multiple comparison method with application to the problem of n rankings of m objects. *Brit. J. Math. Statist. Psychol.*, **18**, 243–254.

Ross, G.J.S. (1990) *Nonlinear Estimation*. New York: Springer-Verlag.

Sackett, D.L., Haynes, R.B., Guyatt, G.H. and Tugwell, P. (1991) *Clinical Epidemiology: A Basic Science for Clinical Medicine*. 2nd edn. Boston: Little Brown.

Sackrowitz, H. and Samuel-Cahn, E. (1999) *P*-values as random variables — expected *P*-values. *Am. Statistician*, **53**, 326–331.

Satterthwaite, F.E. (1946) An approximate distribution of estimates of varaince components. *Biometrics Bulletin*, **2**, 110–114.

Saunders, R. and Laud, P. (1980) The multidimensional Kolmogorov goodness-of-fit test. *Biometrika*, **67**, 237.

Savage, I.R. (1956) Contributions to the theory of rank order statistics I. *Ann. Math. Statist.*, **27**, 590–615.

Scheirer, C.J., Ray, W.S. and Hare, N. (1976). The analysis of ranked data derived from completely randomized factorial designs. *Biometrics* **32**, 429–434.

Scholz, F.W. and Stephens, M.A. (1987) *K*-sample Anderson–Darling tests. *J. Amer. Statist. Assoc.*, **82**, 918–924.

Schuster, E.F. and Gu, X. (1997) On the conditional and unconditional distributions of the number of runs in a sample from a multisymbol alphabet. *Communications in Statistics: Simulation and Computation*, **26**, 423–442.

Scott, A.J., Smith, T.M.F. and Jones, R.G. (1977) The application of time series methods to the analysis of repeated surveys. *Internat. Statist. Rev.*, **45**, 13–28.

Seber, G.A.F. (1982) *The Estimation of Animal Abundance and Related Parameters*. 2nd edn. London: Charles Griffin and Company.

Sekar, C.C. and Deming, W.E. (1949) On a method of estimating birth and death rates and the extent of registration. *J. Amer. Statist. Assoc.*, **44**, 101–115.

Selvin, S. (1995) *Practical Biostatistical Methods*. Belmont, CA: Wadsworth Publishing Company.

Sen, P.K. (1968) Estimates of the regression coefficient based on Kendall's tau. *J. Amer. Statist. Assoc.*, **63**, 1379–1389.

Sen, P.K. (1969) On a class of rank order tests for the parallelism of several regression lines. *Ann. Math. Statist.*, **40**, 1668–1683.

Senn, S.J. (2002) *Crossover Trials in Clinical Research*. 2nd edn. Chichester: John Wiley & Sons.

Shapiro, S.S. and Wilk, M.B. (1965) An analysis of variance test for normality (complete samples). *Biometrika*, **52**, 591–611.

Shapiro, S.S., Wilk, M.B. and Chen, H.J. (1968) A comparative study of various tests for normality. *J. Amer. Statist. Soc.*, **63**, 1343–1372.

Shaughnessy, P.W. (1981) Multiple runs distributions: recurrences and critical values. *J. the Amer. Statist. Assoc.*, **76**, 732–736.

Shirley, E. (1977) A nonparametric equivalent of Williams' test for contrasting existing dose levels of a treatment. *Biometrics*, **33**, 386–389.

Shirley, E.A.C. (1987) Applications of ranking methods to multiple comparison procedures and factorial experiments. *Appl. Statist.*, **36**, 205–213.

Siegel, S. and Castellan, N.J. (1988) *Nonparametric Statistics for the Behavioral Sciences*. 2nd edn. New York: McGraw-Hill.

Siegel, S. and Tukey, J.W. (1960) A nonparametric sum of ranks procedure for relative spread in unpaired samples. *J. Amer. Statist. Assoc.*, **55**, 429–444.

Simonoff, J.S. (2003) *Analyzing Categorical Data* New York: Springer–Verlag.

Simpson, E.H. (1951) The interpretation of interaction in contingency tables. *J. Roy. Statist. Soc. B*, **13**, 238–241.

Smeeton, N.C. (1986) Modelling episodes of mental illness: some results from the Second U.K. National Morbidity Survey. *The Statistician*, **35**, 55–63.

Smeeton, N.C. and Cox, N.J (2003) Do-it-yourself shuffling and the number of runs under randomness for a sample consisting of several categories. *Stata Journal*, **3**, 270–277.

Smeeton, N.C. and Cox, N.J. (2006) Software update: Do-it-yourself shuffling and the number of runs under randomness. *Stata Journal*, **6**, 597.

Smeeton N.C, Rona R.J, Oyarzun M. and Diaz P.V (2006) Agreement between responses to a standardized asthma questionnaire and a questionnaire following a demonstration of asthma symptoms in adults. *Amer. J. Epidemiol.*, **163**, 384–391.

Smeeton, N.C., Rona, R.J., Sharland, G., Botting, B.J., Barnett, A. and Dundas, R. (1999) Estimating the prevalence of malformation of the heart in the first year of life using capture-recapture methods. *Am. J. Epidemiol.*, **150**, 778–785.

Smeeton, N. and Wilkinson, G. (1988) The detection of annual clusters in individual patterns of parasuicide. *Journal of Applied Statistics*, **15**, 179–182.

Smirnov, N.V. (1939) On the estimation of discrepancy between empirical curves of distribution for two independent samples. (In Russian.) *Bull. Moscow Univ.*, **2**, 3–16.

Smirnov, N.V. (1948) Tables for estimating the goodness of fit of empirical distributions. *Ann. Math. Statist.*, **19**, 279–281.

Smith, R.L. and Naylor, J.C. (1987) A comparison of likelihood and Bayesian estimators of the three-parameter Weibull distribution. *Appl. Statist.*, **36**, 358–369.

Snee, R.D. (1985) Graphical display of results of three treatment randomized block experiments. *Appl. Statist.*, **34**, 71–77.

Somers, R.H. (1962) A new asymmetric measure of association for ordinal variables. *Amer. Sociological Rev.* **27**, 799–811,

Spearman, C. (1904) The proof and measurement of association between two things. *Amer. J. Psychol.*, **15**, 72–101.

Sprent, P. (1998) *Data Driven Statistical Methods.* London: Chapman & Hall.

Stuart, A. (1955) A test for homogeneity of the marginal distributions in a two-way classification. *Biometrika*, **42**, 412–416.

Stuart, A. (1957) The comparison of frequencies in matched samples. *Brit. J. Statist. Psychol.*, **10**, 29–32.

Sukhatme, B.V. (1957) On certain two sample nonparametric tests for variances. *Ann. Math. Statist.*, **28**, 188–194.

Swed, F.S. and Eisenhart, C. (1943) Tables for testing randomness of grouping in a series of alternatives. *Ann. Math. Statist.*, **14**, 83–86.

Sweeting, T.J. (1982) A Bayesian analysis of some pharmacological data using a random coefficient regression model. *Appl. Statist.*, **31**, 205–213.

Tarone, R. and Ware, J. (1977) On distribution-free tests for equality of survival distributions. *Biometrika*, **64**, 156–160.

Terpstra, T.J. (1952) The asymptotic normality and consistency of Kendall's test against trend when ties are present in one ranking. *Indag. Math.*, **14**, 327–333.

Theil, H. (1950) A rank invariant method of linear and polynomial regression analysis, I, II, III. *Proc. Nederl. Akad. Wetensch. A*, **53**, 386–392, 521–525, 1397–1412.

Thomas, G.E. (1989) A note on correcting for ties with Spearman's ρ. *J. Statist. Comp. Sim.*, **31**, 37–40.

Thomas, G.E. (2000) Use of the bootstrap in robust estimation of location. *The Statistician*, **49**, 63–77.

Tilling K. and Sterne J.A.C (1998) Capture-recapture methods including covariate effects. *Amer. J. Epidemiol.*, **149**, 392–400.

Tukey, J.W. (1958) Bias and confidence in not quite large samples. Abstract. *Ann. Math. Statist.*, **29**, 614.

van der Waerden, B.L. (1952) Order tests for the two-sample problem and their power, I. *Proc. Kon. Nederl. Akad. Wetensch. A*, **55**, 453–458. Correction, **56**, 80.

van der Waerden, B.L. (1953) Order tests for the two-sample problem and their power; II, III. *Proc. Kon. Nederl. Akad. Wetensch. A*, **56**, 303–310, 311–316.

von Mises, R. (1931) *Wahrscheinlichkeitrechnung und ihre Anwendung in der Statistik und Theoretischen Physik*. Leipzig: F. Deuticke.

Upton, G.J.G. (1992) Fisher's exact test. *J. Roy. Statist. Soc. A*, **155**, 395–402.

Wald, A. and Wolfowitz, J. (1940) On a test whether two samples are from the same population. *Ann. Math. Statist.*, **11**, 147–162.

Walsh, J.E. (1949a) Application of some significance tests for the median which are valid under very general conditions. *J. Amer. Statist. Assoc.*, **44**, 342–355.

Walsh, J.E. (1949b) Some significance tests for the median which are valid under very general conditions. *Ann. Math. Statist.*, **20**, 64–81.

Wasserman, L (2006) *All of Nonparametric Statistics*. New York: Springer.

Whitworth, W.A. (1886) *Choice and Chance*. Cambridge: Deighton Bell.

Wilcoxon, F. (1945) Individual comparisons by ranking methods. *Biometrics*, **1**, 80–83.

Willemain, T.R. (1980) Estimating the population median by nomination sampling. *J. Amer. Statist. Assoc.*, **75**, 908–911.

Williams, D.A. (1986) A note on Shirley's nonparametric test for comparing several dose levels with a zero-dose control. *Biometrics*, **42**, 183–186.

Williamson, P., Hutton, J.L., Bliss J., Blunt J., Campbell, M.J. and Nicholson, R. (2000) Statistical review by research ethics committees. *J. Roy. Statist. Soc. A*, **163**, 5–13.

Yates, F. (1984) Tests of significance for 2×2 contingency tables. *J. Roy. Statist. Soc. A*, **147**, 426–463.

Zelen, M. (1971) The analysis of several 2×2 contingency tables. *Biometrika*, **58**, 129–137.

Index